Craig F. Bohren and Eugene E. Clothiaux

Fundamentals of Atmospheric Radiation

Craig F. Bohren and Eugene E. Clothiaux

Fundamentals of Atmospheric Radiation

An Introduction with 400 Problems

WILEY-VCH Verlag GmbH & Co. KGaA

The Authors of this Book

Craig F. Bohren
Dept. of Meteorology
Pennsylvania State University
bohren@ems.psu.edu

Eugene Clothiaux
Dept. of Meteorology
Pennsylvania State University
cloth@meteo.psu.edu

For a Solutions Manual, lecturers should contact
the editorial department at
vch-physics@wiley-vch.de,
stating their affiliation and the course in which
they wish to use the book.

Library of Congress Card No.: applied for

British Library Cataloging-in-Publication Data:

A catalogue record for this book is available from
the British Library.

**Bibliographic information published by
Die Deutsche Bibliothek**

Die Deutsche Bibliothek lists this publication in the
Deutsche Nationalbibliografie; detailed
bibliographic data is available in the Internet at
http://dnb.ddb.de.

© 2006 WILEY-VCH Verlag GmbH & Co. KGaA,
Weinheim

Printed in the Federal Republic of Germany

Printed on acid-free paper

| **Printing** | betz-druck GmbH, Darmstadt |
| **Binding** | J.Schäffer Buchbinderei GmbH, Grünstadt |

ISBN-13 3-527-40503-9
ISBN-10 3-527-40503-8

To my scientific sparring partners
Bill Doyle, Alistair Fraser, and Akhlesh Lakhtakia
Craig F. Bohren

To my father and mother,
they always choose kindness and curiosity over fixed notions;
to my brothers and sisters,
for sharing the wild ride;
to Jessica,
for her enduring patience and relentless support,
and to our sons Daniel and Joshua,
that they grow in the generosity of spirit of their
grandmothers and grandfathers.
Eugene E. Clothiaux

Contents

Fundamentals of Atmospheric Radiation: An Introduction with 400 Problems. Craig F. Bohren and Eugene E. Clothiaux
Copyright © 2006 Wiley-VCH Verlag GmbH & Co. KGaA, Weinheim
ISBN: 3-527-40503-8

Preface

Like so many textbooks, this one has its origins in the classroom, the fruit of more than 30 years of combined experience teaching courses on atmospheric radiation to graduate and undergraduate students of meteorology. This experience has forced us to recognize that most of our students do not adequately understand the fundamentals of electromagnetic radiation and its interaction with matter. Students come to the classroom with their heads full of mantras, half-truths, or outright errors, and much of our effort has been devoted to trying to convince them that what they think are universal truths are at best approximations or simply wrong. Indeed, all theories are ultimately wrong. And a theory is just scribbles on paper, not reality. Theories can help make sense of reality but they are not reality itself.

We are careful to expose to the clear light of day all assumptions underlying theories, their limitations and ranges of validity. Nothing is intentionally swept under the rug. Because all theories ultimately break down, you must know what underlies them to have a hope of fixing them when they do. Ignorance is not bliss.

James D. Patterson, a retired physics professor, published An Open Letter to the Next Generation in the July 2004 issue of *Physics Today*. This letter is charming and refreshingly honest. Patterson does not brag about his triumphs but instead warns the next generation about mistakes he made in his career. He notes that, "We have to learn basics first, because we need them for all that follows. If we do not learn the basics, we are disadvantaged. A related sin is skipping essential details. Then we do not get to the bottom of things and are not well grounded."

Many of the references at the ends of chapters are to original papers. Again, we quote Patterson: "When we want to know something, there is a tendency to seek a quick answer in a textbook. This often works, but we need to get in the habit of looking at original papers. Textbooks are often abbreviated, second- or third-hand distortions of the facts, and they usually do not convey the flavor of scientific research." We go even further than Patterson and note that whenever you see in a textbook a statement of the form "Einstein [or Newton or some other scientific worthy] said…" replace "said" with "did not say" and what follows is more likely to be true. Even direct quotations are not reliable because so often textbook writers can't be bothered to go to the library (too far to walk) and so pass on what they think they remember that some other textbook writer thinks Einstein (or whoever) might have said. The only sure way to find out what our predecessors said is to read their own words.

We present theories as a hierarchy, each level of which is more encompassing than its predecessors but each correct subject to stated limitations and approximations. Learn at a certain level secure in the knowledge that what you learn need not be unlearned. To go on to the next level is for you to decide. For example, Chapter 5 introduces multiple scattering by

Fundamentals of Atmospheric Radiation: An Introduction with 400 Problems. Craig F. Bohren and Eugene E. Clothiaux
Copyright © 2006 Wiley-VCH Verlag GmbH & Co. KGaA, Weinheim
ISBN: 3-527-40503-8

way of a pile of plates, which can be used to illustrate much of the physics of more complicated multiple-scattering media, such as clouds, and also is a way of introducing concepts and terms in more advanced theories. But there is nothing to be unlearned because what we say is true, subject, as always, to the stated limitations of the theory. You can then move on to the two-stream theory from which you can acquire much of the physical intuition you need to understand multiple scattering. If you wish, you can stop at the end of Chapter 5. You will have mastered something complete unto itself but not the final word (there is no final word). You need not feel ashamed for not knowing the supposedly exact (nothing is exact) equation of radiative transfer or how to solve it. There are plenty of folks who can crunch numbers using this equation but don't understand them or lack the ability to estimate them without resorting to extensive calculations (using someone else's data in someone else's computer program). Even a superficial reading of the history of science conveys the lesson that the best scientists have superb intuition. The number crunchers and formalists occupy the lower ranks. This is even true of mathematicians, who are mistakenly looked upon as logic machines. The good ones know in their bellies what is true. Proofs are needed mostly to convince others. Today, many mathematicians make their livings proving or attempting to prove the conjectures (i.e., flashes of mathematical insights) of their illustrious predecessors.

Understanding should come before number-crunching. Our aim is to give you an intuitive feel for the subject matter, a firm grasp of its foundations, and to show how theories help you understand observations and measurements. Again, Patterson's lament is apposite: "I had been more interested in getting good grades than gaining understanding".

Nowhere in this book will you find condescending and insulting statements of the form "it is trivial to show". Nothing is trivial. We had to work hard for every equation, often arguing for days about "trivial" points. The deeper you delve into a subject, the more subtleties you uncover.

It seems that textbooks are almost required by law to be boring, to be carefully purged of all traces of their human authorship. We occasionally break this law. We tell stories. Some may make you laugh. Others may make you mad (and they certainly will make your professors mad). A word of caution: Peter Pilewskie read some of the first drafts of this book, and told us that he had to be careful not to drink anything while reading because while drinking a soda he happened upon a passage that caused him to convulse with laughter and spew soda over himself and his surroundings.

In an ideal world we'd like this book to read like a racy novel. But even if we were capable of writing one, it would no doubt attract the scorn of what Sinclair Lewis in *Arrowsmith* called "Men of Measured Merriment", by whom we do not mean editors. Our experience has been that the blame for dry, lifeless textbooks lies with their authors, not with censorious and humorless editors. There is a strong sentiment within science that it should be a grim grind, that if you enjoy doing it you are not really working. Many years ago the senior author was a visitor at a university that shall remain nameless. At the time he was working long hours, seven days a week. One day, out of the blue, a red-faced professor marched into his office and blurted, "You! You think you work so hard. You don't work hard because you *enjoy* what you are doing." He was serious. This was no joke. The senior author also was attacked on the floor of the United States House of Representatives because of an article in the *National Enquirer* in which he was quoted as saying that he was having great fun doing research on green thunderstorms. It seems that if you take money from the government for doing research you shouldn't enjoy it (or if you do, pretend that it is disagreeable).

We are much more critical of demonstrable nonsense than is the norm, or even permitted, in textbooks. We reckon that there is a statute of limitations for forgiving textbook writers for errors. When books contain statements that have been known to be false for 50 or 100 years, the time has come to heap ridicule on the heads of those who continue to propagate them. For example, there is no excuse, nor has there been for about 100 years, for continuing to say that the refractive index must be greater than 1 or that there is any necessary relationship between density (mass or number) and refractive index.

In the second volume of his *Recuerdos de mi Vida*, the histologist Santiago Ramón y Cajal notes with some acerbity that "In contrast to shameful custom, the child of traditional laziness, my book was to contain, as solemnly promised in the preface, only original illustrations and conclusions drawn from my own investigations." Although we can't promise that all conclusions in this book are drawn from our own investigations, we can promise that our illustrations are original. We did not write with scissors and paste. We made many measurements solely for this book and designed figures intended to convey ideas as clearly as possible. The instrument used for all spectral measurements was a Photo Research SpectraColorimeter Model PR-650 SpectraScan, which measures radiation from 380 nm to 780 nm in increments of 4 nm with a bandwidth of 8 nm.

We hereby declare this book to be an acronym-free zone. To the extent possible we use no acronyms. They are the bane of scientific writing, making it even more boring and arcane than it would be otherwise. The anonymous author of an article in the April 16, 2005 *Economist* comments on the "delight in creating forced acronyms that plagues many branches of science." A plague acronyms indeed are, and claims that they save space are laughable given that acronym-mongers are invariably sloppy writers who could save much more space by writing more compact sentences. But aside from their ugliness, acronyms are just one more way of creating barriers between those who are in the know and those who are not, cabalistic symbols by which the initiated recognize each other. We are waiting to see a paper (maybe it already has been published) entitled "The effect of SSTs on SSTs."

Wherever possible we give the full names of authors of papers and books we cite. Most scientists do have first names, despite efforts to conceal them, and it is rumored that some even have mothers and fathers. We also spell out in full the titles of journals. Cryptic abbreviations, like acronyms, are yet another way of distinguishing between the in-group and the out-group. Do you know what MNRAS stands for? If not, you are a barbarian, not fit to eat at the same table with the lords of the universe.

A book is supposed to be a conversation between authors and readers. The best way to converse with us is to work the problems. There are almost 400. They are not acts of penance but give you the opportunity to test your mastery of the subject matter (memorization of formulas is not mastery) and they expand on topics touched on briefly if at all in the bodies of the chapters. Many of these problems are questions asked by students or correspondents. We enjoyed answering them. And if you don't enjoy solving problems, you might ask yourself why you are studying science. Scientists solve problems. So get to work. And enjoy yourselves (but frown a lot so that no one will know).

Acknowledgments

Acknowledgements are always problematical because much of what one learns comes not from books and papers but from casual conversations, often in convivial surroundings. More than 25 years ago I spent over a year in the Institute of Atmospheric Physics at the University of Arizona. At the end of each day, Sean Twomey and I, sometimes accompanied by Don Huffman and Phil Krider, would march off to the nearest tavern for a few beers. Although some of our conversation centered on horse racing, Sean's great passion, we also discussed science at length. A year spent in a barroom with Sean Twomey is equivalent to a graduate degree in atmospheric science. I do recall that one topic in this book stems from a story he told me. When he worked for the Commonwealth Scientific and Industrial Research Organization in Australia, manuscripts had to be reviewed by scientists in a division other than that from which the manuscripts originated. Sean blasted a manuscript by a radio astronomer who had committed the blunder of assuming that the sum of exponentials is an exponential. The fuming author called Sean and asked him angrily, "What the hell do you know about radio astronomy?" Sean replied, "Nothing, but I do know something about exponentials."

But my memory is becoming less reliable as I make the inevitable descent into senescence. As the story of the exponentials demonstrates I can remember almost the exact date and place where I acquired some pearls of wisdom, while others are lost in a haze. I neatly handled the problem of acknowledging three colleagues, Bill Doyle, Alistair Fraser, and Akhlesh Lakhtakia, by simply dedicating this book to them. Bill and Akhlesh have been and continue to be my sounding boards on electromagnetic theory. Both have an encyclopedic knowledge of the subject, including its history, and are aware of the many subtleties that don't make their way into textbooks. Although Bill is approaching his 80th year he still retains the enthusiasm one hopes to see in college freshmen. We continue to correspond and talk on the telephone, although now he has to call me at night to spare his office mate from having to listen to our raucous conversations. Akhlesh and I have lunch every few months and call each other with questions at all hours of the day and night.

Alistair Fraser made my 20 years at Penn State a rewarding and fruitful experience. Without him I might not have stayed. Much of what I know about atmospheric optics I learned directly from him or honed what I already knew (or thought I knew). Had it not been for Alistair my academic career almost certainly would have been different. Having such a brilliant scholar and inspiring teacher to work with made it almost inevitable that I would join forces with him. I followed in his footsteps by teaching a unique course of his design, meteorological observations, in which students photograph optical phenomena in the atmosphere and write reports on them. This is the one course that indelibly changes students. They are never the same going out as coming in. And the same can be said about the teacher. Our students were often amazed at how severely Alistair and I criticized each other. We had to explain to them that this was the best way of ensuring that our work was of the highest quality. Alistair is quick to spot logical flaws, a merciless critic of sloppy exposition, a superb interpreter of what can be seen with the naked eye.

At Penn State I also had the good fortune to learn from Herschel Leibowitz, one of the most eminent perceptual psychologists, who would teach me at the breakfast table what physicists should know, but usually don't, about how humans construct a visual world out of raw optical data.

Thanks also to Paul Kay for his criticism of our discussion of color words.

More than 30 years ago my first teacher of radiative transfer was Bruce Barkstrom with whom I collaborated on a paper on radiative transfer in snow on the ground. This was a fruitful and enjoyable collaboration that brought me up to speed on much of what I needed to know.

Although Don Huffman did not contribute directly to this book, other than to provide me with a few references, his lasting influence can be felt on everything I do.

For many years I have corresponded with Warren Wiscombe, who fires questions at me every few months, causing me to refine ideas and correct errors. And this even before email made correspondence much easier.

Ray Shaw was a guiding force behind the discussion of nonexponential attenuation in Chapter 2. Thanks also to Joe Shaw for sending me reprints and to Glenn Shaw for siring Ray and Joe.

Tim Kane directed us to references on optical heterodyning.

If computers and their programs can be "user-friendly", users should have the right to be "computer-unfriendly". As my colleagues know, I am outright computer-hostile. But I am grateful to Harry Henderson and Chuck Pavolski, who responded speedily and graciously to my anguished and profane cries for help when my computers, no doubt sensing my hostility toward them, rebelled against my authority.

To save Tom Kozo possible embarrassment I won't say what he contributed, but he knows.

Manfred Wendisch had the most direct effect on this book. We sent him the first versions of most of the chapters, which he went over with a fine-tooth comb, saving us from many errors, causing us to tidy up terminology and tighten our arguments. He also caused us to take more care to make this book understandable to people whose first language is not English.

Peter Pilewskie critically commented on early versions and independently checked some of our at-first puzzling Monte Carlo calculations in Chapter 6. He also generously allowed us to publish some of his measurements, the only ones in this book we did not make.

When I had some tricky (for me) mathematical questions I turned, as usual, to George Greaves, my former climbing partner, companion on many ascents, some hair-raising, in Iceland and Scotland many years ago.

Others who contributed to this book, if only indirectly by way of the residue of mostly forgotten conversations, are Tom Ackerman, Rich Bevilacqua, Ted Staskiewicz, Tim Nevitt, Cliff Dungey, Raymond Lee, Phil Krider, John Olivero, Denny Thomson, Shermila Singham Carl Ribbing, Larry Woolf, Andy Young, Claes Beckman, Günther Können, Ken Sassen, Dick Bartels, and Fred Loxsom.

Because of my popular science books and writings on atmospheric optics, hardly a week goes by that I don't receive email from someone, somewhere in the world, from senior scientists to elementary school students, asking me questions some of which made their way into this book. To this anonymous army of inquisitive people I am also grateful.

My many students contributed questions, which I tried to answer, and misconceptions, which I tried to dispel. At least half of the problems in this book were taken from examinations and homework problems.

The portable spectrophotometer used for the spectral measurements in this book was purchased through a grant from the National Science Foundation with matching funds from the Penn State University Department of Meteorology.

To date I have written books with three collaborators, with whom I am still friends. So although I am not easy to work with, I am not impossible. Eugene Clothiaux had the hardest row to hoe of all my collaborators. I depended on him for all the heavy work that I am no longer capable of doing. All this while he was struggling up the academic ladder and helping to raise young children. Aside from the intellectual burdens of collaboration, Eugene bore physical burdens that are perhaps unusual. Because I am retired Eugene had to make the trek to my house frequently, carrying books and papers and the latest versions of chapters. His ancient car could not make it all the way up our steep and rutted road (which he calls "the creek bed"), so in all kinds of nasty weather he would park at the house of our neighbors, then trudge up the last quarter-mile, in winter a veritable ice sheet. Now that's dedication!

As usual, my most heartfelt thanks go to Nanette Malott Bohren, my companion of more than 40 years, who had to put up with the mess and stress of yet another book but who carefully pored over draft versions ferreting out logical and typographical errors. Although Nanette has no formal scientific training, she has the amazing ability to spot errors in equations and inconsistent notation.

Craig F. Bohren
Tŷ'n y Coed
Oak Hall, Pennsylvania
July, 2005

My fortune is great in having grown up in the late twentieth century United States. Those Americans living two generations back provided the infrastructure and support that allowed my father to earn a doctorate in physics and my mother a doctorate in math and science education, even though higher education was totally lacking in their families. For a mere $4,000 of my parent's money I was able to study with Jean-Marie Wersinger, George Kozlowski, Charles Brown, Delos McKown and my father at Auburn University as an undergraduate in physics. I was able to parlay this initial investment into a graduate assistantship with Leon Cooper, Mark Bear and Ford Ebner in physics and neural science at Brown University. My luck continued into the 1990s when I received a postdoctoral research fellowship to work with Tom Ackerman, Bruce Albrecht and Denny Thomson at Penn State University. During my years as a research associate and assistant professor at Penn State University, faculty members of the Department of Meteorology were incredibly supportive, to a degree so great that I have dubbed this faculty as King Arthur's Court.

The field of atmospheric radiation is full of feisty, but kind, characters. In all of his years of research in this field Warren Wiscombe has encountered only two scientists whom he has described to me as not only feisty but also a bit nasty. Such individuals are rare in the field of atmospheric radiation and I have yet to meet them – maybe I never will. I view my colleagues much like Klaus Pfeilsticker describes his colleagues in Boulder, Colorado – as his "Boulder Family." I have my ARM Science Team Family, my MISR Science Team Family, my European Union CLOUDMAP2 Family, my Family of Wonderful Graduate Students and my Fellow Members of King Arthur's Court. I have learned, and continue to learn, a tremendous amount from all of these colleagues.

Howard Barker has influenced my thinking about many topics in this book. Ideas from Mark Miller, Pavlos Kollias, and Roger Marchand have no doubt found their way into this book. Tony Clough, during a series of enjoyable dinners dating back to the mid-1990s, has tried time and again to straighten out my thinking on topics in Chapter 2, and I am not sure I have them all straight yet. But I am certain that his and Eli Mlawer's assistance over the years has provided me the best chance of properly running their line-by-line radiative transfer model and monochromatic radiative transfer model, which we used to generate all of the high spectral resolution figures in this book. Rich Bevilacqua provided timely insights to us on retrieving water vapor profiles in the mesosphere at microwave frequencies. I first learned some Monte Carlo methods from Tom Ackerman in the early 1990s, and the first code that I ever used that could be started on one machine and then replicate itself to run on many machines was developed in a collaboration between Elizabeth Post and Tom. In the years since I have learned a great deal from Sasha Marshak (who has devoted time and patience to his many discussions with me), Anthony Davis and Frank Evans about radiative transfer in general and Monte Carlo techniques in particular. Elizabeth Post's original code has undergone radical changes as a result but she would nonetheless recognize the code that remains to this day. Discussions with Qilong Min motivated specific applications in Chapter 6. Of the graduate students I worked with at Penn State those who made a direct contribution to the radiative transfer codes I used for this book include Chuck Pavloski, Seiji Kato, Laura Hinkelman, Daniel Pawlak, Jason Cole and Jonathan Petters. All of the Monte Carlo terrestrial radiation calculations for Chapter 6 were produced by Jason Cole with a Monte Carlo code that he developed during his thesis research. Those with indirect contributions to topics in this book include Jay Mace, Chuck Long, Jim Mather, Andy Vogelmann, Ruei-Fong Lin, Xiquan Dong, Michael Jensen, Urszula Jambor, Adrian George, Kim Fineran, Manajit Sengupta, Greg Schuster and Dave Groff. Students in the atmospheric radiation courses that I taught always provided valuable feedback, with Kelly Cherrey and Jesse Stone's comments being of particular value as this book project came to an end.

For ten of my fourteen years at Penn State I had no idea who Michael Modest was even though I can see his office window from mine. He contributed ideas to Daniel Pawlak and Jason Cole during their study of radiative transfer.

I have never met our editors and technical assistants at Wiley-VCH in Germany – we did everything by email. Nevertheless, their support was wonderful. Andreas Thoss helped us in the early stages and Ulrike Werner helped us reach the end. Uwe Krieg always provided timely support with the Wiley-VCH LaTeX style sheets. I would send emails to them at the end of the day and without exception I would have my answers the following morning. They gave Craig wide latitude in determining the style of this book. After we had missed our third (or was it fourth?) deadline Ulrike told me not to worry, that if I had known what I was getting myself into with the start of this book I never would have done it. She was right, and her patience made ending this book project as pleasant as it could be. While our Wiley-VCH editors took care of our book business, Patrick Cleary's skill, flexibility, and open-mindedness was wonderful on my home front.

I owe a special thanks to Tom Ackerman and my co-author. Over the years I have learned more about atmospheric radiation from these two scientists than from any other person. My hope throughout the 1990s was that Tom and Craig would write a textbook and include me as a co-author if I could perform enough work on their behalf. Such a book might be dubbed the ABCs of atmospheric radiation, or the BAC of atmospheric radiation, but certainly not the CAB of atmospheric radiation. When Alistair Fraser suddenly retired and disappeared from Penn State University in 2001, Craig, a bit rattled from Alistair's departure, asked me to help him write a textbook in which he organized his diverse thoughts and scattered writings on atmospheric radiation. I asked him if I could get Tom to help us out. He said no – that Tom and he would have too many difficulties reaching agreement on content and style throughout the book. When I asked Tom the same question, he agreed with Craig. My hopes went up in smoke. I very much wanted to help Craig but I could not do so without Tom's blessing. I knew that whatever ideas of mine got into a textbook would partly be Tom's. Tom, as always, was amazingly gracious. He told me that I should help Craig and that I should have no worries about ideas of his that got into the book via my contributions. Over the nine years that I worked with Tom his boundless generosity towards me and the fantastic graduate students that he recruited was truly remarkable. He is second to none in this regard. I was indeed fortunate when I first crossed Tom's path in the pastry queue on the Sunday morning of April 21, 1991, during the American Institute of Physics symposium "Global Warming: Physics and Facts" held at Georgetown University.

Craig Bohren lives in a different world from the rest of us. During the course of writing this book, he has received hundreds, if not thousands, of emails from people with no real experience in science, or a bit of informal training, or plenty, or even experts in this field and that. To the best of my knowledge he has answered many, perhaps most if not all, of these emails as he tries to bring understanding to the people who write to him. My guess is that this diversity of his experience over many years has contributed to his strong and forceful statements in his discourse on science. Time and again he has energetically criticized me for writing paragraphs that he describes as incomprehensible. On a day close to the completion of this book, he called me and told me that what I had sent to him made him truly depressed – what I wrote was not clear and he could not make sense of it and it was depressing him to no end. As despair began to sink into me, I had to remember that this was Craig and he takes science communication seriously. He was being blunt because things were not clear to him and he wanted to make them clear. Over the next week he pursued cleaning up my ideas with such vigor. As he put ideas together in a logical and consistent manner I could see his mood lighten and his excitement grow. To me this is quintessential Craig – vigorously criticizing someone, me in this case, to educate as he gains clarity on a topic himself. I have come to appreciate to no end this intellectual sincerity on the part of Craig.

So, when Craig criticizes with passion something or someone in our text, he is doing so to make a point and not to humiliate. Ironically, I know that one of the first people most likely to find shortcomings in our text is going to be Tom Ackerman. I look forward to discussions with Tom in regards to aspects of the text because I know that he will be conversing with me to express his thoughts regarding some point here or there and to educate me as well. My hope is that when my colleagues, other scientists and students find an error they also let me know about it in the spirit of Tom.

Eugene E. Clothiaux
State College, Pennsylvania
July, 2005

1 Emission: The Birth of Photons

This is the first of three foundation chapters supporting those that follow. The themes of these initial chapters are somewhat fancifully taken as the birth, death, and life of photons, or, more prosaically, emission, absorption, and scattering.

In this chapter and succeeding ones you will encounter the phrase "as if", which can be remarkably useful as a tranquilizer and peacemaker. For example, instead of taking the stance that light *is* a wave (particle), then fiercely defending it, we can be less strident and simply say that it is *as if* light is a wave (particle). This phrase is even the basis of an entire philosophy propounded by Hans Vaihinger. In discussing its origins he notes that "The Philosophy of 'As If' ... proves that consciously false conceptions and judgements are applied in all sciences; and ... these scientific Fictions are to be distinguished from Hypotheses. The latter are assumptions which are probable, assumptions the truth of which can be proved by further experience. They are therefore verifiable. Fictions are never verifiable, for they are hypotheses which are known to be false, but which are employed because of their utility."

1.1 Wave and Particle Languages

We may discuss electromagnetic radiation using two languages: wave or particle (photon) language. As with all languages, we sometimes can express ideas more succinctly or clearly in the one language than in the other. We use both, separately and sometimes together in the same breath. We need fluency in both. Much ado has been made over this supposedly lamentable duality of electromagnetic radiation. But no law requires physical reality to be described by a single language. We may hope for such a language, but Nature often is indifferent to our hopes. Moreover, we accept without protest or hand-wringing the duality of sound. We describe sound waves in air as continuous while at the same time recognizing that air, and hence sound, is composed of discrete particles (molecules) in motion.

How do we choose which language to use? Simplicity. Life is short. To understand nature we take the simplest approach consistent with accuracy. Although propagation of sound in air could be described as the motions of molecules, had this approach been taken acoustics would have floundered in a mathematical morass.

In the photon language a beam of radiation is looked upon as a stream of particles called photons with the peculiar property that they carry energy, linear momentum, and angular momentum but not mass. The mass of the photon often is said to be identically zero. But given the near impossibility of measuring zero in the face of inevitable errors and uncertainties, it would be more correct to say that the upper limit of the photon mass keeps decreasing, its present value being about 10^{-24} times the mass of the electron. If it bothers you that a particle

Fundamentals of Atmospheric Radiation: An Introduction with 400 Problems. Craig F. Bohren and Eugene E. Clothiaux
Copyright © 2006 Wiley-VCH Verlag GmbH & Co. KGaA, Weinheim
ISBN: 3-527-40503-8

without mass can carry momentum this is because you are stuck on the notion that momentum is mass times velocity. Sometimes this is true (approximately), sometimes not. Momentum is momentum, a property complete in itself and not always the product of mass and velocity.

Photons are of one kind, differing only in their energy and momenta, whereas waves are of unlimited variety and often exceedingly complex, the simplest kind a plane harmonic wave characterized by a single (circular) frequency ω and direction of propagation (see Secs. 3.3 and 3.4). The dimensions of circular frequency are radians per unit time. You may be more familiar with just plain frequency, often denoted by ν (sometimes f), which has the dimensions of cycles per unit time. The unit of frequency is the hertz, abbreviated Hz, one cycle per second. Because one cycle corresponds to 2π radians, the relation between frequency and circular frequency is simple:

$$\omega = 2\pi\nu. \tag{1.1}$$

All electromagnetic waves propagate in free space (which does not strictly exist) with the same speed c, about $3 \times 10^8 \, \mathrm{m\,s^{-1}}$. A plane harmonic wave in free space can just as well be characterized by its wavelength λ, related to its frequency by

$$\lambda\nu = c. \tag{1.2}$$

You sometimes hear it said that frequency is more fundamental than wavelength. In a sense, this is correct, but wavelength is often more useful. When we consider the interaction of electromagnetic waves with chunks of matter, the first question we must ask ourselves is how large the waves are. Big and small have no meaning until we specify a measuring stick. For electromagnetic radiation the measuring stick is the wavelength. The mathematical expressions describing the interaction of such radiation with matter can be quite different depending on the size of the matter relative to the measuring stick.

How do we translate from wave to photon language? A plane harmonic wave with circular frequency ω corresponds to a stream of photons, each with energy

$$E = h\nu = \hbar\omega, \tag{1.3}$$

where h is Planck's constant ($6.625 \times 10^{-34} \, \mathrm{J\,s}$) and $\hbar = h/2\pi$. The frequency of visible electromagnetic radiation (light) is about 10^{14} Hz, and hence the photons that excite the sensation of vision have energies around 10^{-20} J. This isn't much energy; the kinetic energy of a golf ball as it slices through air is about 10^{13} times greater.

Understanding what happens when an electromagnetic wave is incident from air on the smooth surface of glass, say, is not especially difficult if one uses the wave language. The incident wave excites molecules in the glass to radiate secondary waves that combine to form (approximately) a net reflected wave given by the law of reflection and a net transmitted wave given by the law of refraction. There is no such thing as an absolutely smooth surface, so what is meant is smooth on the scale of the wavelength.

All this makes intuitive sense and causes no perplexity. But now consider what happens when we switch to photon language. If we look upon reflection as the rebound of photons at a surface and transmission as their penetration through it, then why, if all photons are identical, are some reflected and some transmitted? This is indeed puzzling; even more so is why photons should be specularly (by which is meant mirror-like) reflected, because for photons imagined as particles of vanishingly small dimensions, all surfaces are rough.

This is not to say that one couldn't describe reflection and transmission at smooth interfaces in photon language, only that to do so would be exceedingly costly in mental effort. And the reverse sometimes is true. Many years ago one of the authors attended a colloquium entitled "The photoelectric effect without photons." By the photoelectric effect is usually meant the emission of electrons by a surface (often metallic) because of illumination by radiation (often ultraviolet). In photon language the photoelectric effect is simple to describe. When a photon of energy $h\nu$ is absorbed by the surface, the maximum kinetic energy E of the electrons thereby set free is

$$E = h\nu - p, \tag{1.4}$$

where p is the minimum energy an electron loses in breaking free of the surface. A single photon interacting with a single electron gives up its entire energy to that electron, which if sufficient enables the electron to break free of the forces binding it to the metal. According to this equation the energies of the emitted electrons are independent of the incident power whereas the photocurrent (rate and number of emitted electrons) is proportional to it, which accords with experiment. This simple equation, first written down by Einstein in 1905, is one of the keystones of the modern theory of radiation and matter. Yet the speaker at that colloquium years ago, in an effort to describe and explicate the photoelectric effect without photons, assailed the audience with dozens of complicated equations. And even at that, part way through his mathematical tour de force his mind and tongue betrayed him and he blurted out the forbidden word "photon". At that point, your author who was there leapt up from his seat and shouted, "Photons! Photons! You promised no photons."

A mirror illuminated by an incident beam gives rise to a reflected beam. Is this reflected beam redirected incident photons? Alas, we cannot do an experiment to answer this question. To determine if reflected photons are the same as incident photons would require us to be able to identify them. But photons are indistinguishable. We cannot tell one from another. We cannot tag a photon and follow its progress. Thus if you want to believe that reflected photons are the same as incident photons, you may do so. No one can prove you wrong. But you cannot prove you are right. When faced with an undecidable proposition, you may believe whatever you wish. Note that in the wave language we would not likely even ask if the reflected wave is the same as the incident wave.

It is not often acknowledged that there is a third language for talking about light, what might be called the who-gives-a-hoot-what-light-is? language. This is geometrical or ray optics, in which the nature of light isn't addressed at all. Fictitious rays are imagined to be paths along which the energy carried by light is transported, and these paths meander and bifurcate according to simple geometrical laws.

But which language is the more useful? In a letter to *American Journal of Physics*, M. Psimopoulos and T. Theocharis ask the rhetorical questions: "What new discoveries have (i) the particle or photon aspect of light, and (ii) the wave aspect of light, given rise to? Answer: (i) we are not aware of any; (ii) holography, laser, intensity interferometry, phase conjugation." To this list we add radar, all of interferometry, on which much of the science of measurement is based, and interference filters, which have many applications. The view of these authors is extreme, but they also quote the more measured words of Charles Townes, a pioneer in masers and lasers: "Physicists were somewhat diverted by an emphasis in the world of physics on the photon properties of light rather than its coherent aspects." That is, the photon language has been the more fashionable language among physicists, just as French was the fashionable language in the Imperial Russian court. When prestigious and munificent prizes began to be awarded for flushing "ons" (electron, positron, neutron, meson, and so on) from the jungle, shooting them, and mounting their stuffed heads on laboratory walls, the hunt was on, and slowed down only with the demise of the Superconducting Supercollider.

Although the wave language undoubtedly has been and continues to be more fruitful of inventions, the photon language is perhaps more soothing because photons can be incarnated, imagined to be objects we can kick or be kicked by. Waves extending through all space are not so easily incarnated. We can readily conceive of the photon as a thing. And yet an electromagnetic wave is just as much a thing as a photon: both possess energy and momentum (linear and angular) but not, it seems, mass.

1.2   Radiation in Equilibrium with Matter

We often are told that when bodies are heated they radiate or that "hot" bodies radiate. True enough, but it is just as true that when bodies are *cooled* they radiate and that "cold" bodies radiate. *All* matter – gaseous, liquid, or solid – at *all* temperatures emits radiation of *all* frequencies at *all* times, although in varying amounts, possibly so small at some frequencies, for some materials, and at some temperatures as to be undetectable with today's instruments (tomorrow's, who knows?). Note that there is no hedging here: all means all. No exceptions. Never. Even at absolute zero? Setting aside that absolute zero is unattainable (and much lower than temperatures in the depths of the Antarctic winter or in the coldest regions of the atmosphere), even at absolute zero radiation still would be associated with matter because of temperature fluctuations. Temperature is, after all, an average, and whenever there are averages there are fluctuations about them.

Radiation emitted spontaneously, as distinguished from scattered radiation (see Ch. 3), is not stimulated by an external source of radiation. Scattered radiation from the walls of the room in which you read these words may be stimulated by emitted radiation from an incandescent lamp. Turn off the lamp and the visible scattered radiation vanishes, but the walls continue to emit invisible radiation as well as visible radiation too feeble to be perceptible.

We are interested in the spectral distribution of radiation – how much in each wavelength interval – emitted by matter. Consider first the simpler example of an ideal gas in a sealed container held at absolute temperature T (Fig. 1.1). When the gas is in equilibrium its molecules are moving in all directions with equal probability, but all kinetic energies E are not equally probable. Even if all the molecules had the same energy when put into the container, they would in time have different energies because they exchange energy in collisions with each other and the container walls. A given molecule may experience a sequence of collisions in which it always gains kinetic energy, which would give it a much greater energy than average. But such a sequence is not likely, and so at any instant the fraction of molecules with kinetic energy much greater than the average is small. And similarly for the fraction of molecules with kinetic energy much less than the average. The distribution of kinetic energies is specified by

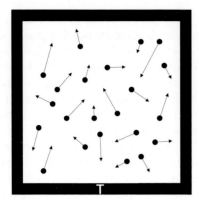

Figure 1.1: At equilibrium, ideal gas molecules in a closed container at absolute temperature T have a distribution of kinetic energies (Fig. 1.2) determined solely by this temperature.

a probability distribution function $f(E)$ which, like all distribution functions, is defined by its *integral* properties, that is,

$$\int_{E_1}^{E_2} f(E) \, dE \tag{1.5}$$

is the fraction of molecules having kinetic energies between any two energies E_1 and E_2. Note that f does not specify *which* molecules have energies in a given interval, only the fraction, or probability, of molecular energies lying in this interval. If f is continuous and bounded then from the mean value theorem of integral calculus

$$\int_{E_1}^{E_2} f(E) \, dE = f(\bar{E})(E_2 - E_1), \tag{1.6}$$

where $\bar{E}$ lies in the interval (E_1, E_2). If we denote E_1 by E and E_2 by $E + \Delta E$ we have

$$f(E) = \lim_{\Delta E \to 0} \frac{1}{\Delta E} \int_{E}^{E+\Delta E} f(x) \, dx. \tag{1.7}$$

Because of Eq. (1.7) $f(E)$ is sometimes called a *probability density*. When the limits of the integral in Eq. (1.5) are the same (interval of zero width) the probability is zero. The probability that a *continuous* variable has exactly a particular value at any point over the interval on which it is defined is zero, as it must be, for if it were not the total probability would be infinite.

A distribution function such as $f(E)$ is sometimes defined by saying that $f(E) \, dE$ is the fraction (of whatever) lying in the range between E and $E + dE$. This is sloppy mathematics because although E represents a definite number dE does not. Moreover, this way of defining a distribution function obscures the fact that f is defined by its integral properties. As we shall see, failure to understand the nature of distribution functions can lead to confusion and error. It would be better to say that $f(E)\,\Delta E$ is *approximately* the fraction of molecules lying between E and $E + \Delta E$, where the approximation gets better the smaller the value of ΔE.

You also often encounter statements that $f(E)$ is the fraction of molecules having energy E *per unit energy interval*. This can be confusing unless you recognize it as shorthand for saying that $f(E)$ must be multiplied by ΔE (or, better yet, integrated over this interval) to obtain the fraction of molecules in this interval. This kind of jargon is used for all kinds of distribution functions. We speak of quantities per unit area, per unit time, per unit frequency, etc., which is shorthand and not to be interpreted as meaning that the interval is one unit wide.

Gases within a sealed container held at constant temperature evolve to an equilibrium state determined solely by this temperature. In this state the distribution function for molecular kinetic energies is the *Maxwell–Boltzmann distribution*

$$f(E) = \frac{2\sqrt{E}}{\sqrt{\pi}(k_{\mathrm{B}}T)^{3/2}} \exp\left(-E/k_{\mathrm{B}}T\right), \tag{1.8}$$

where k_{B}, usually called *Boltzmann's constant*, is $1.38 \times 10^{-23}\,\mathrm{J\,K^{-1}}$, and f is normalized

$$\int_0^\infty f(E)\,dE = 1. \tag{1.9}$$

The limits of integration are symbolic: molecules have neither infinite nor zero kinetic energies; by zero is meant $\ll k_{\mathrm{B}}T$ and by infinite is meant $\gg k_{\mathrm{B}}T$. Because of Eq. (1.9) $f(E)$ is a probability distribution function.

The most probable kinetic energy E_{m} is that for which f is a maximum, the energy at which its derivative with respect to E is zero:

$$E_{\mathrm{m}} = k_{\mathrm{B}}T/2. \tag{1.10}$$

As the temperature of the gas increases so does the most probable kinetic energy of its molecules. Figure 1.2 shows f relative to its maximum as a function of E relative to E_{m}, a universal curve independent of temperature.

What does all this have to do with radiation? Because matter continuously emits radiation, a container with walls so thick that no photons leak from it will fill with a gas of photons (Fig. 1.3). The container is held at a fixed temperature T. At equilibrium the photons in the container, like gas molecules, do not all have the same energy (equivalently, frequency) but are distributed about a most probable value. The distribution function for the energies of photons in equilibrium with matter goes under various names and there are several versions of this function differing by a constant factor. Imagine a plane surface within the container. At equilibrium, the radiation field is isotropic, so regardless of how the surface is oriented the same amount of radiant energy crosses unit area in unit time. We consider only that radiant energy (photons) propagating in a hemisphere of directions either above or below the surface. The energy distribution function (or spectral distribution) is given by the *Planck distribution* (or *Planck function*)

$$P_{\mathrm{e}}(\omega) = \frac{\hbar\omega^3}{4\pi^2 c^2} \frac{1}{\exp\left(\hbar\omega/k_{\mathrm{B}}T\right) - 1}. \tag{1.11}$$

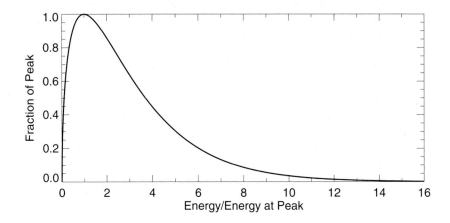

Figure 1.2: Distribution of kinetic energies of an ideal gas at equilibrium shown as a universal function independent of temperature. The kinetic energy relative to that at the peak of the distribution function, however, does depend on temperature.

The integral of this function over any frequency interval is the total radiant energy in that interval crossing unit area in unit time, called the *irradiance* (discussed in more detail in Sec. 4.2).

The Planck function is worthy of respect, if not awe, in that it contains not one, not two, but *three* fundamental (or at least believed to be so) constants of nature: the speed of light in a vacuum c, Planck's constant h, and Boltzmann's constant k_B. You can't get much more fundamental than that.

The most probable photon energy is obtained by setting the derivative of P_e with respect to ω equal to zero; the result is the transcendental equation

$$3(e^x - 1) = xe^x, \tag{1.12}$$

where $x = \hbar\omega/k_B T$, the solution to which (obtained quickly with a pocket calculator) is $x = 2.819$. Thus the most probable photon energy is

$$\hbar\omega_m = 2.819 k_B T. \tag{1.13}$$

Note the similarity of Eq. (1.11) to Eq. (1.8) and Eq. (1.13) to Eq. (1.10), which is not surprising given that both are distribution functions for gases, although of a different kind. The most striking difference between a gas of molecules and a gas of photons is that the number of molecules in a sealed container is conserved (barring chemical reactions, of course) whereas the number of photons is not. As the temperature of the container, which is the source of the photons, increases, the number of photons within it increases. Photons are not subject to the same conservation laws as gas molecules, which are endowed with mass.

At frequencies for which $\hbar\omega \ll k_B T$ Eq. (1.11) can be approximated by

$$P_e(\omega) \approx \frac{k_B T \omega^2}{4\pi^2 c^2}. \tag{1.14}$$

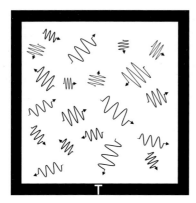

Figure 1.3: An opaque container at absolute temperature T encloses a gas of photons emitted by its walls. At equilibrium, the distribution of photon energies (Fig. 1.4) is determined solely by this temperature.

Folks interested in radiation of sufficiently low frequency (e.g., microwaves) sometimes express radiant power as a temperature. When first encountered this can be jarring until you realize that the Planck function is proportional to absolute temperature at such frequencies.

1.2.1   Change of Variable

We may express the Planck distribution as a function of frequency or wavelength. But in making a change of variables we have to be careful. The physical content of the Planck distribution is contained in its integral. According to the theorem for the change of variables in an integral

$$\int_{\omega_1}^{\omega_2} P_e(\omega)\, d\omega = \int_{\lambda_1}^{\lambda_2} P_e\{\omega(\lambda)\} \frac{d\omega}{d\lambda}\, d\lambda, \tag{1.15}$$

where $\omega(\lambda)$ is the transformation from circular frequency to wavelength and λ_j is the wavelength corresponding to ω_j. The derivative in the integral on the right side of this equation is called the *Jacobian* of the transformation. Equation (1.15) is *not* obtained by canceling the $d\lambda$s, which is merely a way of remembering the theorem. The notation of calculus has evolved so as to make it easy to remember theorems, but notation should not cause us to forget that they all require proofs. No theorem can be proved by purely notational tricks.

According to Eq. (1.15) the Planck function expressed in wavelength terms is

$$P_e(\lambda) = P_e\{\omega(\lambda)\} \frac{d\omega}{d\lambda}, \tag{1.16}$$

where we use the same symbol P_e for both functions even though this is sloppy mathematics. The distinction between a function and its values is often blurred. We write

$$y = f(x) \tag{1.17}$$

to indicate that y is the value the function f assigns to x. Suppose that f is the function "square it": $y = x^2$. If we transform from the variable x to $x = \sqrt{u}$, we obtain the new functional relation $y = u$. This is now a different function, and hence merits its own name (symbol). But to save having to invent more and more symbols, we are sloppy and write $y = f(x) = f\{x(u)\} = f(u)$, when we should write $y = f(x) = g(u)$. We often are even sloppier by confusing the value of the function with the function itself. That is, we write $y = y(x) = y\{x(u)\} = y(u)$. The fundamental rule of mathematical sloppiness is that you are allowed to be sloppy as long as you know how to do things correctly.

Although the Jacobian in Eq. (1.15)

$$\frac{d\omega}{d\lambda} = -\frac{2\pi c}{\lambda^2} \tag{1.18}$$

is negative, this does not mean that the radiant energy in the wavelength interval is negative. The upper limit on the right side of Eq. (1.15) is smaller than the lower limit, which by itself would make the integral negative, but the negative Jacobian makes the integral positive. So we write the Planck function as

$$P_{\mathrm{e}}(\lambda) = P_{\mathrm{e}}\{\omega(\lambda)\} \left| \frac{d\omega}{d\lambda} \right| = P_{\mathrm{e}}\{\omega(\lambda)\} \frac{2\pi c}{\lambda^2} \tag{1.19}$$

and remember to reverse the limits of integration on the right side of Eq. (1.15). The Planck function expressed in wavelength terms is therefore

$$P_{\mathrm{e}}(\lambda) = \frac{2\pi hc^2}{\lambda^5} \frac{1}{\exp\left(hc/\lambda k_{\mathrm{B}}T\right) - 1}. \tag{1.20}$$

For $hc/\lambda k_{\mathrm{B}}T \ll 1$, Eq. (1.20) is approximately

$$P_{\mathrm{e}}(\lambda) \approx \frac{2\pi c k_{\mathrm{B}}T}{\lambda^4}. \tag{1.21}$$

At temperatures around 300 K this equation is a good approximation (within about 1% or less) for wavelengths greater than about 250 μm. As we show in Section 8.1 the spectrum of skylight is approximately proportional to $1/\lambda^4$. As temperature increases without limit, therefore, the Planck function at visible wavelengths has approximately the same spectral dependence as the blue sky. So much for the notion that an exceedingly hot body is "white hot" or that blue is a "cold" color whereas red is a "warm" color.

The two forms of the Planck function presented here have the peculiar property that although the integral over any wavelength interval is equal to the integral over the corresponding frequency interval, the two functions do not peak at the same place. That is, if we find the frequency at which $P_{\mathrm{e}}(\omega)$ is a maximum and transform that frequency into a wavelength, we do not obtain the wavelength at which $P_{\mathrm{e}}(\lambda)$ is a maximum. To find this wavelength, differentiate Eq. (1.20) with respect to λ and set the result equal to zero. This yields the transcendental equation

$$5(e^x - 1) = xe^x, \tag{1.22}$$

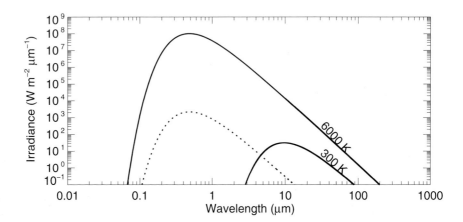

Figure 1.4: Planck function for 6000 K and 300 K. The dashed curve is the irradiance at the top of the atmosphere from a 6000 K blackbody at the Earth–sun distance, which approximates the solar irradiance.

where $x = hc/\lambda k_{\mathrm{B}}T$, the solution to which is $x = 4.961$. From this we obtain *Wien's displacement law* relating temperature to the wavelength λ_{m} at which $P_{\mathrm{e}}(\lambda)$ is a maximum:

$$\lambda_{\mathrm{m}}T = 2902\,\mu\mathrm{m\,K}. \tag{1.23}$$

For $T = 273\,\mathrm{K}$ $(0\,^{\circ}\mathrm{C})$, $\lambda_{\mathrm{m}} = 10.6\,\mu\mathrm{m}$. Equation (1.23) is called a displacement law because it determines how the Planck function is displaced as temperature increases. This displacement is evident in Fig. 1.4, which shows Eq. (1.20) for two temperatures, 6000 K and 300 K. Note also the huge difference in the amount of radiation emitted at these two temperatures.

But if we transform Eq. (1.13) into wavelength terms we obtain a different displacement law

$$\lambda_{\mathrm{m}}T = 5107\,\mu\mathrm{m\,K}, \tag{1.24}$$

where λ_{m} is the wavelength corresponding to the frequency ω_{m} in Eq. (1.13). And this wavelength for 273 K is 18.7 µm, quite a shift from 10.6 µm. Which is correct? They both are. No law requires P_{e} to be plotted versus wavelength. This may be the custom in some fields, but not in others. Many spectroscopists plot spectra as a function of *wavenumber* (inverse wavelength, equivalent to frequency) and would consider doing otherwise an unnatural act. There is, in general, no invariant maximum for a distribution function. This may be unpalatable but it is a fact of life, in the nature of distribution functions. And yet this seems to be a difficult idea to get across. Once, after we had carefully discussed it in class, a student asked in all sincerity, "But where is the *real* maximum of the Planck function?" He thought we knew but were withholding it from the uninitiated, a secret to be revealed only on our deathbeds.

Failure to recognize that the maximum of a distribution function depends on how it is plotted has led and no doubt will continue to lead to errors. In a delightful paper Bernard Soffer

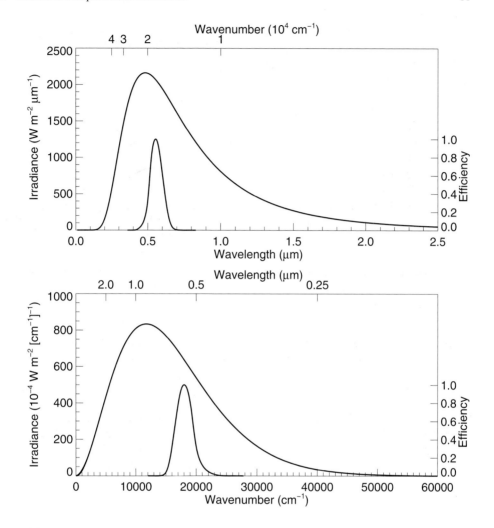

Figure 1.5: Luminous efficiency and Planck distribution at 6000 K (approximately the spectrum of solar radiation). The top figure shows the Planck distribution for wavelength as the independent variable; the bottom figure shows the Planck distribution for wavenumber (frequency). Note that although the peaks of the Planck distribution and luminous efficiency nearly coincide in the top figure, they are appreciably shifted in the bottom figure. These figures are similar to and were inspired by those presented by Soffer and Lynch (1999).

and David Lynch thoroughly demolish the widespread, but unsupportable, notion that the human eye is adapted to respond to the peak of the solar spectrum. If the *luminous efficiency* of the human eye (Sec. 4.1.7), the relative degree to which radiant energy of each visible wavelength is converted into the sensation of brightness, and the solar spectrum are plotted versus wavelength on the same figure, the peaks of the two curves roughly coincide. This is shown in Fig. 1.5, the Planck function for a temperature of 6000 K, which is a good enough approximation to the solar spectrum to make the point, together with the luminous efficiency. This accidental coincidence of peaks has led countless biologists and vision scientists to leap to the unwarranted conclusion that the eye is optimized at the wavelength of the peak of the solar spectrum, a supposed triumph of evolution. We say "accidental" because if wavenumber (frequency) is taken as the independent variable (and why not?), the maxima of the two curves do not coincide, and by an appreciable amount. This is because the luminous efficiency is a *point function*, defined at each wavelength (or frequency), and hence its maximum doesn't change when the independent variable is changed from wavelength to frequency or anything else. But this is not true of the solar spectrum, which, like the Planck function, is a spectral distribution function, its physical content specified by its integral. For its integral properties to be preserved, it has to change shape because equal frequency intervals do not correspond to equal wavelength intervals:

$$\omega_2 - \omega_1 = 2\pi c \left(\frac{\lambda_1 - \lambda_2}{\lambda_1 \lambda_2} \right). \tag{1.25}$$

Soffer and Lynch give other examples of errors resulting from the failure to distinguish between a point function and a distribution function. The moral of this story is that no reliable conclusions can be drawn, in general, from positions of the maxima of distribution functions.

1.2.2   Stefan–Boltzmann Law

The total radiant energy for the Planck function is obtained by integration over all frequencies:

$$\int_0^\infty \frac{\hbar \omega^3}{4\pi^2 c^2} \frac{1}{\exp\left(\hbar\omega/k_B T\right) - 1} \, d\omega. \tag{1.26}$$

Again, the limits of integration here are symbolic: no photon has zero or infinite energy. By transforming the variable of integration from ω to $x = \hbar\omega/k_B T$ this integral becomes

$$\frac{k_B^4 T^4}{4\pi^2 c^2 \hbar^3} \int_0^\infty \frac{x^3}{\exp\left(x\right) - 1} \, dx = \sigma T^4, \tag{1.27}$$

where

$$\sigma = \frac{k_B^4}{4\pi^2 c^2 \hbar^3} \int_0^\infty \frac{x^3}{\exp\left(x\right) - 1} \, dx = 5.669 \times 10^{-8} \, \text{W m}^{-2} \, \text{K}^{-4}. \tag{1.28}$$

This fourth-power law for total (spectrally integrated) irradiance is called the *Stefan–Boltzmann law* and σ is called the *Stefan–Boltzmann constant*.

1.3   Blackbody Radiation

Radiation in equilibrium with matter, as specified by the Planck function, often is called *blackbody radiation*. To understand the origins of this queer (as well as often confusing and misleading) term requires postulating an idealized blackbody, which, like so many bodies in physics, does not exist. A blackbody cannot be excited to radiate, more than it would in isolation, by an external source of radiation of any frequency, direction, or state of polarization. The definition of a blackbody as one that absorbs all radiation incident on it contains a trap for the unwary. Notions about radiation being incident on bodies are valid only when they are much larger than the wavelength. We intuitively expect, based on our everyday experiences with objects large compared with visible wavelengths, that the radiant energy absorbed by an illuminated object is determined by its geometrical area. This expectation breaks down when the body is small compared with the wavelength, a restriction almost never mentioned although Planck recognized it clearly. On page 2 of his *Theory of Heat Radiation*, he says that he always assumes that the "linear dimensions of all parts of space considered, as well as the radii of curvature of all surfaces under consideration, are large compared with the wave lengths of the rays considered." The concept of radiation incident on a body is from geometrical (or ray) optics, which is never strictly valid because all bodies are finite.

When one of the authors was interviewed many years ago for a position at Penn State he was asked if the *emissivity* (see following section) of a body can be greater than 1. He responded, "of course", which may have saved a few lives (and possibly got him a job). At the time some meteorology graduate students were calculating emissivities of small particles at infrared wavelengths, and to their horror and dismay were obtaining values greater than 1. These students were almost suicidal because they had been brainwashed that emissivities cannot be greater than 1. Not so. Supposed upper limits on emissivities hold (approximately) only for objects much larger than the wavelength. In order to determine "all the radiation incident on" a body, what is meant by "incident" has to be well defined, but it is not. There are always departures from geometrical optics, although for sufficiently large bodies these departures may be negligible.

This seemingly heretical assertion about emissivities greater than 1, when cast in the language of antenna engineers, would be considered almost trivial. Any antenna engineer knows that the effective area of a receiver can be much larger than its geometrical area.

Although no strict blackbodies exist, some bodies are approximately black over a limited range of frequencies, directions, and polarization states of the exciting radiation.

Suppose that a hypothetical blackbody is placed inside an opaque cavity the walls of which are held at constant temperature. This blackbody is bathed in equilibrium radiation, which is isotropic (the same in all directions) and unpolarized (see Sec. 7.1). The rate at which radiant energy of all frequencies is absorbed by this blackbody is (subject to the restriction stated by Planck)

$$A \int_0^\infty P_e(\omega) \, d\omega, \tag{1.29}$$

where A is the total surface area of the blackbody. Let $\mathcal{E}_b(\omega)$ be the spectral emittance, the power emitted per unit frequency in a hemisphere of directions per unit area of the blackbody; the total emittance is the integral of the spectral emittance. If the blackbody is in thermal equilibrium (its temperature does not change), the total rates of absorption and emission must be equal:

$$\int_0^\infty P_e(\omega) \, d\omega = \int_0^\infty \mathcal{E}_b(\omega) \, d\omega. \tag{1.30}$$

Because of this equation we might be tempted to set the integrands equal to each other:

$$P_e(\omega) = \mathcal{E}_b(\omega).$$

(1.31)

Although Eq. (1.31) is *sufficient* to ensure that the blackbody is in thermal equilibrium, it is not *necessary* unless supplemented by an additional physical argument. According to the *principle of detailed balance*, Eq. (1.31) is both necessary and sufficient for thermal equilibrium. That is, at each frequency the rate of emission by the blackbody must be equal to the rate of absorption. Underlying this principle is a fundamental symmetry property of nature called *time-reversal symmetry*: the equations of the electromagnetic field have the same form when time is run backwards. To make this clearer consider a simpler example, the equation of motion of a point mass m acted on by a force $\mathbf{F}$ at position $\mathbf{x}$:

$$m\frac{d^2\mathbf{x}}{dt^2} = \mathbf{F}.$$

(1.32)

If $\mathbf{F}$ depends only on the position of m, Eq. (1.32) is unchanged if we make the transformation $t \to -t$, the consequence of which is that if m follows a path to a point where its velocity is $\mathbf{v}$ and then launched from this point with velocity $-\mathbf{v}$, m will retrace its original path.

In the context of the problem of interest here, the principle of detailed balance (time-reversal symmetry) yields the result that absorption and emission are inverse processes. That is, if time were to reverse, absorbed photons would become emitted photons and vice versa.

According to Eq. (1.31) the emission spectrum of a blackbody at temperature T is the Planck spectrum, the spectrum of radiation in equilibrium with matter. Hence such radiation is often called blackbody radiation even though its existence does not hinge on that of a black-body. Defining equilibrium radiation by something that doesn't exist is unsatisfying and also deflects attention from its physical nature.

1.4   Absorptivity and Emissivity

Suppose that a real body is placed inside a cavity held at constant temperature. Such a body is uniformly illuminated by blackbody radiation. The spectral *absorptivity* $\alpha(\omega)$ is defined as that number which when multiplied by the irradiance $P_e(\omega)$ gives the rate of absorption per unit area of radiation per unit frequency by the body. Because blackbody (cavity) radiation is unpolarized and isotropic, α is the absorptivity for such radiation. Real bodies (large compared with the wavelength) are characterized by (dimensionless) absorptivities less than or equal to 1.

We define the spectral *emissivity* $\varepsilon(\omega)$ as that number which when multiplied by the Planck function gives the rate of emission by the real body in all hemispherical directions (again, per unit area and frequency). For the real body to be in thermal equilibrium inside the cavity requires that total absorption be balanced by total emission:

$$\int_0^\infty \alpha(\omega)P_e(\omega)\,d\omega = \int_0^\infty \varepsilon(\omega)P_e(\omega)\,d\omega.$$

(1.33)

Again, a sufficient condition for Eq. (1.33) to be satisfied is

$$\alpha(\omega) = \varepsilon(\omega),$$

(1.34)

and by the principle of detailed balance is also a necessary condition. Thus the spectral emissivity is equal to the spectral absorptivity, which again points to absorption and emission as inverse processes.

Alas, this simple and unambiguous equation has been translated, unnecessarily, by some textbook writers into a mantra, "a good absorber is a good emitter", which is as useless as it is misleading. Whether something is good or bad cannot be determined until criteria for goodness or badness are specified. Steam pipes in power plants are wrapped with aluminum sheeting, and insulation for houses is coated with aluminum foil. Metals such as aluminum, especially when highly polished, have low spectral emissivities over the range of frequencies that encompass most of the integrated Planck function at typical terrestrial temperatures. To reduce the rate of radiative cooling of the contents of pipes (or of anything) they are wrapped with a "bad" emitter, which is "good." Do you wrap hot sandwiches with aluminum foil? If so, you are making use of the "good" properties of a "bad" emitter (low emissivity).

Equation (1.34) often is called *Kirchhoff's law*, but is a restricted form of this law. In general, the absorptivity (and hence emissivity) of a body depends on the direction and state of polarization of the incident radiation as well as its frequency, caveats often omitted. We would not have to worry about this if all bodies were always illuminated by blackbody radiation (unpolarized and isotropic). But alas, this is not true, and real bodies when removed from cavities are illuminated by radiation that usually is not the same in all directions and may be partially polarized (see Sec. 7.1).

1.4.1   Blackbody Radiation without a Blackbody

Blackbody radiation is radiation in equilibrium with matter. We discuss this by way of a container (cavity) at a fixed temperature. You sometimes find textbook treatments of blackbody radiation in which it is stated (or implied) that the walls of the cavity must be black. This is not true. All that is required for a cavity to be filled with blackbody radiation is that the cavity be opaque and have a nonzero emissivity (and hence absorptivity) at all wavelengths. Thus if we fashion a cavity from a material that is neither black nor 100% reflecting at any wavelength, the radiation contained therein is still blackbody radiation. How can this be? How can blackbody radiation be obtained without a blackbody?

To answer this question, consider two opaque parallel plates, large compared with their separation (Fig. 1.6); the spectral emissivity of each plate is ε and both plates are at the same temperature. We want to determine the amount of radiation that crosses unit area (between the plates and parallel to them) in all upward directions. For simplicity we ignore that ε strictly depends on direction and polarization state.

The radiation between the plates is a consequence of radiation emitted *and* reflected (because they are not necessarily black) by them. Because the plates are opaque (no transmission by them) the *reflectivity* (fraction of incident radiation reflected) of a plate is $1 - \alpha$, which from Kirchhoff's law is $1 - \varepsilon$.

Consider first the lower plate. An amount of radiation ε times the Planck function is emitted by this plate. We may omit this function because it is a common factor and simply say that an amount of radiation (per unit frequency) ε is emitted upward by the lower plate. A fraction $1 - \varepsilon$ of this radiation is reflected downward by the upper plate, and then a fraction $1 - \varepsilon$ is reflected upward by the lower plate, which contributes to the upward radiation. Some

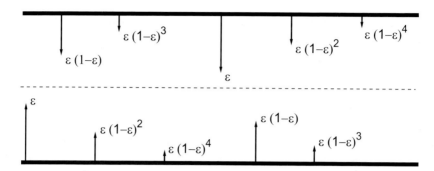

Figure 1.6: The radiation field between two identical opaque plates with emissivity ε is a consequence of emission and reflection by them. The total upward radiation field can be expressed as the sum of upward emitted radiation, downward emitted radiation reflected once, three times, and so on, as well as upward emitted radiation reflected twice, four times, and so on.

of this radiation is reflected twice again to contribute further to the upward radiation, and so on *ad infinitum*. The total contribution of the lower plate to the upward radiation is therefore an infinite series

$$\varepsilon + \varepsilon(1-\varepsilon)^2 + \varepsilon(1-\varepsilon)^4 + \cdots \tag{1.35}$$

which can be written

$$\varepsilon(1 + x + x^2 + \cdots), \tag{1.36}$$

where $x = (1-\varepsilon)^2$. The series in parentheses pops up on so many diverse occasions that its sum is worth knowing.

Consider the finite sum

$$S_n = 1 + x + x^2 + \cdots + x^n. \tag{1.37}$$

Multiply both sides of this equation by x and add 1 to the resulting equation:

$$xS_n + 1 = S_n + x^{n+1} \tag{1.38}$$

and hence

$$S_n = \frac{1 - x^{n+1}}{1 - x}. \tag{1.39}$$

The sum of the infinite series in Eq. (1.36) is the limit of Eq. (1.39) as n goes to infinity. If $|x| < 1$, the limit of x^{n+1} for infinite n is zero. With this restriction on x, the sum S of the infinite series is

$$S = 1 + x + x^2 + \cdots = \frac{1}{1 - x}. \tag{1.40}$$

With this result, the contribution to the upward radiation from emission by the lower plate and multiple reflections between the two plates is

$$\frac{\varepsilon}{1 - (1 - \varepsilon)^2} = \frac{1}{2 - \varepsilon}. \tag{1.41}$$

But we are not finished. The upper plate contributes to the upward radiation indirectly by way of multiple reflections. An amount of radiation ε is emitted downward by the upper plate, and a fraction $1 - \varepsilon$ is reflected upward by the lower plate, thereby contributing to the upward radiation. A fraction of this radiation is reflected downward by the upper plate, then upward by the lower plate. And so on *ad infinitum*. We obtain another infinite series

$$\varepsilon(1 - \varepsilon) + \varepsilon(1 - \varepsilon)^3 + \cdots \tag{1.42}$$

the sum of which is

$$\frac{1 - \varepsilon}{2 - \varepsilon}. \tag{1.43}$$

The total upward radiation is the sum of Eqs. (1.41) and (1.43):

$$\frac{1}{2 - \varepsilon} + \frac{1 - \varepsilon}{2 - \varepsilon} = 1. \tag{1.44}$$

Now multiply this by the Planck function to obtain the amount of radiation in the upward direction between the plates. As far as the radiation field is concerned, it is *as if* the emissivity of the lower (or upper) plate were 1. A cavity with walls that are not black nevertheless fills with blackbody radiation because of emission and multiple reflections. High reflectivity compensates for low emissivity. The only condition we imposed on the cavity is that it be opaque at the frequencies of interest.

Before moving on to averages of emissivities and absorptivities we critically discuss the concept of an average. The following section is applicable to many branches of science.

1.4.2 Averages: A Critical Look

One of the most misleading, if not outright dangerous, yet widespread concepts is that of the unqualified average of a set of values of some physical variable. Not only are there an infinite number of possible averages, not all of them, nor indeed any of them, may be of any use whatsoever. Consider a set of values x_j ($j = 1, 2, \ldots, N$) of anything (temperature, pressure, height, weight, etc.). The *arithmetic average* (or *arithmetic mean*) of this set is defined as

$$\langle x \rangle = \frac{1}{N} \sum_{j=1}^{N} x_j. \tag{1.45}$$

This is one number that in some sense (a point we return to) represents the entire set of values. But a number that might be just as good at representing the set is the *median*, defined as that value of x such that 50% of the values in the set lie above it and 50% below. And then there is the most probable value (or *mode*), defined as the value of x that occurs most frequently. So we already have before us three equally acceptable ways of characterizing a set of values by a single number. But we've hardly scratched the surface of the universe of averages. The arithmetic average is an equally weighted average, and there is no reason why we cannot weight values differently depending on some expectation of their relative importance. That is, we form the average

$$\sum_{j=1}^{N} w_j x_j, \tag{1.46}$$

where the weights satisfy

$$\sum_{j=1}^{N} w_j = 1. \tag{1.47}$$

For the arithmetic average, all the weights are $1/N$. The weights w_j could, but need not, be probabilities of occurrence of the values x_j. Because there is no end to weighting functions, there is no end to possible averages. But matters are even worse. A perfectly respectable (and sometimes useful) average is the *root-mean-square average*, the square root of

$$\sum_{j=1}^{N} w_j x_j^2. \tag{1.48}$$

But why stop at the square root of the average of squares? Why not take the nth root of

$$\sum_{j=1}^{N} w_j x_j^n, \tag{1.49}$$

where n can be *any* number, not just an integer? This is a perfectly good average in the sense that it is one number that characterizes a distribution. We can go further still. Consider any function $f(x)$. Denote the inverse of this function as f^{-1}, defined by

$$f^{-1}\{f(x)\} = x. \tag{1.50}$$

For example, the inverse of the square is the square root; the inverse of the exponential function is the natural logarithm. Now we can generalize the two previous averages as

$$f^{-1}\left\{\sum_{j=1}^{N} w_j f(x_j)\right\}. \tag{1.51}$$

We can generalize this to a continuous distribution of values x:

$$f^{-1}\left\{\int w(x) f(x)\, dx\right\}, \tag{1.52}$$

where

$$\int w(x)\,dx = 1. \tag{1.53}$$

The limits of integration (not shown) are whatever we choose them to be. Given the indefinite number of possible functions f and w, the number of possible averages is boundless. Implicit in Eq. (1.52) is the requirement that the inverse function f^{-1} be single-valued, which is not true for every function. All of these averages are different, in general, and all of them tell different stories, have different uses, or may be useless if not outright dangerous.

First consider a dangerous average. Shoe sizes are distributed over a population. Not everyone wears the same size. Suppose that a shoe manufacturer were to determine the arithmetic average shoe size of a large pool of potential customers and then make shoes only of that size. A lot of customers would go barefoot. Indeed, depending on the distribution of shoe sizes all customers might go without shoes. Suppose, for example, that the population were composed of two groups, those with big feet and those with little feet. The (arithmetic) average shoe size would therefore lie in the intermediate range, and shoes of this size would fit no one. The shoe manufacturer would have lots of unhappy customers and would soon go broke. Here is an example in which one number characterizing a distribution is less than worthless. The shoe manufacturer needs to know the entire distribution of sizes, not just one number obtained from it.

If this example of shoe sizes seems far-fetched consider a region of the planet dominated by marine boundary layer stratus and cirrus. The distribution of cloud-base heights would be peaked around heights in the boundary layer and near the tropopause. The average of these cloud-base heights would produce a mean value far from both peaks in the distribution of heights.

Now let's turn to an average that does have a meaning, although a limited one. Cloud droplets are distributed in size. As we show in Chapter 3 the total power scattered by a cloud droplet depends on its size. Here we can calculate (in principle) an average total scattered power per droplet. The weighting function is the droplet-size probability distribution. The total power scattered by a fictitious cloud of droplets of identical size, all with the same particular average total scattered power per droplet, is the same as that by a real cloud with droplets distributed in size, where the total number of droplets is the same in both clouds. All well and good – but there's a catch (there always is). What size should be associated with the identical droplets in the fictitious cloud? We assumed that there is one and only one average droplet size that gives rise to the average total scattered power per droplet even though this is not always true (see Fig. 3.11). This is not the only problem. Absorption by cloud droplets is, in general, a *different* function of droplet size than scattering by droplets. This means that if we base our calculations of average size on the dependence of absorption on size, we end up, in general, with a different average. And that's not the end of our problems. The angular dependence of scattering is yet another function of droplet size. Although we may be able to determine average sizes that separately have physical meaning for (total) scattered power, absorption, and angular scattering, each one of these averages is different, and thus the question, What is *the* average size of a cloud droplet?, should be greeted with a horselaugh.

Global mean temperature is another example of a dubious concept and for more than one reason. As we have seen, infinitely many mean temperatures are possible, and each one is different. But more important, a single number for an entire planet cannot possibly capture the consequences of temperature changes to human health, wealth, and happiness. As with shoe sizes, one needs the entire distribution, which in this context means everything related to weather: spatial and temporal distributions of temperature, rainfall amount and distribution in time and space, winds, duration, timing, and strength of storms – the list goes on and on. If you live in Minneapolis and were to choose your clothing every day on the basis of the global mean temperature you'd likely be uncomfortable most of the time (or possibly even perish).

1.4.3   Average Emissivity and Absorptivity

Kirchhoff's law was derived for a body illuminated by equilibrium radiation. This naturally leads to the question, Does this law still hold for arbitrary illumination? Although several years ago this question generated some controversy, it now seems safe to say that the dust has settled and we should not hesitate to apply Kirchhoff's law to bodies even when they are not illuminated by equilibrium (blackbody) radiation (see Sec. 2.8). But where we can get into trouble is by misapplying Kirchhoff's law to *averages*.

Total emission by a body at temperature T and with spectral emissivity $\varepsilon(\omega)$ is

$$\int \varepsilon(\omega) P_e(\omega; T)\, d\omega. \tag{1.54}$$

We can't do any damage to this integral by multiplying and dividing it by the same (nonzero) quantity, the integral of P_e, which from Eq. (1.27) is σT^4. Thus we may write Eq. (1.54) as

$$\sigma T^4 \int_0^\infty \varepsilon(\omega) p(\omega; T)\, d\omega = \langle \varepsilon \rangle \sigma T^4, \tag{1.55}$$

where the (normalized) weighting function p is defined by

$$p(\omega; T) = \frac{P_e(\omega; T)}{\int_0^\infty P_e(\omega; T)\, d\omega} \tag{1.56}$$

and the average emissivity is

$$\langle \varepsilon \rangle = \int_0^\infty \varepsilon(\omega) p(\omega; T)\, d\omega. \tag{1.57}$$

The spectral emissivity, like the weighting function, could depend on temperature.

What about absorption? Total absorption by a body is the integrated product of its spectral absorptivity and the spectral distribution of the illumination $F(\omega)$:

$$\int_0^\infty \alpha(\omega) F(\omega)\, d\omega. \tag{1.58}$$

We may define an average absorptivity by

$$\langle \alpha \rangle = \frac{\int_0^\infty \alpha(\omega) F(\omega)\, d\omega}{\int_0^\infty F(\omega)\, d\omega}, \tag{1.59}$$

but this average absorptivity is not, in general, equal to the average emissivity even though $\alpha = \varepsilon$ because the weighting functions for the two averages are, in general, different.

If we assume that the average emissivity Eq. (1.57) is equal to the average absorptivity Eq. (1.59), we should do so knowing that this is not strictly true except for a body at temperature T illuminated by blackbody radiation for this temperature. Equality of the two averages may be a good approximation under some circumstances, such as if the spectral quantities are nearly independent of frequency over the range of frequencies for which the weighting functions have their greatest values.

1.4.4 Brightness and Color Temperature

Suppose we have an instrument that can measure radiant power over some range of frequencies anywhere in the electromagnetic spectrum. For simplicity we assume a narrow field of view for the instrument, but this is not necessary. If we were to point the instrument in a particular direction at a source of radiation, which could, but need not, be a measurably emitting body, the instrument would dutifully measure a radiant power. Now we can ask, What temperature must a blackbody have in order for the instrument reading to be the same? This temperature is called the *brightness temperature* of the source, not to be confused with the ordinary (or thermodynamic) temperature. Even if the radiation measured is mostly or entirely emitted (as opposed to reflected) by a body, its brightness temperature is not the same as its temperature unless we happen to choose a frequency range over which the emissivity of the body is almost 1.

To show that a brightness temperature always exists consider the integral of the Planck function Eq. (1.11) over any range of frequencies:

$$\frac{\hbar}{4\pi^2 c^2} \int_{\omega_1}^{\omega_2} \frac{\omega^3}{\exp\left(\hbar\omega/k_\mathrm{B}T\right) - 1}\, d\omega. \tag{1.60}$$

This integral approaches 0 as $T \to 0$ and ∞ as $T \to \infty$, and its derivative with respect to T is always positive. Thus whatever our instrument reads, we can always find one and only one temperature such that Eq. (1.60) matches it. But keep in mind that this temperature depends on the frequency interval and possibly the direction (unless the source is isotropic). And if the instrument is equipped with a polarizing filter, and we were to rotate it, the brightness temperature might change (unless the source is unpolarized.)

Although the concept of brightness temperature is not restricted, *color temperature* is. The color temperature of a source of (necessarily) visible radiation is the temperature of a blackbody with the same perceived color. As we show in Section 4.3, the gamut of colors accessible to the human observer can be represented as a set of points in a two-dimensional space, whereas the possible colors of blackbodies lie on a curve in this space. Thus blackbodies of all temperatures can match only a small sample of possible colors. Nevertheless, color temperatures can be useful as long as we recognize their limitations. The color temperature of average daylight (sunlight plus skylight) is around 6500 K; that of an ordinary incandescent (tungsten) lamp around 3000 K. Color temperatures of skylight are 10,000–40,000 K. Perhaps it is fortunate that we can't touch the sky. If we could, we'd surely burn our fingers.

1.4.5   A Few Comments on Terminology

No single term, not misleading or faintly ludicrous, for radiation emitted by matter at typical terrestrial temperatures is adequate to distinguish it uniquely and unambiguously from solar radiation. Despite the title of Planck's book (the German title of which may not have the same connotations as its English translation), there is no such thing as "heat radiation" in the sense of a special kind of radiation emitted only by "heated bodies". According to the Planck function, radiation is emitted by *all* bodies regardless of temperature, although to varying degrees at different frequencies. Yves Le Grand calls radiant heat a "meaningless term", and adds, "to say that the sun, for instance, radiates heat is naïve". Moreover, there is no special range of frequencies such that radiation within this range and only within it is capable of raising the temperature of bodies that absorb it. A sufficiently intense source of radiation of almost any frequency can heat a body suitably chosen for its absorption properties at that frequency. For example, lasers at many different frequencies are commercially available for cutting and welding. And although the absorptivity of snow is close to 1 at wavelengths beyond a few micrometers (Fig. 5.16), a single crystal of ordinary salt is transparent from visible wavelengths far into the infrared (to around 60 μm). The heating power of a source of radiation is not uniquely determined by its spectrum but depends as much on its magnitude and the properties of what it illuminates.

Sometimes radiation from the sun is called *shortwave radiation*, whereas that from objects at terrestrial temperatures is called *longwave radiation*. If everyone understands exactly what is meant, no ambiguity arises, but these terms are inherently ambiguous because short and long depend on an arbitrary boundary. The solar spectrum, for example, can be divided into those wavelengths shorter than a certain wavelength and those longer.

If we call the radiation emitted by objects at typical terrestrial temperatures infrared radiation we face some problems. As we show in the following subsection, about half the radiation from the sun lies in the infrared (beyond the red). Moreover, the infrared is a spectral region extending more than three decades, from about 0.7 μm to 1000 μm. Prefixing the qualifiers near, middle, and far to infrared may not help because these terms mean different things to different people. Most important, if we call terrestrial radiation infrared radiation we hobble our thinking. Some people make their living measuring *microwave radiation* emitted by terrestrial objects, and other people make their living measuring microwave radiation from the sun. If we believed in the literal truth of assertions that terrestrial objects emit infrared radiation and that the solar spectrum peters out beyond 2.5 μm we would have to conclude that these people are either charlatans or self-deluded. But they are neither. They simply have risen to the never-ending challenge of measuring today what was not measurable yesterday.

The term *terrestrial radiation* can mislead people into thinking that this is radiation emitted only by solid bodies, ones that you can kick. And, believe it or not, we have stumbled upon conflicting assertions that only solids have emissivities or that only gases have them.

The term *thermal radiation* at least signifies that the emitted radiation is a consequence solely of a body's temperature, in contrast with, say, luminescence, which is more or less independent of temperature, or laser radiation, which has nothing fundamental to do with temperature. But radiation from the sun has just as much right to be called thermal radiation as does radiation emitted by snow, soils, rocks, and cabbages.

The only sure way to avoid confusion and error is to give the source and either a wavelength (or range of wavelengths) or some other clue that will leave no doubts as to what is meant. A good example of ambiguity is the unqualified term infrared imagery, which could mean imagery based on solar infrared radiation scattered by objects or on terrestrial infrared radiation emitted by them. Ambiguity vanishes when the source and wavelength are specified. We call radiation originating from the sun *solar* radiation, that from the Earth *terrestrial* radiation, and give the appropriate wavelengths if necessary.

1.4.6 The Solar Spectrum

In preceding paragraphs we criticized the various terms used to distinguish radiation emitted by the sun from that emitted by objects at typical terrestrial temperatures. But there is no disputing the existence of two distinct spectral regions, only what to call them. Figure 1.4 shows the spectral irradiance of a 6000 K blackbody, with radius equal to that of the sun, at the Earth–sun distance, which approximates the solar irradiance (see Fig. 1.7 and Prob. 1.27), together with that of a 300 K blackbody. There is little overlap between these two spectra. A distinction must be made between the solar irradiance and the irradiance of the sun. At the top of Earth's atmosphere, about 1369 W of solar radiant energy (all wavelengths) crosses each square meter oriented with its normal toward the sun (strictly, this is an average because the Earth–sun distance varies slightly over a year). This quantity, denoted as S_0, is now usually called the *solar irradiance*, formerly the *solar constant*, an inappropriate name given the interest in and effort devoted to measuring its changes. The irradiance of the sun, however, is that *at* the sun. Although the spectra of the two irradiances have the same shape, the magnitude of the solar irradiance is uniformly lower than the irradiance of the sun by the square of the ratio of the sun's radius to the Earth–sun distance. Figure 1.4 also compares the (approximate) solar irradiance with the irradiance of a blackbody at a typical terrestrial temperature. These two spectral irradiances are well separated, whereas the (approximate) irradiance of the sun towers over that of the 300 K blackbody. Where the two curves intersect may be interpreted as the wavelength at which a photon is equally likely to be of solar as of terrestrial origin.

Solar radiation is not blackbody radiation because the sun is not enclosed in an opaque container, but the solar spectrum does approximate that of a blackbody. A detailed solar spectrum is shown in Fig. 1.7, on which the boundaries between ultraviolet, visible, and infrared are marked. About 51% of the solar irradiance lies in the infrared, 42% in the visible, and 7% in the ultraviolet. These numbers, easy to remember, are not exact if for no other reason than that the boundaries between these regions are somewhat arbitrary.

1.4.7 Imaging and Spectral Dependence of Contrast

A good rule of thumb is that most terrestrial objects (e.g., snow, soil, water, vegetation) are approximately black (have spectral emissivities near 1) over the range of frequencies that encompass much of the Planck function for typical terrestrial temperatures. Because of this, the brightness temperature (for infrared frequencies) of terrestrial objects is a good approximation to their thermodynamic temperature. This provides a means for determining temperatures remotely: measure the amount of radiation (in some frequency interval) emitted by an object and convert this radiation into a temperature by way of the Planck function. We can do even

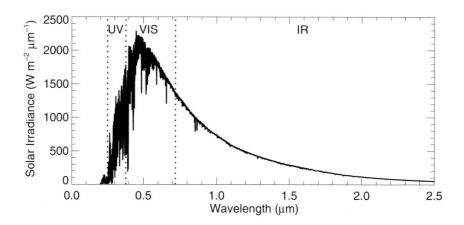

Figure 1.7: Solar spectrum (after Kurucz and Clough).

more. Temperature differences give rise to *contrast*, which allows for the possibility of imaging radiation we cannot see but that an instrument can.

The *image* of any object is a one-to-one transformation: every point on the object corresponds to one and only one point on the image. This transformation is the function of lenses in cameras, in slide projectors, and in your eyes. Remove the lens from a camera and light from objects still illuminates the film, but a single point on the film receives light from many object points. Remove the lens from a slide projector and it still projects light, but no image, onto a screen. A pinhole is the simplest kind of imaging device. One and only one line can be drawn from any point on an object, through the pinhole, to an image point.

Our eyes are imaging devices that respond to different amounts of radiation coming from different directions. We can get about in the world because of these (relative) differences, which is called *contrast*. A whiteout is the absence of contrast. You can experience whiteouts in blizzards or while descending through thick clouds in an airplane (see Sec. 5.2). If you have ever been in a whiteout you know that it can be frightening. You can't tell up from down, right from left. You are lost in a field of radiation the same in all directions.

Our eyes form images using visible radiation, the (indirect) source of which is the sun or lamps. That is, we usually image *scattered* light (we rarely look directly at the sun or at light bulbs). But imaging devices are not restricted to visible frequencies. The radiation emitted by terrestrial objects at different temperatures provides contrast between them, the greater the temperature difference, the greater the contrast. All imaging devices have a contrast threshold below which they cannot distinguish one object from an adjacent one.

Several years ago we got a telephone call from a very frustrated scientist. While poring over infrared images of sea ice he had noticed that the contrast was better in images at shorter wavelengths. He wanted to know why. So he called remote sensing experts and asked them. They all agreed that contrast was indeed better at the shorter infrared wavelengths. But that wasn't his question. He wanted to know *why*, but ran into a blank wall. No one disagreed with his observation but no one could explain it. And the more incomprehension he encountered,

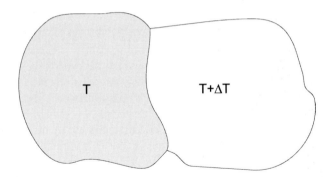

Figure 1.8: Two adjacent objects at different temperatures can be distinguished one from the other if the relative difference in the radiation emitted by them is sufficiently large. This contrast depends on the relative temperature difference $\Delta T/T$ and the chosen frequency of the radiation.

the more frustrated he became. The following is our answer to his question, which is a simple application of the Planck function and also illustrates the importance of beginning with fundamentals when faced with a problem.

Suppose that we have two objects, side by side, at different temperatures T and $T + \Delta T$ (Fig. 1.8). We may define the contrast between these two objects (assumed to be nearly black over the frequency range of interest) as

$$\frac{P_e(T + \Delta T) - P_e(T)}{P_e(T)}. \tag{1.61}$$

The first term in the numerator can be expanded in a Taylor series and truncated to yield an approximation for the contrast:

$$\frac{1}{P_e}\frac{\partial P_e}{\partial T}\Delta T \approx \frac{\Delta T}{T}\frac{x}{1 - e^{-x}}, \tag{1.62}$$

where $x = \hbar\omega/k_B T$. Not surprisingly, the contrast between the two objects depends on their relative temperature difference, but also on frequency by way of the quantity x. The contrast enhancement

$$\frac{x}{1 - e^{-x}} \tag{1.63}$$

is approximately 1 for small x and increases approximately linearly with increasing x. Thus, all else being equal (always an important caveat), greater contrast is obtained for higher frequencies (shorter wavelengths). At typical terrestrial temperatures ($\sim$300 K) $x \gg 1$ at wavelengths in the range $4 - 40\,\mu$m and so the contrast is approximately

$$\frac{\Delta T}{T}\left(\frac{\hbar\omega}{k_B T}\right) = \frac{\Delta T}{T}\left(\frac{hc}{\lambda k_B T}\right). \tag{1.64}$$

If inherent contrast were the *only* criterion for choosing one infrared frequency over another, the highest possible frequency would be the best choice. Of course, at sufficiently high frequencies there might not be enough emitted radiation to image.

1.5   Emission by Clouds

All else being equal, we expect lower air temperatures on mornings after clear nights than after cloudy nights. The correct explanation of this fairly common observation, however, is not widely known. What is widespread is an *incorrect* explanation: radiation emitted by the ground is reflected by clouds. Do clouds reflect radiation? Of course they do. Any fool can see that clouds are bright. And so they may be – at visible wavelengths, which is all that any fool can see. But because we cannot see the kind of radiation emitted by the ground at typical temperatures, we are on thin ice extrapolating from visible to infrared wavelengths. Metals (clean) such as silver and aluminum are highly reflecting at visible wavelengths. As it happens, these metals are just as highly reflecting at infrared, microwave, and even radio wavelengths. But what is true for silver and aluminum is not true for clouds. Although thick clouds may have high reflectivities for visible solar radiation, these same clouds are nearly black at the infrared wavelengths of radiation emitted by the ground (see Fig. 5.15). Our eyes deceive us about clouds. And not only clouds. For example, black and white (to our eyes) paints have nearly identical emissivities, as does black and white skin, over the Planck spectrum at typical terrestrial temperatures.

Suppose that some of the water vapor in clear air condenses into a cloud of water droplets; the total amount of water substance does not change. Suppose also that the temperature (assumed uniform) is the same for the clear as for the cloudy air. What has changed radiatively?

On a partly cloudy spring day we pointed an infrared thermometer at a patch of clear overhead sky. Although the air temperature measured with an ordinary thermometer was $20\,°C$, the temperature recorded by the infrared thermometer was a frigid $-50\,°C$, not even close to air temperature. This is because an infrared thermometer measures brightness temperatures, which are lower or at most equal to terrestrial thermodynamic temperatures. When we shifted the thermometer's field of view from clear sky to an adjacent patch of cloudy sky the brightness temperature shot up to $-3\,°C$. It is not plausible that two adjacent patches of sky differed in temperature by $47\,°C$, so the only possible explanation is that the emissivities of the two patches were different given that the reflectivity of clouds (Fig. 5.15) and air for terrestrial radiation is small. We have done this simple but dramatic experiment many times, always with the same result: clear sky is always radiatively much colder than adjacent cloudy sky. This difference lies mostly in the markedly different spectral emissivities of water vapor and of liquid water. Later that same day, after sundown, as clouds thickened, the overhead brightness temperature had increased to $2–3\,°C$ even though air temperatures had dropped.

On a summer day with broken clouds we pointed an infrared thermometer at the zenith and measured the changing brightness temperature (Fig. 1.9); the state of the sky when the measurements were taken is shown in Fig. 1.10. As clouds flitted in and out of the thermometer's field of view, the brightness temperature fluctuated by about $40\,°C$. One of the *greguerías* of Ramón Gómez de la Serna is a charming way of looking at broken clouds: "*La curiosidad del cielo por ver la tierra abre muchas veces el nublado*" (The sky's curiosity to see the earth often opens clouds).

To further support the hypothesis that the difference between clear and cloudy skies lies in their emissivities, consider Figure 1.11. This figure shows the (calculated) *normal emissivity* over a range of infrared wavelengths (about 12–16 μm) of a uniform layer of moist air at a total pressure of 1 atmosphere, temperature $20\,°C$, and 1 cm of precipitable water. Normal

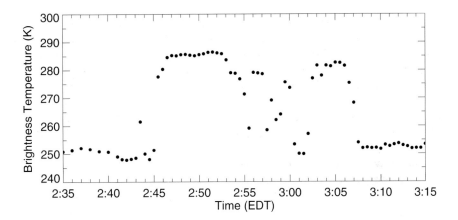

Figure 1.9: Brightness temperature of the zenith sky on a summer day with broken clouds. Measurements were made with the radiation thermometer pointed vertically. See Fig. 1.10 for an all-sky photograph of the cloud cover.

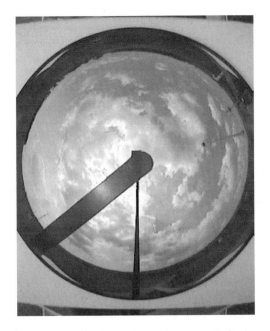

Figure 1.10: All-sky photograph of a partly cloudy sky on a summer day in State College, Pennsylvania. A time-series of the brightness temperatures of the zenith sky is shown in Fig. 1.9.

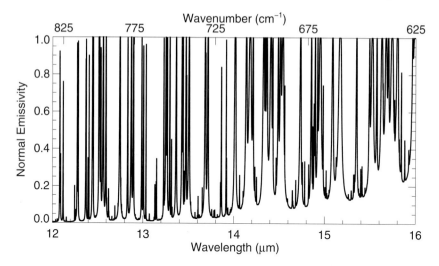

Figure 1.11: Normal emissivity of a uniform layer of moist air at 20 °C, total pressure of 1 atmosphere, and 1 cm of precipitable water. To resolve details only a fairly narrow range of wavelengths is shown; the emissivity spectrum is similar over the entire infrared spectrum.

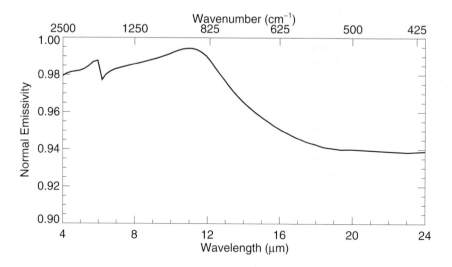

Figure 1.12: Normal emissivity of a layer of pure water sufficiently thick that very little incident radiation is transmitted by it.

emissivity is that in the direction perpendicular to the layer (remember, emissivity, in general, depends on direction). Precipitable water of 1 cm means that if all the water vapor were condensed the result would be a layer of liquid water 1 cm thick. The spectral emissivity of water vapor is a series of sharp peaks and deep valleys as a consequence of the *absorption*

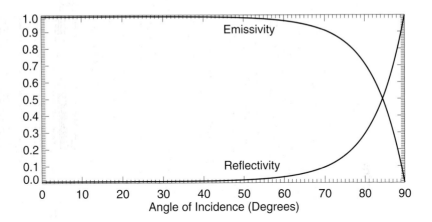

Figure 1.13: Reflectivity for incident unpolarized radiation ($\lambda = 10\,\mu\text{m}$) and the corresponding emissivity for pure water.

lines of water vapor (see Secs. 2.6 and 2.7), narrow spectral regions over which absorption is high. At many wavelengths, the emissivity is close to 1, but at just as many other wavelengths it is smaller, and total emission depends on the *average* emissivity [see Eqs. (1.55) and (1.57)]. Contrast the normal emissivity of water vapor with that of liquid water for a layer sufficiently thick (less than a millimeter) that transmission by it is negligible (Fig. 1.12). This emissivity is a more or less smooth function of wavelength and departs only slightly from 1 over the range 4–24 µm. You don't have to do any calculations to recognize that the average emissivity of the liquid water layer is higher than that of the water vapor. And the emissivity of a thick cloud of water droplets is essentially the same as that of a liquid water layer (see Prob. 2.11). The rate of nocturnal cooling of the ground depends on the difference between its rate of emission and the rate at which it absorbs radiation emitted downward from the sky. The lower this net radiation, all else being equal, the greater the cooling rate. Because emission by clouds, all else being equal, is greater than emission by clear sky, radiative cooling is greater on cloudless nights.

1.5.1   Directional Emissivity

Emissivity depends on the direction of emission (see Sec. 1.4), which is why we qualified the emissivities of interest in the preceding discussion as *normal* emissivities. For an opaque body, the absorptivity for radiation in a particular direction is 1 minus the reflectivity, which by Kirchhoff's law is the emissivity. Because reflectivity depends on direction, so does emissivity. Figure 1.13 shows reflectivity of pure water versus angle of incidence for unpolarized incident radiation ($\lambda = 10\,\mu\text{m}$) and the corresponding emissivity calculated using the Fresnel coefficients (Sec. 7.2). At this wavelength, a layer of water only a few millimeters thick is opaque. If emissivity depends on direction, so must brightness temperature.

The brightness temperature T_b in a direction ϑ is defined by

$$\int \varepsilon(\omega,\vartheta)P_e(\omega,T)\,d\omega + \int\{1 - \varepsilon(\omega,\vartheta)\}P_e(\omega,T_s)\,d\omega = \int P_e(\omega,T_b)\,d\omega, \quad (1.65)$$

where the range of integration is determined by the instrument, $\varepsilon(\omega,\vartheta)$ is the spectral emissivity in the direction ϑ, T is the (thermodynamic) temperature, and T_s is the brightness temperature of the sky (which also depends on direction; see Sec. 2.2). The first integral on the left side of Eq. (1.65) is emission; the second integral is reflection of radiation from the sky. As evidenced by the weak dependence of the (normal) emissivity of water on frequency over a wide range (Fig. 1.12), we can ignore the frequency dependence of the directional emissivity in Eq. (1.65). Because our infrared thermometer responds to a narrow range (about $2\,\mu m$) of wavelengths around $10\,\mu m$, we can approximate Eq. (1.65) by

$$\varepsilon(\vartheta)P_e(\omega,T) \approx P_e(\omega,T_b), \quad (1.66)$$

where the frequency corresponds to $10\,\mu m$. We also assume that reflection is negligible compared with emission, which is not true for near-glancing angles. At this wavelength, and for T around 300 K, the exponential term in the Planck function Eq. (1.11) is much greater than 1, and hence we can approximate Eq. (1.66) by

$$\varepsilon\exp\left(\hbar\omega/k_B T_b\right) \approx \exp\left(\hbar\omega/k_B T\right) \quad (1.67)$$

Take the natural logarithm of both sides and rearrange terms to obtain

$$\frac{\hbar\omega}{k_B T_b} \approx \frac{\hbar\omega}{k_B T} - \ln\varepsilon. \quad (1.68)$$

If we write $T_b = T - \Delta T$ and assume that $\Delta T/T \ll 1$, we can further approximate Eq. (1.68) as

$$\frac{T - T_b}{T} \approx -\frac{k_B T}{\hbar\omega}\ln\varepsilon. \quad (1.69)$$

Thus we predict that at a fixed temperature T, the relative difference between thermodynamic temperature T and brightness temperature T_b for any direction is a linear function of the negative of the natural log of the emissivity in that direction.

To verify the correctness of Eq. (1.69) we measured the brightness temperature of water in a 30 cm × 40 cm pan about 2 cm deep at four elevation angles (60°, 40°, 25°, and 16°). Because an infrared thermometer cannot distinguish emitted from reflected radiation, the water was heated to maximize emission, and measurements were made on a cool, clear day to minimize incident radiation from the sky. The emissivity was obtained from Fig. 1.13. In Eq. (1.69) T is the temperature of the *surface* of the water, which, because of evaporative cooling, is not what one measures by immersing an ordinary thermometer into the water. As a good estimate for T we used the (average) brightness temperature at 60°; the emissivity is close to 1 for this direction (Fig. 1.13). Emission from the overhead sky corresponded to a brightness temperature less than about 218 K on the day measurements were made (see Sec. 2.2 for more on the variation of brightness temperature of the sky with direction). The

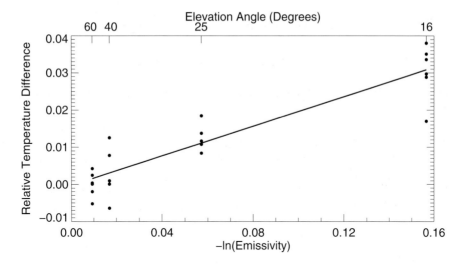

Figure 1.14: Measured differences between thermodynamic temperature and brightness temperature relative to thermodynamic temperature for pure water at four elevation angles with six measurements per angle. The straight line is a linear least-squares fit to the measurements. The chi-square goodness-of-fit statistic is 0.00065 implying a correlation coefficient r^2 of 0.83.

reflectivity of water for radiation from the overhead sky is small (~ 0.01). We made six sets of measurements for the four directions. The average surface temperature of the water was about 338 K, for which the slope in Eq. (1.68) is about 0.22. Figure 1.14 shows a least-squares fit to all the measurements, with slope 0.22, in agreement with the predicted slope even given all the approximations underlying it and the difficulty of making the measurements.

We have more to say about directional emissivity in Section 2.2.

1.6 Emissivity and Global Warming

One cannot open a newspaper or magazine these days without encountering warnings about impending global warming because of the ill-named greenhouse effect. As a consequence of this extraordinary publicity, explanations of the greenhouse effect have been designed more with journalists and politicians in mind than scientists. Even those who publish more frequently in scientific journals than in the popular press either assert that the greenhouse effect is the result of "closing the atmospheric window", thereby "trapping" radiation, or that it is the result of increased emission from the atmosphere. When adherents to both explanations clash, the result is indeed warming, although local rather than global.

If we set aside criticism of the term greenhouse effect (the function of greenhouses is primarily to suppress convection by enclosing a space heated by solar radiation), we are still left with the task of trying to square two apparently irreconcilable physical explanations.

To do so we turn to a simple example of radiative equilibrium, that for a system consisting of two uniform slabs, one above the other (Fig. 1.15). The absolute temperature of the top slab

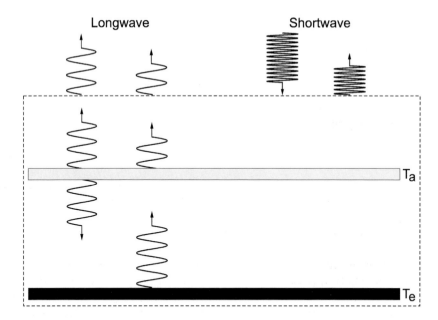

Figure 1.15: Radiative equilibrium between two uniform slabs, infinite in lateral extent. The top slab is transparent to incident shortwave radiation, whereas both slabs emit and absorb longwave radiation. The bottom slab is black to longwave radiation.

is T_a; that of the bottom slab is T_e. The absorptivity of the top slab averaged over the spectrum of the radiation emitted by the bottom slab (assumed to be a blackbody) is α. We assume that the emissivity ε of the top slab (averaged over its emission spectrum) is equal to α, which is not strictly true unless the two temperatures are equal or the emissivity and absorptivity are independent of frequency. We also assume that the top slab does not reflect radiation from the bottom slab, and hence the transmissivity τ (fraction of incident radiation transmitted) of the top slab is $1 - \alpha = 1 - \varepsilon$. The top slab is transparent to radiation from a source outside the system illuminating it from above, where S is the net *irradiance* (net radiant energy crossing unit area in unit time) of the source. The source spectrum hardly overlaps the emission spectra of the slabs.

Consider first the two slabs taken as a single system. If it is in radiative equilibrium, the net radiant energy input to it must be balanced by the radiation emitted by the bottom slab and transmitted by the top slab plus the radiation emitted by the top slab:

$$S = \sigma T_e^4 (1 - \varepsilon) + \varepsilon \sigma T_a^4. \tag{1.70}$$

A radiative energy balance applied to the top slab yields

$$2\varepsilon \sigma T_a^4 = \alpha \sigma T_e^4 = \varepsilon \sigma T_e^4. \tag{1.71}$$

The term on the left is total emission by the top slab, which emits both upward and downward (hence the factor 2); the term on the right is absorption by the top slab of radiation emitted by the bottom slab.

We can solve these two equations to obtain

$$T_e^4 = \frac{S}{\sigma(1 - \varepsilon/2)} \tag{1.72}$$

for the radiative equilibrium temperature of the bottom slab. There are no other modes of energy transfer; the slabs are suspended in a vacuum.

Because of the assumption $\alpha = \varepsilon = 1 - \tau$, we can rewrite Eq. (1.72) as

$$T_e^4 = \frac{2S}{\sigma(1 + \tau)}. \tag{1.73}$$

Note that the temperature of the slabs is a consequence of an external source, but for fixed S the equilibrium temperature depends on the radiative properties (ε or, equivalently, τ) of the top slab.

What does this all have to do with the atmosphere, specifically global warming? Take the top slab to represent the atmosphere, the bottom slab to represent the ground. The total amount of solar radiant energy intercepted by Earth is $S_0 \pi R_e^2$, where R_e is its radius. The total surface area, however, is $4\pi R_e^2$, and so the solar irradiance spread uniformly over the entire globe is $(1369/4)\,\mathrm{W\,m^{-2}}$. But a fraction of this radiant energy is reflected, as evident from satellite photos, which (at visible wavelengths) show a mostly dark Earth (the oceans) brightened here and there by clouds. An estimate for the total fraction of reflected solar radiation, the planetary *albedo* a, is 0.30 (30%). This yields a net solar irradiance $S = S_0(1 - a)/4 = 240\,\mathrm{W\,m^{-2}}$ (albedo is more or less synonymous with reflectivity, although albedo is usually applied to radiation reflected in a hemisphere of directions whereas reflectivity may apply to radiation reflected in a particular direction or in a hemisphere of directions). This is the net solar radiant energy input to Earth spread out evenly over the entire planet. If we take this value for S in Eq. (1.72) we obtain different temperatures T_e depending on ε. For $\varepsilon = 1$, $T_e = 303\,\mathrm{K}\,(30\,°\mathrm{C})$, whereas for $\varepsilon = 0$, $T_e = 255\,\mathrm{K}\,(-18\,°\mathrm{C})$. These temperatures are at least in line with typical air temperatures near Earth's surface, neither ridiculously higher nor lower.

Although nitrogen and oxygen are the numerically dominant atmospheric gases, they are not radiatively dominant. The contribution to the emissivity (over the Planck spectrum for typical terrestrial temperatures) of Earth's atmosphere is mostly from comparatively small amounts of certain *infrared-active* gases (see Ch. 2), water vapor being by far the most abundant, although still less than about 1% of the atmosphere. Carbon dioxide is another infrared-active gas, with an abundance of around 360 ppm. This gas would be present in the atmosphere even if Earth were devoid of human inhabitants, but has been increasing since the beginning of the Industrial Revolution, presumably because of increased burning of fossil fuels.

The emissivity of the present atmosphere is, say, 0.8, which if substituted in Eq. (1.72) with $S = 240\,\mathrm{W\,m^{-2}}$ yields $T_e = 289\,\mathrm{K}\,(16\,°\mathrm{C})$. We note that this is *not* the global average temperature. In the first place, there is no such thing as *the* average temperature (or *the* average anything) but rather infinitely many possible averages depending on the function of temperature averaged and how it is weighted. And what exactly does T_e correspond to? Is it the temperature of the ground? If so, this is not the air temperature, which varies with height, especially near the ground. All we can say for certain is that $T_e = 289\,\mathrm{K}$ is the radiative equilibrium temperature of a slab that absorbs about $240\,\mathrm{W\,m^{-2}}$ of radiant energy, above which is another slab with $\varepsilon = 0.8$ but which does not absorb any solar radiation. In addition, the lower slab is black to the radiation emitted by the upper slab, and radiation is the *only* form of energy transfer (this is what is meant by *radiative* equilibrium).

It would be a bit of a stretch to say that the simple system consisting of two slabs shown in Fig. 1.15 is a model of the atmosphere. It is at best an analogue, useful for helping us understand some basic physics, possibly to frame testable hypotheses, even to estimate relative changes if used judiciously. For example, if ε of the atmosphere is increasing because of increased amounts of infrared-active gases, this suggests that temperatures in the lower atmosphere could increase.

Equation (1.73) is the basis for interpreting global warming as the result of "closing the window". As the transmissivity of the (analogue) atmosphere decreases, the radiative equilibrium temperature T_e increases. Equation (1.72) is the basis for interpreting global warming as the result of increased emission. As the emissivity increases, so does the radiative equilibrium temperature. Which interpretation is correct? One interpretation cannot be right and the other wrong if they are based on the same theory. The two interpretations are merely two different ways of saying the same thing.

Although one interpretation cannot be right and the other wrong, one may be less misleading, more felicitous than the other. We prefer the increased emission interpretation for a few reasons. According to this interpretation we are warmed at the surface of Earth by *two* sources of radiation: the sun and the atmosphere. With this interpretation the atmosphere is actually doing something (emitting) whereas according to the other interpretation it only prevents something from happening. Moreover, the notion that the atmosphere traps radiation is at best a bad metaphor, at worst downright silly. In the emission interpretation the atmosphere is a source of radiation, not a photon trap that corrals wayward photons and sends them back to Earth just as a truant officer returns wayward children to school. A truant officer can return children to school because they are distinguishable, whereas photons are not. If this doesn't bother you, what about the fact that the spectrum of the radiation emitted by the ground to the atmosphere is not the same as the spectrum of radiation emitted to the ground by the atmosphere?

To further bolster the emission interpretation, consider the following thought experiment. Quickly paint the entire globe with a highly conducting metallic paint, thereby reducing the emissivity of the surface to near zero. Because the surface no longer emits radiation, none can be "trapped" by the atmosphere. Yet the atmosphere keeps radiating as before, oblivious to the absence of radiation from the surface (at least initially; as the temperature of the atmosphere drops, its emission rate drops). Of course, if the surface doesn't emit radiation but continues to absorb solar radiation, the surface temperature rises and no equilibrium is possible until the emission spectrum shifts to regions for which the emissivity is not zero.

A more physically relevant quantity than temperature is the downward emission of radiation, which from Eqs. (1.71) and (1.72) is

$$F_\downarrow = \frac{\varepsilon S}{2 - \varepsilon}. \tag{1.74}$$

This equation is Earth's radiation budget in a nutshell. It tells us that emission from the atmosphere to the surface is a consequence of a radiant energy *transformation*. Solar radiation (S) is transformed, because of absorption by the surface, into longer wavelength radiation

$F_\downarrow$ from the atmosphere. But this transformation cannot occur unless the atmosphere has a nonzero emissivity. As ε increases, the downward radiation increases, and the rate of increase is greatest at $\varepsilon = 1$. S is the *net* incoming solar radiation to the planet, which is affected by the radiative properties of clouds and their spatial extent as well as the output of the sun and the Earth–sun distance. Although the slabs in our analogy were uniform, the atmosphere is not. The sky can be cloudy or clear or both at the same time, and we have seen that the emissivity of clouds is greater than that of clear air. This points to the opposing roles of clouds in reducing $F_\downarrow$ because of greater reflection (reduced S) and increasing it because of greater emission.

The total amount of radiant energy of all wavelengths absorbed by the bottom (surface) slab is

$$F_\downarrow + S = \frac{S}{1 - \varepsilon/2}, \tag{1.75}$$

which lies between S $(\varepsilon = 0)$ and $2S$ $(\varepsilon = 1)$.

An average (or better yet, effective) temperature that does have an unambiguous physical meaning is the effective radiative equilibrium temperature T_{eff} of Earth defined as the temperature of a blackbody with a total emission equal to the net solar radiation received by Earth averaged over its entire surface:

$$\sigma T_{\text{eff}}^4 = S = \frac{S_0}{4}(1 - a). \tag{1.76}$$

For $S = 240 \, \text{W m}^{-2}$, $T_{\text{eff}} = 255 \, \text{K}$. This is the equivalent blackbody temperature an observer on the moon would infer for Earth looked upon as an infrared sun. Just as we on Earth say that the sun is equivalent to a 6000 K blackbody (based on the solar irradiance), an observer on the moon would say that Earth is equivalent to a 255 K blackbody (based on the terrestrial irradiance). Note that the effective temperature defined by Eq. (1.76) in no (direct) way depends on the emissive properties of Earth's atmosphere.

In general, energy (or power) is a more relevant physical quantity than temperature. Energies are additive, temperatures are not; energy is conserved, temperature is not. Energy fluxes drive atmospheric processes. But W m^{-2} is banned from American newspapers, both because it is an SI quantity and because it is much too scientific for readers of even the most pretentious newspapers in the land. Similarly, instead of energy fluxes we get the wind-chill temperature, which obscures the fact that energy fluxes, not temperatures, kill people by hypothermia. Canadian meteorologists at one time, perhaps still, gave energy fluxes (rates of energy transfer from bodies), and we recall once seeing a value of $1000 \, \text{W m}^{-2}$ during the winter, a number that makes you shiver if you understand energy fluxes.

The total amount of radiation emitted by the atmosphere over a given time and a given area is a definite number. It may be difficult to measure but it is not an average.

Equation (1.74) contains a trap, not a radiation trap but one for the unwary, and it has caught some big fish. At first glance it would seem that the worst that could happen would be for ε to reach 1, and hence the maximum irradiance from the atmosphere to the surface would be pegged at S. The addition of any more infrared-active gases would have no effect. This dubious conclusion illustrates the perils of pushing an analogy too far. We made a simplifying assumption likely to become invalid as the concentration of infrared-active gases increases

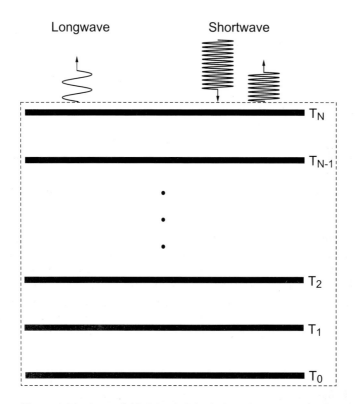

Figure 1.16: A set of N slabs, infinite in lateral extent, each of which is black to longwave radiation. All the slabs except the bottom one are transparent to incident shortwave radiation.

without limit, namely, that the atmosphere can be represented as a single slab (temperature). Let's see how matters change if we consider a set of N slabs, each of which is black to longwave radiation (Fig. 1.16). As before all slabs are transparent to shortwave radiation except the bottom one. Radiative energy balances for the entire system, then for each slab in succession from the top downward yield

$$S = \sigma T_N^4, \ 2\sigma T_N^4 = \sigma T_{N-1}^4, \ 2\sigma T_{N-1}^4 = \sigma T_N^4 + \sigma T_{N-2}^4, \ \ldots \tag{1.77}$$

from which it follows that

$$\sigma T_{N-j}^4 = (j+1)S \quad (j = 0, 1, \ldots, N). \tag{1.78}$$

The downward radiation to the bottom slab (surface)

$$F_\downarrow = NS \tag{1.79}$$

therefore increases without limit as N increases. Of course, this simple analogy would at some point break down, but it does show that downward radiation to the surface could increase with increasing concentration of infrared-active gases, accompanied by higher atmospheric (tropospheric) temperatures.

Does the atmosphere "act like a blanket"? No, not really. Blankets, like almost all insulation, suppress convection. This is why so many insulating materials bear a family resemblance. What do wool, down, cork, felt, hair, glass wool, foam, earth, snow, etc. have in common? They all are porous materials. When subjected to temperature differences, air moves, therefore transporting energy. But the air in pores, if they are sufficiently small, doesn't move much. So the function of blankets is to suppress the movement of air by enclosing it in small pores. If you want a blanket to also reduce net radiation, wrap the blanket with aluminum foil, which has a low emissivity. This is why the insulation in houses or around pipes is coated with foil. But note that this is just the opposite of what happens in the atmosphere. As the emissivity of the atmosphere increases, we expect downward radiation from it to increase (all else being equal). This is yet another reason why assertions about the "atmosphere acting like a blanket" are absurd.

Now a parting shot about the term "greenhouse effect". The only difference between a greenhouse and an ordinary house is where their furnaces are located: outside (the sun) for a greenhouse, inside for an ordinary house. This is why the walls of a greenhouse are (partly) transparent to solar radiation. These same walls are more or less opaque to the infrared radiation emitted by the interior. But so are the walls of ordinary houses, and no one says that they "trap" radiation. You can demonstrate for yourself just exactly what houses do by opening all the doors and windows of your house on a cold and blustery winter day. The walls "trap" just as much radiation as when the house was shut tightly, but now the furnace is heating the great out of doors.

References and Suggestions for Further Reading

The quotation in the opening paragraph is from Hans Vaihinger, 1935: *The Philosophy of 'As if'*, 2nd ed. Barnes & Noble, p. xlii.

We owe the insight about the duality of sound to Fritz Bopp, 1957: The principles of the statistical equations of motion in quantum theory, in *Observation and Interpretation in the Philosophy of Physics*, S. Körner, Ed., Dover, p. 190.

For the latest experimental limits on the photon mass see Jun Luo, Liang-Cheng Tu, Zhong-Kun Hu, and En-Jie Luan, 2003: New experimental limits on the photon rest mass with a rotating torsion balance. *Physical Review Letters*, Vol. 90, pp. 081801 (1–4). The authors cite a string of papers, published during the past 33 years, reporting ever-decreasing upper limits on the photon mass. If it were known to be identically zero, many physicists have been wasting their time and the taxpayers' money.

For an English translation of Einstein's 1905 paper giving a theoretical interpretation of the photoelectric effect (the official reason for his 1921 Nobel prize) see the collection edited by Henry A. Boorse and Lloyd Motz, 1966: *The World of the Atom*, Vol. I, Basic Books, pp. 544–57, an invaluable two-volume collection of and commentaries on the most important papers on atomism from antiquity to the age of quarks and other strange particles.

Despite its age, the treatise by Arthur Llewelyn Hughes and Lee Alvin Dubridge, 1932: *Photoelectric Phenomena*, McGraw-Hill is still worth reading, especially the first two chapters on the history of Eq. (1.4) and its experimental verification, which required considerable effort and ingenuity. The authors assert (p. 5) that there "are but few more interesting developments in all of physics than the growth of the surface photoelectric effect from an obscure physical phenomenon to one of profound theoretical significance, and, finally, to one of surpassing commercial importance."

Heinrich Hertz is usually credited with the (1887) experimental discovery of the (surface) photoelectric effect, although it is much less well known that he invented the thermodynamic diagrams used by meteorologists (Elizabeth Garber, 1976: Thermodynamics and meteorology. *Annals of Science*, Vol. 33, pp. 51–65.)

The history of attempts to interpret the photoelectric effect without photons is recounted by Roger H. Stuewer, 1970: Non-Einsteinian interpretations of the photoelectric effect. *Minnesota Studies in the Philosophy of Science*, Vol. 5. *Historical and Philosophical Perspectives of Science*, Roger H. Stuewer, Ed., pp. 246–63, University of Minnesota Press. Section V is a refutation of the oft-repeated mantra that it is impossible to understand the photoelectric effect without invoking photons.

Arguments for the greater usefulness of the wave language for light are in a letter by M. Psimopolous and T. Theocharis, 1986: "...To see it as it is...to know it as it isn't..." *American Journal of Physics*, Vol. 54, p. 969.

Charles H. Townes's statement about physicists being "diverted" is in his 1984 paper Ideas and stumbling blocks in quantum electronics. *IEEE Journal of Quantum Electronics*, Vol. 20, pp. 547–50.

For an excellent critical discussion of the photon concept, including its long and controversial history, see Richard Kidd, James Ardini, and Anatol Anton, 1989: Evolution of the modern photon. *American Journal of Physics*, Vol. 57, pp. 27–35.

For a simple experimental demonstration that unilluminated bodies we cannot see may emit enough visible light to be detected, see Craig F. Bohren,1987: *Clouds in a Glass of Beer*, John Wiley & Sons, pp. 74–5.

Minimum temperatures for a blackbody to be visible are given by Leo Levi, 1974: Blackbody temperature for threshold visibility. *Applied Optics*, Vol. 13, p. 221.

Derivations of the Maxwell–Boltzmann distribution, Eq. (1.8), are given in treatises on the kinetic theory of gases such as Earle H. Kennard, 1938: *Kinetic Theory of Gases*, McGraw-Hill, Ch. II; Leonard B. Loeb, 1961: *The Kinetic Theory of Gases*, Dover, Ch. IV; and Sir James Jeans, 1982: *An Introduction to the Kinetic Theory of Gases*, Ch. IV. Loeb (pp. 130–8) and Jeans (pp. 124–30) discuss some of the early experimental verifications of this distribution. Deriving it by detailed consideration of molecular energy exchanges by collisions is not trivial, but plausible arguments (not a rigorous derivation) for the distribution of molecular speeds (from which follows that for energies) is given by Craig F. Bohren and Bruce A. Albrecht, 1998: *Atmospheric Thermodynamics*, Oxford University Press, pp. 61–4.

The Planck distribution, Eq. (1.11), is not easy to derive. If it were, Planck would not be almost a household name. One of the best derivations we've seen is by David Bohm, 1989: *Quantum Theory*, Dover, Ch. 1. To follow this derivation requires a good grounding in electromagnetic theory, statistical mechanics, and classical mechanics.

Nowadays we blithely write down the Planck distribution forgetting that the theoretical and experimental path to it was long and arduous. Imagine how difficult it was to make absolute spectral measurements more than 100 years ago. The history of the Planck distribution is recounted in Hans Kangro, 1976: *Early History of Planck's Radiation Law*, Taylor & Francis. For a history of the role that blackbody radiation played in the evolution of quantum mechanics see Thomas S. Kuhn, 1978: *Black-Body Theory and the Quantum Discontinuity 1894–1912*, Oxford University Press.

The equilibrium distribution of molecular energies in a closed container can come about only because of interactions (energy exchanges) between molecules within the gas and between gas molecules and the walls of the container. The Maxwell–Boltzmann distribution applies strictly to an ideal gas in which molecules exert no forces on each other (or the time over which they do exert forces is zero, which also is physically unrealistic). Photons do not interact with each other, so the equilibrium distribution of a photon gas can come about only because of interactions of photons with the walls of the container. Radiation in a perfectly reflecting container (which does not exist) would never evolve to the equilibrium distribution if it did not initially have this distribution. These points are made by F. E. Irons, 2004: Reappraising Einstein's 1909 application of fluctuation theory to Planckian radiation. *American Journal of Physics*, Vol. 72, pp. 1059–67.

Planck's *The Theory of Heat Radiation*, an English translation of the second edition of which (1913) was published by Dover in 1959, is full of insights and qualifiers that have been forgotten over the years. For example, Planck recognized that "the surface of a body never emits rays, but rather it allows parts of the rays coming from the interior to pass through" (p. 4), that a "finite amount of energy. . . is emitted only by a finite. . . volume, not by a single point" (p. 5), that all matter is inhomogeneous at some scale (p. 8), that "in nature there is no such thing as absolutely parallel light or an absolutely plane wave front" (p. 14), and that the maximum of a distribution function depends on how it is expressed (p. 16).

Proof of the theorem for the change of variables in integration is given by R. Creighton Buck, 1978: *Advanced Calculus*, 3rd ed., McGraw-Hill, pp. 182–3. This excellent textbook is rigorous without being oppressive. Buck recognizes that theorems must be proven by a sequence of logical steps, not by notational tricks. But he does recognize that notation "serves as a guide in the correct application of theorems."

The consequences of the failure to recognize that a distribution function does not have a unique maximum are discussed by Bernard H. Soffer and David K. Lynch, 1999: Some paradoxes, errors, and resolutions concerning the spectral optimization of human vision. *American Journal of Physics*, Vol. 67, pp. 946–58.

A discussion of the principle of detailed balance in the context of proving Kirchhoff's law is given by F. Reif, 1965: *Fundamentals of Statistical and Thermal Physics*, McGraw-Hill, pp. 382–4.

A concise but good discussion of time-reversal symmetry is given by P. C. W. Davies, 1976: *The Physics of Time Asymmetry*, University of California Press, pp. 22–7. The time-reversal symmetry of fundamental dynamical laws (e.g., Newton's laws, electromagnetic theory, quantum mechanics) is to be distinguished from what Davies calls the "asymmetry of the world with respect to time" (i.e., we all grow older, alas, even though growing younger is not forbidden by dynamical laws).

We are by no means the first to ridicule the notion that there are "good" and "bad" emitters. In an obituary notice for John Henry Poynting (whose eponymous vector appears in Chapter 4), Oliver Lodge noted that Poynting's "rebellion against an excessive anthropomorphism. . . some substances being praised as good radiators while others are stigmatized as bad. . . though doubtless more than half humourous was in itself wholesome" (*Collected Scientific Papers by John Henry Poynting*, 1920, Cambridge University Press, p. xii).

Kirchhoff's law in Section 1.4 is not his original version, for which see William Francis Magie, 1965: *A Source Book of Physics*, Harvard University Press, pp. 354–60, a superb collection of excerpts from more than 100 papers of great historical significance preceded by short biographies of their authors. For example, the empirical (i.e., obtained by curve fitting) discovery of the radiation law, Eq. (1.27), later derived by Boltzmann, is given in Stefan's own words on pp. 378–81.

For a brief biography of Gustav Robert Kirchhoff and the significance of his work see the entry by L. Rosenfeld in *Dictionary of Scientific Biography*, Vol. VII, pp. 379–83. Other than original papers, the *DSB* is the first place to look if you want to know what our illustrious predecessors really did and said. The last place to look is in textbooks, which are notorious spreaders of rumors, half-truths, and outright errors. Ask any historian of science. Better yet, read Tony Rothman's (2003) delightful *Everything's Relative and Other Fables from Science and Technology*, John Wiley & Sons.

The controversy over the validity of Kirchhoff's law is reviewed by H. P. Baltes, 1976: On the validity of Kirchhoff's law of heat radiation for a body in a nonequilibrium environment. *Progress in Optics*, Vol. 13, pp. 1–25.

The term *emittance* is sometimes used for what we call emissivity. Indeed, it has been suggested that emissivity be used for bodies that are "pure and smooth", emittance for bodies that are not. For a criticism of this suggestion see the letter by William L. Wolfe, 1982: A proclivity for emissivity. *Applied Optics*, Vol. 21, p. 1. Wolfe's parting shot is "On reflection, I like reflectivity and emissivity". So do we, but for a contrary view see Joseph C. Richmond's rebuttal following Wolfe's letter. If the term *emittance* is to be used at all it is best reserved as an abbreviation for *emitted* irradi*ance* (or radi*ance*). Note in Chapter 4 that radiometric and photometric dimensional quantities (radiance, irradiance, luminance, etc.) all end in *ance*, and hence in the same spirit emittance ought to be emissivity times the Planck function.

The assertion on page 22 that radiant heat is a "meaningless term" is from Yves Le Grand, 1957: *Light, Colour, and Vision*, John Wiley & Sons, p. 4.

Luminescence (mentioned in Sec. 1.4.5) is the term first used by Eilhardt Wiedemann (1888) for "all those phenomena of light which are not solely conditioned by the rise in temperature." Luminescence, or "cold light", is contrasted with incandescence, or "hot light". Examples of luminescence are the dim light of phosphorous, the light of fireflies, and the light emitted by substances excited by various kinds of electromagnetic radiation or subatomic particles. By light is usually meant visible or near-visible.

Fluorescence originally was applied to light that ceases immediately upon removal of its source of excitation, in contrast with phosphorescence, which persists. But this distinction is not absolute, long-lived fluorescence merging continuously into short-lived phosphorescence. Moreover, what is meant by "long-lived" and "short-lived" depends on the sensitivity of instruments. Some observable consequences of fluorescence are presented in Section 4.1.7.

In all examples of luminescence electrons in atoms and molecules are excited by some means into higher energy levels. At a later time these electrons decay to lower energy levels, with the attendant emission of photons having energies equal to the differences in the energy levels. In fluorescence, the time between excitation and decay may range from a second to as low as 10^{-9} s, which is still a long time compared with the inverse frequency of visible radiation.

For a history of luminescence see the monumental treatise by Edmund Newton Harvey, 1957: *A History of Luminescence From the Earliest Times Until 1900*, American Philosophical Society. See especially Chapter XI on fluorescence. This term was coined by Sir George Gabriel Stokes, a central figure in Chapter 7. For Stokes's abstract of his lengthy paper on fluorescence see Magie, pp. 344–52.

The solar spectrum in Figure 1.7 is derived from a spectrum developed by Robert L. Kurucz of the Harvard-Smithsonian Center for Astrophysics. The best reference for this spectrum is his unpublished article The Solar Irradiance by Computation (November 25, 1997). We obtained our copy from Kurucz. Tony Clough averaged Kurucz's spectrum to 1 cm^{-1} resolution. We started with Clough's spectrum and averaged it further to 20 cm^{-1} resolution. The best way to obtain the Kurucz solar irradiance values is by downloading the Line-by-Line Radiative Transfer Model (LBLRTM) developed by Tony Clough and his co-workers. Both the original Kurucz spectrum and Clough's 1 cm^{-1} resolution spectrum come bundled with LBLRTM, which is available through Atmospheric and Environmental Research (AER), Tony Clough's employer.

For an older set of tables of the solar irradiance at intervals of 10 nm or 20 nm over most of the spectrum, which may be more convenient for calculations and yet still adequate for many purposes, see M. P. Thekaekara and A. J. Drummond, 1971: Standard values for the solar constant and its spectral components. *Nature Physical Science*, Vol. 229, pp. 6–9.

For simple experiments to demonstrate that emission by aluminum foil is quite different from that of bodies painted white or black, and that despite appearances to the contrary, some (visibly) black and white bodies are nearly identical at infrared wavelengths, see Craig F. Bohren, 1991: *What Light Through Yonder Window Breaks?*, John Wiley & Sons, Ch. 7. See also Richard A. Bartels, 1990: Do darker objects really cool faster? *American Journal of Physics*, Vol. 58, pp. 244–8.

The *greguería* on page 26 is from Ramón Gómez de la Serna, 1994: *Greguerías*, Clasicos Castlia, p. 69.

Problems

1.1. We note in Section 1.2.1 that the Planck distribution can be expressed as a function of frequency or wavelength. But this doesn't exhaust the possibilities: any single-valued function of frequency is a suitable independent variable. Moreover, other distributions related to the Planck distribution are also possible. For example, the spectral number density n (number per unit volume) of photons in a cavity at a fixed temperature is proportional to the Planck function divided by the photon energy (see Prob. 4.51), and n can be expressed as photons per unit frequency $n(\omega)$ or per unit wavelength $n(\lambda)$. Do so and then find the corresponding displacement laws for the maxima of these distributions.

1.2. The inverse of a function might not be single valued (i.e., the inverse might not exist). To convince yourself of this, and to obtain a better understanding of what is meant by the inverse of a function, sketch a simple function with a multi-valued inverse. This requires thinking carefully about what is meant by a mathematical function and its inverse. A simple sketch is all that is needed.

1.3. All raindrops are not identical but are distributed in size. The *Marshall–Palmer distribution*

$$N(D) = N_o e^{-\Lambda D} \tag{1.80}$$

often has been used for raindrops, where N_o and Λ are constants. $N(D)$ has the property that

$$\int_{D_1}^{D_2} N(D)\, dD \tag{1.81}$$

is the number of raindrops per unit volume with diameters between D_1 and D_2. From this distribution one can obtain the mean diameter, call it $\langle D \rangle$. But the scattering of sunlight (see Sec. 3.5) by raindrops is proportional to D^2, so as far as light scattering is concerned the root-mean-square diameter, the square root of the average of D^2, is of physical significance.

The mass of a raindrop is proportional to the cube of its diameter, so the relevant average diameter here is the cube root of the average of D^3. The contribution of a drop to rainfall rate depends on the product of its volume and its terminal velocity, which is approximately proportional to diameter. Here the relevant mean diameter is the fourth root of the average of D^4. Finally, the radar backscattering cross section (see Sec. 3.5.2) is proportional to the sixth power of diameter.

Find the relative values of the root-mean-sixth, root-mean-fourth, root-mean-cube, root-mean-square, and mean diameters for the Marshall–Palmer distribution (for D ranging from 0 to ∞). What do you conclude from this?

1.4. A radiation thermometer infers temperatures of objects (at typical terrestrial temperatures) by measuring infrared radiation from them. We have done the following demonstration. We put some water and ice cubes in a shiny, thin-walled metal container with an open top (a tin can with the label removed will do) and stirred the water thoroughly. Then we pointed the thermometer at the water surface. The temperature reading was close to 0 °C, but when we immediately pointed the thermometer at the side of the container, the temperature reading increased. Then we replaced the ice–water mixture with hot water. The temperature of the water measured at its top surface was a certain value, but when we pointed the thermometer at the side of the container, the temperature reading decreased. Explain.

1.5. Estimate plausible minimum and maximum values (W m^{-2}) at Earth's surface of radiation emitted downward by the atmosphere.

1.6. Try the following demonstration. Hold your arms out from your sides, level with your shoulders, hands open, for about 10 seconds. Then suddenly bring your arms together so that your palms are a centimeter or less apart *facing each other but not touching*. Hold your hands in this position for perhaps 10 seconds or more. Describe what you experience and explain it.

1.7. For the previous problem students have mentioned radiation from the air. By simple physical arguments you should be able to show that radiation from air in a room is negligible.

HINT: A good estimate for the infrared brightness temperature of the sky is about 250 K.

1.8. Estimate the temperature of the filament of a 100 W electric light bulb. State all assumptions. Estimate the uncertainty in your estimate because of uncertainties in the quantities needed for your calculations.

HINT: If you can find a burned-out bulb, smash it (carefully) and examine the filament. If you can't find such a bulb, try to find a clear light bulb so that you can estimate the dimensions of a filament. Be sure to ask yourself if your estimate is reasonable (300 K is not, nor is 30,000 K).

1.9. If ozone in the upper atmosphere is a "good absorber" of solar ultraviolet radiation, why isn't it also a "good emitter" of such radiation, thereby undoing the good it is said to do?

1.10. Suppose that you have a friend, a Texan, say, who tells you that in Texas it once was so hot that his spit boiled when it hit the pavement. Is it possible for water to boil when placed on a surface heated by sunlight (no lenses please)? What is wanted here is a simple, quantitative argument showing that spit boiling on hot pavement either is possible or is not. If possible, is it plausible?

HINT: At the top of the atmosphere, the amount of solar radiation incident on every square meter (pointed toward the sun) is about 1369 W, but, of course, is less at the surface.

1.11. The following was taken from a book on radar systems: "It is known from the theory of blackbody radiation that any body which absorbs energy radiates the same amount of energy that it absorbs." Discuss.

1.12. How might heights of clouds above the surface of Earth be estimated from observations made from a satellite well above them? The satellite is not equipped with a laser or any other source of radiation. State all assumptions.

1.13. What fraction of electricity bills for lighting with incandescent lamps is wasted because they emit invisible as well as visible radiation? You may take the color temperature of incandescent lamps to be 2500 K and assume constant emissivity of the filament.

HINT: You can do this problem analytically by making judicious approximations or you can write a computer program for numerically integrating the relevant integral. If you are in a hurry, you can content yourself with an upper limit on this fraction.

1.14. Estimate the maximum temperature of the illuminated surface of the moon. You may take $1369\,\mathrm{W\,m^{-2}}$ as the solar irradiance at the top of Earth's atmosphere. State all assumptions. After you have made your estimate, try to find out how it squares with any measurements you can find.

1.15. Sketch the spectral absorptivity, from ultraviolet well into the infrared, of the best solar energy collector. By "best" is meant that the temperature it reaches is as high as possible. Take a stab at estimating this temperature. You may take the solar spectrum to extend from $0.25\,\mu\mathrm{m}$ to $2.5\,\mu\mathrm{m}$ and $1000\,\mathrm{W\,m^{-2}}$ as an estimate for the maximum solar irradiance at the ground.

You can't determine this temperature on the back of an envelope but will have to write a computer program. At the very least, write down the equation that has to be solved and give some thought to how you might do so.

1.16. Several years ago a newspaper advertisement described a new kind of radiant heating system. The radiating panels were said to be "hot to the touch but they do not burn." The advertisement went on to say "the panels radiate long infrared rays which are most readily absorbed by the objects, including the human body. In fact, they are the same rays that warm us from the sun." Discuss.

1.17. When one of the authors and his wife lived in Wales they had in their bedroom a heating panel similar to that described in Problem 1.16. Now they have a small electric heater in their bedroom, and it seems to do about as good a job as the panel. This heater has 6 strips, each about 20 cm long and 0.5 cm wide. When the heater is operating, the strips glow red. On the basis of this information you should be able to estimate the size of the panel in Wales. You may take the panel to be square. State all assumptions and, as always, be sure that your answer makes sense.

1.18. Why do temperature inversions (temperature increasing with height) often form at night immediately above the tops of clouds?

1.19. In the simple two-slab analogue to the Earth–atmosphere system, the top slab representing the atmosphere was assumed to emit the same amount of radiation to space as to the ground. Is this true for the real atmosphere? Why or why not?

1.20. If a warm object is wrapped with aluminum foil the rate of cooling of the object decreases (by "warm" is meant that the temperature of the object is higher than that of its surroundings). You can verify this by heating some water, pouring it into a jar, then measuring the rate at which the water temperature decreases with time. Do this with a bare jar and then with one wrapped in aluminum foil.

Aluminum foil has a lower emissivity than glass, and hence the foil reduces emission. But at the same time, the foil has a lower absorptivity, which means that the foil decreases the absorbed radiation from the surroundings. Because the foil giveth and the foil taketh away why isn't its net effect zero? What is the essential condition for the cooling rate to decrease because of wrapping the jar with foil? Another version of this question is, Under what conditions would wrapping an object with foil increase the rate of cooling?

You may assume that Kirchhoff's law is valid even for spectrally averaged emissivity and absorptivity.

1.21. The following statement is in an opening paragraph in a paper (Carl G. Ribbing, 1990: Beryllium oxide: a frost preventing insulator. *Applied Optics*, Vol. 15, pp. 882–84) on low emissivity coatings for dew and frost prevention: "The possibility for wavelength-selective condensation prevention is based on the existence of atmospheric transmittance windows. It is only by emission of radiation in these wavelength intervals that a body can cool to a substantially lower temperature than the immediately surrounding air. . ."

Atmospheric windows are spectral regions in the terrestrial infrared where the atmosphere has a comparatively high transmissivity.

Discuss the statement enclosed in quotation marks. It can be restated as follows: To develop coatings with spectral emissivitivies that reduce net radiative cooling, one need consider only the atmospheric window regions of the terrestrial infrared spectrum.

What implicit assumptions underlie this statement? Do you agree with it? You can give either mathematical or physical arguments or both.

HINT: It may help to divide the terrestrial infrared spectrum into two regions: window and non-window. Don't hesitate to make simplifying assumptions that enable you to concentrate on what is essential rather than get bogged down in details.

1.22. One of the authors lived in Arizona on a former dude ranch in the desert outside Tucson. One year he painted the roof of his cottage (not air-conditioned) with a metallic paint. The aim was to increase the reflectivity of the roof for solar radiation, thereby reducing the temperature inside the cottage. If your aim is to keep the interior of a house cooler in a region of intense solar illumination would you paint the roof of the house with metallic or white paint? Explain your choice. This is an open-ended question without a right or wrong answer. For measurements relevant to this problem see Dena G. Russell and Richard A. Bartels, 1989: The temperature of various surfaces exposed to solar radiation: an experiment, *The Physics Teacher*, Vol. 27, pp. 179-81.

1.23. Photons have energy and momentum (linear and angular). We gave an expression for the energy $h\nu$ of a photon but not its momentum. But from this expression you should be able to guess the expressions for photon linear momentum and angular momentum (to within dimensionless constants) by simple dimensional arguments. Don't look up the answer in a physics book. If you know the answer, don't do this problem because it won't teach you anything.

1.24. Estimate the maximum irradiance of a towering inferno (e.g., a huge forest fire) relative to that of the sun on a clear day at noon.

1.25. The September 30, 1995 issue of *Science News* contained the following in an article on engineering solutions to global warming: "One of the more expensive options would be to install a giant solar deflector, built from materials on the moon, at a point 1.5 million miles from Earth in the direction of the sun. Thinner than a human hair, the diaphanous sheet would stretch 2,000 kilometers across and deflect 2 percent of the radiation headed toward Earth. Estimated cost: $1 trillion to $10 trillion."

You might think that thousands of hours of computer time went into computing the value "2 percent". In fact, you can obtain this number (approximately) in a few minutes without a calculator. The purpose of the "diaphanous sheet" is to compensate for the increased emissivity of the atmosphere as a consequence of doubling the concentration of CO_2. Assume that the increase in the radiative equilibrium temperature of the simple slab model of the Earth–atmosphere system at the surface because of this increased CO_2 is 1.5 K and that the present radiative equilibrium temperature is 288 K. This is all you need in order to do this problem. Huge computers, elaborate global circulation models and radiative transfer codes are not necessary. The analysis underlying a project estimated to cost more than $1 trillion can be done on the back of an envelope – if you know what you are doing.

1.26. Estimate the total rate of emission of radiation from a (naked) human. A good estimate for skin temperature is $33°C$. The surface area S (in cm^2) of a human can be estimated from the DuBois formula

$$S = 71.84\, W^{0.425}\, L^{0.725},$$

where W is body mass in kg and L is height in cm. This is the total surface area, about 80% of which is the effective radiating area. To good approximation human skin and hair have an emissivity close to 1 over the relevant range of infrared wavelengths. The Dubois formula is in Max Kleiber, 1975: *The Fire of Life*, 2nd ed., Robert E. Krieger, p. 184. This very readable book is filled with fascinating data such as the weight loss of the fasting dog Oscar (p. 28) and the time evolution of the rectal temperature of pigs in air at different temperatures (p.177).

Express your result in kilocalories (food calories) per day. One calorie is 4.184 J and a kilocalorie is 1000 cal. On the basis of this calculation what do you conclude about your necessary daily food intake if there were no radiation from the environment? Suppose that there were no such radiation. What strategy would humans likely have had to adopt in order to survive? What class of animals would we resemble?

Now determine your *net* infrared radiation, in kcal/day, in an environment at room temperature ($20°C$). Estimate the fraction of your daily food intake (in kcal) that is needed just to balance net radiation.

How does clothing alter your numbers and conclusions? To answer this you either have to estimate the temperature drop from skin to clothing or measure this drop with an infrared thermometer.

You often hear the assertion (but not from us) that "most body heat is lost from your head." Can you make any sense out this, its implication being that your head is some kind of super-radiator? Yet the emissivity of skin and hair on your head is no different from that on the rest of your body, and head temperatures are not greatly different from those on other areas of your body. Moreover, you have more hair on your head than other parts of your body (unless you are bald). You can make your answer quantitative by estimating the contribution of your head area to total body area.

1.27. What is the temperature of a blackbody with an emission spectrum that peaks at the same wavelength as the solar spectrum (outside Earth's atmosphere)? Suppose that the sun were replaced by a blackbody of the same radius and distance to Earth. What is the temperature of this blackbody such that the solar irradiance is its presently accepted value (about $1369 \, \text{W m}^{-2}$)? The angular width of the sun is about $0.5°$.

1.28. For what angle is the reflectivity of water at $10 \, \mu\text{m}$ such that emission by it at $100°C$ equals reflection of clear-sky radiation of the same wavelength? You may take the radiative temperature of the sky to be 250 K. This problem is related to the directional emissivity experiment discussed in Section 1.5.1.

1.29. We show in Section 1.2.1 that the wavelength at which the Planck function has a maximum (for a given temperature) is not an absolute quantity but depends on the independent variable chosen, which is arbitrary. Mark Heald (2003) in an article in *American Journal of Physics* (Where is the "Wien peak"?, Vol. 71, pp. 1322–23) agrees but notes that a peak wavelength can be defined unambiguously as a *median wavelength*: the wavelength that divides the Planck function into two regions of equal area (integrated irradiance). First, convince yourself that this median wavelength is independent of the variable in the Planck function and why. Then determine the displacement law for the median wavelength.

HINT: The second part of this problem requires numerically evaluating an integral.

1.30. Brightness temperature is equal to thermodynamic temperature if the emissivity of the object of interest is 1 at the wavelengths detected by the radiation thermometer and the detected radiation is solely that emitted by this object. Assume that the second condition is satisfied and derive a simple expression for the relative difference between brightness temperature (at, say, a wavelength of $10 \, \mu\text{m}$ or thereabouts) and thermodynamic temperature as a function of the departure of the emissivity from 1 for typical terrestrial temperatures.

HINT: Expand the brightness temperature in a Taylor series around $\varepsilon = 1$.

1.31. The result of the previous problem might lead you to believe that errors in brightness temperatures because of departures of emissivity from 1 are large. Measurements of the brightness temperature of objects are always made in a particular environment. In what kind of environment would the error be greatest? In what kind of environment would the error be least (possibly zero) even if the emissivity of the object of interest were appreciably less than 1? To help you answer these questions consider measuring the brightness temperature of a wall inside a house. Even though the emissivity of a wall probably is less than 1, the radiation thermometer reading is more or less correct. Why?

1.32. Because photons possess linear momentum, light can exert *radiation pressure*, that is, transfer momentum to illuminated objects. Determine the radiation pressure of direct solar radiation illuminating a surface with reflectivity 1 at all solar wavelengths. Take the irradiance to be $1000 \, \text{W m}^{-2}$. Express your answer relative to sea-level atmospheric pressure. What is the radiation pressure on a surface that is black at all solar wavelengths?

HINTS: Problems 1.1 and 1.23 should be helpful. The relation between radiation pressure and irradiance is so simple that you could obtain it (to within a constant factor) by dimensional analysis.

1.33. It has been stated countless times that no body can emit more radiation in any wavelength interval than a blackbody at the same temperature. Yet there are plenty of examples in which bodies can emit thousands of times more than blackbodies. Give a few simple examples. You don't have to go far to find them.

1.34. We have hit on a sure-fire weight loss scheme, the Bohren and Clothiaux Radiation Diet. According to Einstein's famous mass energy equation, which now is such a commonplace that it is a staple of cartoons, if a body loses energy its mass decreases. So we are planning construction of a world-wide chain of weight-loss rooms. The walls of a room are black to the kind of infrared radiation emitted at body temperature and cooled to the temperature of dry ice ($-78.5\,°C$) to maximize net radiation from clients in the room. While sitting in it they will radiate their excess weight without exercise or diet at a cost of only $10 an hour. We are offering exclusive franchises to readers of this book for the absurdly low fee of only $50,000. How can you lose? Send us money now and take advantage of this advance offer.

1.35. Estimate how long it would take in a completely darkened room to photograph an object at room temperature. Assume that with the fastest film available and the aperture as large as possible (smallest f-stop) you can photograph the coils of an electric stove (estimated temperature of, say, 1500 K) with an exposure time of 1/100 s. All that is wanted here is a crude estimate. Would it take hours, weeks, months, years, centuries?

1.36. Green vegetation is more absorbing of solar radiation with wavelengths less than about 700 nm than of radiation with greater wavelengths (note the steeply rising radiance of grass beyond about 690 nm in Fig. 4.13). Can you think of reasons why plants might have evolved to have this spectral absorption property?

1.37. Radiation literally could be trapped in a container with perfectly reflecting (100%) walls. Photons injected into such a container would rattle around forever. But perfection is not of this world. Estimate how much departure from perfection (what reflectivity < 100%) could be tolerated in a container with linear dimensions of order 10 cm such that radiant energy trapped within it would decrease by no more than, say, $1/e$ of its initial value in an hour.

1.38. Photons emitted by a body cannot be interrogated as to their origins: luminescence or thermal emission. All that can be measured is total emission. But it is possible to determine if part of the emitted radiation is a consequence of luminescence and even how much. If total emission (at a given frequency) exceeds that of a blackbody, the excess can be attributed to luminescence (almost by definition). But a body can emit less than a blackbody (at the same temperature) and yet still be luminescent. What do you need to know in order to determine the contribution of luminescence to total emission? Obtain an expression for this contribution.

1.39. Show that in principle one can measure the temperature of a blackbody by measuring only the ratio of radiation emitted by it at two different wavelengths. How should these wavelengths be chosen and how does your choice depend on the expected range of temperatures?

HINT: Sketching a few functions should help.

1.40. Show that the form of the Stefan–Boltzmann law, Eq. (1.27), can be derived by applying the first and second laws of thermodynamics to a photon gas (in a cavity) given two results for such a gas: its pressure p is equal to one-third its energy density u, which depends on T only.

HINT: Consider a reversible expansion (or compression) of the photon gas. The total internal energy of the gas is uV, where V is its volume.

1.41. What is the temperature of the air near the surface and just above the top layer in the two-slab model developed in Section 1.6?

HINT: Place a thin layer in each location and assume that it does not affect the overall energy balance of the two-slab model. This question is motivated by Section 3.8 in Dennis L. Hartmann, 1994: *Global Physical Climatology*, Academic Press, pp. 61–3.

1.42. Meteorology students often are taught that the maximum daytime temperature (of what we are left to guess) occurs when the net radiation balance (again, of what we must guess) is zero. Critically discuss this.

1.43. Peter W. Huber and Mark P. Mills, 2005: *The Bottomless Well*, Basic Books, p. 6, assert that the amount of solar energy reaching the "surface from above" is "roughly ten thousand times as much as humanity consumes in the form of fossil fuels... Green plants seize and temporarily store a tiny fraction of it. During the night, the dark side of the Earth radiates all the rest, along with the geothermal heat, back out into the black depths of the cosmos." Critically discuss.

1.44. Why is the photon language of almost no use to most radio engineers? There is a simple yet fundamental reason that has essentially nothing to do with phase differences. A related question, which is also a hint, is what must the photon number density be such that a detector with area A senses a beam as a more or less discrete set of photons?

2 Absorption: The Death of Photons

If emission may be called, somewhat poetically, the birth of photons, absorption may be called their death, although their spirit (energy) lives on in whatever absorbs them. The previous chapter is devoted mostly but not exclusively to emission because it would be artificial to try to divorce it completely from absorption given that they are inverse processes. In this chapter the emphasis is on absorption.

We purposely used the term absorption by a body in the previous chapter to avoid conveying that radiation is absorbed at or by surfaces. It is not, if only because they do not exist. Surfaces are mathematical idealizations of transitions in material properties over distances smaller than we can perceive with our eyes or fingers or measuring instruments. Perhaps *interface* better conveys this sense of a sharp transition. If radiation incident at the interface between a negligibly absorbing medium and an absorbing medium is attenuated by absorption over a very short distance, we might say that the surface (interface) absorbed the radiation. But it would be more accurate to say that absorption occurred *because* of the interface and over a short distance (on some scale) within the medium. We begin by making no assumption about the characteristic distance over which absorption occurs: it could be meters or kilometers or micrometers or nanometers.

2.1 Exponential Attenuation

Consider attenuation by absorption of a monodirectional beam of monochromatic radiation by an optically homogeneous medium. By *optically homogeneous* is meant homogeneous on the scale of the wavelength; no medium is absolutely homogeneous (continuous). By *monodirectional* is meant that the radiation is confined to a narrow set of directions, and by *monochromatic* is meant that it is confined to a narrow set of wavelengths. Again, a truly monodirectional, monochromatic beam does not exist. Even a laser beam has a finite angular divergence and range of wavelengths.

The beam is directed along the x-axis and its *irradiance*, radiant energy crossing unit area per unit time (Sec. 4.2) at $x = 0$ is F_0. We can imagine the region from 0 to an arbitrary distance x to be subdivided into N identical slices each of thickness $\Delta x = x/N$ (Fig. 2.1). Upon transmission from 0 to Δx the irradiance of the incident beam is reduced (attenuated) from F_0 to F_1 because of absorption within Δx. It is reasonable to assume that if Δx is sufficiently small attenuation is proportional to Δx and to F_0, where the proportionality constant (*absorption coefficient*) is denoted by κ:

$$F_0 - F_1 = F_0 \kappa \Delta x. \tag{2.1}$$

Fundamentals of Atmospheric Radiation: An Introduction with 400 Problems. Craig F. Bohren and Eugene E. Clothiaux
Copyright © 2006 Wiley-VCH Verlag GmbH & Co. KGaA, Weinheim
ISBN: 3-527-40503-8

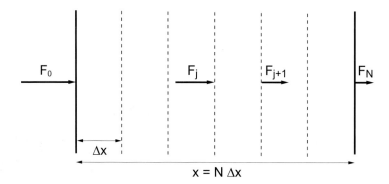

Figure 2.1: An absorbing, plane-parallel medium can be imagined to be made up of N slices, each of thickness Δx. A monodirectional beam with irradiance F_0 is incident on this medium. The transmitted irradiance F_j at a distance $j\Delta x$ into the medium decreases with increasing j because of absorption in all preceding layers.

It is more convenient to write this as

$$F_1 = F_0(1 - \kappa\Delta x).\tag{2.2}$$

This is the irradiance at a distance Δx from the origin. At a distance $2\Delta x$, by the same argument, the irradiance is

$$F_2 = F_1(1 - \kappa\Delta x) = F_0(1 - \kappa\Delta x)^2.\tag{2.3}$$

An implicit assumption in going from Eq. (2.2) to Eq. (2.3) is that transmission by the two slabs is independent, and hence transmission by each can be multiplied to obtain transmission by the two combined (this assumption is scrutinized in Sec. 2.4). The pattern now should be clear: after transmission over a distance $x = N\Delta x$, the irradiance is

$$F_N = F_0(1 - \kappa x/N)^N.\tag{2.4}$$

Now take the limit as N becomes indefinitely large keeping κx constant:

$$F = \lim_{N\to\infty} F_0(1 - \kappa x/N)^N,\tag{2.5}$$

where we omit the subscript on F. Do you recognize this limit? It is one of the many ways of defining the exponential function:

$$e^{\xi} = \lim_{n\to\infty}(1 + \xi/n)^n.\tag{2.6}$$

From Eqs. (2.5) and (2.6) we obtain the law of exponential attenuation (by absorption)

$$F = F_0 \exp(-\kappa x).\tag{2.7}$$

This law is also valid for attenuation by scattering if multiple scattering is negligible (see Sec. 5.2.3).

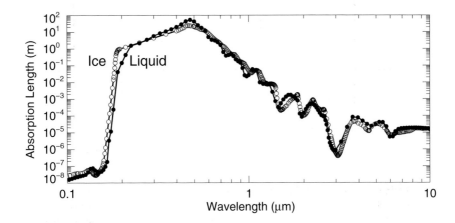

Figure 2.2: Absorption length (inverse absorption coefficient) of pure ice and liquid water from UV to IR. The data for liquid water were taken from Querry *et al.* (1991), those for ice from Warren (1984).

The various names by which this law is called exemplify *Stigler's law of eponymy*: "No scientific discovery is named after its original discoverer." The law of exponential attenuation, often called Lambert's law, was first stated in Pierre Bouguer's *Essay on the Gradation of Light* (1729), although we could find no evidence that he established it experimentally. Most chemists call it Beer's law, which is wide of the mark given that Bouguer preceded Beer by more than 100 years, Beer did not discover an exponential law of attenuation with distance, and, in fact, did not explicitly state *any* exponential law. The most we can say is that by reworking Beer's data one can unearth what he did not: an exponential attenuation law for solutions of fixed thickness but variable concentration of the absorbing solute.

Because κx is dimensionless, κ must have the dimensions of inverse length, and hence $1/\kappa$ must have the dimensions of length. The *e-folding length* is the distance over which a monodirectional beam is attenuated by a factor $1/e$. Because this term is a bit of a mouthful we prefer *absorption length* for $1/\kappa$. As a general rule, whenever any physical quantity can be expressed as a length it is wise to do so. Lengths are easier to get a feel for, more so than time, even mass. We can both touch and see lengths.

The absorption coefficient (absorption length) of a material depends on wavelength, often varying by as much as a factor of 10^{10}, an example of which is as near as a faucet. Figure 2.2 shows the huge range of absorption lengths for liquid water and ice from 0.1 μm to 10 μm. The absorption length for (pure) liquid water and ice over the visible spectrum, shown on a linear scale in Fig. 5.12, is greatest in the blue, least in the red. From this figure it is evident why water in a drinking glass is not noticeably colored: the dimensions of glasses are small compared with the absorption length of water over the visible spectrum. This curve also shows that transmission of white light over several meters through water is sufficient to attenuate much more of the long-wavelength components than the short. Indeed, with increasing distances the only component that survives corresponds to the greatest absorption length. Water is intrinsically blue and needs no impurities to make it so. In Section 5.3.1 we explore the consequences of this to the observed colors of natural ice bodies such as glaciers, icebergs, ice caves, icefalls, and even holes in snow.

2.1.1   Absorptivity and Absorption Coefficient: A Tenuous Connection

Absorptivity and absorption coefficient are not the same. In the first place, the former is dimensionless whereas the latter has the dimensions of inverse length, which itself ought to signal caution. More to the point, the connection between them is sometimes tenuous at best. Consider, for example, radiation incident on bodies sufficiently thick that transmission by them is negligible. We now can attach a more precise meaning to "sufficiently thick": much thicker than the absorption length at the wavelength of the radiation. With this assumption, the absorptivity of the body is 1 minus its reflectivity. How does the reflectivity of the body depend on its absorption coefficient? For many materials over many wavelength intervals, reflectivity changes hardly at all even with huge increases in absorption coefficient. And if there is a change, it is likely to result in a *decrease* in absorptivity (see Prob. 7.20). For example, the absorption coefficient of metals such as silver and aluminum is usually huge compared with that of insulators such as quartz and salt, a million times or more, especially at visible and near-visible wavelengths. And yet reflectivities of metals are high, and hence their absorptivities are lower than those of insulators. Finally, there is this important distinction to be kept in mind: absorptivity is a property of a *body* whereas absorption coefficient is a property of a *material*.

2.1.2   Absorptance and Absorbance: More Room for Confusion

As if distinguishing between absorptivity and absorption coefficient were not difficult enough, we also have to keep these terms separate from the near homophones *absorptance* and *absorbance*. Although absorptance is sometimes used as a synonym for absorptivity, this is not recommended given that we try to restrict terms ending in "ance" to amounts of radiant power. For example, emittance, which can be looked upon as shorthand for *emitted* irradi*ance*, is radiant power per unit area. Similarly, absorptance can be looked upon as shorthand for absorbed irradiance.

Absorbance, a term widely used by chemists, is the negative logarithm (base 10) of the transmissivity of a sample of an absorbing material (usually liquid) in a container (cell). Because a transmissivity less than 1 is a consequence both of reflection by the container and absorption by its contents, the apparent absorbance can be nonzero even with an empty cell or one filled with a negligibly absorbing liquid. To correct for reflection, the absorbance of the cell is subtracted from the apparent absorbance to obtain that of the sample. To good approximation the transmissivity of the sample in the cell often is

$$T = T_0 \exp(-\kappa h), \tag{2.8}$$

where h is the sample thickness, κ its absorption coefficient, and T_0 the transmissivity of the cell without the sample in place. Take the negative logarithm of both sides of Eq. (2.8) to obtain

$$-\log T + \log T_0 = \kappa h \log e = 0.434 \kappa h. \tag{2.9}$$

The left side of this equation is absorbance corrected (approximately) for reflection by the cell. With this correction, absorbance measured by chemists is, except for a constant factor, the absorption optical thickness (κh) of the sample (see Sec. 5.2).

2.1.3 The Sum of Exponentials is not an Exponential

The exponential attenuation law Eq. (2.7) strictly holds only for monochromatic radiation because κ depends on frequency. Any real source is distributed over frequency, and hence the integrated transmitted irradiance is

$$\int F_0(\omega) \exp(-\kappa x)\, d\omega, \tag{2.10}$$

where $F_0(\omega)$ is the *spectral irradiance* (irradiance per unit frequency interval) at $x = 0$. The limits of integration can be anything, and for simplicity we do not express κ as a function of frequency. Although each spectral component of the incident beam is attenuated exponentially with distance, the integrated beam is not. And this is true even if the incident irradiance does not depend on frequency. This basic property of exponential attenuation has sometimes been forgotten, resulting in errors.

To show that the sum (integral) of exponentials is *not*, in general, an exponential, we assume that

$$\exp(-Kx) = \int \exp(-\kappa x)\, d\omega, \tag{2.11}$$

where K is independent of x and ω. Differentiate both sides of this equation with respect to x to obtain

$$K = \frac{\int \kappa \exp(-\kappa x)\, d\omega}{\int \exp(-\kappa x)\, d\omega}. \tag{2.12}$$

The right side of this equation, the average of κ weighted by a normalized exponential, depends on x, in general, and hence we contradict our original assumption that K is independent of x, which therefore must be false. This is a proof by contradiction: assume something is true, explore the consequences, and when a contradiction results, the original assumption must have been false. This, by the way, is a variation on a theme in Section 1.4.2: the average of a function is not necessarily the function of the average.

When κ is independent of frequency over the range of interest, the integrated irradiance does decrease as a simple exponential. And when $\kappa x \ll 1$ for the frequency range and distances of interest we can approximate the exponential in Eq. (2.10) by the first two terms in its Taylor series expansion to obtain the following approximation for the transmitted irradiance:

$$\int F_0(\omega)(1 - \kappa x)\, d\omega. \tag{2.13}$$

This can be written as

$$F_0(1 - \langle \kappa \rangle x), \tag{2.14}$$

where the integrated irradiance at $x = 0$ is

$$F_0 = \int F_0(\omega)\, d\omega \tag{2.15}$$

and the average absorption coefficient is

$$\langle\kappa\rangle = \frac{\int F_0(\omega)\kappa\, d\omega}{\int F_0(\omega)\, d\omega}. \tag{2.16}$$

We obtained Eq. (2.14) by approximating an exponential by the first two terms in a Taylor series, but we can do the reverse, approximate the first two terms in a series by an exponential:

$$1 - \langle\kappa\rangle x \approx \exp(-\langle\kappa\rangle x), \tag{2.17}$$

which yields approximate exponential attenuation for the integrated irradiance:

$$F \approx F_0 \exp(-\langle\kappa\rangle x). \tag{2.18}$$

But in general, this equation is not correct.

2.1.4 Attenuation in a Nonuniform Medium

Nature is not so cooperative as to provide us only with media having uniform properties. The absorption coefficient κ can vary from point to point. The medium depicted in Fig. 2.1 can be nonuniform and subdivided into N equal slices so thin that in each of them the absorption coefficient is nearly constant. Transmission by the jth slice is

$$F_{j+1} \approx F_j \exp(-\kappa_j \Delta x), \tag{2.19}$$

where F_j is the irradiance incident on the jth slice, F_{j+1} is the irradiance transmitted by this slice, κ_j is the absorption coefficient $\kappa(x)$ at some point in the interval (x_j, x_{j+1}), and $\Delta x = x/N$. It follows from this that transmission over the distance x is

$$F = F_0 \exp\left(-\sum_{j=1}^{N} \kappa_j \Delta x\right). \tag{2.20}$$

The limit of the sum in this equation is the integral

$$\int \kappa(x)\, dx = \lim_{N\to\infty} \sum_{j=1}^{N} \kappa_j \Delta x, \tag{2.21}$$

and hence the exponential attenuation law for a nonuniform medium (spatially varying absorption coefficient) is

$$F = F_0 \exp\left\{-\int \kappa(x)\, dx\right\}. \tag{2.22}$$

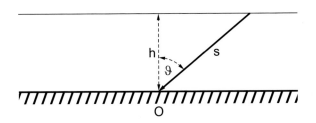

Figure 2.3: The path length s through a slab, and hence its absorptivity, is greater the more the path is slanted from the vertical.

2.2   Directional Emissivity of the Atmosphere

We note in Section 1.4 that emissivity depends on direction, and also show calculations of the spectral normal emissivity of a uniform layer of moist air. And in Section 1.5 we show how the emissivity of a layer of water depends on direction. With the law of exponential attenuation in hand we can go a step further and show how the emissivity of the atmosphere depends on direction.

Although Earth is round, its radius is about 1000 times greater than the thickness of that part of the atmosphere containing most of the infrared-active gases. Because of this we can pretend that the atmosphere is a planar slab infinite in lateral extent. For simplicity we assume that the absorption coefficient κ of the atmosphere does not depend on altitude. That this is not true does not affect our general conclusions.

Because the reflectivity of the atmosphere for terrestrial infrared radiation is negligible, the absorptivity of the atmosphere is 1 minus its transmissivity, which over any path is given by the exponential attenuation law:

$$\exp(-\kappa s), \tag{2.23}$$

where s is the path length. If κ varies along the path, Eq. (2.23) is replaced by the path integral of κ as in Eq. (2.22). For a slab atmosphere of thickness h and a path making an angle ϑ with the vertical (zenith), $s = h/\cos\vartheta$ (Fig. 2.3), and hence the emissivity (absorptivity) of the atmosphere in any direction ϑ is

$$\varepsilon = 1 - \exp(-\kappa h/\cos\vartheta). \tag{2.24}$$

Here is an example in which absorptivity (emissivity) does increase with increasing absorption coefficient. Emissivity is least overhead and increases to 1 toward the horizon. But increased emissivity toward the horizon is not the only reason why *emission* increases in this direction. Emission depends on the temperature of the emitting body as well. The temperature of the atmosphere is not uniform. In the troposphere, which contains most of the atmosphere, temperature usually decreases with height.

Radiation from the atmosphere at the surface, in a given direction, comes from all points on an atmospheric path along that direction. But radiation emitted at each point must be transmitted over some distance through the atmosphere to the observation point O (Fig. 2.3). The greater this distance, the less radiation transmitted. This is evident from Eq. (2.24), according to which emissivity does not increase indefinitely with increasing path length but

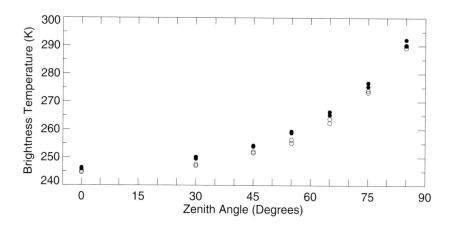

Figure 2.4: Brightness temperature as a function of zenith angle on a clear summer day. The first measurement always was made at zenith ($0°$), then for increasing angles to about $85°$ ($5°$ above the horizon). This was done four times: twice for the radiation thermometer pointed east-northeast (solid circles), twice pointed west-southwest (open circles).

asymptotically approaches 1. All points on the atmospheric path contribute to radiation at the observation point O. But those closest to it contribute most for two reasons: the concentration of emitting gases, and hence κ, decreases with height (distance from the observation point), and the shorter the path from emitting gases to the observation point, the greater the fraction of emitted radiation transmitted to it.

Even if the atmosphere were uniform in temperature, emitted radiation (at the ground) would be least overhead and increase toward the horizon. Because temperature usually decreases with height, this directional variation of emission is even greater. We leave it as a problem (Prob. 5.19) to show that the decrease of temperature with height in Earth's atmosphere is a minor contributor to the total variation of emission with direction.

If emission does indeed increase from zenith to horizon, so should the brightness temperature of the atmosphere. To demonstrate this, we measured brightness temperatures at various zenith angles (Fig. 2.4) from a valley floor near State College on a clear summer day. As predicted by Eq. (2.24) brightness temperature increases with increasing zenith angle.

2.3   Flux Divergence

If more radiant energy enters a region than leaves it, radiant energy must be converted into other forms within the region. This transformation is usually manifested by a temperature increase but could be manifested in other ways (e.g., photosynthesis).

Consider a monochromatic, monodirectional beam propagating along the x-axis in an optically homogeneous medium characterized by an absorption coefficient κ. A region between x and $x + \Delta x$ is bounded by planes of area A. The difference between the radiant energy entering this region and that leaving it is

$$AF_0 \exp(-\kappa x) - AF_0 \exp\{-\kappa(x + \Delta x)\}, \tag{2.25}$$

where F_0 is the irradiance at $x = 0$. Divide this difference by the volume of the region

$$\frac{AF_0 \exp(-\kappa x) - AF_0 \exp\{-\kappa(x + \Delta x)\}}{A\Delta x} \tag{2.26}$$

and take the limit as $\Delta x \to 0$ to obtain the rate of energy conversion (transformation) per unit volume around a point:

$$-F_0 \frac{d}{dx} \exp(-\kappa x) = -\frac{dF}{dx} = F_0 \kappa \exp(-\kappa x). \tag{2.27}$$

Thus the rate of energy transformation per unit volume is the negative of the spatial derivative of the irradiance, often called the *flux divergence*. The negative flux divergence is proportional to the local rate of temperature change under the assumption that radiant energy transformation results only in temperature increases. Note that the flux divergence is a product of two functions, one of which increases with increasing κ, the other of which decreases; the maximum of Eq. (2.27) occurs for $\kappa x = 1$ for fixed $x > 0$.

Suppose that we want to heat an object by illuminating it with radiation. As a concrete example, take the radiation to lie in the microwave region. How should we choose the frequency? The frequency of maximum absorption coefficient would give the highest heating rate but it might be localized near the surface of the object. For the object to be heated more or less uniformly, the absorption length $(1/\kappa)$ should be comparable with the linear dimensions of the object. The frequency used in microwave ovens is chosen so that the absorption length of water (mostly the water in food enables it to be heated in microwave ovens) is about equal to the linear dimensions of typical objects heated in them (all microwave ovens are about the same size). Many years ago on an examination we gave the absorption properties of water over a wide range of frequencies and asked students to pick the frequency of a microwave oven (we didn't know). Some students fell into the trap of picking the frequency of maximum absorption coefficient, but others recognized that an absorption length of, say, a few mm, would result in braised but not well-cooked food. Following the exam, we consulted the experts in our electronics shop and, to our delight, learned that the frequency chosen on the basis of an absorption length for water of around 10–20 cm is the frequency used in ovens.

2.3.1   The Sum of Exponentials is not an Exponential: Another Example

From now on we use the term *plane-parallel* medium to mean one confined between two parallel planes infinite in lateral extent. The properties of the medium may vary from point to point but only in the direction perpendicular to the planes. We sometimes call this direction the vertical direction even though it need not coincide with gravity. For our purposes, radiation is oblivious to gravity.

Suppose that a plane-parallel, uniform, absorbing medium is illuminated by *isotropic* radiation (Fig. 2.5). By isotropic is meant that the radiation (specifically, the radiance [Sec. 4.1.2]) is the same in all directions in a hemisphere. The total irradiance (Sec. 4.2) is the weighted,

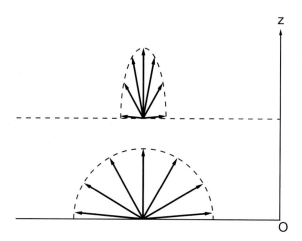

Figure 2.5: An initially isotropic source of radiation, incident at $z = 0$ on an absorbing medium, becomes more sharply peaked toward the z-direction with increasing z.

by cosine and solid angle, sum of contributions from all directions. Any particular direction is specified by ϑ, the angle with the vertical. At any depth z into the medium, radiation along a particular path is attenuated by the factor $\exp(-\kappa z/\cos\vartheta)$. At this depth, the irradiance is again the weighted sum of contributions from radiation in all directions. Although radiation from each direction is the same at the illuminated boundary of the medium ($z = 0$), this is not true at $z > 0$. Thus the initially isotropic radiation field does not remain isotropic. In the limit of indefinitely large κz, the emerging radiation would be monodirectional in the vertical direction, although of vanishingly small irradiance. Moreover, although the irradiance is the sum (integral) of exponentials and depends only on z, it is not an exponential function of z.

2.4 Absorption Cross Section

Determining the absorption coefficient of liquids and solids from the absorption properties of their individual molecules is not an easy task because they are sufficiently close together that they interact strongly. This is evident from Figs. 1.11 and 1.12, which show that the spectral emissivity of liquid water bears little resemblance to that of water vapor. Beginning with the latter it is not an easy step to the former. Interactions between water molecules in the liquid phase all but destroy their individuality. For gases and suspensions of particles, however, we do have a hope of determining absorption coefficients beginning with the properties of a single molecule or particle.

By *particle* we mean a bound collection of molecules sufficient in number that it has macroscopic properties such as temperature and pressure. There is no such thing as the temperature or pressure of a molecule. Even the radius of a molecule is a nebulous quantity: every method for measuring molecular diameters yields a different result. A particle may itself be composed of a material with an absorption coefficient, but there is no such thing as the absorption coefficient of a single molecule. All molecules of the same substance are essentially identical, but every particle is unlike every other particle. Like temperature and pressure, absorption coefficient is a statistical quantity, an average over an ensemble of many molecules.

We expect the absorption coefficient of a gas to depend on the concentration of its molecules. After all, the inverse of the absorption coefficient is the absorption length, and it would hardly make sense if a gas at one concentration had the same absorption length as the same gas at a higher concentration. The term *concentration* instead of *density* is used here to emphasize that absorption of electromagnetic radiation is not fundamentally dependent on mass. Electromagnetic waves exert forces on charges, not masses, which just go along for the ride. When you use the unqualified term density, make sure that you are clear whether you mean mass density (mass per unit volume) or number density (molecules per unit volume). We use number density and concentration to mean more or less the same thing.

Suppose that an isolated molecule is illuminated by a monodirectional, monochromatic beam of irradiance F. This molecule absorbs energy, by which is meant it transforms radiant energy into other forms, at a rate W_a, which is proportional to F. The dimensions of W_a are power, whereas those of F are power per unit area. Thus the proportionality factor σ_a between the two

$$W_a = \sigma_a F \tag{2.28}$$

must have the dimensions of length squared (area). For this reason σ_a is given the name *absorption cross section*. This is the *effective area* of the molecule for removing energy from the incident beam, and should not be confused with the geometrical cross-sectional area of the molecule even if such an area had a precise meaning, which it does not. The diameters of the kinds of molecules that inhabit the atmosphere are about 3×10^{-8} cm, which corresponds to geometrical cross-sectional areas of about 7×10^{-16} cm^2. But absorption cross sections of molecules are usually much smaller than this value.

Equation (2.28) holds equally well if the illuminated object is a particle rather than a molecule. We have more to say about absorption cross sections of particles in Section 2.9.

Cross sections of various kinds are fundamental in several areas of physics, common to the kinetic theory of gases, neutron physics, high-energy particle physics, and optics.

Suppose that a thin slab of gas with area A and thickness Δx, populated by N molecules per unit volume with absorption cross section σ_a, is illuminated by a monodirectional, monochromatic beam directed perpendicular to the faces of the slab (Fig. 2.6). The total number of molecules in the slab is $NA\Delta x$, and hence the total effective area for removing radiant energy from the beam is $\sigma_a NA\Delta x$, where we assume that no molecule overlaps (or shadows) another in the direction of the beam. The decrease in the transmitted power over the distance Δx is

$$A\Delta F = -F\sigma_a NA\Delta x, \tag{2.29}$$

where F is the incident irradiance. From this it follows that

$$\Delta F = -FN\sigma_a \Delta x = -F\kappa \Delta x, \tag{2.30}$$

where

$$\kappa = N\sigma_a \tag{2.31}$$

is the absorption coefficient from Eq. (2.1).

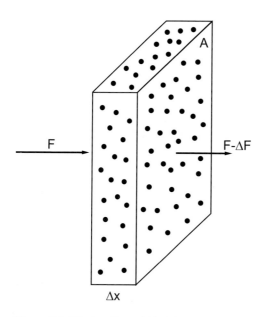

Figure 2.6: The irradiance F incident on a slab randomly populated by absorbing molecules is attenuated by an amount proportional to the total cross-sectional area (projected effective area) of all the molecules in the slab.

We implicitly assumed that all the molecules are identical. Even if they are, however, they are unlikely to all be oriented identically, and so the absorption cross section in Eq. (2.31) should be interpreted as an average. Because of the additivity of κ:

$$\kappa = \sum_j N_j \sigma_{aj}, \quad N = \sum_j N_j, \tag{2.32}$$

where j indicates a particular molecular orientation and σ_a in Eq. (2.31) is the orientational average

$$\frac{1}{N} \sum_j N_j \sigma_{aj}. \tag{2.33}$$

If all orientations are equally likely (random orientation), all cross sections are equally weighted.

Another assumption underlying Eqs. (2.31)–(2.33) is that the consequences of interference (coherence) are negligible (see Secs. 3.4 and 5.1). If not, to determine the irradiance transmitted at any x, we could not skip the intermediate step of determining the amplitude and phases of the electric and magnetic fields, from which irradiances follow.

The absorption cross section also depends on the polarization state (see Ch. 7) of the beam if the molecule is asymmetric. But if the molecules are randomly oriented, the absorption coefficient of the gas is independent of polarization.

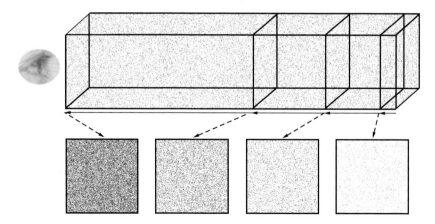

Figure 2.7: The square tube is randomly filled with 25,000 particles (black squares). What an observer would see (looking to the right) at different points along this tube is depicted in the squares below the tube.

Everything we say about absorption by a gas is equally applicable to a suspension of particles. All that need be done is replace the absorption cross section of a molecule with that of a particle (see Sec. 2.9).

Equation (2.7) for exponential attenuation, with the absorption coefficient of a gas (or suspension of particles) given by Eq. (2.31), contains hidden assumptions worth bringing into the open. Even if coherence is negligible, the possibility remains that one molecule or particle can affect absorption by others. Suppose, for example, that a tube contains a fixed number of absorbers each with the same absorption cross section. How they are distributed in space determines attenuation by them. If *all* the absorbers happen to line up one behind the other, their total projected absorption area is just the absorption cross section of a single absorber. But if they are distributed so that *no* absorber is behind another, their total projected absorption area is much greater. Attenuation is vastly different for these two spatial distributions even if the total number of absorbers in the tube is the same. All absorbers lined up in a row is physically unrealistic (except in a crystal), but this extreme distribution signals that how the projected absorption areas of the absorbers overlap determines attenuation by them.

To probe further we did computations for two possible arrangements of absorbers represented by black squares, henceforth called particles. A large square with area 4 units, representing the cross-sectional area of a square tube, was filled with identical particles with cross-sectional area 0.0001 units according to two different prescriptions. In one prescription the positions of the particles within the large square were randomly chosen, added to the large square one after another, and the total cumulative particle area recorded (Fig. 2.7). The total area of overlap of an added particle with all other particles in the tube was subtracted from the cumulative particle area to produce the total projected area. In the other prescription, the particles were not allowed to overlap. If the position of a particle was such that it overlapped any other, it was relocated to remove all overlap, and hence the resulting distribution of particles was not perfectly random.

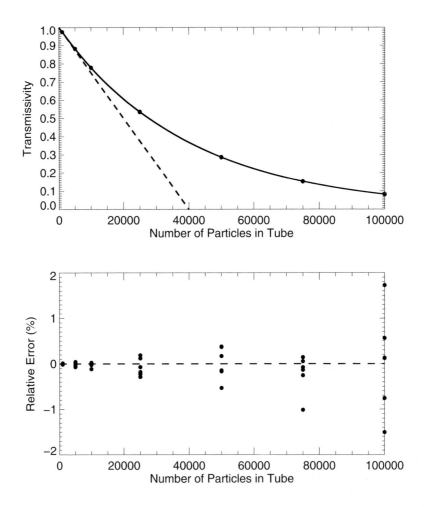

Figure 2.8: The top figure shows transmissivity versus number of particles for two different prescriptions: overlapping particles (solid circles) and non-overlapping particles (dashed curve). Each solid circle on the solid curve (exponential attenuation) is the result of six sets of calculations for 1000, 5000, 10,000, 25,000, 50,000, 75,000, and 100,000 particles. Because the variation over the six sets is not resolvable in the upper part of the figure, the lower part shows the ratio of the difference between each calculated transmissivity and exponential decreasing transmissivity, relative to exponential decreasing transmissivity.

The quantity 1 minus the total projected area of particles divided by the cross-sectional area of the tube is the transmissivity, shown for both overlapping and non-overlapping particles in Fig. 2.8. For a sufficiently small number of particles, the transmissivity is the same, but as this number increases, the two curves diverge. For the overlapping particles, the transmissivity

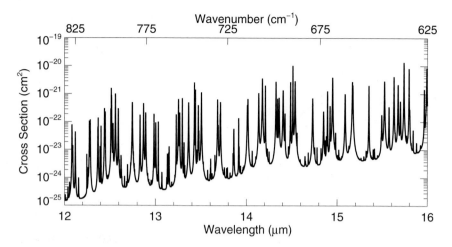

Figure 2.9: Absorption cross section of a water molecule for a temperature of $20°\text{C}$ and a total pressure of 1 atmosphere.

closely follows the exponential law $\exp(-NA_{\mathrm{p}})$, where N is the total number of particles in the tube and A_{p} is the cross-sectional area of a particle relative to the cross-sectional area of the tube. Each dot on the solid curve $[\exp(-NA_{\mathrm{p}})]$ in Fig. 2.8 is the result of six sets of calculations, each of which gave slightly different results, although they are not resolvable. To resolve the differences, the lower part of the figure shows the relative deviation of each calculation, defined as the ratio of the difference between the computed transmissivity and that predicted by the exponential law relative to the exponential law.

For the non-overlapping particles, transmissivity decreases with particle number more rapidly than exponential, a straight line $1 - NA_{\mathrm{p}}$. And if all the particles were on a line, the transmissivity would be $1 - A_{\mathrm{p}}$ for all $N > 0$. From this we conclude that Eq. (2.7) is applicable to a perfectly random distribution of absorbers. If the positions of absorbers are correlated, the exponential law is called into question, although we know of no experimental evidence that this ever happens to an appreciable extent with atmospheric molecules or particles.

Now we are better able to understand the origins of the spectral emissivity curve for a uniform layer of moist air (Fig. 1.11). We obtained this curve using Eq. (2.24) for $\vartheta = 0$ with κ given by Eq. (2.31). Thus the frequency dependence of the emissivity originates in that of the absorption cross section of the water molecule (Fig. 2.9). We can write the normal emissivity of a uniform layer of moist air of thickness h as

$$\varepsilon_{\mathrm{n}} = 1 - \exp(-Nh\sigma_{\mathrm{a}}), \qquad (2.34)$$

with

$$Nh = \frac{\rho_{\mathrm{w}}d}{m_{\mathrm{w}}} = 3.3 \times 10^{22}d, \qquad (2.35)$$

where ρ_{w} is the density of liquid water, m_{w} the mass of a water molecule, and d the depth of a liquid water layer that would result if all the water vapor molecules in the moist air were condensed to liquid. The quantity $Nh\sigma_a$ is the absorption optical depth τ_a (see Sec. 5.2), which for $d = 1$ cm is $3.3 \times 10^{22}\, \sigma_a$, with the cross section in cm^2.

The absorption cross section of water vapor (Fig. 2.9) varies by more than a factor of 1000 over the range of infrared frequencies shown. At the peaks, the cross section is sufficiently large that the corresponding emissivity is almost 1, at the troughs, much less than 1 (Fig. 1.11). The absorption cross section is not strictly a property of a molecule but depends on interactions with neighboring molecules, even in the gas phase. The positions of the peaks do not change so much as their widths, which is not especially difficult to account for (as we do subsequently). What Fig. 2.9 shows is the absorption coefficient divided by N, which defines the effective absorption cross section, a property of the molecule and its environment, which for our calculations is a total pressure of 1 atm and a temperature of 20 °C. To interpret the frequency dependence of the cross section we must dig deeper. To do so we need the right shovel, the rudiments of complex variables, which we turn to next.

2.5   The ABCs of Complex Variables

Before proceeding we digress briefly on the rudiments of complex variables, the minimum necessary to understand what follows. For many readers this will be a review.

The history of mathematics is to a large extent the history of the evolution of the concept of number, from the integers, to rational numbers (ratios of integers), to irrational numbers (such as $\sqrt{2}$) and finally to complex numbers. We may define a complex number z as the ordered pair

$$z = (x, y), \tag{2.36}$$

where x and y are just old-fashioned numbers. By ordered pair is meant that (x, y) is not the same as (y, x); x is called the *real* part of z, y the *imaginary* part, sometimes written

$$x = \Re\{z\}, \; y = \Im\{z\}. \tag{2.37}$$

The terms "real" and "imaginary" are archaic, relics from a time when our ancestors were nervous about admitting such mathematical unicorns as complex numbers into the mathematical zoo. The rule for addition of complex numbers follows naturally

$$z_1 + z_2 = (x_1 + x_2, y_1 + y_2) \tag{2.38}$$

whereas that for multiplication does not:

$$z_1 z_2 = (x_1 x_2 - y_1 y_2, x_1 y_2 + x_2 y_1). \tag{2.39}$$

That is, even if you were innocent of any knowledge of complex numbers, if asked to add them, you'd likely come up with Eq. (2.38). Any complex number $(x, 0)$ with zero imaginary part can be written simply as x. We also have from the addition rule [Eq. (2.38)]

$$(x, y) = (x, 0) + (0, y) \tag{2.40}$$

and from the multiplication rule [Eq. (2.39)]

$$(0, y) = (0, 1)(y, 0).\tag{2.41}$$

If the symbol i denotes $(0, 1)$, any complex number can be written simply as

$$z = x + iy.\tag{2.42}$$

The quantity i has the peculiar property that its square is -1,

$$i^2 = -1,\tag{2.43}$$

which is why it was deemed "imaginary." When a proper (real) number, positive or negative, is squared the result is always a positive number.

The ordered pair (x, y) can be represented geometrically as a point in a plane, called the complex plane, with rectangular coordinates x and y or, equivalently, as a displacement vector in this plane spanning the points $(0, 0)$ and (x, y). We can transform rectangular coordinates to plane polar coordinates:

$$x = a \cos \vartheta, \; y = a \sin \vartheta,\tag{2.44}$$

where a is the length of the vector representing the complex number z and ϑ is the angle between this vector and the x-axis in the counterclockwise sense. Thus any complex number can be written

$$z = a \cos \vartheta + ia \sin \vartheta,\tag{2.45}$$

where a is called the *modulus* of z and ϑ its *argument*. The *complex conjugate* of any complex number $z = (x, y) = x + iy$, often written as z^*, is $(x, -y) = x - iy$. The product of a complex number with its complex conjugate is always non-negative:

$$zz^* = x^2 + y^2 = a^2 = |z|^2 .\tag{2.46}$$

We may define

$$\exp(i\vartheta) = \cos \vartheta + i \sin \vartheta.\tag{2.47}$$

If we differentiate the right side of Eq. (2.47) with respect to ϑ we get the same result as when we take the derivative of the left side in the usual way treating i as a constant:

$$\frac{d}{d\vartheta}(\cos \vartheta + i \sin \vartheta) = -\sin \vartheta + i \cos \vartheta = i(\cos \vartheta + i \sin \vartheta) = \frac{d}{d\vartheta}\exp(i\vartheta).\tag{2.48}$$

And similarly for derivatives of all orders as well as integrals. Multiplication yields

$$\begin{aligned}\exp(i\vartheta_1)\exp(i\vartheta_2) &= \cos \vartheta_1 \cos \vartheta_2 - \sin \vartheta_1 \sin \vartheta_2 + i(\cos \vartheta_1 \sin \vartheta_2 + \cos \vartheta_2 \sin \vartheta_1)\\ &= \cos(\vartheta_1 + \vartheta_2) + i \sin(\vartheta_1 + \vartheta_2)\\ &= \exp\{i(\vartheta_1 + \vartheta_2)\}.\end{aligned}\tag{2.49}$$

Equations (2.48) and (2.49) tell us that whenever we encounter the complex number $\cos \vartheta + i \sin \vartheta$ we can write it succinctly as $\exp(i\vartheta)$, and then treat this entity just as we would the exponential function of a real variable. Now we can write an arbitrary complex number in *polar form* as

$$z = a \exp(i\vartheta),$$ (2.50)

which for our purposes usually will be the most convenient form. Addition of complex numbers can be interpreted geometrically as the addition of vectors in the complex plane, whereas multiplication of complex numbers

$$z_1 z_2 = a_1 a_2 \exp\{i(\vartheta_1 + \vartheta_2)\}$$ (2.51)

can be interpreted as stretching (or shrinking) and rotating vectors. In particular, multiplying any complex number by i doesn't change its length (modulus) but rotates it (counterclockwise) by $\pi/2$ because $\exp(i\pi/2) = i$.

The real part of the sum of complex numbers is the sum of its real parts

$$\Re\{z_1 + z_2\} = \Re\{z_1\} + \Re\{z_2\}$$ (2.52)

whereas this is, in general, not true of the product of complex numbers:

$$\Re\{z_1 z_2\} \neq \Re\{z_1\}\Re\{z_2\}.$$ (2.53)

Equation (2.52) is the simplest example of a linear operation. Differentiation is another. That is, if f_1 and f_2 are arbitrary differentiable functions of the real variable t then

$$\frac{d}{dt}\{f_1 + f_2\} = \frac{df_1}{dt} + \frac{df_2}{dt}.$$ (2.54)

And this is true for all orders of differentiation as well as integration (i.e., the integral of a sum is the sum of integrals). Moreover, we can interchange the order of taking the real part and any other linear operation. For example, if f is a complex-valued function of the real variable t, then

$$\Re\left\{\frac{df}{dt}\right\} = \frac{d}{dt}\Re\{f\}.$$ (2.55)

Equation (2.55) is just one example of a general rule:

$$\Re\{Lf\} = L\Re\{f\},$$ (2.56)

where L is *any* linear operator (e.g., differentiation and integration). A consequence of Eq. (2.56) is that if a physical quantity is governed by a linear equation, we can find its complex-valued solution, then take its real part.

Make no mistake about it, all physical quantities are real numbers. So at this point you may be wondering why anyone would be so crazy as to find complex solutions to physical problems. Although lugging around what appears to be excess baggage (i.e., the imaginary part of a solution), only to toss it on the scrap heap, may seem to be the height of foolishness, the small effort entailed in mastering the rudiments of complex variable theory is an enormous labor-saving device. Although this can be appreciated fully only with experience, a few stories support our claim.

Determining reflection and transmission because of an optically smooth, infinite planar boundary between two arbitrary optically homogeneous and infinite media is a problem in electromagnetic theory, the governing equations for which are linear. For any state of polarization of a plane wave incident on an interface at any angle, we can find the reflected and transmitted electric and magnetic fields relative to the incident fields. The ratios that result from these solutions are complex quantities because we obtained the complex solution to the field equations, called the *Fresnel coefficients* (see Sec. 7.2). They are compact, easy to remember, and take only a line or two to write down. The corresponding real reflectivity and transmissivity, which are ratios of irradiances, are obtained by multiplying these complex ratios by their complex conjugates, which yields real numbers. The Fresnel coefficients are readily programmed using a computer language that allows for operations on complex numbers. The resulting program takes a few lines. We once rewrote the Fresnel coefficients so that the reflectivity and transmissivity were expressed entirely in terms of real variables. The derivation took about 20 pages of close-packed algebra, many hours of work, lots of checking and rechecking. And the result was a cumbersome set of complicated equations, difficult to remember. Moreover, the computer program required to obtain numbers from these equations was at least ten times longer than a program based on complex arithmetic. And we began with the complex solution to the problem. Had we insisted from the outset on not using complex solutions, we might still be trying to obtain the real Fresnel coefficients. Keep in mind that this reflection and transmission problem is perhaps the simplest in electromagnetic theory. A more complicated planar problem is reflection and transmission by a single film and, more complicated yet, by multilayer films. Solving any of these problems without using complex solutions to the field equations would be a nightmare.

A vastly more complicated problem is determining the scattered field when an arbitrary homogeneous sphere is illuminated by a plane wave (discussed in Sec. 3.5). The resulting solution is an infinite series of terms, each of which is a complicated complex-valued function. Many years ago we received a telephone call from a man who proposed to take this complex solution and express it entirely in real terms so that he could then write a computer program using a language that could not do complex arithmetic. We asked him if he had committed some grievous sin for which he felt the need to atone by undertaking this terrible penance. He laughed. He was determined. A week later he called back, chastened: "You were right. It's impossible." And again, he began with a solution that had been obtained by way of complex-valued functions. Had he attempted to solve the scattering problem in a state of willful ignorance of complex numbers, he probably still would be slaving away.

The moral of these stories is that anything you can do with complex solutions to linear problems you can do without them but often you would be crazy to try. The same can be said for vectors. The payoff for learning the rudiments of vector analysis is that you can solve problems that without vectors would be horrible. Vectors are not necessary but they sure are useful. Take a look at old books on hydrodynamics and electromagnetic theory, ones from which vectors are absent, and you'll see what we mean.

Before proceeding, we need to issue just one word of caution. The only time we risk getting into trouble using complex solutions to (real) problems is when we need to perform nonlinear operations (e.g., multiplication) on solutions. To do this we first have to take real parts.

2.6   Interpretation of the Molecular Absorption Coefficient

We could content ourselves with accepting the absorption coefficient as a quantity obtained by transmission measurements and ask no questions about its origin. But such questions arise when we want to know why the absorption coefficient for the same material can be so different at different frequencies or why the absorption coefficient at a given frequency can be so different for different materials. To answer such questions we begin with the simplest possible example, a harmonic oscillator of mass m and positive charge e acted on by a time-harmonic (monochromatic) electromagnetic wave of frequency ω. If $\mathbf{E}$ is the electric field of this wave, the force on the oscillator is $e\mathbf{E}$. You may wonder why, if light is an electro*magnetic* wave, we ignore the force exerted by the magnetic field. The reason is that this force is negligible compared with the electric force if the velocity of the oscillator is small compared with the free-space speed of light.

A time-harmonic electric field has the form $\mathbf{E}_0 \cos \omega t$. According to Eq. (2.47)

$$\exp(-i\omega t) = \cos \omega t - i \sin \omega t, \tag{2.57}$$

and hence the electric field is the real part of the complex electric field:

$$\mathbf{E}_0 \cos \omega t = \Re\{\mathbf{E}_0 \exp(-i\omega t)\}. \tag{2.58}$$

This electric field is the complex representation of the real field of interest. As noted in the previous section, we may deal with complex representations of real quantities as long as we restrict ourselves to linear operations. Note that we could change i to $-i$ in Eq. (2.58) and the result would be the same. There are two conventions for the sign of the complex representation of time-harmonic fields. It doesn't matter which convention you choose as long as you stick with it. If you switch conventions in mid stream you may drown.

The position of the oscillator relative to the origin of some fixed coordinate system is $\boldsymbol{\xi}$. We assume that the oscillator has an equilibrium (no force) position $\boldsymbol{\xi}_0$ when not acted on by an external electric field. We also assume that the oscillator is acted on by a restoring force proportional to the displacement from equilibrium and by a dissipative force proportional to its velocity. Thus the equation of motion for the oscillator is

$$m\frac{d^2\boldsymbol{\xi}}{dt^2} = -K(\boldsymbol{\xi} - \boldsymbol{\xi}_0) - b\frac{d\boldsymbol{\xi}}{dt} + e\mathbf{E}, \tag{2.59}$$

where K and b are constants. If we denote by $\mathbf{x}$ the displacement from equilibrium, Eq. (2.59) can be written

$$\frac{d^2\mathbf{x}}{dt^2} = -\omega_0^2\mathbf{x} - \gamma\frac{d\mathbf{x}}{dt} + \frac{e}{m}\mathbf{E}, \tag{2.60}$$

where $\omega_0^2 = K/m$ and $\gamma = b/m$. Because **x** satisfies a linear differential equation we can allow **x** to be complex and take its real part if needed.

The time-harmonic electric field is given by Eq. (2.58). After a sufficiently long time **x** is also time-harmonic with the same frequency as the electric field: $\mathbf{x} = \mathbf{x}_0 \exp(-i\omega t)$; this is the steady-state solution. When this assumed form of the solution is substituted in Eq. (2.60) we obtain for the complex amplitude

$$\mathbf{x}_0 = \frac{e}{m} \frac{\mathbf{E}_0}{\omega_0^2 - \omega^2 - i\gamma\omega}. \tag{2.61}$$

The instantaneous rate P at which work is done on the oscillator by the external field is

$$P = \mathbf{F} \cdot \frac{d\boldsymbol{\xi}}{dt} = e\mathbf{E} \cdot \frac{d\mathbf{x}}{dt}. \tag{2.62}$$

Here we have to be careful. Taking the dot (or scalar) product of vectors is not a linear operation. So we have to find the real displacement and take its scalar product with the real field. If we multiply and divide Eq. (2.61) by the complex conjugate of the denominator in this equation we obtain

$$\mathbf{x} = \frac{e}{m} \mathbf{E}_0 \frac{\exp\{-i(\omega t - \phi)\}}{\sqrt{(\omega_0^2 - \omega^2)^2 + \gamma^2\omega^2}}, \tag{2.63}$$

where

$$\sin\phi = \frac{\gamma\omega}{\sqrt{(\omega_0^2 - \omega^2)^2 + \gamma^2\omega^2}}, \quad \cos\phi = \frac{\omega_0^2 - \omega^2}{\sqrt{(\omega_0^2 - \omega^2)^2 + \gamma^2\omega^2}}. \tag{2.64}$$

The quantity ϕ is the *phase difference*, or simply the phase, between the amplitude of the oscillator and the force that drives it. When the phase difference is not zero, the maximum displacement of the oscillator does not occur at the same time as that of the electric field.

Now take the time derivative of the displacement [Eq. (2.63)] and multiply its real part by the real part of the electric field to obtain

$$P = \frac{e^2}{m} E_0^2 \frac{\omega}{\sqrt{(\omega_0^2 - \omega^2)^2 + \gamma^2\omega^2}} \{\cos^2 \omega t \sin\phi + \sin\omega t \cos\omega t \cos\phi\}. \tag{2.65}$$

This is the *instantaneous* rate at which work is done on the oscillator. But more often we are interested in the *time average* of P, denoted by $\langle P \rangle$, because detectors cannot respond instantaneously to the high frequencies of visible and near-visible radiation. For example, for a detector to respond instantaneously, even to comparatively sluggish 100 μm radiation, would require a response time of about 10^{-11} s.

Because we were critical in Section 1.4 of how the term "average" is bandied about without qualification, we had better make it clear that by time average here we mean a uniformly weighted average over the averaging interval. That is, if $f(t)$ is a time-dependent function, its average between times t_1 and t_2 is

$$\langle f \rangle = \frac{1}{t_2 - t_1} \int_{t_1}^{t_2} f \, dt. \tag{2.66}$$

The time average of the square of the cosine over a time interval much larger than the period is approximately 1/2 and the average of the product of the sine and cosine is zero. Thus from Eq. (2.65) we have

$$\langle P \rangle = \frac{e^2}{2m} E_0^2 \frac{\omega \sin \phi}{\sqrt{(\omega_0^2 - \omega^2)^2 + \gamma^2 \omega^2}}. \tag{2.67}$$

To do work on the oscillator requires energy. Electromagnetic fields are ascribed properties such as energy and momentum (but apparently not mass), which can be thought of as carried by photons. Thus $\langle P \rangle$ may be interpreted as the rate W_a at which energy is absorbed from the incident electromagnetic field by the oscillator. For $\langle P \rangle$ to be nonzero requires a nonzero phase difference ϕ, which in turn requires that dissipation (γ) be nonzero. Because the irradiance of the incident (or exciting) electromagnetic wave is proportional to the square of the amplitude of its electric field, it follows from Eq. (2.28) that $\langle P \rangle / E_0^2$ is, except for a constant factor, the frequency-dependent absorption cross section of the oscillator.

We may combine Eqs. (2.64) and (2.67) to obtain

$$\langle P \rangle = W_a = \frac{e^2}{2m} E_0^2 \frac{\gamma \omega^2}{(\omega_0^2 - \omega^2)^2 + \gamma^2 \omega^2}. \tag{2.68}$$

Maximum absorption occurs at ω_0, the frequency at which the derivative of W_a with respect to ω is zero. This frequency is often called the *resonant* or *natural frequency*, the frequency at which the oscillator, subject to no external forces, would oscillate forever in the absence of dissipation if displaced from its equilibrium position and released. W_a and hence the absorption cross section at the resonant frequency are proportional to $1/\gamma$. To remind ourselves that we are interested in the frequency dependence of absorption cross sections we can rewrite Eq. (2.68) as

$$\sigma_a = \gamma \sigma_{am} \frac{\gamma \omega^2}{(\omega_0^2 - \omega^2)^2 + \gamma^2 \omega^2}, \tag{2.69}$$

where σ_{am} is the maximum absorption cross section. In the high-frequency limit ($\omega \gg \omega_0$) the absorption cross section decreases as $1/\omega^2$; in the low-frequency limit ($\omega \ll \omega_0$) it increases as ω^2 provided that for both limits $\gamma \ll \omega_0$. The frequencies at which absorption falls to one-half its maximum value are obtained from Eq. (2.69):

$$\omega_{1/2} = \sqrt{\omega_0^2 + \gamma^2/4} \pm \gamma/2. \tag{2.70}$$

Equation (2.69) is the simplest example of an *absorption line*, a range of frequencies over which absorption is comparatively high. The *width* γ of the line, the range of frequencies for which absorption is at least one-half the maximum, follows from Eq. (2.70).

We are almost always interested in lines that are narrow in the sense that the width is much less than the resonant frequency ($\gamma \ll \omega_0$), in which instance Eq. (2.69) is approximately

$$\sigma_a = \frac{\gamma \sigma_{am}}{4} \frac{\gamma}{(\omega_0 - \omega)^2 + \gamma^2/4} \tag{2.71}$$

and Eq. (2.70) is approximately

$$\omega_{1/2} = \omega_0 \pm \gamma/2. \tag{2.72}$$

The frequency-dependent function multiplying the invariant quantity $\gamma\sigma_{am}/4$ is sometimes called the *Lorentz line shape*. Because its width at half maximum is γ, whereas its peak is $4/\gamma$, this suggests that the area

$$\gamma \int_0^\infty \frac{d\omega}{(\omega_0 - \omega)^2 + \gamma^2/4} \tag{2.73}$$

underneath the profile is independent of γ. By a transformation of variable the integral becomes

$$\gamma \int_{-\omega_0}^\infty \frac{d\mu}{\mu^2 + \gamma^2/4}. \tag{2.74}$$

The antiderivative of the integrand is

$$\frac{2}{\gamma} \tan^{-1} \frac{2\mu}{\gamma}, \tag{2.75}$$

and hence Eq. (2.74) is approximately 2π provided that $\gamma \ll \omega_0$, which confirms our expectation that the integral of the absorption cross section is independent of γ. As the width of the line decreases its peak increases in such a way that the area underneath the line is constant.

What does all this have to do with molecules? By definition a molecule is a *bound* object: its constituent electrons and nuclei hang together and their range of motion is limited. Similarly, the harmonic oscillator is bound by the spring constant (restoring force) K. Although molecules are electrically neutral (unless ionized), they are composed of equal numbers of oppositely charged electrons and protons. Our molecular oscillator has charge e, so lurking in the background is an equal and opposite charge $-e$. We can take the motion of e to be relative to $-e$. Two equal and opposite charges separated by a distance d is called a *dipole*; its *dipole moment* $\mathbf{p}$ is e times the position vector of the positive relative to the negative charge. Thus our oscillator is a dipole with an oscillating (or vibrating) dipole moment $\mathbf{p} = e\mathbf{x}$. Similarly, molecules have dipole moments, either *induced* by the exciting field or *permanent* as a consequence of their structure or both. Although molecules are electrically neutral, their centers of positive and negative charge may not coincide, either because an external electric field pushes the centers of positive and negative charge in opposite directions or because this charge separation is built into the molecule. Here we consider only induced dipole moments; that is, $\mathbf{x}$ vanishes when the external electric field vanishes. Moreover, $\mathbf{p}$ lies along the direction of the field, which therefore exerts no torque on the dipole. But a permanent dipole moment in an electric field experiences a torque and hence rotates. In general, a dipole both vibrates and rotates.

A harmonic oscillator driven by an external electric field leads to a simple classical model of the frequency dependence of absorption of electromagnetic radiation by individual molecules. To good approximation a molecule can be considered to be a dipole, and this is true even for sufficiently small particles (in Sec. 3.4 we elaborate on what is meant by "sufficiently").

Before proceeding we pause to contemplate the form of the resonant frequency

$$\omega_0^2 = \frac{K}{m},$$ (2.76)

which tells us something about the origin of all oscillations. This resonant frequency is the ratio of a restoring force (K) to inertia (m). This is a general result. Oscillations of any mechanical system usually can be traced to a struggle between a restoring force and inertia. For example, a simple pendulum oscillates because of the opposition between the restoring force of gravity and the inertia of the pendulum. Oscillations of a plucked string (see Sec. 3.3) occur because of opposition between the restoring force provided by the tension in the string and its inertia. Whenever you encounter an expression for the resonant frequency of a mechanical system, examine it carefully and try to interpret it as a consequence of a struggle between a restoring force and inertia. For an atom the restoring force is the attraction between the opposite charges of the induced dipole and the inertia is from the mass of the electrons associated with the dipole (Sec. 3.4.9).

2.6.1   Why the Obsession with Harmonic Oscillators?

At first glance, the amount of space devoted in textbooks to harmonic oscillators seems out of proportion to their importance. Yet the reason for this emphasis is that under fairly general conditions *any* mechanical system can be considered to be a collection of independent harmonic oscillators.

Consider N interacting point masses. For simplicity, we specify the positions of the point masses by their rectangular cartesian coordinates, denoted by ξ_i ($i = 1, 2, \ldots, 3N$). Suppose that the point masses exert position-dependent forces on each other that can be derived from a potential V, and suppose further that the system has an equilibrium configuration ($\xi_1^0, \xi_2^0, \ldots$). When the coordinates take these values the net force on any point mass is zero. We expand the potential in a Taylor series about the equilibrium position:

$$V = V_0 + \sum_i \left(\frac{\partial V}{\partial \xi_i}\right)_0 (\xi_i - \xi_i^0) + \frac{1}{2}\sum_i\sum_j \left(\frac{\partial^2 V}{\partial \xi_i \partial \xi_j}\right)_0 (\xi_i - \xi_i^0)(\xi_j - \xi_j^0) + \ldots \quad (2.77)$$

By definition the equilibrium position is that for which all interaction forces vanish:

$$\left(\frac{\partial V}{\partial \xi_i}\right)_0 = 0.$$ (2.78)

V_0 is a constant, so can be omitted because its derivatives do not contribute to the forces. If we truncate the series in Eq. (2.77) at the quadratic term, we obtain the potential in the *harmonic approximation*:

$$V \approx \frac{1}{2}\sum_i\sum_j K_{ij}(\xi_i - \xi_i^0)(\xi_j - \xi_j^0),$$ (2.79)

where

$$K_{ij} = K_{ji} = \left(\frac{\partial^2 V}{\partial \xi_i \partial \xi_j}\right)_0.$$ (2.80)

This approximation, also called the small-amplitude approximation, is better the smaller the departure from the equilibrium position.

Denote by x_i the displacement $\xi_i - \xi_i^0$. By the definition of a potential, the force component F_j is

$$F_j = -\frac{\partial V}{\partial \xi_j},\tag{2.81}$$

which from Eq. (2.79) is

$$F_j = -\sum_i K_{ji}(\xi_i - \xi_i^0) = -\sum_i K_{ji}x_i.\tag{2.82}$$

With this equation and the assumption of a damping force

$$-b_i\frac{d\xi_i}{dt} = -b_i\frac{dx_i}{dt}\tag{2.83}$$

associated with each velocity component, the equations of motion for the system of point masses are

$$m_j\frac{d^2\xi_j}{dt^2} = m_j\frac{d^2x_j}{dt^2} = -\sum_i K_{ji}x_i - b_j\frac{dx_j}{dt} \quad (j = 1, 2, \ldots, 3N).\tag{2.84}$$

Although there are $3N$ distinct coordinates, there are only, at most, N distinct masses. Because these equations are linear with constant coefficients we can express them in compact form by defining a column matrix $\mathbf{x}$, the $3N$ elements of which are x_j:

$$\mathbf{M}\frac{d^2\mathbf{x}}{dt^2} = -\mathbf{K}\mathbf{x} - \mathbf{B}\frac{d\mathbf{x}}{dt},\tag{2.85}$$

where $\mathbf{K}$ is a symmetric $3N \times 3N$ matrix with elements K_{ij}, $\mathbf{M}$ is a diagonal matrix with elements m_j, and $\mathbf{B}$ is a diagonal matrix with elements b_j. Multiply both sides of Eq. (2.85) by the inverse matrix $\mathbf{M}^{-1}$, a diagonal matrix the elements of which are the inverse masses:

$$\frac{d^2\mathbf{x}}{dt^2} = -\mathbf{M}^{-1}\mathbf{K}\mathbf{x} - \mathbf{M}^{-1}\mathbf{B}\frac{d\mathbf{x}}{dt} = \mathbf{K}'\mathbf{x} - \mathbf{B}'\frac{d\mathbf{x}}{dt}.\tag{2.86}$$

Now assume that we can find a linear transformation

$$\mathbf{x} = \mathbf{A}\mathbf{u}\tag{2.87}$$

such that both $\mathbf{A}^{-1}\mathbf{K}'\mathbf{A}$ and $\mathbf{A}^{-1}\mathbf{B}'\mathbf{A}$ are diagonal, where $\mathbf{A}^{-1}$ denotes the inverse transformation. With the transformation Eq. (2.87), Eq. (2.86) becomes

$$\frac{d^2\mathbf{u}}{dt^2} = -\mathbf{\Omega}\mathbf{u} - \mathbf{\Lambda}\frac{d\mathbf{u}}{dt},\tag{2.88}$$

with

$$\mathbf{A}^{-1}\mathbf{K}'\mathbf{A} = \mathbf{\Omega}, \quad \mathbf{A}^{-1}\mathbf{B}'\mathbf{A} = \mathbf{\Lambda},\tag{2.89}$$

where Ω and Λ are diagonal matrices. Thus the $3N$ coupled equations of motion [Eq. (2.85)] become $3N$ uncoupled equations

$$\frac{d^2 u_j}{dt^2} = -\omega_j^2 u_j - \Gamma_j \frac{du_j}{dt} \ (j = 1, 2, \ldots, 3N), \tag{2.90}$$

where ω_j^2 are the elements of Ω and Γ_j are the elements of Λ.

Do these equations of motion look familiar? They should. They are the equations of motion of $3N$ *independent* harmonic oscillators. The frequencies ω_j are called the *normal* (or *characteristic*) *frequencies* of the system and the coordinates u_j are called the *normal coordinates*. Each independent oscillation is called a *normal mode*, and the general motion of the entire system is, from Eq. (2.87), a superposition of normal modes.

By purely mathematical trickery we reduced a system of interacting point masses to a system of non-interacting harmonic oscillators. This reduction requires for its validity that the system have a position of equilibrium the departures from which are sufficiently small. With these assumptions, *every* mechanical system behaves as if it were a set of independent harmonic oscillators of different frequencies. The system can range from a single water molecule composed of two hydrogen atoms and an oxygen atom to an entire building.

We once attended a colloquium given by a civil engineer interested in the forces that turbulent winds exert on buildings. He claimed that changes in materials and designs had altered the natural frequencies of buildings. We don't know much about turbulence, even less about structural engineering, but all we needed to understand the gist of his talk was Eqs. (2.76) and (2.90).

Reduced to its simplest elements, a building (or any structure) is a system of independent harmonic oscillators (e.g., weights and springs). The natural frequencies of this system depend on the stiffness of the structure (K) and the density of materials (m). If either or both are changed, the natural frequencies change. The time-varying turbulent wind field can be decomposed into components with different frequencies, and the aim of the designer is to be sure that these frequencies, especially the dominant ones, do not coincide with the natural frequencies of the structure. As we have seen, the amplitude of an oscillation depends on the frequency of the driving force.

2.7   Classical versus Quantum-Mechanical Interpretation of Absorption

What once was, and perhaps still is, the standard treatise on atomic spectra, by Condon and Shortley, fills 432 pages of text. Herzberg's treatises fill 581 pages for diatomic molecules, 538 pages for polyatomic molecules, and 670 pages for the electronic spectra of polyatomic molecules. Townes and Schawlow devote 648 pages to microwave spectroscopy. And the physical strain of just lifting these nearly 3000 pages is as nothing compared to the mental strain of absorbing them. Add to these tomes the dozens if not hundreds of books on quantum mechanics and modern physics. Spectroscopy is the science of details. Indeed, this is the source of the almost religious awe in which quantum mechanics is held by physicists. A theory capable of ordering – if not explaining – so many details is akin to magic.

This somewhat depressing inventory of details, systematized by a theory that, as Richard Feynman has so often been quoted as saying, "nobody understands", signals that what follows can only scratch the surface of a vast subject. Given the few pages we can devote to it, our aim must be the modest one of conveying the bare bones of what students of atmospheric science should know about the infrared spectra of molecules that inhabit the atmosphere.

According to classical mechanics, the mechanics of Newton, a mechanical system can have a continuous set of energies (kinetic plus potential). This seems so self-evidently true that it is rarely stated explicitly and even more rarely questioned. But in reality, energy is like dollar bills: you can have one bill in your wallet or two or three, but never 1.523. Dollar bills are *quantized*. And so is energy, but this is not evident until we consider mechanical systems on a scale not directly accessible to our senses. We do not live at this microscopic (or atomic) scale, so we have no right to expect that the physics of ordinary macroscopic objects is valid there. This is much like moving from one country to another. In one society, certain rules of behavior and customs are taken for granted. But when you enter a different society, you sometimes discover that many of the familiar rules no longer apply. In India, Pakistan, and Sri Lanka, people eat with their fingers. To do otherwise would seem unnatural to the inhabitants of these countries. But in Western countries, eating with your fingers is considered to be extremely impolite. Children who do this are told in no uncertain terms to desist.

The differences between the rules of behavior of macroscopic and microscopic objects are considerably greater than the differences between those of the inhabitants of New York penthouses and the inhabitants of huts in the Amazon jungle.

If the laws of quantum mechanics must be taken on faith, consider that so must the laws of classical mechanics. Why $F = ma$? You learn this first as an axiom, and so what follows it may seem strange. But if you had learned quantum mechanics first, $F = ma$ might seem strange. Whatever you learn first sets the standard for what is normal.

Consider first a harmonic oscillator with natural frequency ω_0. According to classical mechanics this oscillator can have a continuous range of energies depending on its initial amplitude. But according to quantum mechanics the energies of this harmonic oscillator are quantized, having only the discrete set of energies

$$E_n = \hbar\omega_0 \left(n + \frac{1}{2} \right), \tag{2.91}$$

where $n = 0, 1, 2, \ldots$. This result seems to contradict common sense, but it really doesn't because common sense is based on our experience with macroscopic objects. The difference in energy between physically realizable (allowed) adjacent energy states of the harmonic oscillator is

$$\Delta E = \hbar\omega_0. \tag{2.92}$$

The oscillators we encounter at the macroscopic level might have natural frequencies as high as 100 Hz. This corresponds to an energy difference of 6.63×10^{-32} J. To get a feeling for how much energy this is, consider how much energy the eyebrow of a flea has when it falls from a flea on the back of a dog. A flea, which can be seen without a microscope, has dimensions of a millimeter or so, and hence the total volume of the flea is of order $10^{-9}\,\mathrm{m}^3$. The characteristic linear dimension of a flea's eyebrow is at least 100 times smaller than the overall dimensions of the flea. Thus the volume of the flea's eyebrow is of order $10^{-15}\,\mathrm{m}^3$. We estimate the density of flea flesh to be that of water, $1000\,\mathrm{kg\,m}^{-3}$, which gives a mass of $10^{-12}\,\mathrm{kg}$ for the flea's eyebrow. Suppose that the flea resides on the back of a Great Dane, about 1 m high at the shoulders. The potential energy of the flea's eyebrow, relative to that at the ground, is about $10^{-11}\,\mathrm{J}$. As small as this potential energy is, it is still 10^{21} times greater than the difference between adjacent energy levels of a harmonic oscillator with natural frequency 100 Hz. Although the energies of macroscopic oscillators are quantized in principle, the spacing of energy *levels*, as they are called, is so small macroscopically that in practice the levels are continuous. We are forced to come to grips with the discreteness of energy levels only when we consider systems with very high natural frequencies. All else being equal, natural frequencies increase with decreasing mass [Eq. (2.76)], and hence the consequences of discrete energies are not negligible at the atomic scale.

With this preamble, consider absorption of electromagnetic energy by a single isolated oscillator from the classical and quantum-mechanical points of view. According to the classical analysis in Section 2.6, the rate at which power is absorbed by an oscillator from a time-harmonic electromagnetic wave of given amplitude depends on its frequency ω. Absorption is sharply peaked in a narrow range of frequencies, called an absorption line (or band), centered on the natural frequency of the oscillator. The width of the line is a consequence of damping of the oscillator.

Now consider the same process from a quantum-mechanical point of view. The incident monochromatic electromagnetic wave is considered to be a stream of photons, each with energy $\hbar\omega$. Absorption of electromagnetic energy is a consequence of absorption of photons. If the oscillator absorbs a photon, the energy of the oscillator must increase. But this increase can be only one of a set of discrete values. Unless the energy of the photon is equal to the difference between two energy levels of the oscillator, it cannot absorb the photon. This accounts for the narrowness of absorption lines.

2.7.1   Molecular Energy Levels

An electrically neutral atom consists of a nucleus composed of uncharged neutrons and positively charged protons bound together and surrounded by negatively charged electrons equal in number to the protons. Almost the entire mass of an atom resides in its nucleus, which is smaller by a factor of about 1000 than the atom as a whole. That is, the electrons, on average, are at distances from the nucleus large compared with its size.

A molecule is a collection of two or more atoms bound together. They can vibrate about their equilibrium positions and the molecule as a whole can rotate. A vibrating object has vibrational kinetic and potential energy; a rotating object has rotational kinetic energy. The center of mass of the molecule can move and it can interact with neighboring molecules. And electrons have potential and kinetic energies. These various modes of motion of a molecule can be decomposed *approximately* into *translational* (position and velocity of the center of mass of the molecule), *rotational*, *vibrational*, and *electronic* modes, and hence the total energy of the molecule is

$$E = E_{\mathrm{trans}} + E_{\mathrm{rot}} + E_{\mathrm{vib}} + E_{\mathrm{elec}}. \tag{2.93}$$

That this decomposition is only approximate is evident from the rotations and vibrations of two mass points connected by a spring. The rotational kinetic energy of this system depends on its moment of inertia, which in turn depends on the separation between the mass points. But if they vibrate, their separation changes, and hence so does the moment of inertia. Thus the two types of motion, rotation and vibration, are not completely decoupled. Although Eq. (2.93) often is a good approximation, don't be surprised when it fails to explain what is observed.

Each of the separate contributions to the total energy E in Eq. (2.93) is quantized except the translational kinetic energy of the center of mass of a molecule. A truly isolated molecule is an idealization. We are almost always faced with an ensemble of many molecules. Even in a low density gas, interactions between molecules are not completely negligible. Strictly speaking, translational energy is also quantized but the level separation is so small that it cannot be observed. To the extent that Eq. (2.93) is a good approximation we can therefore speak of the electronic energy levels of a molecule, its vibrational energy levels, and so on. The lowest allowed energy of any kind is called the *ground state*. All higher energy levels, or states, are called *excited states*. When a molecule absorbs a photon it is said to be excited into a higher-energy state or to undergo a transition from one energy state to another of higher energy. A molecule in an excited state can then spontaneously drop to a lower energy state accompanied by the emission of a photon equal in energy to the difference in energy levels, which underscores our previous assertions about absorption and emission being inverse processes. The Lorentz line shape [Eq. (2.71)] is just as valid quantum-mechanically as it is classically, but the terms in it are interpreted quite differently. Classically, ω_0 is the natural frequency of an oscillator and γ is a factor in a viscous damping term; quantum-mechanically, $\hbar\omega_0$ is the difference between two energy levels and $1/\gamma$ is the *lifetime* of the transition between them (i.e., the average time the molecule exists in the higher energy state).

The energy of a photon is

$$\hbar\omega = h\nu = \frac{hc}{\lambda} = hc\tilde{\nu}. \tag{2.94}$$

Except for the factor hc, a universal constant, the energy of a photon is inversely proportional to the wavelength of the associated wave. Thus we may express photon energies as inverse wavelengths, which are called *wavenumbers*, usually with units cm^{-1}. Why wave*number*? Because it is the number of waves in unit length. Spectroscopists are careless in the use of the symbol ν: sometimes it denotes frequency, sometimes wavenumber (inverse wavelength). And to make matters worse, the units of wavenumber are usually cm^{-1}, and so if hc is in SI units and $\tilde{\nu}$ in cm^{-1}, photon energy is $100hc\tilde{\nu}$. We try to be consistent and use ν for frequency, $\tilde{\nu}$ for wavenumber. A word of caution: $2\pi/\lambda$, often written as k or q, is sometimes called wavenumber.

A wavelength of $10\,\mu m$ corresponds to a wavenumber of $1000\,cm^{-1}$. Wavelengths in the middle of the visible spectrum correspond to wavenumbers of around $20,000\,cm^{-1}$. Sometimes you will encounter statements about a photon energy being so many wavenumbers, 2000, say. Unless stated otherwise, this probably means $2000\,cm^{-1}$.

Another quantity with dimensions of energy that keeps cropping up in all kinds of problems is $k_B T$, which we can divide by hc so as to express it in the same units as wavenumbers. At typical terrestrial temperatures (300 K say), $k_B T/hc$ is around $200\,cm^{-1}$.

Consider now a gas of molecules in thermal equilibrium at temperature T. These molecules can collide and exchange energy. Indeed, a molecular collision is *defined* as an interaction between two or more molecules in which the energy of each molecule after the interaction is different from that before (total energy being conserved). The most probable kinetic energy of a gas molecule is $k_B T/2$ (see Sec. 1.2), so this is how much energy can be exchanged in a typical collision. Some molecules have greater energies than the most probable, some less, the probability of energy E being proportional to $\exp(-E/k_B T)$. (We omit the other factor dependent on energy because it does not vary so strongly). This quantity, called the *Boltzmann factor*, crops up in all kinds of problems. Indeed, it would not be an exaggeration to say that it, not love, is what makes the world go round. Rates of chemical reactions are determined by the Boltzmann factor. Our bodies are complicated, finely-tuned engines in which many chemical reactions are continuously taking place, their rates determined by the Boltzmann factor. A small change in temperature changes not only the absolute rates of these reactions but their relative rates as well. This is why our bodies constantly attempt to keep our deep core temperature at around 37 °C. An increase in this temperature of only 1–2 °C is sufficient to send us to bed wracked with pain and fever. An increase of a few more degrees might be fatal. When you cook food you are exploiting the Boltzmann factor. An egg will rot before it hardens at room temperature, but immerse it in water at 100 °C and it is hard-boiled in minutes. A rough rule of thumb used by chemists is that chemical reaction rates double with every 10 °C increase in temperature. A temperature increase from 20 °C to 100 °C corresponds to 8 doubling times, a factor of 256.

If a gas is in thermal equilibrium its molecules are distributed in their energy states. The ratio, on average, of the number N_j of molecules having energy E_j and the number of molecules N_i having energy E_i is approximately

$$\frac{N_j}{N_i} \approx \frac{\exp(-E_j/k_B T)}{\exp(-E_i/k_B T)} = \exp\{-(E_j - E_i)/k_B T\}. \tag{2.95}$$

Typically, the separation between electronic energy levels is around $10,000\ \mathrm{cm}^{-1}$. Suppose that E_i corresponds to the ground electronic state and that E_j corresponds to the first excited electronic state. Because the difference ΔE between these two energy states is around $10,000\ \mathrm{cm}^{-1}$, the ratio of the number of molecules with electrons in the first excited state to the number in the ground electronic state at 300 K is of order $\exp(-10,000/200) = \exp(-50) \approx 10^{-25}$. At typical terrestrial temperatures almost all molecules in a gas are in their ground electronic state. To populate excited electronic states would require a ten-fold or more increase in absolute temperature. Typical separations between adjacent vibrational energy levels are about $1000\ \mathrm{cm}^{-1}$. Again, this is large compared with $k_B T$ for ordinary temperatures, and hence most gas molecules at these temperatures are in their vibrational ground states. Typical separations between rotational energy states are 10–$100\ \mathrm{cm}^{-1}$, which are comparable with $k_B T$. Thus at ordinary temperatures many molecules are in excited rotational states.

Merely because molecules vibrate and rotate does not necessarily mean that they radiate (emit). Absorption, as we have seen, is the inverse of emission. It is perhaps easier to discuss spontaneous emission because it occurs in the absence of an external exciting field. One of the results of classical electromagnetic theory is that accelerated charges radiate electromagnetic waves. A corollary of this is that for an electrically neutral charge distribution (e.g., molecule) to radiate, its dipole moment must change with time, either in magnitude or direction or both. A molecule with a permanent dipole moment is said to be *polar*. A common example is the water molecule (H_2O), which owes its permanent dipole moment to its asymmetry: it is composed of two atoms of one kind (hydrogen) and one atom of another (oxygen), which all do not lie on a line. As a consequence of its permanent dipole moment, a rotating water molecule radiates (emits).

The two most abundant molecules in the atmosphere are the diatomic molecules nitrogen (N_2) and oxygen (O_2). Both of these molecules are *homonuclear*, which is just a fancy way of saying that they are composed of identical atoms. Such molecules cannot have a permanent dipole moment. To have such a moment would require the center of positive charge to be associated with one atom and the center of negative charge to be associated with the other. But symmetry rules this out: both atoms are identical, and hence one cannot be positively and the other negatively charged. Although nitrogen and oxygen molecules can rotate, in so doing they do not radiate (much). These molecules also can vibrate. But again symmetry rules out the possibility of a changing dipole moment during this vibration. The two atoms must enter into the vibration in a symmetric way, and hence the dipole moment associated with one atom is equal and opposite to that associated with the other.

We have seen that the motions of a molecule can be expressed as a sum of normal modes, each with a characteristic frequency. These frequencies lie in the infrared and a radiating mode is called *infrared active*. A mode that does not radiate is called *infrared inactive*. The terms infrared active and inactive, which are familiar to infrared spectroscopists, are preferable to the popular but misleading term "greenhouse gas." Water vapor is infrared active; nitrogen and oxygen, for the most part, are infrared inactive. Greenhouse gases are produced by resident cats with digestive problems.

The normal modes of vibration of the water molecule are shown schematically in Fig. 2.17. There are three modes: the O-H (symmetric) stretching mode, the H-O-H bending mode, and the O-H (asymmetric) stretching mode. The corresponding normal frequencies are $\tilde{\nu}_1 = 3657.1 \, \text{cm}^{-1}$ ($2.73 \, \mu\text{m}$), $\tilde{\nu}_2 = 1594.8 \, \text{cm}^{-1}$ ($6.27 \, \mu\text{m}$), and $\tilde{\nu}_3 = 3755.9 \, \text{cm}^{-1}$ ($2.66 \, \mu\text{m}$). The O-H stretching modes are so named because the vibration occurs approximately along the O-H bond.

Keep in mind that these normal frequencies are for the isolated water molecule, water in the gaseous phase. When water vapor condenses to form liquid water, the positions of the absorption bands shift. And when liquid water freezes to form ice, there is yet another shift in the absorption bands. Nevertheless, the normal frequencies of the isolated water molecule are good guides to the approximate positions of absorption bands in the condensed phases. This is evident in Fig. 2.2. Note the large dips in the absorption length around 3 µm and 6 µm. These are the consequence, even in the condensed phases, of vibrations of the isolated water molecule, which does not lose its identity completely when it condenses.

According to quantum mechanics, the vibrational energy levels of the water molecule are quantized:

$$E_n = h\nu_i \left(n + \frac{1}{2} \right), \tag{2.96}$$

where ν_i is any one of the three normal mode frequencies of the water molecule. On the basis of this equation alone we would expect absorption bands not only at the fundamental frequency ν_i but at overtones as well, integral multiples of the fundamental frequency: $2\nu_i, 3\nu_i, \ldots$. But this expectation is not quite borne out by experience. Equation (2.96) does not tell the entire story. We have to account for *selection rules*, rules that tell us which energy transitions are allowed. The different energy states described by Eq. (2.96) are also states of different angular momentum. The photon carries quantized angular momentum. Thus when a photon is absorbed by a molecule, and it undergoes a transition to a higher energy level, the transition must be such that angular momentum is conserved. Thus if one unit of angular momentum is annihilated (so to speak) when a photon is absorbed, the molecule must increase its angular momentum, and hence energy state, by the same amount. And similarly for emission: when a photon is emitted, the one unit of angular momentum created upon the birth of the photon must be compensated for by the same decrease in the angular momentum of the molecule. Remember that angular momentum is a vector, and hence the angular momentum of the emitted or absorbed photon and the change in angular momentum of the molecule must add vectorially to zero. This requirement of angular momentum conservation leads to the selection rule $\Delta n = \pm 1$. That is, transitions are allowed only for integral changes in the quantum number n in Eq. (2.96).

This selection rule is not absolute because it is based on the harmonic approximation (Sec. 2.6.1). Because the forces between atoms in the water molecule are not exactly harmonic, the previous section rule can be violated. That is, transitions corresponding to *overtones* of the fundamental frequencies and even *combinations* of different frequencies (e.g., $\nu_1 + 2\nu_2, \nu_1 - \nu_2$, etc.) are possible. These transitions are very weak, but not so weak as to be unobservable.

Although absorption drops precipitously from infrared to visible, it does not reach zero. Moreover, absorption over the visible spectrum is least in the blue–green and rises toward the red in both the liquid and solid phases. (This is evident in Fig. 2.2 but even more so in Fig. 5.12.) Water is a weak blue dye, and this is intrinsic, not the result of some vague impurities. This intrinsic selective absorption by water leads to observable consequences: the blue of the sea, of crevasses in glaciers, ice caves, and frozen waterfalls (see Sec. 5.3.1). What is the molecular mechanism for this blueness?

Overtones of fundamental vibration frequencies in the infrared and combinations of these frequencies make their presence felt, weakly, but observably in the visible. The experimental proof of the vibrational origin of the visible absorption spectrum of water was obtained several years ago by Chuck Braun and Sergei Smirnov, who published their results in a delightful paper, "Why is water blue?", one of those papers you must read before going to your grave. Braun and Smirnov measured the visible and near-visible absorption spectra of ordinary water (H_2O) and heavy water (D_2O). The vibration spectrum of a molecule depends on the masses of its constituent atoms [see Eq. (2.76)]. Because the nuclear mass of heavy water is greater than that of ordinary water we expect a shift (*isotope shift*) toward lower frequencies in the vibration spectrum of D_2O relative to that of H_2O. And this shift was indeed observed. In ordinary water there is a peak at about 750 nm, which shifts to about 1000 nm for heavy water. Braun and Smirnov did not calculate the magnitude of the isotope shift because of the greater mass of deuterium relative to that of hydrogen, so we did using Eq. (2.76) and found good agreement with the measured shift.

Rising absorption in the red gives ordinary water its bluish color, whereas absorption by heavy water is flat throughout the visible spectrum and begins to rise only well into the infrared.

When questions about visible absorption by water arise, chemists seem to be divided into two groups: those who adamantly deny that water has a visible absorption spectrum and those who admit that it does but are just as adamant that it is a consequence of hydrogen bonding.

You may think that we are joking about the first group, yet we could tell you many stories. One will suffice. Several years ago one of the authors gave a talk in Italy on colors in nature for an audience mostly of biological scientists. Afterwards, a photochemist accused him of being a criminal and pervert for telling innocent biologists that water has a visible absorption spectrum.

In chemistry, hydrogen bonding is the universal solvent… for ignorance. Whenever a puzzle arises and there is any liquid water in the neighborhood, hydrogen bonding dissolves the puzzle in the same way that Alexander untied the Gordian knot with his sword. Hydrogen bonding plays the same role that friction does in the undergraduate physics laboratory. Liquid water, however, is blue not *because* of hydrogen bonding but *despite* it. Hydrogen bonding in liquid water shifts all the infrared absorption bands, including their overtones and combinations, to lower frequencies compared with the gas phase.

Braun and Smirnov are of the opinion that the color of water is the only example of a color in nature that results from vibrational rather than electronic transitions. It is commonly believed that all colors in nature can be traced to electronic excitations. Water provides an example to the contrary.

Carbon dioxide is a linear, symmetric molecule. By linear is meant that its bonds lie on a straight line. By symmetric is meant that it is composed of a carbon atom flanked on either side by oxygen atoms. This symmetry implies that the carbon dioxide molecule does not have a permanent dipole moment, which precludes this molecule from having what is called a *rotational* absorption band: an absorption band associated with rotation of the molecule unaccompanied by vibrations. But the carbon dioxide molecule can vibrate in such a way that it has a changing dipole moment. This in turn implies that this molecule, when vibrating in one of its vibrational modes, also can rotate to give a changing dipole moment. So carbon dioxide has *vibration-rotational* bands.

Carbon dioxide has a bending mode with frequency $\tilde{\nu}_2 = 667.4\,\text{cm}^{-1}$ (15 μm) and an asymmetric stretching mode with frequency $\tilde{\nu}_2 = 2349.2\,\text{cm}^{-1}$ (4.25 μm). The symmetric stretching mode is infrared inactive. It is the vibration-rotational band near 15 μm that is the major player in the global warming scenario associated with increased carbon dioxide in the atmosphere.

Rotational energies also are quantized. For example, those of the *rigid rotator*, a massless rigid rod of length r with one end fixed, the other end attached to a mass m, and free to rotate are given by

$$E_J = \frac{\hbar^2 J(J+1)}{2I}, \; J = 0, 1, 2, 3, \ldots, \tag{2.97}$$

where $I = mr^2$ is the moment of inertia about the rotation axis. As with transitions of the harmonic oscillator, those of the rigid rotator are restricted by a selection rule: $\Delta J = \pm 1$.

Although molecules can rotate, and hence have rotational energy levels, they are never exactly describable as rigid rotators. Like the harmonic oscillator, the rigid rotator is an idealization, and hence this selection rule is not absolute. When we say with seeming absoluteness that a transition is forbidden, we mean that the transition probability is very small compared with that for an allowed transition but not identically zero. Quantum mechanics is a most permissive theory, allowing improbable events that according to classical mechanics are impossible.

You may wonder why we jumped directly to the quantum-mechanical rigid rotator instead of making a more gradual transition by way of the classical rotator. This is what we did for the harmonic oscillator, beginning with its classical equation of motion and then stating, but not proving, how the discrete energy levels of the quantum-mechanical oscillator are related to the classical natural frequency of the oscillator. Why didn't we follow the same approach for the rigid rotator? And why were we unable to find any authors who do or even bring up the subject?

As it happens, analysis of the classical harmonic oscillator is considerably less difficult than analysis of the quantum-mechanical oscillator. When we turn to the rigid rotator, this happy state of affairs is reversed: the quantum-mechanical rotator is more amenable to analysis. The classical rigid rotator has no natural frequency. It can spin at any angular speed. A simple pendulum does have a natural frequency for small-amplitude oscillations, but this frequency depends on the magnitude of an *external* force: gravity. Thus even if we could solve the classical equation of motion for a driven rigid rotator, we wouldn't acquire any fruitful analogies or equations we could carry with us into the quantum domain. And similarly for electronic energy levels. But with only a few exceptions, these energy levels are not of great relevance to the atmosphere. One exception is the Chappuis bands of ozone, which play an important role in the color of the zenith twilight sky (see Sec. 8.1.3).

2.8   Absorption by Molecules: The Details

The previous section laid out the bare bones of molecular absorption and its physical interpretation. In this section we put some flesh on these bones, a meal that requires considerable chewing and digestion. At the very least students of atmospheric science should be familiar with Figs. 2.12 and 2.13 and understand their implications. Everything in this section is aimed at explaining the strange shapes of the curves in these figures.

A beam of radiation transmitted by a gas undergoes attenuation by absorption if photons in the beam have energies equal to differences in two molecular energy levels for which the selection rules are obeyed. Molecules make transitions from lower to upper levels. But the reverse is also possible. Some molecules can be *induced* or *stimulated* to make transitions to a lower level, energy conservation therefore requiring that photons be emitted. These photons have the same energy and direction as those that induced them. Although this process may seem strange at first glance, remember that any process in which energy is conserved is possible. *Spontaneous emission* of photons by gas molecules, in contrast to this *induced emission*, is, of course, also possible. Indeed, by emission without qualification, we usually mean spontaneous emission.

Induced emission and absorption are opposite processes, the former increasing the radiant energy in a beam, the latter decreasing it, but all we can observe is the net result. Thus the apparent (observed) absorption coefficient κ is less than it would be in the absence of induced emission. And we also might have to correct for spontaneous emission because a detector downstream of a source necessarily receives induced and spontaneous radiation from the gas as well as source radiation not absorbed by it.

2.8.1 Absorption versus Spontaneous and Induced Emission

Estimating κ in the presence of induced emission is a difficult but worthwhile task. Let E_ℓ and E_u ($E_u > E_\ell$) be two molecular energy levels, for which radiative transitions between levels are allowed. The energy levels of a molecule depend on its structure and on the forces exerted on it by its neighbors (no molecule lives alone) and these forces depend on the ever-changing distances between molecules. These interaction forces, often called collisions, result in perturbations of the energy levels of the isolated molecule, and hence in their difference, often expressed as a frequency $(E_u - E_\ell)/h$. This frequency for the *isolated* molecule is denoted by ν_0. The distribution of frequencies about ν_0, called a *line shape*, is narrow but its width is never identically zero (see Fig. 2.18). Because they are in motion, molecules are illuminated by radiation of frequency slighted shifted from that of the source, and emit radiation the frequency of which is also shifted (see Sec. 3.4.6). This *Doppler broadening* of the spectral line (centered on ν_0) is equivalent to perturbing energy level differences but is a consequence of relative motion rather than inter-molecular forces. In the lower atmosphere Doppler broadening of spectral lines is usually less than that resulting from interactions, often called *collisional broadening* or *pressure broadening*, a misleading term because the broadening has nothing fundamental to do with pressure. Inter-molecular force broadening is a more descriptive, but also more awkward, term.

Keep in mind that treating a gas as a collection of identical molecules each with the same but fuzzy or perturbed energy levels is an approximation. The only rigorous way to determine the energy levels of a gas would be to treat it as a system of molecules each interacting with all the others. This system as a whole has a set of energy levels. Alas, this is an insoluble many-body problem. So we settle for less and adopt the fiction that each molecule has the same set of fuzzy energy levels, the fuzziness specified by the line width. This is a good approximation for a gas because its molecules interact weakly, but fails completely when the gas is compressed into a liquid or a solid, for which the inter-molecular forces are dominant rather than perturbations. The absorption spectrum of sodium gas, for example, bears no resemblance to that of sodium metal.

N is the total number of molecules per unit volume, N_u in an upper state, N_ℓ in a lower state. Divide frequency space into M bins centered on frequency ν_0. Denote by n_{um} the number density of molecules with an upper energy level such that the difference between this level and a lower level lies in the mth bin. The rate of change of n_{um} because of spontaneous emission is assumed to be given by

$$\left(\frac{dn_{um}}{dt}\right)_s = -R_{s,um}n_{um}, \tag{2.98}$$

where $R_{s,um}$ is the rate constant for spontaneous emission, for generality assumed to be possibly different for every transition, and s denotes spontaneous. Equation (2.98) has the form of a first-order chemical kinetics equation based on the plausible assumption that the instantaneous rate at which something changes is proportional to the amount of that something at any instant. This is also the equation of radioactive decay. The frequency of the photon emitted is that associated with the mth bin. If $f_{\text{u}m}$ is the probability that a molecule has upper and lower states such that the energy difference lies in the mth bin, we can write Eq. (2.98) as

$$\left(\frac{dn_{\text{u}m}}{dt}\right)_{\text{s}} = -R_{\text{s,u}m} f_{\text{u}m} N_{\text{u}}. \tag{2.99}$$

The rate of change of $n_{\text{u}m}$ because of induced emission must depend on the radiation environment of the molecule, and hence the rate equation must contain the number density $n_{\text{p},m}$ of photons with energies (frequencies) in the mth bin:

$$\left(\frac{dn_{\text{u}m}}{dt}\right)_{\text{e}} = -R_{\text{e,u}m} f_{\text{u}m} N_{\text{u}} \frac{n_{\text{p},m}}{\Delta\nu}, \tag{2.100}$$

where e denotes induced emission and $\Delta\nu$ is the bin width. Note that the dimensions of the rate constant in this equation are different from those in Eq. (2.99).

The number density of lower states $n_{\ell m}$ decreases because of absorption of photons and, like induced emission, depends on the radiation environment. Thus the rate equation analogous to Eq. (2.100) is

$$\left(\frac{dn_{\ell m}}{dt}\right)_{\text{a}} = -R_{\text{a},\ell m} f_{\ell m} N_{\ell} \frac{n_{\text{p},m}}{\Delta\nu}, \tag{2.101}$$

where a denotes absorption and $f_{\ell m}$ is the probability that a molecule has lower and upper states such that the energy difference lies in the mth bin. The quantity $R_{\text{a},\ell m}$ is essentially the *Einstein absorption coefficient*, $R_{\text{e,u}m}$ the *Einstein induced emission coefficient*, and $R_{\text{s,u}m}$ the *Einstein spontaneous emission coefficient*. Einstein assumed in 1916 that the transition rates Eq. (2.100) and Eq. (2.101) are proportional to the spectral density $n_{\text{p},m}/\Delta\nu$, later proven by quantum mechanics when it had developed into a sufficiently detailed theory. We are content to accept Einstein's assumption. Ultimately, its justification is agreement with measurements.

Because only the product of a rate constant and an occupation probability occurs in Eqs. (2.99)–(2.101), we can fold any dependence of the rate constants on bin into the occupation probabilities and write

$$\left(\frac{dn_{\text{u}m}}{dt}\right)_{\text{s}} = -R_{\text{s}} f_{\text{s,u}m} N_{\text{u}}, \tag{2.102}$$

$$\left(\frac{dn_{\text{u}m}}{dt}\right)_{\text{e}} = -R_{\text{e}} f_{\text{e,u}m} N_{\text{u}} \frac{n_{\text{p},m}}{\Delta\nu}, \tag{2.103}$$

$$\left(\frac{dn_{\ell m}}{dt}\right)_{\text{a}} = -R_{\text{a}} f_{\text{a},\ell m} N_{\ell} \frac{n_{\text{p},m}}{\Delta\nu}. \tag{2.104}$$

The sum of Eqs. (2.102) and (2.103), times a constant factor, is the rate at which molecules lose energy by emission, whereas Eq. (2.104), times the negative of that same constant factor, is the rate at which molecules gain energy by absorption. In equilibrium losses must balance gains and hence

$$\frac{\overline{N}_u}{\overline{N}_\ell} = \frac{R_a f_{a,\ell m}(n_{p,m}/\Delta\nu)}{R_s f_{s,um} + R_e f_{e,um}(n_{p,m}/\Delta\nu)},$$
(2.105)

where the bars indicate equilibrium values. We assume that this energy balance holds bin by bin, that is, we invoke the principle of detailed balance (see Sec. 1.3). From this equation follows

$$\frac{n_{p,m}}{\Delta\nu} = \frac{(R_s f_{s,um}/R_e f_{e,um})}{(\overline{N}_\ell R_a f_{a,\ell m}/\overline{N}_u R_e f_{e,um}) - 1}.$$
(2.106)

In equilibrium the ratio of number densities is given approximately by

$$\frac{\overline{N}_u}{\overline{N}_\ell} = \frac{g_u \exp(-E_u/k_B T)}{g_\ell \exp(-E_\ell/k_B T)} = \frac{g_u}{g_\ell} \exp(-h\nu_0/k_B T),$$
(2.107)

where g_u and g_ℓ are the multiplicities of states, the number of different ways a molecule can have the same energy. We say approximately because we approximate a system of weakly interacting molecules with definite, but unknown, energy levels by a collection of non-interacting molecules with fuzzy energy levels. You have seen a version of Eq. (2.107) before in Section 1.2, the Maxwell–Boltzmann distribution of molecular kinetic energies, which are not quantized. The same kind of probability distribution also holds for the quantized (internal) molecular energy levels.

In equilibrium the number density of photons in the mth bin is proportional to the Planck irradiance $P_e(\nu_m)$ at frequency ν_m (see Prob. 5.41):

$$\frac{n_{p,m}}{\Delta\nu} = \frac{4P_e(\nu_m)}{ch\nu_m}.$$
(2.108)

If we set Eq. (2.108) equal to Eq. (2.106) and use Eq. (2.107) we obtain

$$P_e(\nu_m) = \frac{(ch\nu_m R_s f_{s,um}/4R_e f_{e,um})}{(g_\ell R_a f_{a,\ell m}/g_u R_e f_{e,um})\exp(h\nu_0/k_B T) - 1}$$
(2.109)

$$= \frac{b_m}{c_m \exp(h\nu_0/k_B T) - 1}.$$

This must be equal to the Planck function [Eq. (1.11)], with, of course, $\omega = 2\pi\nu$ transformed to ν:

$$P_e(\nu_m) = \frac{2\pi h\nu_m^3}{c^2} \frac{1}{\exp(h\nu_m/k_B T) - 1}$$
(2.110)

$$= \frac{a_m}{\exp(h\nu_m/k_B T) - 1},$$

from which it follows that

$$\frac{a_m}{\exp(h\nu_m/k_B T) - 1} = \frac{b_m}{c_m \exp(h\nu_0/k_B T) - 1}.$$
(2.111)

If we assume that a_m, b_m and c_m are independent of temperature (they contain only ratios of quantities that depend on temperature in approximately the same way), then the only way Eq. (2.111) can be satisfied for arbitrary T is if

$$a_m = b_m = \frac{ch\nu_m R_{\mathrm{s}} f_{\mathrm{s,um}}}{4 R_{\mathrm{e}} f_{\mathrm{e,um}}} = \frac{2\pi h \nu_m^3}{c^2}, \tag{2.112}$$

and

$$c_m \exp(h\nu_0/k_{\mathrm{B}}T) = \frac{g_\ell R_{\mathrm{a}} f_{\mathrm{a},\ell m}}{g_{\mathrm{u}} R_{\mathrm{e}} f_{\mathrm{e,um}}} \exp(h\nu_0/k_{\mathrm{B}}T) = \exp(h\nu_m/k_{\mathrm{B}}T). \tag{2.113}$$

We can solve these equations to obtain

$$R_{\mathrm{s}} f_{\mathrm{s,um}} = \frac{g_\ell}{g_{\mathrm{u}}} R_{\mathrm{a}} f_{\mathrm{a},\ell m} \frac{8\pi \nu_m^2}{c^3} \exp[-h(\nu_m - \nu_0)/k_{\mathrm{B}}T], \tag{2.114}$$

$$R_{\mathrm{e}} f_{\mathrm{e,um}} = \frac{g_\ell}{g_{\mathrm{u}}} R_{\mathrm{a}} f_{\mathrm{a},\ell m} \exp[-h(\nu_m - \nu_0)/k_{\mathrm{B}}T]. \tag{2.115}$$

Thus all three rate coefficients are related and depend only on molecular properties and temperature. Although we obtained Eqs. (2.114) and (2.115) by considering a gas in equilibrium with radiation, we assume that these relations hold even when Eq.(2.105) does not, which we turn to next.

Consider the radiance L in a gas. We define radiance much more carefully and thoroughly in Section 4.1. For the moment we define $L_m \Delta\nu \Delta\Omega$ as the amount of radiant energy in the mth bin crossing unit area normal to the beam per unit time and confined to a set of directions with solid angle $\Delta\Omega$. Over a distance Δz the radiance increases because of spontaneous and induced emission and decreases because of absorption. We assume that scattering (see Ch. 3) is negligible. Let A be the area of a rectangle to which the z-axis is perpendicular. The number of molecules in the volume $A\Delta z$ is the number density times this volume. The rate of energy decrease depends on the various transition rates in Eqs. (2.102)–(2.104):

$$\Delta L_m \, \Delta\nu \, A \, \Delta\Omega = A\Delta z h \nu_m \left[-\left(\frac{dn_{\mathrm{um}}}{dt}\right)_{\mathrm{s}} \frac{\Delta\Omega}{4\pi} - \left(\frac{dn_{\mathrm{um}}}{dt}\right)_{\mathrm{e}} + \left(\frac{dn_{\ell m}}{dt}\right)_{\mathrm{a}} \right]. \tag{2.116}$$

The first two terms on the right side are gains, the third a loss. We assume that photons are spontaneously emitted with equal probability in all directions, which accounts for the factor $\Delta\Omega/4\pi$. For the general radiation field, $n_{\mathrm{p},m}$ in Eqs. (2.103) and (2.104) is *not* that given by Eq. (2.108) but instead is obtained from

$$ch\nu_m n_{\mathrm{p},m} = L_m \, \Delta\Omega \, \Delta\nu. \tag{2.117}$$

If we substitute Eqs. (2.102)–(2.104) in Eq. (2.116) and use Eqs. (2.107), (2.110), (2.114), (2.115) and (2.117) we obtain, in the limit $\Delta z \to 0$,

$$\frac{dL_m}{dz} = -\kappa_{\ell \mathrm{u},m} L_m + \frac{N_{\mathrm{u}}}{\overline{N}_{\mathrm{u}}} \overline{\kappa}_{\ell \mathrm{u},m} \frac{P_{\mathrm{e}}}{\pi}, \tag{2.118}$$

where

$$\kappa_{\ell u,m} = \frac{N_\ell R_a}{c} \frac{f_{a,m}}{\Delta \nu} \left[1 - \frac{\overline{N_\ell}}{N_\ell} \frac{N_u}{\overline{N_u}} \exp(-h\nu_m/k_B T) \right]. \tag{2.119}$$

Now take the limit as the bin width goes to zero:

$$\kappa_{\ell u} = \frac{N_\ell R_a f}{c} \left[1 - \frac{\overline{N_\ell}}{N_\ell} \frac{N_u}{\overline{N_u}} \exp(-h\nu/k_B T) \right], \tag{2.120}$$

where

$$f = \lim_{\Delta \nu \to 0} \frac{f_{a,m}}{\Delta \nu} \tag{2.121}$$

is a probability density. We suppress the bin index and treat frequency as a continuous variable. We also omit the subscript ℓ on f_a, it being understood that absorption is always from a lower to a higher energy state.

According to Eq. (2.31) an absorption coefficient is a number density times an absorption cross section. If we divide Eq. (2.120) by N_ℓ we obtain the absorption cross section for those molecules with energy level E_ℓ that make a transition to E_u:

$$\sigma_{\ell u} = \frac{R_a f}{c} \left[1 - \frac{\overline{N_\ell}}{N_\ell} \frac{N_u}{\overline{N_u}} \exp(-h\nu/k_B T) \right]. \tag{2.122}$$

If the number densities of molecules with upper and lower energy levels are those corresponding to thermal equilibrium [i.e., those given by Eq. (2.107)] at temperature T, the equation of transfer Eq. (2.118) becomes

$$\frac{dL}{dz} = -\overline{\kappa}_{\ell u} L + \overline{\kappa}_{\ell u} \frac{P_e}{\pi}, \tag{2.123}$$

where

$$\overline{\kappa}_{\ell u} = \frac{\overline{N_\ell} R_a f}{c} \left[1 - \exp(-h\nu/k_B T) \right]. \tag{2.124}$$

The multiplier of L is the absorption coefficient, that of P_e/π the emission coefficient, and in Eq. (2.123) the two are equal, which is a form of Kirchhoff's law (see Sec. 1.4). The validity of this law does not require a gas at temperature T to be bathed only in blackbody radiation corresponding to this temperature but rather that the number of molecules in the upper and lower energy states be the thermal equilibrium values.

The absorption coefficient Eq. (2.124) decreases with increasing temperature because $\overline{N_\ell}$ decreases whereas $\overline{N_u}$ increases: there are fewer molecules with the lower energy state to absorb photons and more molecules with the upper energy to be induced to emit photons (equivalent to negative absorption). Although we usually think of emission, the Planck function times an emission coefficient, or emissivity, as being a consequence of spontaneous emission, the absorption coefficient, and hence emission coefficient, is affected by induced emission.

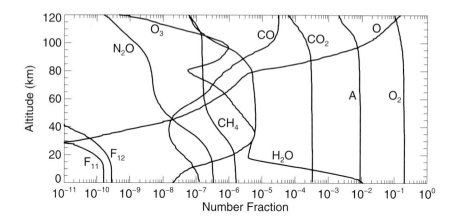

Figure 2.10: Average concentration profiles (expressed as a fraction of the total number density) of various atmospheric molecules. F11 and F12 denote the chlorinated fluorocarbons Freon-11 and Freon-12.

We implicitly assumed molecules with only two energy levels between which transitions are allowed. But there are many such pairs of levels, and hence the absorption coefficient is the sum of many transitions

$$\kappa = \sum_{\ell u} \overline{\kappa}_{\ell u} \tag{2.125}$$

with the corresponding absorption cross section defined by

$$\sigma_{\mathrm{a}} = \frac{\kappa}{N} = \sum_{\ell u} \frac{N_{\ell}}{N} \sigma_{\ell u}, \tag{2.126}$$

where N is the total molecular number density. All the different lines are separate but not completely: their tails overlap, and the sum of all this overlap gives a continuum background on which is superimposed a series of sharp lines.

We may write Eq. (2.120) so as to emphasize that the absorption coefficient is the product of the line shape f and the *line strength*

$$S_{\ell u} = \frac{N_{\ell} R_{\mathrm{a}}}{c} \left[1 - \frac{\overline{N}_{\ell}}{N_{\ell}} \frac{N_{\mathrm{u}}}{\overline{N}_{\mathrm{u}}} \exp(-h\nu/k_{\mathrm{B}} T) \right]. \tag{2.127}$$

At normal terrestrial temperatures $\exp(-h\nu/k_{\mathrm{B}} T)$ is usually $\ll 1$ for vibrational and, especially, electronic transitions, but not necessarily for rotational transitions. At such temperatures an appreciable fraction of molecules may be in excited (higher than the ground state) rotational energy states.

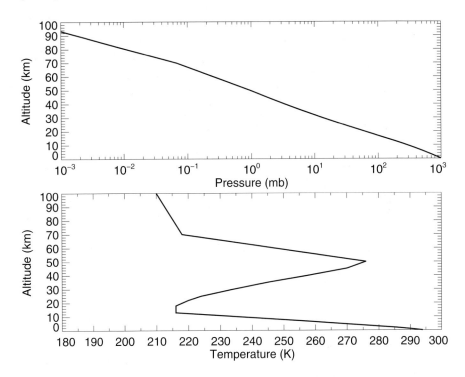

Figure 2.11: Typical mid-latitude summer pressure and temperature profiles. After McClatchey *et al.* (1972).

2.8.2   Absorption by Atmospheric Molecules

The basic determinant of transmission by Earth's atmosphere is the absorption coefficient Eq. (2.125), which depends on concentration profiles of atmospheric gases, their electronic, vibrational, and rotational energy levels, and the variation of temperature and pressure with height.

Several years ago students in a graduate seminar course surveyed the scientific literature for ranges of concentrations of atmospheric gases. The average values they obtained are shown in Fig. 2.10. A more recent perusal of the literature indicates that these values are typical. Although molecular nitrogen (not shown) is the most abundant gas, it is not a major player in absorption by the lower atmosphere except in an indirect way by virtue of interaction (collisional broadening) with its less abundant but radiatively active neighbors.

The absorption coefficient is proportional to molecular concentration, which varies with height. Less evident is an additional variation with height because the line shape depends on total pressure and, to a lesser extent, temperature. Vertical pressure profiles are monotonously similar whereas temperature profiles exhibit more latitudinal and seasonal variation. For the calculations shown here we used typical mid-latitude, summer pressure and temperature profiles (Fig. 2.11).

Cross Sections and Transmissivities

Figure 2.12 shows the absorption cross section at sea level (1013 mb and 294 K) over six wavelength decades for the dominant radiatively active atmospheric molecules. The absorption cross section of a molecule is defined as the ratio of its absorption coefficient to its number density [Eq. (2.126)]. We computed the absorption coefficient with a *line-by-line* radiative transfer code. By this is meant that the spectrum is divided into such small intervals (e.g., wavenumber intervals 0.001 cm^{-1} or less) that the finest spectroscopic details are captured. Absorption lines are narrowest for the lowest pressures (see Fig. 2.19), so these lines dictate the interval width; the necessity of line-by-line calculations for accurate results is evident from the huge variation in absorption, as much as four decades, over a small range of wavenumbers. Although wavenumber is proportional to energy level differences, to show how cross sections vary from the ultraviolet (0.1–0.4 μm) through the visible (0.4–0.7 μm) into the infrared (0.7 μm to 1 mm) and out to the microwave (1 mm to 10 cm) spectral regions, wavelength is easier to visualize.

Three fundamental lessons follow from Fig. 2.12: the magnitudes of absorption cross sections vary over a huge range, more than 10 decades; the spectra of the molecules are quite different; and the spectra are extraordinarily complicated. To determine how these spectra determine transmission by the atmosphere we need absorption coefficients for each molecular species at many altitudes, the total absorption coefficient at a height being the sum over species. The exponential of the negative of this absorption coefficient integrated along a vertical path from sea level to high in the atmosphere is the transmissivity [see Eq. (2.22)]. The result is shown in Fig. 2.13, in six separate panels so as to clearly display the marked differences between spectral regions. We also include the consequences of scattering by atmospheric molecules, the total transmissivity being the product of the absorption and scattering transmissivities (assuming single scattering; see Ch. 5).

Ozone absorbs almost all solar radiation shortward of 0.3 μm. With the exception of ozone and, to a lesser extent oxygen, atmospheric molecules are negligibly absorbing over the visible, and hence the smooth decrease in transmissivity from about 0.6 μm to 0.3 μm is a consequence of scattering by atmospheric molecules (mostly nitrogen because it is the dominant species). Absorption from 0.6 μm to 0.8 μm is dominated by oxygen, and from 0.8 to 5 μm by water vapor helped by two appreciable peaks attributable to carbon dioxide. Within the spectral region 0.3–3 μm there are several broad regions over which the transmissivity exceeds 0.7. Most of the solar spectrum lies in this region, and hence much solar radiation penetrates clear skies and is transmitted to Earth's surface.

From 4 μm to 1 mm absorption by water vapor, with contributions from carbon dioxide and ozone, is often so strong that the transmissivity is nearly zero over broad ranges within this region. One important exception is 8–12 μm, where transmissivity often exceeds 0.6. Emission by Earth peaks in this region, sometimes called the *window region* because transmission of radiation from the surface is high. From 0.1 mm to 1 mm, transmission is essentially zero, and beyond about 1 mm solar and terrestrial radiation do not contribute much to Earth's radiation budget. But from 1 mm to 10 cm atmospheric transmissivity generally increases, approaching 1 at 10 cm. No wonder weather radars operate at 10 cm!

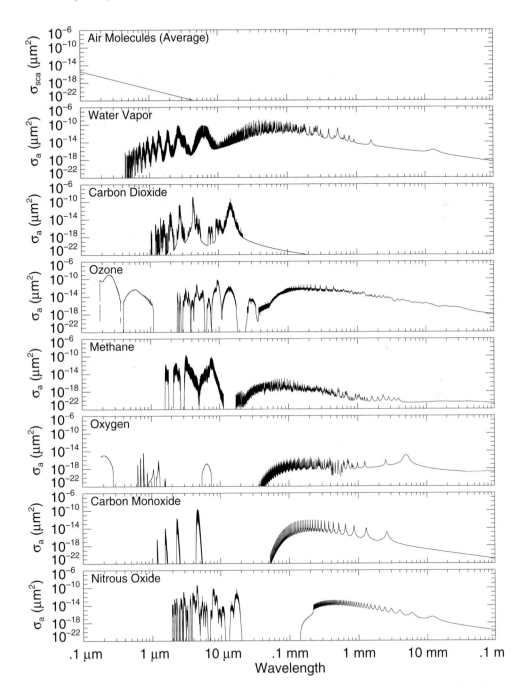

Figure 2.12: Absorption cross sections for most of the strongly absorbing atmospheric gases (1013 mb and 294 K) and average scattering cross section for air molecules (top panel).

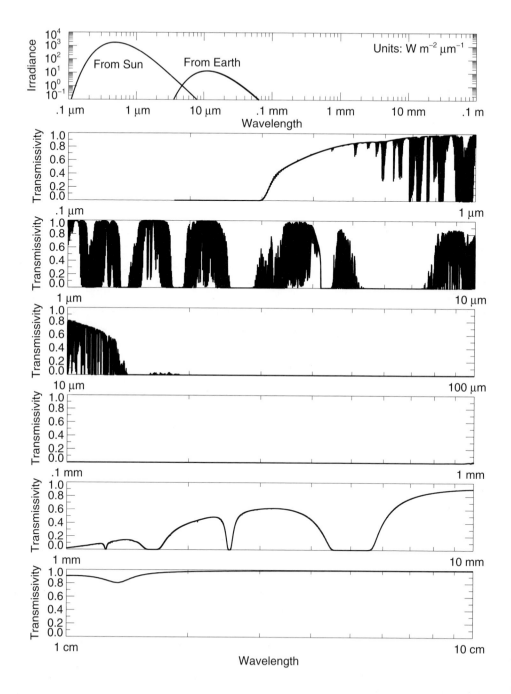

Figure 2.13: Transmissivity along a vertical path through the entire atmosphere. Each panel is one wavelength decade. The top panel shows approximate solar and terrestrial irradiances.

State Populations and the Structure of Absorption

To understand the detailed structure of the absorption cross sections in Fig. 2.12 we have to probe deeper into the energy states of molecules, the statistical distribution of these states, and line broadening.

The probability that a molecule of a gas in equilibrium at temperature T has a quantized energy E_s is

$$p(E_s) = \frac{g_s \exp\left(-E_s/k_{\mathrm{B}}T\right)}{\displaystyle\sum_{s=0}^{\infty} g_s \exp\left(-E_s/k_{\mathrm{B}}T\right)}, \tag{2.128}$$

where g_s, as we note following Eq. (2.107), is the multiplicity (or density or degeneracy, an unfortunate term give its unsavory connotations, or number) of states with energy E_s. You have seen a density of states before perhaps without realizing it. For example, in the continuous distribution function for molecular kinetic energies [Eq. (1.8)] $\sqrt{E}$ multiplying the Boltzmann factor $\exp(-E/k_{\mathrm{B}}T)$ is the density of states. This becomes more evident if we transform from E to speed v, in which instance the density of states is proportional to v^2. If molecular translational states are uniformly distributed in a three-dimensional velocity space, the number of states between v and $v + \Delta v$ is proportional to v^2, and all these states correspond to approximately the same energy. And similarly for the Planck function Eq. (1.11). If we divide it by $\hbar\omega$, we obtain a quantity proportional to the spectral distribution of the number density of photons, which divided by the integral over all frequencies is the probability distribution. Here the density of states is proportional to ω^2. The factor multiplying this density of states is not the Boltzmann factor, except for photon energies $\hbar\omega \gg k_{\mathrm{B}}T$, because photons don't obey the same statistical law as molecules.

The simplest molecule for which absorption is shown in Fig. 2.12 is carbon monoxide (CO), and so we would be wise to attempt to understand its spectrum before considering more complicated molecules. Carbon monoxide has a single vibrational mode with frequency ν_1; the corresponding vibrational energy levels are

$$E_n = h\nu_1\left(n + \frac{1}{2}\right) = hc\tilde{\nu}_1\left(n + \frac{1}{2}\right), \quad n = 0, 1, \ldots, \tag{2.129}$$

where $\tilde{\nu}_1 = 2143.3\,\mathrm{cm}^{-1}$. If we treat the carbon monoxide molecule as a rigid rotator its rotational energy levels are approximately

$$E_J = BhcJ\left(J + 1\right), \quad J = 0, 1, \ldots, \tag{2.130}$$

where we take $B = 1.9563\,\mathrm{cm}^{-1}$ for the molecule in its lowest (ground) vibrational state. Although a truly rigid rotator cannot vibrate, the simple energy levels in Eqs. (2.129) and (2.130) result in a cross section spectrum that agrees with much more detailed line-by-line calculations.

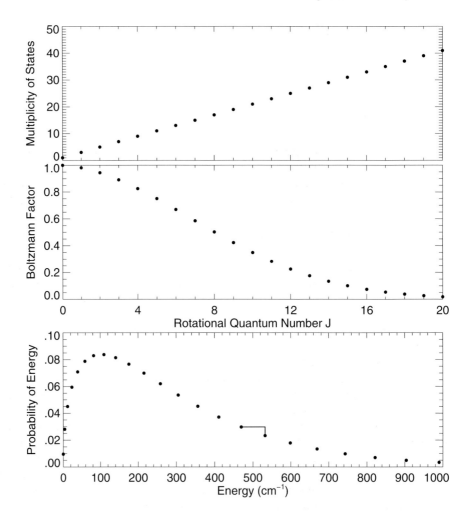

Figure 2.14: Multiplicity of states of a rigid rotator for different rotational quantum numbers J (top). Boltzmann factor for the carbon monoxide molecule at $T = 294$ K (middle). Probability that this molecule has a particular energy (bottom); the energy (wavenumber) difference indicated by the two connected dots corresponds to a transition from $J = 15$ to $J = 16$.

The multiplicity of vibrational states is $g_n = 1$, whereas that of the rotational states is $g_J = 2J + 1$. Thus the probability that a CO molecule has the vibrational energy E_n is

$$p(E_n) = \frac{\exp\{-hc\tilde{\nu}_1(n+1/2)/k_\mathrm{B}T\}}{\displaystyle\sum_{n=0}^{\infty}\exp\{-hc\tilde{\nu}_1(n+1/2)/k_\mathrm{B}T\}} \tag{2.131}$$

and the probability that it has the rotational energy E_J is

$$p(E_J) = \frac{(2J+1)\exp\{-B\,h\,c\,J(J+1)/k_{\mathrm{B}}T\}}{\displaystyle\sum_{J=0}^{\infty}(2J+1)\exp\{-B\,h\,c\,J(J+1)/k_{\mathrm{B}}T\}}. \tag{2.132}$$

At normal terrestrial temperatures, almost all CO molecules are in their ground ($n = 0$) vibrational state (see the top panel in Fig. 2.16), which is why the simple rotational energy level structure [Eq. (2.130)] yields good results.

Figure 2.14 shows the multiplicity of states g_J as a function of rotational quantum number J, the corresponding Boltzmann factor $\exp(-E_J/k_{\mathrm{B}}T)$ for $T = 294$ K, and the probability $p(E_J)$. Because g_J increases monotonically with increasing J whereas the Boltzmann factor decreases, the maximum of $p(E_J)$, unlike that of $p(E_n)$, does not correspond to the ground state.

Energy level transitions of a linear harmonic oscillator with rigid rotations, resulting in emission or absorption of radiation, must obey specific selection rules: the rotational quantum number J must change by 1 and the vibrational quantum number n can change by *no more* than 1. Every allowed energy level difference corresponds to a photon energy. For example, the bottom panel of Fig. 2.14 shows an energy level difference of about 60 cm^{-1} for an allowed transition from $J = 15$ to $J = 16$, which corresponds to a photon of this energy. The top panel in Fig. 2.15 shows the absorption cross section $\overline{\kappa}_{\ell u}/N$ for each rotational transition. These cross sections are a discrete set for frequencies at the peak of each absorption line. In reality, lines are broadened and partly overlap. The cross section [Eq. (2.126)] taking this into account is shown in the middle panel in Fig. 2.15. These calculations require the line shape (discussed in the following subsection) and the rate constant R_{a} for rotational transitions of carbon dioxide. For a rigid rotator theoretical estimates are available and we used those in Eqs. 3.9 and 3.38 (converted to SI units) of Goody and Yung (1989). These relatively simple calculations are in remarkably good agreement with line-by-line calculations (bottom panel) that include all whistles and bells. But the results are not identical. The very fine features in the line-by-line calculation are a consequence of the different isotopes of carbon, which result in CO molecules with different masses. The simple theory also neglects anharmonic terms, which is a fancy way of saying that there is no such thing as an exactly harmonic oscillator. Note that in Section 2.6.1 we referred to the harmonic approximation. Although we can ignore terms higher than quadratic in the expansion of the potential, Nature cannot.

The Sierra Nevada mountain range in California (also in Spain) owes its name to its long, jagged profile seen at a distance against the horizon sky: *sierra nevada* is literally "snow-covered saw." Sierras are abundant in Fig. 2.12, for example, the *Sierra Rotatoria* of carbon monoxide stretching eastward across the spectrum from about 50 μm. Far to the west lie four *picachos* (to borrow another geographical term from the Spanish), isolated peaks jutting from the surrounding plane, to which we now turn our attention, especially the *Gran Picacho Vibratorio* at about 4.67 μm. These picachos, less imaginatively called bands with their associated lines, are a consequence of transitions of the carbon monoxide molecule from its $n = 0$ to $n = 1$ vibrational state accompanied by a change $\Delta J = \pm 1$ of its rotational state.

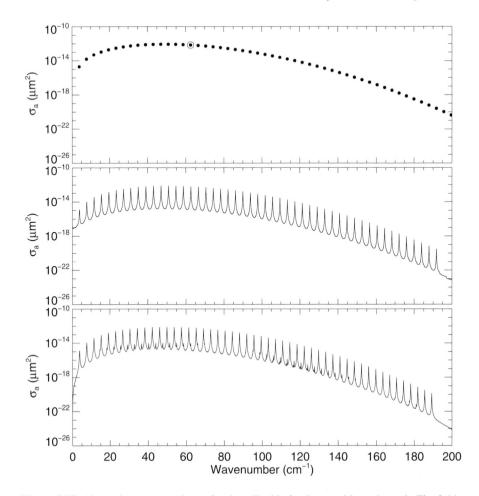

Figure 2.15: Absorption cross sections of carbon dioxide for the transitions shown in Fig. 2.14. The black dot enclosed by a circle (top panel) corresponds to a photon energy (wavenumber) equal to the difference between the $J = 15$ and $J = 16$ energy levels. These discrete cross sections are calculated at line centers and do not include overlap. The middle panel illustrates all of each broadened line and overlapping of lines. The bottom panel shows detailed line-by-line calculations. All calculations are for $T = 294$ K at a pressure of 1013 mb.

The probability that a molecule has a vibrational energy E_n *and* a rotational energy E_J is

$$p(E_n, E_J) = p(E_n)p(E_J), \tag{2.133}$$

with $p(E_n)$ given by Eq. (2.131), $p(E_J)$ by Eq. (2.132). The probability $p(E_n)p(E_J)$ is shown in the top panel of Fig. 2.16 for $n = 0$ and $n = 1$, each for a range of J. The horizontal and vertical lines indicate energy differences corresponding to the $n = 0$ to $n = 1$ transition with J changing from 15 to 14 and to 16. The set of higher probabilities ($n = 0$) shown in this figure duplicates that shown in the lower panel of Fig. 2.14. Added to Fig. 2.16 is a set of probabilities ($n = 1$) that are much lower (because $E_{n=1} > E_{n=0}$).

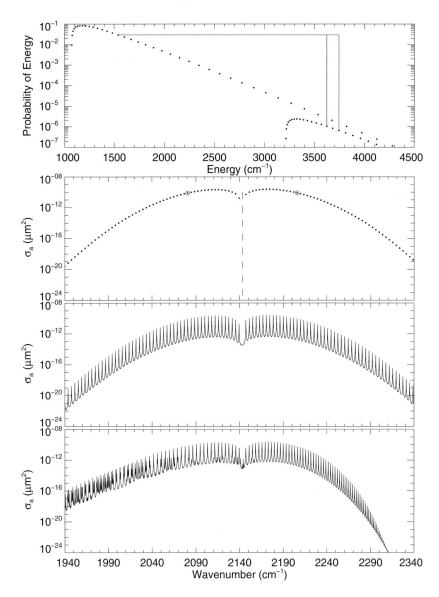

Figure 2.16: Top panel: Probability of a vibration-rotational energy level of carbon monoxide for vibrational quantum numbers $n = 0$ and $n = 1$ and a range of rotational quantum numbers J. Horizontal and vertical lines indicate energy level differences corresponding to the $n = 0$ to $n = 1$ transition with J changing from 15 to 14 and to 16. Second panel from top: Absorption cross sections with no overlapping or broadening of lines for a set of allowed energy differences for two branches: one for an increase in J by 1, one for a decrease by 1. Third panel from top: Absorption cross sections with overlapping and broadening of lines according to simple theory. Bottom panel: Absorption cross sections from line-by-line calculations.

The second panel from the top in Fig. 2.16 shows the absorption cross section with no overlapping or broadening of lines for a set of allowed energy differences, and hence photon energies, for two branches: $\Delta J = 1$ and $\Delta J = -1$. This set of points is similar to that shown in Fig. 2.15 with its mirror image added. That is, the points in Fig. 2.16 are nearly symmetric about the dashed vertical line, which corresponds to the forbidden transition $\Delta J = 0$. The third panel in Fig. 2.16 is the counterpart of the second panel in Fig. 2.15. Because there is no simple theory of the rate constant for $n = 0$ to $n = 1$ transitions, we used the same line strengths as for the rotational band, then multiplied the resulting cross sections by a scale factor so that their peak value matched that obtained from the line-by-line calculation (bottom panel in Fig. 2.16). The lack of obvious resemblance between the cross section spectrum in Fig. 2.16 and the *Gran Picacho Vibratorio* in Fig. 2.12 is a result of the quite different scales.

As evidenced by comparing the third and fourth panels in Fig. 2.16, the relatively simple theory is adequate to explain the general form of the lowest energy vibration-rotational band of carbon monoxide (and other atmospheric gases). That agreement between simple theory and detailed line-by-line calculations is not as good for vibration-rotational bands as for rotational bands (with a molecule in its lowest vibrational state) is to be expected because the higher the vibrational state, the less valid the rigid rotator model of a molecule.

What about the lower picachos to the west of *Gran Picacho Vibratorio*? These are inexplicable by theory based on a strictly harmonic oscillator. Anharmonic terms (i.e., what was discarded in the expansion of the potential) give rise to transitions between $n = 0$ and $n = 2, 3, 4$ states, and hence frequencies that are integral multiples (overtones) of the fundamental frequency. The picachos for carbon monoxide are equally spaced in wavenumber, each about 100 times lower than its neighbor to the east. We briefly discuss overtones and combination bands of the water molecule in Section 2.7.1. Carbon monoxide has only overtones because of its single normal mode frequency. In Fig. 2.15 and 2.16 (bottom panels) absorption cross sections at the peaks are perhaps a thousand times greater than in the troughs. But the cross section never drops to zero.

With one exception, the absorption spectra in Fig. 2.12 are, for the most part, attributable to rotational and vibration-rotational transitions. The fundamental normal mode frequencies for all molecules (except oxygen) in Fig. 2.12 are shown in Fig. 2.17. By taking combinations and overtones of these fundamental frequencies, one can estimate the positions of the vibrational peaks shortward in wavelength of pure rotational lines. The exception is ozone at wavelengths less than about 1 μm. For this molecule electronic transitions are responsible for absorption from approximately 0.4 μm to 1 μm (*Chappuis* bands), from 0.3 μm to 0.35 μm (*Huggins* bands), from 0.2 μm to 0.3 μm (*Hartley* bands), and at wavelengths less than 0.2 μm. The relevance of the Chappuis bands to the color of the twilight sky is discussed in Section 8.1.3. At sufficiently short wavelengths (high energies) photons destroy molecules rather than merely promote them into higher energy states. A molecule is an association of atoms, so dissolution of the association is called dissociation, or more precisely, photo-dissociation to indicate that the breakup is mediated by photons.

Oxygen is different from the other six gases of Fig. 2.12 in that its relatively weak rotational structure is a consequence of its permanent *magnetic* dipole moment (see Sec. 7.3). The vibration-rotational band centered on 6.43 μm is a result of electric quadrupole transitions, and the sets of bands between 0.9 μm and 1.4 μm and between 0.6 μm and 0.9 μm are vibration-rotational bands superimposed on electronic transitions, which also are also re-

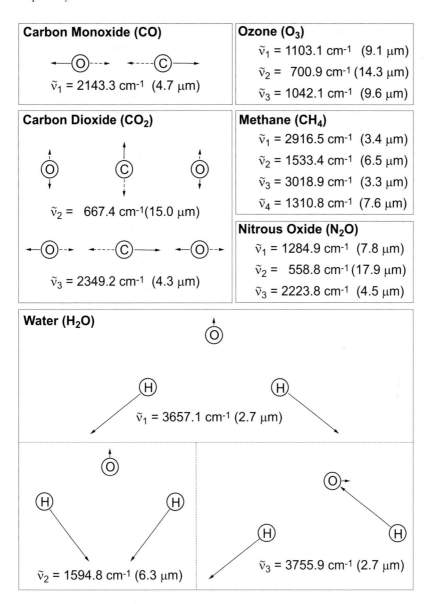

Figure 2.17: Schematic normal mode vibrations of carbon monoxide, carbon dioxide, and water vapor and normal mode frequencies for ozone, methane, and nitrous oxide. The frequencies for carbon monoxide are from Goody and Yung (1989, pp. 210–11), for carbon dioxide from Goody and Yung (1989, pp. 204–07) and Herzberg(1945, pp. 66, 272–6), for water from Goody and Yung (pp. 198–201) and Herzberg (pp. 171, 280–93), for ozone from Goody and Yung (pp. 207–10) and Herzberg (pp. 285–7), for methane from Goody and Yung (p. 211) and Herzberg (pp. 306–9) and nitrous oxide from Goody and Yung (p. 210) and Herzberg (pp. 277–8).

sponsible for absorption between 0.2 μm and 0.26 μm (*Herzberg* band and continuum) and between 0.13 μm and 0.2 μm (*Schumann–Runge* bands and continuum).

Line Shapes

Absorption lines are broadened by the line shape function. We derive one line shape in Section 2.6, the Lorentz line shape, from the classical equation of motion of a harmonic oscillator. The form of this line shape carries over into the quantum domain but the terms in it are interpreted differently. For example, the frequency ω_0 in Eq. (2.60) is a resonant frequency whereas its quantum-mechanical interpretation is an energy level difference divided by $\hbar$. If we transform from circular frequency ω to wavenumber $\tilde{\nu}$ in Eq. (2.73) and multiply by a factor so that the integrated line shape is normalized, we obtain

$$f_c(\tilde{\nu}) = \frac{\alpha_c}{(\tilde{\nu} - \tilde{\nu}_0)^2 + \alpha_c^2}, \tag{2.134}$$

where the half-width $\alpha_c = \gamma/4\pi c$ and $\tilde{\nu}_0$ is the wavenumber at the center of the line. The subscript c denotes collisional broadening, a consequence of inter-molecular, in contrast with intra-molecular, forces. Equation (2.134) also applies to *natural* line broadening, the irreducible line broadening of absorption lines of an isolated molecule at rest. The natural line broadening half-width is so small for the electronic, vibrational, and rotational transitions of interest in atmospheric applications that we need not consider it further.

At 1013 mb and 294 K the collision-broadened half-width α_c is typically 0.03–0.10 cm^{-1} for vibration-rotational transitions. For example, a water vapor line at 1085.436 cm^{-1} has a *foreign-broadened* half-width $\alpha_f = 0.0755$ cm^{-1} and a *self-broadened* half-width $\alpha_s = 0.3580$ cm^{-1}. These terms are self-explanatory: a molecule interacts with its fellow countrymen and with foreigners. The total half-width is the number-weighted average of the two half-widths. High in the atmosphere collisional broadening of the water vapor line is dominated by foreign broadening because water vapor decreases with height more rapidly than do nitrogen and oxygen (Fig. 2.10). Foreign broadening is (approximately) proportional to pressure.

The Lorentz line shape for this transition for the mid-latitude summer profile (Fig. 2.11) is shown in Fig. 2.18 at the surface (1013 mb, 294 K) and high in the atmosphere (10 mb, 238 K). At high altitudes the concentration of water vapor is so low (Fig. 2.10) that broadening of this line is dominated by foreign broadening. If collisional broadening is a consequence of molecular interactions, we expect broadening to decrease with increasing altitude, and the line shapes in Fig. 2.18 support this expectation. (Note the great difference in wavenumber scales between low and high altitudes.) This figure also shows two other line shapes, which we turn to next.

Doppler broadening of lines is a result of relative motion between a source of photons and molecules that absorb them. We discuss the Doppler shift in detail in Section 3.4.6 and derive an expression for the distribution of the frequencies of scattered radiation as a consequence of random molecular motions of an ideal gas [Eq. (3.80)]. Given this expression, and taking into account that the Doppler shift for absorption is half that for scattering, we obtain the Doppler

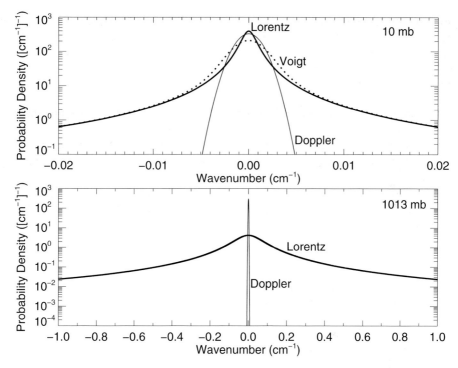

Figure 2.18: Line shapes for a water vapor line high in the atmosphere (10 mb, 238 K) and at the surface (1013 mb, 294 K). Wavenumbers are relative to that at the peak. The Voigt line shape is a convolution of the Lorentz and Doppler line shapes.

line shape

$$f_D(\tilde{\nu}) = \frac{1}{\alpha_D \sqrt{\pi}} \exp\left\{ -\left(\frac{(\tilde{\nu} - \tilde{\nu}_0)}{\alpha_D} \right)^2 \right\}, \tag{2.135}$$

where the Doppler half-width $\alpha_D = (\tilde{\nu}_0/c)\sqrt{2k_B T/m}$. Doppler line shapes for the water vapor line also are shown in Fig. 2.18.

Determining the line shape because of both collisional and Doppler broadening is not trivial. To understand why, consider that we approximate absorption by gas molecules as transitions between two sets of energy levels, where the energy level difference (frequency) is distributed according to the Lorentz line shape. Each one of these energy level differences is Doppler shifted by molecular motion. Because of this, the total line shape is a *convolution* of the two line shapes, called the *Voigt* line shape:

$$f(\tilde{\nu}) = \int_{-\infty}^{\infty} f_c(\tilde{\nu}')f_D(\tilde{\nu} - \tilde{\nu}')\,d\tilde{\nu}' = \int_{-\infty}^{\infty} f_c(\tilde{\nu} - \tilde{\nu}')f_D(\tilde{\nu}')\,d\tilde{\nu}', \tag{2.136}$$

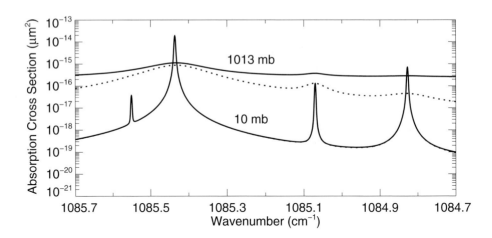

Figure 2.19: Line-by-line calculations of water vapor absorption at two pressure altitudes. The dashed curve shows a calculation from which continuum absorption is omitted. At 10 mb, omission of the continuum results in a curve indistinguishable from that shown.

where the integration limits are symbolic not literal. Equation (2.136) is the line shape for *most* temperatures and pressures, but in certain limits reduces approximately to the Doppler or Lorentz line shape. For example, low in the atmosphere (Fig. 2.18) the Doppler line shape is nearly a spike compared with the much broader Lorentz line shape. When the two are convolved, the result is almost the Lorentz line shape. The other extreme is high in the atmosphere, where the Lorentz line shape is a spike compared with the much broader Doppler line shape. The two profiles at 10 mb in Fig. 2.18 convolve to a line shape that is neither one nor the other.

Figure 2.19 shows four water vapor lines for two pressures. At the lower pressure (10 mb) all lines are distinguishable, but at the higher pressure (1013 mb) collisional broadening is so great that two lines are completely lost and two are demoted from proud peaks to humble dunghills. This figure also shows the results of line-by-line calculations with and without *continuum* absorption; the difference is, however, imperceptible at the lower pressure. Continuum absorption is that added to bring theory and measurements into agreement, usually by theoretically or empirically adjusting lines so that they are accurate far from their centers. By positing stable aggregates of two (dimers) or three (trimers) or more water molecules, additional, but weak, absorption is expected (and has been found experimentally) in infrared regions of the spectrum.

Line shapes (e.g., Lorentz and Voigt) obtained by modeling the weak interaction between a single infrared-active molecule and only one other molecule are accurate near line centers. But line shapes far from centers are much more difficult to model. In spectral regions far from all line centers, in which only the wings of lines overlap, accurate treatment of these wings is essential for estimates of absorption.

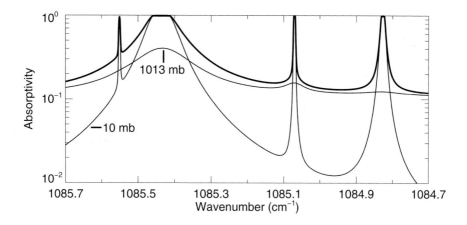

Figure 2.20: Water vapor absorptivities of a two-layer model atmosphere, one layer at 1013 mb and 294 K, one layer at 10 mb and 238 K. The thicker dark curve is the absorptivity of the two layers combined.

Implications for Remote Sensing of Water Vapor

The dependence of collisional broadening on pressure (altitude) and line strength on concentration provides a way to remotely determine vertical water vapor profiles using transmission measurements. To show this we consider a simple two-layer atmosphere. Each layer has a uniform pressure and temperature, the lower one at 1013 mb and 294 K, the upper one at 10 mb and 238 K. The physical thicknesses are chosen to make the absorptivity, averaged over the spectral region shown in Fig. 2.19, of the layers the same.

The absorptivity of each layer is quite different because of collisional broadening (Fig. 2.20). Although all that can be measured from the ground is the total absorptivity of the two layers together, absorptivity at the peaks is dominated by the upper layer, between peaks by the lower layer. Thus a measurement of total absorptivity (transmission) at sufficiently high resolution (< 0.01 cm^{-1}) yields information about both layers, in particular the amount of water vapor in each because of the dependence of the line strength [Eq. (2.127)] on concentration.

The mesosphere, that part of Earth's atmosphere between about 50 km and 80 km, is a difficult place to visit: too high for balloons, too low for satellites. And yet for many years water vapor profiles in the mesosphere have been inferred from ground measurements of high-resolution radiance spectra around a 22.2 GHz water vapor line. Near this frequency even a cloudy atmosphere is not opaque, pressure broadening occurs to high altitudes, and the line shape can be resolved.

Low in the atmosphere (e.g., the troposphere) collisional broadening flattens the 22.2 GHz line (Fig. 2.21). At these altitudes emission does not vary appreciably near line center, and hence if all the water vapor in a vertical column above 40 km were to vanish, the radiance spectrum would be flat (Fig. 2.21). But from 40 km to 80 km collisional broadening decreases,

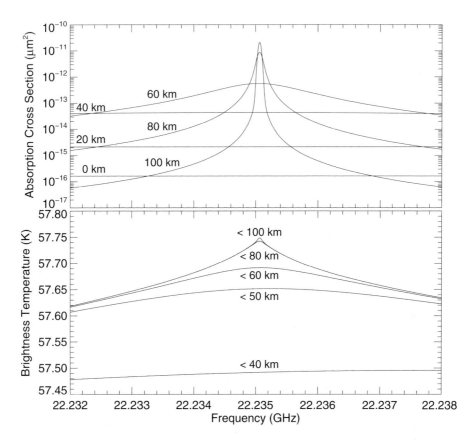

Figure 2.21: Water vapor absorption cross section spectra every 20 km from the surface to 100 km (top panel). Brightness temperatures that would be measured at the surface if all the water vapor above the indicated altitude were not present (bottom panel). Note the switch from wavenumber to frequency, and radiance expressed as a brightness temperature (Sec. 1.4.4), the variables favored by those who measure microwave radiance.

the absorption cross section near line center increases, and hence so does emission relative to that in the wings. By 80 km, Doppler broadening dominates, and the line shape no longer changes apart from small changes with temperature. As a consequence, surface radiances are not sensitive to the location of water vapor above 80 km. Only below this altitude, but above the troposphere, are surface radiances across the line sensitive to the location of water vapor. The radiance near line center increases relative to that in the wings as the water vapor is concentrated at higher altitudes within this range. Spatial resolution of retrieved water vapor profiles is about a pressure scale height.

Water vapor retrievals are not limited to the mesosphere, and have been obtained for the (upper) stratosphere as well. Any middle-atmosphere constituent with sufficient concentration and a rotational microwave line can be probed by this approach, for example, CO, ClO, HNO_3, and N_2O.

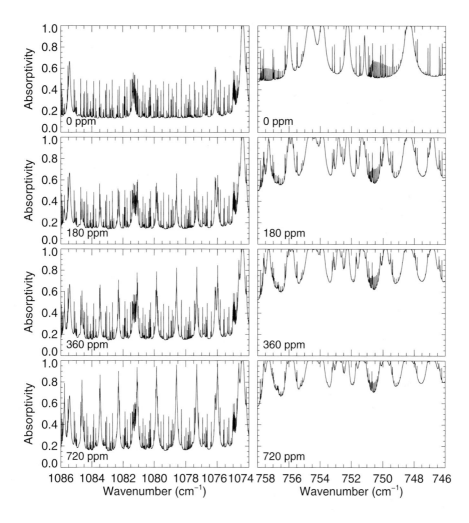

Figure 2.22: Absorptivity along a vertical path, for two spectral regions, for a mid-latitude summer temperature and pressure profile. The concentration of carbon dioxide (ppm = parts per million) for each absorptivity is indicated on the panels.

Line Absorptivity and Saturation

We now turn to absorption by a uniform layer of thickness h of a gas with a single, isolated absorption line. As previously, we divide the line into M equally spaced (in wavenumber) bins. Suppose that n_m photons in the mth bin are incident on the layer. The total line transmissivity is the ratio of the total transmitted to the total incident photons (reflection is assumed negligible):

$$T = \frac{\sum\limits_{m=1}^{M} n_m \exp(-\kappa_m h)}{\sum\limits_{m=1}^{M} n_m}, \tag{2.137}$$

where κ_m is the absorption coefficient for the mth bin. If n_m is uniform

$$T = \frac{1}{M} \sum\limits_{m=1}^{M} \exp(-\kappa_m h). \tag{2.138}$$

The *line absorptivity* is

$$A = 1 - T = 1 - \frac{1}{M} \sum\limits_{m=1}^{M} \exp(-\kappa_m h) = \frac{1}{M} \sum\limits_{m=1}^{M} A_m, \tag{2.139}$$

where $A_m = 1 - \exp(-\kappa_m h)$ is the absorptivity for the mth bin. Thus A is the average absorptivity of the line.

Consider first the limit of a weak line ($\kappa_m h \ll 1$ for all m):

$$A \approx \frac{1}{M} \sum\limits_{m=1}^{M} \kappa_m h. \tag{2.140}$$

If we assume that the exponential in Eq. (2.127) is negligible, Eq. (2.140) becomes

$$A \approx \frac{hS}{M} \sum\limits_{m=1}^{M} f_m = \frac{hS}{\tilde{\nu}_2 - \tilde{\nu}_1} \sum\limits_{m=1}^{M} f_m \Delta\tilde{\nu} = \frac{hS}{\tilde{\nu}_2 - \tilde{\nu}_1}, \tag{2.141}$$

where $\Delta\tilde{\nu}$ is the bin width and $(\tilde{\nu}_1, \tilde{\nu}_2)$ are what we agree to call the (somewhat arbitrary) limits of the line. Because the line strength S is proportional to the number density of molecules, the line absorptivity in the weak line limit increases linearly with number density. But this can't go on indefinitely. In the strong line limit ($\kappa_m h \gg 1$ for all m), $A \to 1$. The line is saturated in the sense that a further increase in number density does not change the line absorptivity.

The spectral absorptivity, over two fairly broad spectral regions, corresponding to a mid-latitude summer profile (Figs. 2.10 and 2.11) but with varying amounts of uniformly mixed carbon dioxide, is shown in Fig. 2.22. The consequences of line saturation are evident for the two spectral regions. Figure 2.23 is a closer look at a narrow spectral region within each of the broad regions in Fig. 2.22. To show the behavior of carbon dioxide itself we use the absorption optical thickness with *no* carbon dioxide to determine what the absorptivity would be if carbon dioxide were the *only* absorbing gas. The line absorptivity for the two spectral regions of Fig. 2.23 is shown as a function of carbon dioxide concentration in Fig. 2.24.

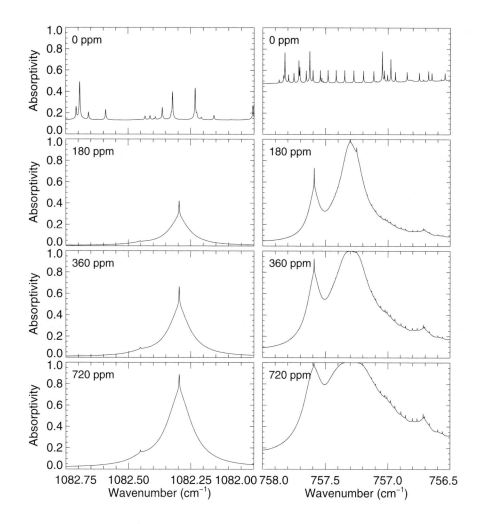

Figure 2.23: Absorptivity along a vertical path for two narrow regions within the spectral regions in Fig. 2.22. The concentration of carbon dioxide is indicated on the panels. Each curve is what the absorptivity would be if carbon dioxide were the only absorbing gas.

For the spectral region $1082-1082.75$ cm^{-1} the line absorptivity increases almost linearly with concentration (weak line limit), whereas for the spectral region $756.5-758$ cm^{-1}, the line absorptivity increases linearly with concentration at first but then increases with ever decreasing slope, gradually reaching the asymptotic value (strong line limit) 1.

If a carbon dioxide line occurs in a spectral region where lines from other molecular species are already saturated, carbon dioxide has little effect on absorptivity (Fig. 2.22). The converse of this is that carbon dioxide, indeed any infrared-active gas, has its greatest effect in regions where lines from other molecules are not saturated. Saturation of absorptivity does

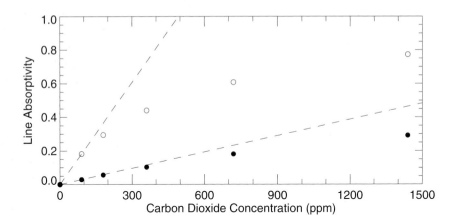

Figure 2.24: Line absorptivity of an atmosphere with carbon dioxide the only absorbing gas for the spectral regions 1082–1082.75 cm^{-1} (closed circles) and 756.5–758 cm^{-1} (open circles) as a function of carbon dioxide concentration (parts per million). Dashed lines represent linear fits to the results at the two lowest carbon dioxide concentrations.

not necessarily imply saturation of consequences. Emission from the atmosphere to Earth depends on emissivity (absorptivity). But it also depends on the atmospheric temperature profile. Although one might be tempted to think that increasing carbon dioxide can only eventually saturate all lines with no other effect, resulting in an upper limit on infrared radiation from the atmosphere, this assumes (probably incorrectly) that the temperature profile does not change. It is as unreasonable to expect an increase in carbon dioxide in the atmosphere to eventually result in some asymptotic value for infrared atmospheric irradiance [Eq. (1.74)] as to expect it to increase indefinitely [Eq. (1.79)].

2.9   Absorption by Particles

Although the form of Eq. (2.28) is indifferent to whether the illuminated object is a molecule or a particle, we write the symbol C_{abs} for the absorption cross section of a particle to signal that particles and molecules are different. Implicit in the definition of cross sections is that the irradiance of the incident illumination be constant over lateral dimensions large compared with the size of the particle. As with molecules the absorption cross section of a particle depends on its orientation (unless it is an isotropic sphere) and the state of polarization of the illumination (again, unless it is an isotropic sphere). The absorption cross section of a particle often is normalized by its geometrical cross sectional area G projected onto a plane perpendicular to the illumination. The resulting dimensionless quantity

$$Q_{\mathrm{abs}} = \frac{C_{\mathrm{abs}}}{G} \tag{2.142}$$

is called an *efficiency* or *efficiency factor* for absorption. The advantage of this normalization, namely, not having to fret about units, is outweighed by several disadvantages. In the first place, we expect quantities called efficiency factors to be less than or at most equal to 1 (100%), whereas it is not unusual to encounter overachieving particles with Q_{abs} *greater* than 1, sometimes appreciably greater, at some wavelengths. Also, some particles don't have well-defined or easily determined geometrical cross sections (e.g., soot aggregates, the kinds of particles found in smokes), and molecules most definitely do not. And the normalization in Eq. (2.142) is arbitrary. Why not normalize by total surface area or the square of mean chord length or ...? To be fair Q_{abs}, at least for a sphere, does have a physical meaning. At a given wavelength it is the emissivity of the particle, which underlies the assertion in Section 1.3 about emissivities greater than 1. But the efficiency factors for scattering and extinction (Sec. 3.5) have no physical meaning.

Both molecules and particles have well-defined absorption cross sections, in principle measurable; both molecules and particles have masses, which again are measurable. So the only really meaningful normalized cross section, by means of which we can compare molecules and particles, is cross section per unit mass. But if a particle has a definite volume v (a molecule does not), another normalized cross section that does have a physical meaning is the *volumetric* cross section, the cross section per unit particle volume. The absorption coefficient of a dilute suspension of N identical particles is

$$\kappa = NC_{abs} = NvC_{abs}/v = fC_{abs}/v, \tag{2.143}$$

where $f = Nv$ is the volume fraction of particles in the suspension (i.e., the fraction of the total volume containing something you can kick). If any quantity deserves to be called an efficiency it is the volumetric absorption cross section C_{abs}/v. Equation (2.143) can be generalized to a suspension of non-identical particles:

$$\kappa = f \left\langle \frac{C_{abs}}{v} \right\rangle, \tag{2.144}$$

where

$$\left\langle \frac{C_{abs}}{v} \right\rangle = \frac{\sum_j f_j \left(C_{abs,j}/v_j \right)}{\sum_j f_j} = \frac{1}{f} \sum_j f_j \left(C_{abs,j}/v_j \right) \tag{2.145}$$

and the subscript j denotes anything that makes one particle different from another – shape, size, orientation, or composition.

We now turn to another advantage of expressing cross sections of particles as volumetric cross sections. Consider the simplest imaginable particle, an optically homogeneous slab of uniform thickness d and surface area A with dimensions much larger than λ illuminated at normal incidence by a beam with irradiance F. The law of exponential attenuation [Eq. (2.8)] strictly applies only to a medium without boundaries. Boundaries add the complication of reflections, with the attendant possibility of interference (see Sec. 3.4). We assume that the optical properties of the slab are sufficiently similar to those of the surrounding medium that reflections are negligible and that the slab is sufficiently thick compared with the wavelength that interference is negligible. With these assumptions the rate W_a at which energy is absorbed by the slab is

$$W_a = FA\{1 - \exp(-\kappa_b d)\},\tag{2.146}$$

where we write the bulk absorption coefficient of the slab material as κ_b to distinguish it from the absorption coefficient [Eq. (2.144)] of a dilute suspension of particles composed of that material. If we further assume that $\kappa_b d \ll 1$ (weak absorption), Eq. (2.146) becomes

$$W_a \approx FAd\kappa_b.\tag{2.147}$$

From Eq. (2.28) the absorption cross section of the slab is therefore $Ad\kappa_b$ and its cross section per unit volume is

$$C_{abs}/v \approx \kappa_b.\tag{2.148}$$

At the other extreme $\kappa_b d \gg 1$ (strong absorption) and Eq. (2.146) yields

$$C_{abs}/v \approx 1/d.\tag{2.149}$$

For some materials Equations (2.148) and (2.149) for a slab particle are good estimates for the volumetric absorption cross section of some particles in the limits of weak and strong absorption. A further advantage of considering volumetric absorption by a particle is that it can be compared with the bulk absorption coefficient of its parent material. Sometimes the absorption spectrum of a particle is similar to that of its parent material, but sometimes exhibits not even a trace of a family resemblance.

If the distribution of sizes and shapes of particles in a suspension, all of identical composition, is such that they can fill all space (suspensions of identical spheres cannot), κ given by Eq. (2.144) should approach κ_b as f approaches 1. And indeed it does *if* the volumetric absorption cross section is given by Eq. (2.148), but definitely does not *if* it is given by Eq. (2.149). For strong absorption the volumetric absorption cross section of particles is dominated by their size and not the bulk absorption coefficient of their parent material. This signals caution in applying Eq. (2.144) to dense suspensions of particles. Unfortunately, we can't give precise criteria for what is meant by "dense", although we touch on this subject in Sections 5.1 and 5.3. Depending on the size of the particles relative to the wavelength, when a suspension of them becomes sufficiently dense, it is probably more realistic to look upon it as a slightly porous medium ($f \approx 1$), the absorption coefficient of which is approximately

$$\kappa = f\kappa_b.\tag{2.150}$$

This expression does have the correct asymptotic behavior as $f \to 1$. Thus Eq. (2.144) can be used with confidence for $f \ll 1$, and Eq. (2.150) for $f \approx 1$, but the values of f at which the transition from the one to the other begins and ends cannot be specified precisely.

For what materials and particles is Eq. (2.148) not a good approximation? Suspensions of small gold and silver particles (colloidal gold and silver) when illuminated by white light display upon transmission vivid colors completely unrelated to the appearance of the bulk

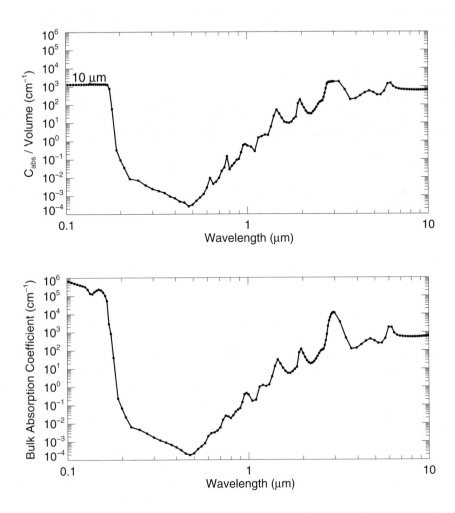

Figure 2.25: Volumetric absorption cross section spectrum of a 10 μm-diameter water droplet (top) compared with the bulk absorption spectrum of liquid water (bottom). For smaller diameters volumetric absorption in the UV more closely follows the bulk absorption coefficient; for larger diameters, peaks in the volumetric absorption spectrum beyond about 2 μm are flattened.

metals. The absorption spectrum of silver, in particular, is as nearly featureless as a dry lake bed, and although that of gold is more interesting, which is why gold is golden, absorption spectra of particles of these metals bear no resemblance whatsoever to those of their parent materials.

Alas, gold and silver particles are not commonly found in the atmosphere but particles of insulating crystalline materials (e.g., quartz, ammonium sulfate) are, and at infrared wavelengths the absorption spectra of such particles do not always dutifully follow those of their

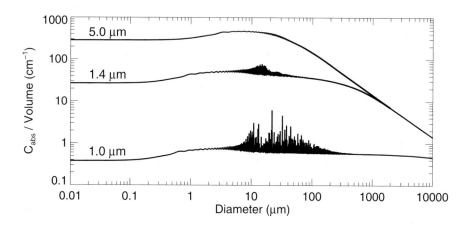

Figure 2.26: Volumetric absorption cross section for water droplets of varying diameter at the three wavelengths shown.

parent materials. But the differences are not so striking as they are for metallic particles, a matter of shifts of absorption peaks rather than their appearance seemingly from nowhere.

Water, however, is an example of a material for which Eqs. (2.148) and (2.149) often are good approximations. The volumetric absorption cross section of a 10 μm-diameter water droplet (Fig. 2.25), calculated using the exact theory for a sphere (Sec. 3.5), is similar to the bulk absorption coefficient of water from visible to infrared. For droplets smaller than about 10 μm (not shown), the volumetric absorption cross section approaches more closely the bulk absorption coefficient in the ultraviolet (< 0.2 μm), whereas for droplets larger than about 10 μm the peaks in the infrared become more and more flattened. This flattening of the absorption spectrum of a particle is expected on physical grounds. As the bulk absorption coefficient of a particle increases, absorption is concentrated more and more in the outer layers of the illuminated particle, and what changes is not *how much* energy is absorbed but *where*.

Figure 2.26 shows the volumetric absorption cross section as a function of droplet diameter at three different wavelengths. For sufficiently small diameters, the cross section is approximately the bulk absorption coefficient and independent of diameter, as predicted by Eq. (2.148). With increasing diameter, the cross section decreases as the inverse of diameter and is independent of the bulk absorption coefficient. Moreover, the diameter at the transition from a constant to a decreasing value is smaller the greater the bulk absorption coefficient, which increases with increasing wavelength. All this is consistent with Eq. (2.149). The exceedingly fine structure in the absorption cross section at $\lambda = 1.0$ μm is a consequence of interference (see Sec. 3.4). Note that this interference structure does not occur until the diameter is greater than the wavelength, and is increasingly suppressed with increasing bulk absorption coefficient. Equations (2.148) and (2.149) for a slab particle were obtained under the assumption of negligible interference. If we had included interference we would have obtained oscillations in the absorption cross section of a slab, interpreted as a consequence of interference between multiply-reflected waves. The greater the value of $\kappa_b d$, the more these waves are damped (attenuated), and the more the interference pattern disappears.

2.9.1   Molecules and Particles: Similarities and Differences

As far as absorption (or scattering) is concerned a molecule is a particle of zero dimensions. Although molecules do indeed have extension in space they are fuzzy. In any interaction of electromagnetic radiation with matter, the relevant measuring stick is the wavelength, against which molecules are quite small, even for wavelengths well into the ultraviolet. The separate parts of molecules therefore radiate in unison. A corollary of this is that absorption by molecules and by small (compared with the wavelength) particles ought to be similar. And indeed they are, with some notable exceptions. The absorption spectrum of water vapor at infrared frequencies exhibits many narrow, closely spaced rotational lines (Fig. 2.12), whereas these lines vanish completely in bulk water, which we interpreted as collisional broadening taken to its extreme, and hence vanish from the absorption spectra of water droplets of all sizes. Although vibrational bands are broadened and shifted in going from vapor to the condensed phases (liquid and ice), they still are prominent in the infrared absorption spectrum of water (Fig. 2.25) and hence in the absorption spectra of small water droplets. Where molecules and particles go their separate ways is when particles become comparable with or greater than the wavelength. This additional degree of freedom (size) results in particle absorption spectra that may bear no resemblance to that of individual molecules or even the condensed phases of these molecules. A single particle larger than the wavelength and weakly absorbing can exhibit a series of narrow, closely spaced lines in its absorption spectrum, reminiscent of rotational bands in molecular spectra but arising from a completely different cause: interference.

   Another difference between molecules and particles is that we have a hope of calculating the energy levels of molecules, and hence the frequency dependence of molecular cross sections, by quantum mechanics. Difficult, yes, but not impossible. Calculating the cross sections of particles, aggregations of many closely-packed interacting molecules, by quantum mechanics is essentially impossible. This would be like trying to forecast the weather using quantum mechanics – in principle not impossible, but in practice not advisable. To calculate absorption cross sections of particles we must use classical electromagnetic theory and obtain the electromagnetic properties of the materials of which they are composed from measurements. This is how the volumetric absorption cross sections for water droplets shown in Figs. 2.25 and 2.26 were obtained.

References and Suggestions for Further Reading

Examples of Stigler's law of eponymy, drawn mostly from mathematics, are in Stephen M. Stigler, 1999: *Statistics on the Table*, Harvard University Press, Ch. 14. Another name for this "supremely important law of the history of science" is "the Infinite Chain of Priority: Somebody Else Always Did It First" (Tony Rothman, 2003: *Everything's Relative and Other Fables from Science and Technology*, John Wiley & Sons, p. xiii.).

Pierre Bouguer's *Optical Treatise on the Gradation of Light* (1760) was translated, with an introduction and notes, by William Edgar Knowles Middleton, 1961, University of Toronto Press. For a sketch of Bouguer's life and work see Middleton's entry in *Dictionary of Scientific Biography*.

For Lambert's magnum opus see Johann Heinrich Lambert, 2001: *Photometria*, Illuminating Engineering Society of North America, translated from the Latin by David DiLaura, who, in the lengthy historical introduction (p. lxxxi), acknowledges that "Bouguer... was the first to recognize exponential decay in absorbing media."

For what Beer did (and did not) do see Heinz G. Pfeiffer and Herman A. Liebhafsky, 1951: The origins of Beer's law. *Journal of Chemical Education*, Vol. 28, pp. 123–5.

For a history of the exponential attenuation law see Dorothy R. Malinin and John H. Yoe, 1961: Development of the laws of colorimetry: a historical sketch. *Journal of Chemical Education*, Vol. 38, pp. 129–31. Despite its title, this paper is *not* about colorimetry (see Sec. 4.3). Its authors add yet another claimant to the law of exponential attenuation, that of F. Bernard, who in the same year as Beer (1852) "arrived at a similar relationship between absorptive capacity and concentration."

The absorption length for liquid water in Fig. 2.2 is from Table 1 in Marvin R. Querry, David M. Wieliczka, and David J. Segelstein, 1991: Water (H_2O), pp. 1059–77 in Edward D. Palik, Ed., 1991: *Handbook of Optical Constants of Solids*, Vol. II, Academic Press. That for ice is from the compilation by Stephen G. Warren, 1984: Optical constants of ice from the ultraviolet to the microwave. *Applied Optics*, Vol. 23, pp. 1206–25.

We have to be clear as to exactly what nonexponential attenuation, a topic of current interest, means. To us Eq. (2.22) is not an example of nonexponential attenuation but rather of nonuniform attenuation; the underlying law is still exponential. Nor is integrated attenuation over an absorption line (Sec. 2.1.3) nonexponential attenuation. Exponential attenuation may hold locally (point by point) but not globally (integrated). To us nonexponential attenuation (at a single wavelength) is a consequence of spatial correlations among absorbers (or scatterers) in a statistically homogeneous medium. This is the sense in which nonexponential attenuation is meant by Alexander B. Kostinski, 2001: On the extinction of radiation by a homogeneous but spatially correlated random medium. *Journal of the Optical Society of America A*, Vol. 18, pp. 1929–33, Alexander B. Kostinski, 2002: On the extinction of radiation by a homogeneous but spatially correlated random medium: reply to comment. *Journal of the Optical Society of America A*, Vol. 19, pp. 2521–25, and Raymond A. Shaw, Alexander B. Kostinski, and Daniel D. Lanterman, 2002: Super-exponential extinction of radiation in a negatively-correlated random medium. *Journal of Quantitative Spectroscopy and Radiative Transfer*, Vol. 75, pp. 13–20 in which one can find citations of relevant papers. Note in particular that Fig. 3 (dot diagrams similar to our Fig. 2.7) in the second paper by Kostinski shows at a glance the differences between positive correlation (sub-exponential), no correlation (exponential) and negative correlation (super-exponential), a graphic realization of the old adage that one picture is worth a thousand words.

Careful examination of Fig. 2.2 (see also Prob. 5.31) reveals small differences between absorption by liquid and solid water at wavelengths around 1 μm. These differences are the basis of methods for remotely determining cloud composition (ice or liquid water). See Peter

Pilewskie, 1989: Cloud Phase Discrimination by Near-Infrared Remote Sensing. Ph.D. dissertation, University of Arizona. This is a model of what a dissertation should be (graduate students take note). It is short (107 pages), blessedly free of acronyms and non-nutritious fluff, the figures are few, uncluttered and comprehensible at a glance, and, as evidenced by the following paper, it has not grown stale with age: Wouter H. Knap, Piet Stammes, and Robert B.A. Koelemeijer, 2002: Cloud thermodynamic-phase determination from near-infrared spectra of reflected sunlight. *Journal of the Atmospheric Sciences*, Vol. 59, pp. 83–96.

For journal articles on this method see Peter Pilewskie and Sean Twomey, 1987: Discrimination of ice from water in clouds by optical remote sensing. *Atmospheric Research*, Vol. 21, pp. 113–22; Peter Pilewskie and Sean Twomey, 1987: Cloud phase discrimination by reflectance measurements near 1.6 and 2.2 μm. *Journal of the Atmospheric Sciences*, Vol. 44, pp. 3419–21; and Peter Pilewskie and Sean Twomey, 1992: Optical remote sensing of ice in clouds. *Journal of Weather Modification*, Vol. 24, pp. 80–82. In this paper the authors present spectral measurements of transmission as well as reflection by clouds, and point out that transmission samples an entire cloud whereas reflection samples only its outer layers.

For a clear treatise on complex variables see Ruel V. Churchill, 1960: *Complex Variables and Applications*. McGraw-Hill.

The following are the bibliographic details for the formidable reading list at the beginning of Section 2.7: Edward U. Condon and George H. Shortley, 1967: *The Theory of Atomic Spectra*. Cambridge University Press; Gerhard Herzberg, 1950: *Molecular Spectra and Molecular Structure*: I. *Spectra of Diatomic Molecules*, 2nd ed. D. Van Nostrand; Gerhard Herzberg, 1945: *Molecular Spectra and Molecular Structure*: II. *Infrared and Raman Spectra of Polyatomic Molecules*. D. van Nostrand; Gerhard Herzberg, 1966: *Molecular Spectra and Molecular Structure*: III. *Electronic Spectra and Electronic Structure of Polyatomic Molecules*. D. Van Nostrand; Charles H. Townes and Arthur L. Schawlow, 1975: *Microwave Spectroscopy*. Dover.

The paper by Charles L. Braun and Sergei N. Smirnov, 1993: Why is water blue? *Journal of Chemical Education*, Vol. 70, pp. 612–14, could have been published in an archival journal (loosely translated from the Latin, archival means "no one reads this stuff") but instead its authors chose a journal devoted to exposition. The result is a paper that is a pleasure to read. When you finish it you only wish it were longer so that the pleasure will linger.

The temperature and pressure profiles in Fig. 2.11 were taken from R. A. McClatchey, R. W. Fenn, J. E. A. Selby, F. E. Volz, and J. S. Garing, 1972: Optical Properties of the Atmosphere, 3rd ed. AFCRL-72-0479, Air Force Cambridge Research Laboratories.

The spectral calculations in this chapter were done using the latest version of LBLRTM, a line-by-line code developed by Tony Clough and his collaborators. See, for example, Shepard A. Clough, Michael J. Iacono, and Jean-Luc Moncet, 1992: Line-by-line calculations of atmospheric fluxes and cooling rates: application to water vapor. *Journal of Geophysical Research – Atmospheres*, Vol. 97, pp. 15761–85. The spectroscopic parameters in this code come from HITRAN, an acronym for high-resolution transmission molecular absorption database. The version most recently published (in a journal) is Laurence S. Rothman *et al.*, 2003: The HITRAN 2000 molecular spectroscopic database: edition of 2000 including updates through 2001. *Journal of Quantitative Spectroscopy and Radiative Transfer*, Vol. 82, pp. 5–44. The 2004 database eventually will be published in the same journal. We depart from our usual practice of giving (when known) the full names of all authors because this paper has 31 of them.

For an English translation of Einstein's 1916 paper Emission and Absorption of Radiation in Quantum Theory see Alfred Engel's 1997 translation *The Collected Papers of Albert Einstein, Vol. 6, The Berlin Years: Writings, 1914–1917*, pp. 212–16. Einstein did not use the terms spontaneous and induced emission, absorption, or photons in this paper. What we call induced emission and absorption Einstein associated with the "work done by the electric field on the resonator", the "energy change due to incident radiation."

For a discussion of the Einstein coefficients see, for example, Robert Martin Eisberg, 1961: *Fundamentals of Modern Physics*. John Wiley & Sons, p. 458–60.

The treatise by Richard M. Goody and Yuk L. Yung, 1989: *Atmospheric Radiation: Theoretical Basis*. Oxford University Press has been invaluable to us in writing this chapter. Textbooks can serve up nonsense when their authors leave their areas of expertise and carelessly discuss subjects with which they are not familiar. But they can also serve up gems when their authors concentrate on clear exposition of subjects with which they are familiar. Goody and Yung's discussions of absorption by atmospheric molecules falls into the latter category. Their sections on thermal emission, vibration-rotational spectra, and line shapes served as roadmaps that we came back to time and again as we wrote our sections on absorption, induced emission and spontaneous emission.

Schematic diagrams of the broadening of spectral lines because of interactions are standard fare in solid state physics treatises. See, for example, Figs. 1 and 2 on page 225 of Frederick Seitz, 1987: *The Modern Theory of Solids*. Dover. Seitz notes that "If we have a large number of free atoms that are infinitely separated from one another and are stationary, their electronic levels are discrete so that the energy levels of the entire system are discrete. Thus the width of the emission lines is determined entirely by natural broadening. The atoms interact, however, if they are brought within a finite distance of one another, and this interaction broadens these levels of the entire system..."

The history of the development of line shapes with far wings that accurately account for continuum water vapor absorption is an interesting one. To learn more about this see Shepard A. Clough, Frank X. Kneizys and R. W. Davies, 1989: Line shape and the water vapor continuum. *Atmospheric Research*, Vol. 23, pp. 229–41, and Qiancheng Ma and Richard H. Tipping, 1991: A far wing line shape theory and its application to the water continuum absorption in the infrared region. 1. *J. Chem. Phys*, Vol. 95, pp. 6290–6301. For measurements of water dimer concentrations in the atmosphere and estimates of the weak absorption by them see

Klaus Pfeilsticker, Andreas Lotter, Christine Peters, and Hartmut Bösch, 2003: Atmospheric detection of water dimers via near-infrared absorption. *Science*, Vol. 300, pp. 2078–80.

For a discussion of retrieving water vapor profiles in the mesosphere see Richard M. Bevilacqua, 1982: An Observational Study of Water Vapor in the Mid-Latitude Mesosphere. Doctoral Thesis, The Pennsylvania State University. Parts of this thesis are more readily available in Richard M. Bevilacqua, John J. Olivero, Philip R. Schwartz, Christopher J. Gibbins, Joseph M. Bologna, and Dorsey J. Thacker, 1983: An observational study of water vapor in the mid-latitude mesosphere using ground-based microwave techniques. *Journal of Geophysical Research*, Vol. 88, pp. 8523–34.

For more on absorption by particles see the treatises on scattering by particles cited in the references at the end of Chapter 3. Scattering and absorption by particles go hand in hand.

Problems

2.1. Given the assumption that the frequency of a microwave oven is chosen so that the absorption length in water is comparable with the size of objects to be heated in it, you should be able to estimate how long it takes to heat a cup of coffee from room temperature to what we call the *McDonald's point*, the temperature (85 °C) to which McDonald's *used* to heat its coffee. The power of a typical microwave oven is 1000–1500 W. The specific heat capacity of liquid water is about 4200 J kg^{-1} K^{-1}, its density 1000 kg m^{-3}.

HINT: There is no need to make this problem complicated. All that is wanted is an estimate. The first time we did this problem we made crude approximations, guessed at some of the relevant physical parameters, and yet still came up with a number that made sense.

2.2. Morning fog often disappears soon after sunrise, and as a consequence you often hear it said that the sun "burned off the fog". Because this explanation seems plausible, its correctness is rarely questioned. But can sunlight really vaporize fog droplets in less than an hour? You should be able to answer this question quantitatively. State all assumptions.

   The density of liquid water is about 1000 kg m^{-3} and about 2.5×10^6 J is required to evaporate 1 kg of liquid water. Over the solar spectrum the absorption cross section per unit volume of a cloud droplet is approximately equal to the bulk absorption coefficient of pure water. At Earth's surface, most of the solar irradiance lies at wavelengths less than about 2.5 μm.

HINT: To determine the exact rate at which a cloud droplet absorbs solar radiation would require considerable computation. But this isn't necessary if all that is wanted is to determine if this rate is sufficient to evaporate a droplet in an hour or less. First set up the problem as exactly as you can. Then make approximations, always erring toward overestimating the rate of absorption. If after making such approximations you obtain an evaporation time much greater than an hour, more detailed calculations will not change your conclusion.

2.3. Physical quantities can be represented as complex variables provided they are subjected to linear operations only. Nevertheless, for some purposes one may multiply the complex representations of real quantities without introducing any error. Show that the time average of the product of two time-harmonic quantities with complex representations $A \exp(-i\omega t)$ and $B \exp(-i\omega t)$, where A and B are complex numbers, is

$$\frac{1}{2}\Re\{AB^*\},$$

where the asterisk denotes the complex conjugate.

2.4. A radiation thermometer can be used to measure the temperature of the ocean (from above its surface, of course). But temperature varies with depth in the ocean, so what temperature does the radiation thermometer measure? Figure 2.2 will help you answer this question.

2.5. Estimate the highest natural frequency of a (macroscopic) mechanical oscillator that you are likely to encounter in your everyday lives.

2.6. Find an expression for the resonant frequency of two one-dimensional oscillators of different masses m_1 and m_2 connected by a spring with constant K. Find the normal modes of motion of this system and give them a simple physical description.

2.7. With Eq. (2.47) you can express $\cos\vartheta$ and $\sin\vartheta$ in terms of $\exp(i\vartheta)$ and $\exp(-i\vartheta)$, then derive all those trigonometric identities you may have slaved over in high-school trigonometry. Begin by showing that $\sin^2\vartheta + \cos^2\vartheta = 1$, then move on to expressions for the cosine and sine of the sum of angles, and so on.

2.8. Define the transmissivity of a slab as the ratio of the transmitted to the incident radiant energy. Is the transmissivity of a given slab for a diffuse source greater than, equal to, or less than that for a monodirectional source? The spectra of the two sources are identical. In answering this question, first ignore reflection. How does your answer depend on the angle of incidence of the monodirectional source?

Suppose that you do not ignore reflection. In which way, if any, does this change your conclusions about the relative transmissivities of a slab for monodirectional and diffuse radiation?

2.9. Almost everyone these days knows that a layer of ozone in the stratosphere absorbs ultraviolet (UV) radiation from the sun. Suppose that above this ozone layer there is a layer of scattering particles as a consequence of volcanic eruptions. One might expect this particle layer together with the ozone layer to reduce the solar UV irradiance at the surface even more than does the ozone layer alone. It has been suggested, however, that the particle layer could result in a greater UV irradiance at the surface. Where (latitude) and when (season) might this be true? Explain your answer. You may take the particle layer to be of uniform thickness and infinite in lateral extent. The trivial solution to this problem is obtained by taking the sun to be below the horizon.

HINT: You do not need to know anything more about scattering by particles than that it exists. This problem, which is related to the previous one, should be easier if you draw a diagram.

2.10. We stated without proof in Section 1.5 that the infrared emissivity of a layer of water is similar to that of a cloud of water droplets. Now you have the tools necessary to show this. A typical liquid water path value for a thick cloud is of order 1 cm. That is, if all the water in the cloud were compressed into a uniform layer it would be about 1 cm thick. The diameter of a typical cloud droplet is about 10 μm. The absorption cross section per unit volume of such a droplet (Fig. 2.25) will help you to show that the infrared emissivity of a cloud, like that of a layer of water, is close to 1 over a broad range of wavelengths.

2.11. The quantity κh in Section 2.2 is the (normal) absorption optical thickness of the (slab) atmosphere (see Sec. 5.2 for a discussion of optical thickness). Strictly, the normal optical thickness is the integral of κ along a radial path through the entire atmosphere. Estimate the absorption optical thickness of the clear atmosphere for the wavelength range (10 μm) of the infrared thermometer used to make the brightness temperature measurements shown in Fig. 2.4. What is the corresponding emissivity? Assume that the (absolute) temperature of the atmosphere is constant. This is not a bad assumption given that water vapor is the major contributor to the absorption optical thickness, the scale height (e-folding distance) for the decrease of water vapor concentration is 1–2 km, and over this distance the absolute temperature decreases by only about 5%. At 10 μm and for temperatures around 300 K, the exponential in the Planck function is much greater than 1.

2.12. In Section 2.9 we derive approximate expressions for the absorption cross section per unit volume of a normally illuminated slab particle in the limits of weak ($\kappa_b d \ll 1$) and strong absorption ($\kappa_b d \gg 1$), where d is the thickness of the slab. Show that, subject to the same underlying assumptions, these expressions are valid for a particle of more general shape. Take the particle to be homogeneous (no holes) and bounded by a surface such that no line in the direction of the beam intersects it in more than two points. For such a particle, d in Eq. (2.148) is the average distance through the particle along the direction of the incident beam; d in Eq. (2.149) is A/v, where A is the cross-sectional area of the particle projected onto the beam and v is its volume.

HINTS: A sketch is essential. Consider a thin tube, its axis parallel to the beam, intersecting the particle. Determine the net amount of incident radiant energy absorbed by this tube (energy in minus energy out). Then add up (integrate) the total for all the tubes into which the particle can be considered to be composed.

2.13. This is Problem 2.2 stated a different way. Show that the radiative equilibrium temperature of a typical cloud droplet exposed to direct sunlight cannot be more than a small fraction of a degree greater than what its temperature would be in the shade (all else being equal). Figure 2.25 is essential for this problem.

2.14. Our local newspaper published a front-page article about auroras that were on seen two successive nights in our county. According to the article, "Auroras...occur when charged particles from the sun interact with atoms of nitrogen and oxygen in the upper atmosphere, giving off particles called photons that appear red or green.... Light is created when the atoms in the atmosphere attempt to cool down." Please discuss.

2.15. We state in Section 2.7.1 that we calculated the measured spectral shift in the visible absorption spectrum of water vapor as a consequence of the different masses of heavy water (D_2O) and ordinary water (H_2O). With a few hints and having done Problem 2.6 you should be able to duplicate our calculation. Braun and Smirnov attribute the absorption peak in the visible at 700 nm and in the infrared at 1000 nm to the combination-overtone $\nu_1 + 3\nu_3$, where ν_1 is the symmetric OH vibrational frequency and ν_3 is the antisymmetric OH vibrational frequency. Both frequencies are nearly equal (within a few percent).

2.16. Suppose that we have a gas in equilibrium at temperature T of N molecules that can exist in only one of two states, upper with energy E_u and lower with energy E_ℓ. Show that at sufficiently low temperature essentially all of the molecules are in the lower state and that the number of molecules in this state decreases with increasing temperature. Also show that at sufficiently high temperature *not* all of the molecules are in the upper state.

HINT: You need Eq. (2.95).

2.17. In 1852 Beer published an attenuation law for absorbing aqueous solutions according to which the diminution λ in *amplitude* of a beam upon transmission through a solution over a path of length D (in decimeters) is given by $\lambda = \mu^D$. Show that this leads to an exponential law for attenuation of irradiance. What we call transmissivity is λ^2. For more on this see Heinz G. Pfeiffer and Herman A. Liebhansky, 1951: The origins of Beer's law. *Journal of Chemical Education*, Vol. 28, pp. 123–125. These authors argue that Beer did not recognize that concentration and length are symmetric variables, that is, the absorption coefficient κ (in modern notation) of an absorbing solution is proportional to the concentration of the absorbing solute, and hence it is the product of concentration and path length that determines the transmissivity. What would have been the form of Beer's law if Beer had recognized this symmetry?

2.18. We are often told in textbooks on classical mechanics that the energy of a body is defined only to within an additive constant. In fact, similar statements are made even in treatises on quantum mechanics. For example, on page 5 of Herzberg's *Molecular Spectra and Molecular Structure I. Spectra of Diatomic Molecules* he states "The energy E of the atom contains an arbitrary additive constant." Yet such statements are never, to our knowledge, made about the energy of a photon. It is $h\nu$. Period. Why the difference?

2.19. The Maxwell–Boltzmann distribution [Eq. (2.8)] for molecular kinetic energies E contains the factor $\sqrt{E}$, which makes it difficult to determine analytically the fraction of molecules with kinetic energies greater than a particular value. But there is a ploy we can resort to if we are willing to settle for an exactly integrable inexact theory. We note in Section 2.8 that when the M–B distribution for energy E is transformed to a distribution for speed v [see also Prob. 3.13] the factor v^2 multiplying the exponential reflects the density of states. The number of points (states) in velocity space corresponding to (approximately) the same kinetic energy is proportional to v^2 because the total volume in velocity space between v and $v + \Delta v$ is proportional to v^2. Knowing this, you should be able to determine the distribution function for the kinetic energies of molecules constrained to lie in a *plane*. With this result you can then determine, analytically, the fraction of these molecules with kinetic energies greater than a given value. Then answer the following question. For a given energy $\overline{E}$ appreciably greater than $k_B T$, how does the fraction of molecules with energies greater than $\overline{E}$ change if the *mean* kinetic energy is changed by only a factor of, say, 2? To answer this question is to obtain insight into why rates of chemical reactions depend so strongly on temperature. This problem was inspired by Cyril N. Hinshelwood, 1940: *The Kinetics of Chemical Change*, Oxford University Press, p. 5–6, an uncommonly good writer (and eminent scientist), whose (1951) *The Structure of Physical Chemistry*, Oxford University Press is, in places, lyrical (see, e.g., the opening section of the first chapter).

2.20. A chemical reaction occurs when two (or more) stable molecules unite to form another stable molecule. If the reacting molecules are stable then their potential energies must be minima. And if the product is stable, it must have a potential energy minimum. The system, so to speak, makes a transition from one valley to another, and in so doing crosses a mountain. The energy required to cross this mountain, which comes from collisions, is the activation energy. Typical activation energies are around 30 times $k_B T$ at room temperature. With this result and the previous problem you should be able to obtain a rule of thumb of chemists we mention in Section 2.7.1: increasing temperature by $10\,^\circ C$ doubles the rate of chemical reactions. As with the previous problem, this one was inspired by Cyril Hinshelwood's book on chemical kinetics (p. 41).

2.21. How does the absorption cross section of atmospheric molecules compare with their geometrical cross section? Keep in mind that the diameter of a molecule is not a precisely defined quantity.

2.22. We note that some of the differences in Fig. 2.15 between the the results of simple and line-by-line calculations for the rotational levels, and hence absorption cross section, of carbon monoxide are a consequence of different isotopes of carbon. What about the isotopes of oxygen? Estimate the difference in the energy levels, and hence photon energies, for the most abundant isotopic forms of CO. Compare your results with Fig. 2.15. Also, estimate the relative contributions (at the peak of a line) of these isotopes to the average absorption cross section for naturally occurring carbon monoxide. Compare your result with the line-by-line calculations.

HINT: To do this problem you will have to dig up atomic masses and abundances of carbon and oxygen isotopes.

2.23. What is the fundamental difference between stimulated emission and scattering (in Sec. 1.2 we describe scattered radiation as being "stimulated" by an external source)?

2.24. At approximately what pressure altitude would the line shape for the water vapor line shown in Fig. 2.18 be dominated by Doppler broadening?

HINT: No calculations are needed.

2.25. In the discussion of Fig. 2.18, which shows line shapes of a water vapor band, we note that at the surface the foreign-broadened half-width is 0.0755 cm^{-1} whereas the self-broadened half-width is 0.3580 cm^{-1} (for the same temperature and pressure), a factor of almost 5 greater. Why the big difference?

2.26. Show that if the Doppler and Lorentz line shape functions [Eqs. (2.135) and (2.134), respectively] are normalized, so is their convolution [Eq. (2.136)].

2.27. Figure 2.23 shows absorptivity over narrow spectral regions by an atmosphere containing different concentrations of carbon dioxide. A line-by-line computer program contains all absorbing gases, most notably water vapor. How does one correct the total absorptivity in order to obtain that for an atmosphere with only carbon dioxide as the absorbing gas? What assumption has to be made?

2.28. Show that the effect of an increase in the concentration of an infrared-active molecule on emission from the atmosphere (i.e., on emissivity) is least, all else being equal, in spectral regions where the transmissivity is least. For simplicity, consider only the normal emissivity and assume a uniform atmosphere of finite thickness.

2.29. James Howard Kunstler, on page 139 of *The Long Emergency: Surviving the Converging Catastrophes of the Twenty-First Century*, asserts that "methane freed into the atmosphere is a ten times more effective greenhouse gas than carbon dioxide." Discuss this assertion.

2.30. Problem 1.5 asks you to estimate plausible maximum and minimum values at Earth's surface of radiation emitted downward from the atmosphere. Why is it easier to estimate an upper limit than a lower limit?

3 Scattering: The Life of Photons

Between birth (emission) and death (absorption) photons are scattered. That is, their direction of propagation changes. The zigzag course of a photon as it wends its way through matter is, so to speak, its biography. In this chapter we take a more catholic view of scattering than is the norm. Unification long has been an aim of physics because it enables our limited human minds to comprehend a universe of bewildering complexity. So among our tasks in this chapter is to remove artificial walls that should have been torn down long ago in the cause of intellectual freedom just as the Berlin Wall was torn down in the cause of political freedom.

3.1 Scattering: An Overview

Why is light scattered? No single answer will satisfy everyone, yet because scattering by particles is amenable to treatment mostly by classical electromagnetic theory, our answer lies within this theory.

Although palpable matter may appear continuous and often is electrically neutral, it is composed of discrete electric charges. Light is an oscillating electromagnetic field, which can excite these charges to oscillate. Oscillating charges radiate electromagnetic waves, a fundamental property of such charges with its origins in the finite speed of light. These radiated electromagnetic waves are scattered waves, excited by a source external to the scatterer. Incident waves from the source excite secondary waves from the scatterer, and the superposition of all these waves is what is observed. The secondary waves are said to be *elastically* scattered if their frequency is that of the source (*coherently* scattered also is used).

Scientific knowledge grows like the accumulation of bric-a-brac in a vast and disordered closet in a house kept by a sloven. Few are the attempts at ridding the closet of rusty or broken or obsolete gear, at throwing out redundant equipment, at putting things in order. For example, spurious distinctions still are made between reflection, refraction, scattering, interference, and diffraction despite centuries of accumulated knowledge about the nature of light and matter.

Why do we think of specular (mirror-like) reflection as occurring *at* surfaces rather than *because* of them whereas we usually do not think of scattering by particles in this way? One reason is that we can see and touch the surfaces of mirrors and ponds. Another reason is the dead hand of traditional approaches to the laws of specular reflection and refraction.

The empirical approach arrives at these laws as purely geometrical statements about what is observed, and a discreet silence is maintained about underlying causes (always a safe course). The second approach is by way of continuum electromagnetic theory: reflected and refracted waves satisfy the partial differential equations of the electromagnetic field (the Maxwell equations). Perhaps because this approach, which yields the amplitudes and phases

Fundamentals of Atmospheric Radiation: An Introduction with 400 Problems. Craig F. Bohren and Eugene E. Clothiaux
Copyright © 2006 Wiley-VCH Verlag GmbH & Co. KGaA, Weinheim
ISBN: 3-527-40503-8

of waves, entails imposing conditions at boundaries, reflected and refracted waves are mistakenly thought to originate from boundaries rather than from all the matter they enclose. This second approach comes to grips with the nature of light but not of matter, which is treated as continuous. The third approach is to recognize explicitly that reflection and refraction are consequences of scattering of waves by discrete matter. Although this scattering approach was developed by Paul Ewald and Carl Wilhelm Oseen early in the last century, it has diffused with glacial slowness. When the optically smooth interface between optically homogeneous media is illuminated, the reflected and refracted waves are superpositions of vast numbers of secondary waves excited by the incident wave. Moreover, every molecule, not just those at or near the interface, contributes to the total. Thus reflected and refracted light is, at heart, an interference pattern of light scattered by discrete molecules (see Sec. 7.2.1 for more about this). The fourth approach is to recognize the discreteness of both matter and radiation fields. This is the method of quantum electrodynamics, presumably the most rigorous but, alas, nearly impossible to apply except for very simple systems (e.g., the hydrogen atom).

No optics textbook would be complete without sections on interference and diffraction, a distinction without a difference: there is no diffraction without interference. Moreover, diffraction is encumbered with many meanings: a synonym for scattering; small deviations from rectilinear propagation; wave motion in the presence of an obstacle; scattering by a flat obstacle; any departure from geometrical (ray) optics; scattering near the forward direction; and scattering by a periodic array. A term with so many meanings has no meaning. Even the etymology of diffraction is of little help, coming from a Latin root meaning to break, the origin of fraction, fracture, fractal, and fracas.

There is no fundamental difference between diffraction and scattering. Scattering by a sphere (see Sec. 3.5.1) is sometimes called diffraction by a sphere. For many years we have offered a million-dollar prize to anyone who can devise a detector that distinguishes between scattered and diffracted waves, accepting the one but rejecting the other. So far no one has collected, and the money continues to draw interest in a numbered Swiss bank account.

The only meaningful distinction is that between approximate theories. What are called diffraction theories often obtain answers at the expense of obscuring if not completely distorting the physics of the interaction of light with matter. For example, an illuminated slit in an opaque screen may be the mathematical source but it is not the physical source of a so-called diffraction pattern. Only matter in the screen can give rise to secondary waves that superpose to yield the observed pattern. Yet generations of students have been taught that empty space is the source of the radiation from a slit. To befuddle them even more, they also have been taught that two slits give an interference pattern whereas one slit gives a diffraction pattern. But every pattern of scattered light called a diffraction pattern is a consequence of interference. And every theory bearing the label diffraction is a wave theory because only such theories can account for interference.

A variation on the bogus notion that empty space is the source of electromagnetic waves is the oft-repeated mantra that a changing electric field "produces" a magnetic field, and a changing magnetic field "produces" an electric field. Not true. Electric fields are produced by charges, magnetic fields by charges in motion (currents). There are always material *sources* of electromagnetic fields; they do not arise out of empty space.

If we can construct a mathematical theory (called for no compelling reason diffraction theory) that enables us to avoid having to consider the nature of matter, all to the good. But this theory and its quantitative successes should not blind us to the fact that we are pretending. Sometimes the pretense cannot be maintained, and when this happens a finger is mistakenly pointed at "anomalies", whereas what is truly anomalous is that a theory so devoid of physical content could ever give adequate results.

A distinction must be made between a physical process and the superficially different theories used to describe it. There is no fundamental difference between specular reflection and refraction by films, diffraction by edges or slits, and scattering by particles. All are consequences of light exciting matter to radiate. The only difference is in how this matter is arranged in space and the approximate theories sufficient for a quantitative description of the scattered light. Different terms for the same physical process are incrustations deposited during the evolution of our understanding of light and its interaction with matter.

3.2   Scattering by a Dipole

The electric charges in matter are acted on by electric fields. Indeed, they are defined as forces per unit charge. When exposed to the oscillatory electric field of an electromagnetic wave, charges in matter are set into oscillation. The fields of these charges therefore change but not instantaneously. This change is propagated outward from the charges in the form of waves traveling at the speed of light. The simplest example of an electrically neutral system is a dipole.

In Section 2.6 we considered the forces acting on a damped oscillator (dipole). We discovered that a dipole, when acted on by a time-harmonic electric field, oscillates at the same frequency as this field. But if the dipole oscillates, it must radiate, and radiation carries energy. So our analysis was incomplete. We didn't account for a dipole losing energy by radiating. This can be accounted for by saying that it is as if the dipole were acted upon by another dissipative force called the *radiative reaction*. Work per unit time done by this force is the rate of energy radiated by the dipole. The radiative reaction is proportional to the second time-derivative of velocity, and hence the equation of motion of a dipole for which the radiative reaction is included is

$$m\frac{d^2\mathbf{x}}{dt^2} = -K\mathbf{x} - b\frac{d\mathbf{x}}{dt} + w\frac{d^3\mathbf{x}}{dt^3} + e\mathbf{E}_\mathrm{o}\exp(-i\omega t), \tag{3.1}$$

where w is a constant.

As in Section 2.6 we are interested in the steady-state solution to this equation,

$$\mathbf{x} = \mathbf{x}_\mathrm{o}\exp(-i\omega t). \tag{3.2}$$

Substituting Eq. (3.2) into Eq. (3.1) yields

$$\mathbf{x}_\mathrm{o} = \frac{e}{m_\mathrm{o}}\frac{\mathbf{E}_\mathrm{o}}{\omega_\mathrm{o}^2 - \omega^2 - i\gamma\omega}, \tag{3.3}$$

where

$$\gamma = \gamma_\mathrm{a} + \gamma_\mathrm{s}\omega^2/\omega_\mathrm{o}^2 \tag{3.4}$$

and

$$\gamma_{\mathrm{a}} = \frac{b}{m}, \quad \gamma_{\mathrm{s}} = \omega_{\mathrm{o}}\frac{w}{m}. \tag{3.5}$$

Equation (3.3) has the same form as Eq. (2.61), so we can immediately write the expression for the time-averaged rate at which work is done (power) on the dipole by the incident electric field, which falls naturally into two parts:

$$\langle P \rangle = \langle P_{\mathrm{a}} \rangle + \langle P_{\mathrm{s}} \rangle, \tag{3.6}$$

where

$$\langle P_{\mathrm{a}} \rangle = \frac{e^2}{2m}E_{\mathrm{o}}^2\frac{\omega^2\gamma_{\mathrm{a}}}{(\omega_{\mathrm{o}}^2 - \omega^2)^2 + \gamma^2\omega^2}, \tag{3.7}$$

$$\langle P_{\mathrm{s}} \rangle = \frac{e^2}{2m}\frac{E_{\mathrm{o}}^2}{\omega_{\mathrm{o}}^2}\frac{\omega^4\gamma_{\mathrm{s}}}{(\omega_{\mathrm{o}}^2 - \omega^2)^2 + \gamma^2\omega^2}. \tag{3.8}$$

We interpret $\langle P_{\mathrm{a}} \rangle$ as the power absorbed and $\langle P_{\mathrm{s}} \rangle$ as the power scattered by the dipole. Absorbed power is a consequence of transforming radiant energy into other forms. Scattered power is a consequence of incident radiation in one direction exciting radiation in all directions. $\langle P_{\mathrm{s}} \rangle$ is the time-average of the power radiated (scattered) in all directions by the dipole. As required, $\langle P_{\mathrm{a}} \rangle$ vanishes as γ_{a} (dissipation) vanishes, as does $\langle P_{\mathrm{s}} \rangle$ when γ_{s} (radiative reaction) vanishes. Scattering is not a transformation of radiant energy into other forms but rather into different directions; radiant energy is conserved.

In the scattering considered here the frequency of oscillation of the dipole, and hence the frequency of the radiation it scatters, is equal to the frequency of the incident radiation (*elastic scattering*). But all scattering is not elastic. Indeed, there is no such thing as truly elastic scattering if for no other reason than that matter is in motion, so the frequency of the radiation scattered by moving matter is Doppler-shifted (see Sec. 3.4.6) from that of the incident radiation. The term *quasi-elastic scattering*, which sometimes is used, signals that strict elastic scattering is a mathematical fiction.

If the frequency ω of the incident electric field is much less than the resonant frequency ω_{o} (frequency at the center of the absorption band) and the width of the band γ is much less than ω_{o}

$$\langle P_{\mathrm{s}} \rangle \approx \frac{e^2}{2m}\frac{E_{\mathrm{o}}^2}{\omega_{\mathrm{o}}^6}\gamma_{\mathrm{s}}^2\omega^4. \tag{3.9}$$

This is the much-celebrated scattering law first derived by Lord Rayleigh in 1871. Frequency to the fourth power is proportional to wavelength to the inverse fourth power. We have more to say about the observable consequences of this law in Section 8.1.

We obtained Rayleigh's scattering law in the following form: the radiation scattered in all directions by a dipole, for a given irradiance (proportional to E_{o}^2), is proportional to the fourth power of the frequency of the incident radiation that excites this scattered radiation. But we also recognize that Rayleigh's law is an approximation, not valid at frequencies near an absorption band. That is, strict ω^4 scattering by a dipole is yet another mathematical fiction, a useful approximation provided we recognize its limitations.

Equations (3.7) and (3.8) also show that scattering and absorption are not completely independent of each other. Absorbed power depends on γ, which has a contribution from scattering (γ_s) and absorption (γ_a), and similarly for scattered power.

Just as we defined the absorption cross section in Section 2.4 we can define the scattering cross section of a molecule (dipole) σ_s by

$$\langle P_s \rangle = W_s = \sigma_s F, \tag{3.10}$$

where W_s is the radiant power scattered in all directions by the molecule and F is the irradiance of the beam that excites this scattered radiation. Unlike the absorption cross section, the scattering cross section can be apportioned into contributions in different directions. No scatterer of electromagnetic radiation scatters exactly the same in all directions (Sec. 7.3), and scattering can be highly directionally asymmetric depending on the size of the scatterer relative to the wavelength (Sec. 3.5.3).

We use the term dipole and molecule almost as synonyms because for our purposes every molecule is a dipole, although not conversely. If sufficiently small, a particle can be a dipole oscillator. To explain why requires a firm understanding of interference, which in turn requires a firm understanding of waves, which we turn to next.

3.3   Waves on a String: The One-Dimensional Wave Equation

The easiest path to understanding waves is by way of the simplest example: waves on a string under tension (Fig. 3.1). In its equilibrium position the string lies along the x-axis, but if the string is displaced in the y-direction it will vibrate because of opposition between the inertia of the string and a restoring force provided by the tension. Consider a segment of string lying between x and $x + \Delta x$. If σ is the mass per unit length of the string (assumed constant), the equation of motion of this segment is

$$\sigma \Delta x \frac{\partial^2 y}{\partial t^2} = F_y, \tag{3.11}$$

where F_y is the y-component of the force on Δx, a consequence of the tension, assumed constant. This assumption requires that the lateral displacement of the string be small. We also assume that the string always lies in the xy-plane. The total force acting on Δx by the string on both sides of it is

$$F_y = T \sin\{\vartheta(x + \Delta x)\} - T \sin\{\vartheta(x)\}, \tag{3.12}$$

where T is the tension in the string and ϑ is the angle between the string and the x-axis. Unless the tension is very small, the force of gravity is negligible. Because of the assumption of small displacements

$$\sin \vartheta \approx \tan \vartheta = \frac{\partial y}{\partial x}. \tag{3.13}$$

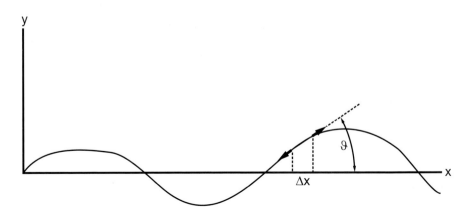

Figure 3.1: The (small) y-displacement of a uniformly tense string stretched along the x-axis varies in space and time. The angle ϑ between the tangent to the string at any point and the x-axis is shown greatly exaggerated.

If Eqs. (3.11)–(3.13) are combined we obtain

$$\sigma \Delta x \frac{\partial^2 y}{\partial t^2} = T\left(\frac{\partial y}{\partial x}\right)_{x+\Delta x} - T\left(\frac{\partial y}{\partial x}\right)_x. \tag{3.14}$$

Divide both sides of this equation by Δx and take the limit as $\Delta x \to 0$:

$$\sigma \frac{\partial^2 y}{\partial t^2} = T\frac{\partial^2 y}{\partial x^2}. \tag{3.15}$$

We also can write this equation as

$$\frac{1}{v^2} \frac{\partial^2 y}{\partial t^2} = \frac{\partial^2 y}{\partial x^2}, \tag{3.16}$$

where

$$v = \sqrt{T/\sigma}. \tag{3.17}$$

Note that v has the dimensions of speed, and hence is called the *phase speed* (about which we say more in this section and Sec. 3.5.1). Equation (3.16) is a *wave equation*, specifically, a one-dimensional scalar wave equation. A wave is anything that satisfies a wave equation. If you know something about one kind of wave you know something about all kinds of waves. Begin with the simplest kind, understand it, and you are on your way to understanding more complicated waves (e.g., three-dimensional vector waves).

Multiply both sides of Eq. (3.15) by the time derivative of y:

$$\sigma \frac{\partial y}{\partial t} \frac{\partial^2 y}{\partial t^2} = \frac{\partial}{\partial t} \left\{ \frac{1}{2}\sigma\left(\frac{\partial y}{\partial t}\right)^2 \right\} = T\frac{\partial y}{\partial t}\frac{\partial^2 y}{\partial x^2}. \tag{3.18}$$

The left side of Eq. (3.18) is the time rate of change of the kinetic energy (per unit length) of the string. This is a clue that we are headed toward an energetic form of the equation of motion. Before obtaining this equation we pause to reflect that although multiplying the equation of motion by partial derivatives and rearranging terms cannot change its physical content, it can result in a more felicitous form. To proceed we note that

$$\frac{\partial}{\partial x}\left(\frac{\partial y}{\partial x}\frac{\partial y}{\partial t}\right) = \frac{\partial y}{\partial t}\frac{\partial^2 y}{\partial x^2} + \frac{\partial^2 y}{\partial x \partial t}\frac{\partial y}{\partial x} \tag{3.19}$$

and

$$\frac{\partial}{\partial t}\left(\frac{\partial y}{\partial x}\right)^2 = 2\frac{\partial y}{\partial x}\frac{\partial^2 y}{\partial t \partial x}. \tag{3.20}$$

With the assumption that the order of differentiation is irrelevant,

$$\frac{\partial^2 y}{\partial x \partial t} = \frac{\partial^2 y}{\partial t \partial x}, \tag{3.21}$$

Equations (3.18)–(3.21) can be combined to yield

$$\frac{\partial}{\partial t}\left\{\frac{1}{2}\sigma\left(\frac{\partial y}{\partial t}\right)^2 + \frac{1}{2}T\left(\frac{\partial y}{\partial x}\right)^2\right\} = T\frac{\partial}{\partial x}\left(\frac{\partial y}{\partial t}\frac{\partial y}{\partial x}\right). \tag{3.22}$$

This equation resembles energy conservation equations of the kind encountered in the dynamics of systems of point particles. The only difference is that the system of interest here is continuous rather than discrete, as evidenced by the partial derivatives. By now we should suspect that the second term in brackets on the left side of Eq. (3.22) is some kind of potential energy per unit length of string.

A short segment of string with x-coordinates x and $x + \Delta x$ has length Δx when the string lies along the x-axis. In general, the length of this segment is Δs, where

$$\Delta s = \sqrt{(\Delta x)^2 + (\Delta y)^2} = \Delta x\sqrt{1 + \left(\frac{\Delta y}{\Delta x}\right)^2}. \tag{3.23}$$

If the slope of the string is everywhere much less than unity [an assumption made previously in Eq. (3.13)], the change in length of the segment is

$$\Delta s - \Delta x \approx \Delta x\left\{1 + \frac{1}{2}\left(\frac{\Delta y}{\Delta x}\right)^2\right\} - \Delta x. \tag{3.24}$$

Divide both sides of this equation by Δx and take the limit as $\Delta x \to 0$ to obtain the change in length per unit length

$$\frac{1}{2}\left(\frac{\partial y}{\partial x}\right)^2. \tag{3.25}$$

This quantity multiplied by the tension T is the work done (per unit length) in stretching the string, and is the second term in brackets on the left side of Eq. (3.22). Hence we are justified in calling this term a potential energy (per unit length).

If we integrate Eq. (3.22) from x_1 to x_2 we obtain

$$\frac{d}{dt}\int_{x_1}^{x_2}\left\{\frac{1}{2}\sigma\left(\frac{\partial y}{\partial t}\right)^2 + \frac{1}{2}T\left(\frac{\partial y}{\partial x}\right)^2\right\}dx = T\left(\frac{\partial y}{\partial x}\frac{\partial y}{\partial t}\right)_2 - T\left(\frac{\partial y}{\partial x}\frac{\partial y}{\partial t}\right)_1. \tag{3.26}$$

The integral is the total energy (kinetic plus potential) of the string between x_1 and x_2. Thus the time rate of change of the energy of this system is equal to a quantity evaluated at one boundary of the string less that same quantity evaluated at the other boundary. It is therefore natural, if not inescapable, to call this quantity an *energy flux*, the rate at which energy is transported across the boundary of the system from its surroundings (the rest of the string). If we define the energy flux vector by

$$T\frac{\partial y}{\partial x}\frac{\partial y}{\partial t}\mathbf{e}_x, \tag{3.27}$$

where $\mathbf{e}_x$ is a unit vector in the positive x-direction, the energy flux into the system is the scalar product of this vector with the outward normals to the system on its boundary, which is just the two points x_1 and x_2.

3.3.1   Solutions to the Wave Equation

Any twice-differentiable function $f(u)$ of the single variable u can be transformed into a solution to Eq. (3.16) by setting

$$u = x \pm vt. \tag{3.28}$$

That the function of two variables so obtained is a solution is readily verified by repeated use of the chain rule for differentiation

$$\frac{\partial}{\partial x}f\{u(x,t)\} = \frac{df}{du}\frac{\partial u}{\partial x}, \tag{3.29}$$

and so on. Note that f is arbitrary subject only to the requirement that it be twice differentiable, and hence there are infinitely many possible solutions to the one-dimensional, scalar wave equation.

The function $f(x-vt)$ has a simple geometrical interpretation, which helps us to visualize, and hence understand waves. At $t = 0$, $y = f(x)$ describes the shape (displacement) of the string. At some later time Δt, the shape of the string will be different but yet the same because the argument of f, and hence its value, will be the same at $x + \Delta x$ if

$$x = x + \Delta x - v\Delta t, \tag{3.30}$$

from which

$$\frac{\Delta x}{\Delta t} = v. \tag{3.31}$$

Thus the waveform $f(x)$ can be looked upon as propagating (without change of shape) in the positive x-direction with constant speed v.

The energy flux vector corresponding to this wave is

$$-Tv\left(\frac{df}{du}\right)^2\mathbf{e}_x. \tag{3.32}$$

At the left boundary of the system, the outward normal is $-\mathbf{e}_x$, and hence the energy flux into the system from left to right is positive, as required of a wave propagating in the positive x-direction. At the right boundary, the wave transports energy out of the system.

The function $f(x + vt)$ corresponds to a wave propagating in the negative x-direction. Because the wave equation is linear, a sum of solutions is also a solution. That is, if y_1 and y_2 are solutions, so is $a_1 y_1 + a_2 y_2$, where a_1 and a_2 are arbitrary constants. Thus it is possible for waves to propagate on a string in both directions simultaneously. Indeed, two equal but opposite (in sign) pulses propagating in opposite directions can meet, annihilate each other briefly when they exactly overlap, then proceed unscathed carrying no memories of or scars from their meeting.

3.3.2   Sinusoidal Wave Functions

One of the simplest possible solutions to the one-dimensional wave equation is a sinusoid:

$$y = a\cos(kx - \omega t), \tag{3.33}$$

where the wavenumber $k = 2\pi/\lambda$ and frequency ω are arbitrary subject to the requirement that $v = \omega/k$. We noted previously (Sec. 2.7.1) that wavenumber could mean $1/\lambda$ or $2\pi/\lambda$; and frequency, unqualified, could be ν or $\omega = 2\pi\nu$. The argument of the cosine, $kx - \omega t$, is called the *phase* of the wave, and Eq. (3.33) describes what is often called a *plane harmonic wave*. Plane because the *surfaces of constant phase*

$$kx - \omega t = \text{const.} \tag{3.34}$$

are planes, harmonic because the time dependence is sinusoidal (harmonious). A harmonic wave is sometimes referred to as monochromatic even when it is not an electromagnetic wave at visible frequencies.

According to Eq. (3.34) the phase at (x, t) will be the same at $(x+\Delta x, t+\Delta t)$ if $\Delta x/\Delta t = \omega/k = v$, hence the name phase speed for v, the speed of surfaces of constant phase (planes perpendicular to the x-axis). This is not, however, the speed of any physical object but rather that of mathematical surfaces.

From Eqs. (3.27) and (3.33) it follows that the (instantaneous) power transmitted by the plane harmonic wave at a left boundary of the string is

$$T\omega k a^2 \sin^2(kx - \omega t). \tag{3.35}$$

Suppose that ω is so high that an energy flux detector responds not to instantaneous but to time averages of Eq. (3.35):

$$T\omega k a^2 \langle\sin^2(kx - \omega t)\rangle, \tag{3.36}$$

where

$$\langle\sin^2(kx - \omega t)\rangle = \frac{1}{t_2 - t_1} \int_{t_1}^{t_2} \sin^2(kx - \omega t)\, dt. \tag{3.37}$$

Take the time interval $t_2 - t_1$ to be large compared with $1/\omega$. Transforming the variable of integration to $t' = t + kx/\omega$ is equivalent to setting $x = 0$ in Eq. (3.37). The integrals of $\sin^2 \omega t$ and $\cos^2 \omega t$ over many periods are approximately equal. Because $\sin^2 \omega t + \cos^2 \omega t = 1$ the sum of these two integrals is 1. Thus the integral Eq. (3.37) is approximately $1/2$ and the time-averaged energy flux is

$$\frac{1}{2} T \omega k a^2. \tag{3.38}$$

Because the wave equation is linear, we can deal with complex representations of real wave functions (see Sec. 2.5). That is, the wave function in Eq. (3.33) is the real part of the complex wave function

$$y_c = a \exp(ikx - i\omega t). \tag{3.39}$$

The time-averaged energy flux associated with this wave is therefore

$$\frac{1}{2} T \omega k \, |y_c|^2 \,. \tag{3.40}$$

3.4   Superposition and Interference

As we noted previously, because of the linearity of the wave equation, the sum (superposition) of solutions is also a solution. Superposition of waves leads to the possibility of *interference*, by which we mean the following. Suppose that plane waves with the same frequency but differing in phase are superposed somehow. The complex representations of these waves are

$$\psi_1 = a_1 \exp(ikx - i\omega t + i\varphi_1), \ \ \psi_2 = a_2 \exp(ikx - i\omega t + i\varphi_2), \tag{3.41}$$

where a_1 and a_2 are real constants. The time-averaged power (energy flux) transmitted by the sum of these two waves is (omitting a constant factor), from Eq. (3.40),

$$a_1^2 + a_2^2 + 2a_1 a_2 \cos \Delta\varphi, \tag{3.42}$$

where $\Delta\varphi = \varphi_2 - \varphi_1$ is the phase difference between the two waves, a_1^2 is the power that would be transmitted by ψ_1 alone, and a_2^2 the power that would be transmitted by ψ_2 alone. The additional term is the *interference* term. The cosine of the phase difference (or phase shift) lies between -1 and 1, and hence the transmitted power lies between $(a_1 - a_2)^2$ and $(a_1 + a_2)^2$. For simplicity assume that the two waves have equal amplitude a; with this assumption the transmitted power lies between 0 and $4a^2$. The first possibility is often called *destructive interference* and the second *constructive interference*. Destructive and constructive interference do not exhaust the gamut of possibilities: these two terms apply to extremes (and for waves of equal amplitude at that). Depending on the phase difference, the power transmitted by two interfering waves of equal amplitude may be zero or four times the power transmitted by each separately or anything in between. This peculiar property of wave interference is not shared by particles. For example, the kinetic energy transmitted by two billiard balls is the sum of the kinetic energies transmitted by each acting separately.

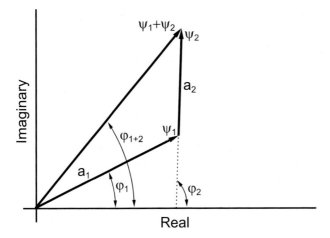

Figure 3.2: Addition of two complex numbers depicted graphically as the addition of two vectors in a plane (the complex plane).

Destructive interference should not be confused with *annihilation* of radiation, nor constructive interference with *creation* of radiation. Equation (3.42) could be interpreted (incorrectly) as implying that two beams can be superposed to create or annihilate radiant energy. But this equation applies strictly to the superposition of two plane waves, and such waves do not exist. If interference results in an increase in radiant power in some directions, this must be compensated for by a decrease in other directions. Radiation can be annihilated (absorbed) or created (emitted) only by matter.

The superposition and interference of waves can be depicted graphically as addition of vectors in the complex plane (see Sec. 2.5). For example, the two waves in Eq. (3.41), omitting the common factor $\exp(ikx - i\omega t)$, are so depicted in Fig. 3.2. A glance at this figure conveys that the resultant of two waves depends on their phase difference. Waves and other time-harmonic quantities represented by vectors in the complex plane are the *phasors* much beloved of electrical engineers (if you have to talk to them – and sometimes this is unavoidable – it helps to know their lingo). The geometrical depiction of waves as phasors helps us understand interference of waves. If the time-harmonic factor $\exp(-i\omega t)$ is included, we have a dynamical picture of a wave as a phasor rotating in the complex plane with constant angular speed ω. Having exposed some of the jargon of electrical engineers we further note that they often use the positive time-harmonic convention and with j in place of i. This is because they fear that their bread-and-butter, electric current, usually represented as i, might be confused with the square root of -1. Anyone who writes time-harmonic quantities as $\exp(j\omega t)$ is almost certain to be an electrical engineer. By their symbols and conventions ye shall know them.

An essential requirement for interference is *coherence*, a concept so fundamental that we devote a section to it. But first we consider superposition of waves with different frequencies.

3.4.1   Superposition of Waves with Different Frequencies

Two time-harmonic waves with different frequencies are separately time-harmonic but their sum is not (yet another variation on the theme that the sum of exponentials is not an exponential; see Secs. 2.1.3 and 2.3.1). Consider two plane harmonic waves on a string propagating in the positive x-direction with the same phase velocity but different frequencies and wavenumbers:

$$A_1 \cos(k_1 x - \omega_1 t), \ A_2 \cos(k_2 x - \omega_2 t). \tag{3.43}$$

For simplicity we determine the power transmitted by the sum of these two waves at $x = 0$. The result, after some manipulation, is

$$C_1 \sin^2 \omega_1 t + C_2 \sin^2 \omega_2 t + C_3 \cos(\omega_1 - \omega_2)t + C_4 \cos(\omega_1 + \omega_2)t, \tag{3.44}$$

where C_j is a constant. To derive this equation we used the identity

$$\sin x \sin y = \frac{1}{2} \cos(x - y) - \frac{1}{2} \cos(x + y). \tag{3.45}$$

Equation (3.44) is the instantaneous (not the time-averaged) power, the sum of four terms: two with the frequencies of the separate waves, one with a frequency equal to the difference of frequencies and one with a frequency equal to their sum. When we take the time average, the terms oscillating with the sum and difference of frequencies vanish (provided that the period of the lower frequency term is small compared with the averaging time). As far as time-averaged power transmission is concerned, therefore, two waves of different frequencies do not interfere: the total transmitted power is the sum of the powers transmitted by each wave as if it alone were propagating.

This result underlies the description of polychromatic light as a mixture of waves with different frequencies: the total transmitted power is the sum of transmitted powers for each component of the mixture. Denote by $\Delta\omega$ the difference in the frequencies of two superposed waves. For the time average of the third term in Eq. (3.44) to be negligible requires the averaging time to be large compared with $1/\Delta\omega$. At the frequencies of visible and near-visible light, $1/\omega$ is of order 10^{-14} s. Thus even if $\Delta\omega$ is as small as 10^{-9}, the averaging time need be greater than only about 10 μs.

3.4.2   Coherence

Interference is possible only because of *coherence*. To cohere means to stick together, usually two or more similar objects. Two waves are said to be coherent, for example, if there is a definite and fixed phase relation between them. In a sense, the waves stick together. Equation (3.42) is a coherent superposition of two waves if the phase difference $\Delta\varphi$ is fixed. Now suppose that this phase difference varies in time much more slowly than the (instantaneous) amplitude of either wave but much more rapidly than the time response of a detector. If $\Delta\varphi$ ranges at least over all values from $-\pi$ to π, the time average of $\cos(\Delta\varphi)$ is zero. The two waves are said to be *incoherent*, and the total power transmitted by them is the sum of the individual transmitted powers.

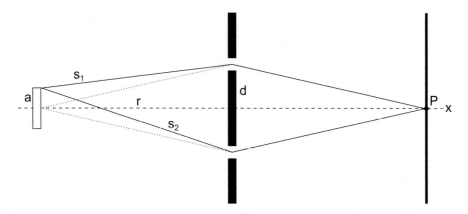

Figure 3.3: An opaque screen with two holes in it separated by a distance $2d$ illuminated by an extended source (disc) of radius a. Each point on the disc, which radiates independently of every other point, is imagined to excite two waves, one from each hole (in reality, it is the screen that radiates). These two waves combine at P, the phase difference between them determined by the path difference $s_2 - s_1$.

Coherent waves must arise from coherent sources. An array of sources is coherent if all its elements bear a fixed spatial relationship to each other. The elements may move and still be coherent, as long as they move in unison. Only a coherent array of sources can give rise to an interference pattern, by which is meant a distribution of radiant power that is not the sum of the radiant powers from each source. Two identical flashlights, for example, do not yield an interference pattern. Shine them onto the same spot on a wall and this illuminated spot is twice as bright as it would be with only one flashlight; the flashlights are mutually incoherent.

Suppose that we illuminate a coherent array of scatterers. Is this sufficient to give an interference pattern? Not unless the array is illuminated by a source of light that itself is coherent to some degree.

Sunlight is sometimes said to be incoherent. Laser light is often said to be coherent. But complete incoherence and coherence are extremes never realized in nature. Every light source is coherent to a degree, and even lasers are incoherent to a degree. One way of specifying the degree of coherence of a light source is by its *lateral coherence length*. To determine this length as simply as possible, consider the two-slit interference problem found in elementary physics textbooks (Fig. 3.3). An opaque screen with two slits (holes) in it is illuminated on one side of the screen, and an observing screen is placed on the other side. Some textbooks show another screen with a single slit or hole between the source (e.g., a light bulb) and the opaque screen with the slits, although the purpose of this mysterious single hole may not be explained. The screen with the two holes in it is a coherent array because the holes are fixed and do not move.

Consider an extended source of light, an incandescent disc, for example, of radius a centered on the axis of the system. The distance between the screen and this source is r. Every point on the source radiates independently of every other point. Waves originating from different points bear no fixed and definite phase relationship. And yet an interference pattern still is possible if the source is sufficiently small. The holes are separated by a distance $2d$ and the center of the coordinate system lies equidistant between them; the coordinates of the two holes are $(0, d, 0)$ and $(0, -d, 0)$. The source (disc) is centered on the x-axis.

Consider first the source point $(-r, 0, 0)$. In the usual textbook treatment of two-slit interference (criticized in Sec. 3.1), a wave emitted by a source point illuminates the two holes, which magically become sources of outgoing waves (despite the embarrassing fact that this is physically impossible). These two waves combine (interfere) at observation points on the observing screen. The radiant power at these points depends on the phase shift between the two waves, which in turn depends on the different paths they traversed. For simplicity take the observation point P to lie on the axis. With this assumption the difference in paths from the holes to P is zero. The distance from the source point $(-r, 0, 0)$ at the center of the disc to either hole in the screen is the same. Thus the two waves that interfere at P are exactly in phase (no phase shift).

But now consider a source point at the edge of the disc $(-r, a, 0)$. The distance from this point to the nearest hole is

$$s_1 = \sqrt{r^2 + (d-a)^2} \tag{3.46}$$

and the distance to the farthest hole is

$$s_2 = \sqrt{r^2 + (d+a)^2}. \tag{3.47}$$

If d and a are much less than r, the path difference is approximately

$$\Delta s = s_2 - s_1 \approx \frac{2ad}{r}, \tag{3.48}$$

which corresponds to a phase difference [see Eq. (3.33)]

$$\Delta \varphi = 2\pi \left(\frac{\Delta s}{\lambda} \right) = \frac{2\pi}{\lambda} \frac{2ad}{r}. \tag{3.49}$$

This phase difference, the greatest possible, is for points lying on the rim of the incandescent disc. The phase difference is zero for the source point at the center of the disc. Hence all phase differences lie between 0 and $\Delta \varphi$.

If the maximum phase difference is much less than 2π, all points on the source will give essentially the same interference pattern at the observation point. This condition will certainly be satisfied if

$$d \ll \frac{\lambda r}{2a}. \tag{3.50}$$

The solid angle Ω (see Sec. 4.1.1) subtended by the source at the holes is $\pi a^2 / r^2$, and hence Eq. (3.50) can be written

$$d \ll \frac{\lambda}{\sqrt{\Omega}}. \tag{3.51}$$

The quantity $\lambda/\sqrt{\Omega}$, called the *lateral coherence length*, is a property of the source. To determine if illumination of a coherent array can yield an interference pattern we need to know the linear dimensions of the array relative to the coherence length. The condition in Eq. (3.51) is a bit strong. We still can get an interference pattern even if d is only less than the coherence length. The mystery of the screen with a single hole placed between an extended source and a screen with two slits now disappears. The function of this first hole is to reduce the angular spread of the light illuminating the slits. Nowadays the two slits are likely to be illuminated by lasers, but before they existed interference patterns were possible using ordinary incandescent light bulbs illuminating a small hole to provide a source with a lateral coherence length larger than the slit spacing.

Note that all points of the source are mutually incoherent: the wave from one point is completely uncorrelated with waves from all other points. Yet this source can still yield interference patterns. Waves from different points do not interfere with each other but rather with themselves (by way of the secondary waves they excite in the slit screen), and if the coherence length is large compared with the slit spacing, each of these separate interference patterns is approximately the same.

At visible wavelengths the coherence length of sunlight is about 50 μm, and hence sunlight can give rise to interference patterns. One example is the *corona* (Sec. 8.4.1). The coherent array that produces this pattern is a single cloud droplet, with dimensions of order 10 μm, less than the coherence length of sunlight.

Although a single cloud droplet can produce an interference pattern, an array of droplets cannot. That is, waves from different droplets do not interfere. A typical number density of droplets in a cloud is 200 cm^{-3}, which corresponds to an average distance between droplets of about 2000 μm, much greater than the coherence length of sunlight. Even if cloud droplets were fixed in space (or at least fixed relative to each other), the waves from different droplets would not interfere, and the power scattered by N droplets still would be N times the power scattered by one.

A striking demonstration of how the coherence properties of an array of scatterers and that of the light illuminating them combine to yield different patterns can be done using a laser with a beam spreader.

An ordinary sheet of white paper in sunlight appears uniformly bright, but that same paper illuminated by a laser exhibits a speckle pattern (*laser speckle*), a mottled pattern of dark and bright regions. The paper is a coherent array: all the fibers in it are fixed in place. And the coherence length of the laser is huge compared with the separation between fibers. Now illuminate a glass of milk with this same laser. The speckle pattern disappears: the milk is uniformly bright. Milk is a suspension of fat globules, the average separation between which is much smaller than the lateral coherence length of the laser. But the globules are in constant motion, and hence do not form a coherent array. Trying to obtain an interference pattern with a suspension of jiggling fat globules would be like trying to obtain a two-slit interference pattern with slits moving rapidly and randomly toward and away from each other.

3.4.3 Distinction between a Theory and an Equation

We criticized the simple theory of the two-slit interference pattern which is based on the physically incorrect assumption that empty space (slits) is a source of waves. Why, then, does this theory give correct results (sometimes)? Consider first an opaque screen with no slits and illuminated on one side. No light is transmitted. Why? If you accept the superposition principle for electromagnetic waves, you cannot believe that the incident wave is destroyed. It exists everywhere just as it did without the screen in place. But the screen gives rise to secondary waves excited by the primary (incident) wave, and the superposition (interference) of all these waves is what is observed. With the screen in place, interference is destructive everywhere behind it: no *net* wave is transmitted. If it bothers you that the incident wave still exists in the space on the dark side of the screen, consider a standard problem in electrostatics. A conducting shell is placed in an external electric field. Inside the shell the total field is zero (electric shielding). Why? The external field induces a charge distribution on the outer surface of the shell (with zero total charge), within which the total field is the sum of the field of this charge distribution *and* the external field. These two fields are equal and opposite inside the shell so their sum is zero. But no one, to our knowledge, claims that the external field ceases to exist within the shell. To do so would contradict the superposition principle. What is true for electrostatics must also be true for electrodynamics.

An important distinction here (and throughout science), but not made often enough, is that between a theory and an equation. They are not the same. A theory of the two-slit interference pattern based on the literal assumption that empty space is the source of waves is incorrect, but the resulting equation (sometimes) is correct. Who cares as long as "you get the right answer"? Sometimes you *don't* get the right answer, and if your theory is wrong and you don't know why you won't be able to fix it.

Suppose that we remove material from the screen by punching two holes in it. If the interference pattern without the holes results in destructive interference everywhere on the dark side of the screen because of secondary waves excited in the screen material, removing a bit of this material ought to change the pattern. And this is what happens: light is observed at points on the dark side of the screen where previously there was none. To good approximation, it is *as if* the holes themselves were the mathematical sources of waves, whereas their physical source is the entire screen with two holes in it. This approximation cannot be universally valid because it is insensitive to the screen thickness and material and polarization state (Ch. 7) of the incident light. A diffraction grating is, in essence, a periodic array of many slits, and it has been long known that the elementary theory of the grating, which is fundamentally no different from two-slit theory, fails to account for what are often called Rayleigh–Wood anomalies.

In the photon language our description of illumination of an opaque screen is quite different. We would say that incident *photons* do not exist on the dark side of the screen, which does not contradict the assertion that the incident *wave* does. Where confusion may arise is when we ask if there is any incident *light* on the dark side. The answer depends on the language we choose, and hence what we mean by light. And this answer becomes garbled when we speak and think in both languages simultaneously (possibly without realizing it). If it seems enigmatic that an incident wave *can* penetrate solid matter, it is just as enigmatic that incident photons *cannot*. The size of a nucleus, where most of an atom's mass is concentrated, is smaller than the atom by at least a factor of 10^{-4}, which in turn means that the (volume) fraction of an atom occupied by mass is less than 10^{-12}. Despite its apparent solidity, matter is almost entirely empty space very sparsely populated by charges (electrons and protons). Thus it is not difficult to understand – or at least accept – why waves can penetrate such nearly empty space but not why photons cannot.

3.4.4 Scalar Waves in Three Dimensions

To proceed further we must consider waves in more than one dimension. A generalization of the one-dimensional wave equation [Eq. (3.16)] is

$$\nabla^2 \Psi = \frac{1}{v^2} \frac{\partial^2 \Psi}{\partial t^2}, \tag{3.52}$$

where Ψ is some (unspecified) physical quantity of interest. Equation (3.52) is the governing equation of acoustic waves in a fluid, for example.

The simplest three-dimensional solution to this equation is a plane harmonic wave

$$\Psi = a \exp(i\mathbf{k} \cdot \mathbf{x} - i\omega t), \tag{3.53}$$

where the *wavevector* $\mathbf{k}$ is arbitrary subject to the requirement that its magnitude k satisfy $\omega^2/k^2 = v^2$. That Eq. (3.53) is a solution to Eq. (3.52) can be verified by direct substitution. The surfaces of constant phase for the wave in Eq. (3.53), specified by

$$\mathbf{k} \cdot \mathbf{x} - \omega t = k_x x + k_y y + k_z z - \omega t = \text{const.}, \tag{3.54}$$

are planes to which the wavevector is perpendicular. Thus we can visualize Eq. (3.54) as planes of constant phase propagating in the direction $\mathbf{k}$ with phase speed v.

Any time-harmonic function $\Psi(\mathbf{x}, t) = a(\mathbf{x}) \exp(-i\omega t)$ is a solution to Eq. (3.52) provided that

$$\nabla^2 a + k^2 a = 0. \tag{3.55}$$

If, for example, a varies with the radial coordinate r only, Eq. (3.55) becomes

$$\frac{1}{r^2} \frac{d}{dr} \left(r^2 \frac{da}{dr} \right) + k^2 a = 0. \tag{3.56}$$

Given Eq. (3.56), the function $u = ar$ satisfies

$$\frac{d^2 u}{dr^2} + k^2 u = 0, \tag{3.57}$$

solutions to which are

$$u = A \exp(\pm ikr), \tag{3.58}$$

and hence

$$\Psi = \frac{A}{r} \exp(\pm ikr - i\omega t) \tag{3.59}$$

is a solution to Eq. (3.52). Equation (3.59) describes *spherical waves*, outgoing (from the origin of coordinates) for the plus sign, incoming for the minus sign. The surfaces of constant phase are spheres, as are the surfaces of constant amplitude ($A/r = \text{const.}$).

Plane and spherical waves do not, of course, exhaust all the possibilities, but they are the simplest waves, and hence the starting point for understanding more complicated waves. Both of these types of wave considered here are perfectly coherent in the following sense. The coherence time of a single wave is the greatest time interval for which the phase difference is more or less constant; the coherence length along the direction of propagation is the greatest space interval for which the phase difference is more or less constant. The coherence length is the product of the phase speed of the wave and the coherence time. Thus Eqs. (3.53) and (3.59) describe waves with infinite coherence times, which means that such waves are idealizations never realized in nature. Even a laser beam does not have an infinite coherence time. To do so it would have to have been turned on at the beginning of time and shine until the end of time. So by an infinite coherence time is meant a time much greater than the period (inverse frequency) of the wave.

3.4.5   Acoustic Waves

We noted that Eq. (3.52) is the governing equation for acoustic waves. Waves on a string can be looked upon as one-dimensional acoustic waves in that they are governed by this equation in one dimension. And waves on the strings of musical instruments excite three-dimensional acoustic waves in the air surrounding them.

Although our primary interest is (vector) electromagnetic waves, acoustic waves in fluids are scalar waves and hence simpler. For this reason we often draw analogies between acoustic and electromagnetic waves. Although the two are similar, they are also different, most notably in the way they usually are detected, including by humans. Detectors of light, such as our eyes and photomultiplier tubes, are power detectors: the detected signal is the time-averaged power because of the very short period (inverse frequency) of light waves relative to the detector response. But the *instantaneous amplitude* of sound waves is detected by the human ear.

Suppose that two time-harmonic waves of different frequency but equal amplitude are superposed:

$$a \cos \omega_1 t + a \cos \omega_2 t. \tag{3.60}$$

By using the identities

$$\cos^2(x/2) = \frac{1 + \cos x}{2}, \quad \sin^2(x/2) = \frac{1 - \cos x}{2}, \tag{3.61}$$

$$\cos\left(\frac{x \pm y}{2}\right) = \cos\frac{x}{2}\cos\frac{y}{2} \mp \sin\frac{x}{2}\sin\frac{y}{2}, \tag{3.62}$$

we can write Eq. (3.60) as

$$2a \cos\left\{\frac{(\omega_1 - \omega_2)t}{2}\right\} \cos\left\{\frac{(\omega_1 + \omega_2)t}{2}\right\}. \tag{3.63}$$

If the two frequencies are relatively close to each other we may interpret this equation as describing a wave of frequency equal to the average of the two frequencies, called the *carrier frequency*, with amplitude varying (modulated) at a frequency equal to half the difference of the two frequencies. This frequency difference is called the *beat frequency*. Two harmonic sound waves of different frequency when superposed give rise to the phenomenon of *beats*. And the human ear often can hear these beats.

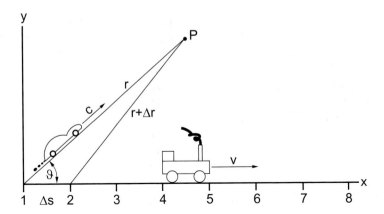

Figure 3.4: Letters are thrown from a train with constant speed v to the drivers of cars waiting at equally spaced stations. These cars carry the letters at constant speed c to a distant point P. In general, the frequency at which letters are sent is not the same as the frequency at which they are received.

But beats are not the sole province of acoustics. In radio engineering producing beats is called *heterodyning*, from the Greek *hetero* meaning different and *dynamis* meaning power. At optical frequencies, mixing of waves of two different frequencies to produce a measurable beat frequency is called *optical beating*, *optical mixing*, or *optical heterodyning*. This has atmospheric applications, most notably the measurement of wind velocities by exploiting the *Doppler effect*, which we turn to next.

3.4.6 The Doppler Effect

The Doppler effect, briefly stated, is the difference between the frequency of transmission of something – call it a signal – and the frequency of its reception because of relative motion between the source and receiver of the signal. There is nothing inherent in the Doppler effect that requires the signal to be transmitted by an acoustic, electromagnetic, or any other kind of wave. To show this, consider the following example inspired by Thomas P. Gill's excellent book, *The Doppler Effect*: a train carrying letters is moving with constant speed v along a straight section of track (Fig. 3.4). Stations are spaced at regular intervals Δs. At each station a car is sitting with its engine running. As the train passes a station, a letter is thrown to the driver of the car, who then drives at constant speed c to a distant point P where the letter is delivered. The time between sending letters is constant, as is the time between receiving them. But the sender of the letters and their recipient at P disagree about these time intervals.

The time t_1' at which the first letter is received is greater than the time t_1 at which it was sent because of the travel time over the distance r between the first station and P:

$$t_1' = t_1 + \frac{r}{c}. \tag{3.64}$$

And the time t_2' at which the second letter is received is greater than the time t_2 at which it was sent:

$$t_2' = t_2 + \frac{r + \Delta r}{c}, \tag{3.65}$$

where $r + \Delta r$ is the distance between the second station and P. The time interval between sending the two letters is Δt, that between receiving them is $\Delta t'$:

$$\Delta t = t_2 - t_1, \ \Delta t' = t_2' - t_1'. \tag{3.66}$$

Combine Eqs. (3.64)–(3.66) to obtain

$$\Delta t' = \Delta t + \frac{\Delta r}{c}. \tag{3.67}$$

This equation is the Doppler effect in a nutshell: $\Delta t' \neq \Delta t$ for a finite speed of transmission.

Let the origin of the xy-coordinate system be at the first station. The coordinates of P are $(r \cos \vartheta, r \sin \vartheta)$. We can find Δr from

$$r + \Delta r = \sqrt{(r \cos \vartheta - \Delta s)^2 + (r \sin \vartheta)^2} = r \sqrt{1 - \frac{2 \Delta s \cos \vartheta}{r} + \left(\frac{\Delta s}{r}\right)^2}. \tag{3.68}$$

If the separation between stations is small compared with r, we can expand the right side of Eq. (3.68) in a power series and retain only the first two terms:

$$\Delta r \approx -\Delta s \cos \vartheta + \frac{1}{2} \frac{(\Delta s)^2}{r}. \tag{3.69}$$

The separation between stations is

$$\Delta s = v \Delta t. \tag{3.70}$$

Equations (3.67), (3.69), and (3.70) yield

$$\frac{\Delta t'}{\Delta t} = 1 - \frac{v}{c} \cos \vartheta + \frac{1}{2} \frac{v}{c} \frac{\Delta s}{r}. \tag{3.71}$$

The limit of this quotient as Δs approaches zero (zero separation between stations) is

$$\lim_{\Delta s \to 0} \frac{\Delta t'}{\Delta t} = 1 - \frac{v \cos \vartheta}{c}. \tag{3.72}$$

The frequency of transmission of letters is $\nu = 1/\Delta t$; their frequency of reception is $\nu' = 1/\Delta t'$. Thus the ratio of frequencies (in the limit of zero separation between stations) is

$$\frac{\nu'}{\nu} = \left(1 - \frac{v \cos \vartheta}{c}\right)^{-1}. \tag{3.73}$$

If $v \ll c$ (slow train, fast car), this ratio is approximately

$$\frac{\nu'}{\nu} \approx 1 + \frac{v \cos \vartheta}{c}. \tag{3.74}$$

Note that $\cos \vartheta$ can be positive or negative. When it is positive the transmitter (source) is approaching the receiver (observer) and the frequency of reception ν' is *greater* than the frequency of transmission ν. This frequency shift is sometimes called a *blue shift* because the frequency of visible light would be shifted upward (toward the blue) when the distance between the transmitter and receiver is decreasing. When $\cos \vartheta$ is negative, the transmitter is receding from the receiver and the frequency of reception is *less* than the frequency of transmission. This frequency shift is sometimes called a *red shift* because the frequency of visible light would be shifted downward (toward the red). But a blue shift has nothing to do with the blue sky, nor a red shift with red sunsets. The terms blue shift and red shift are used even for signals propagated by waves that do not evoke sensations of color (e.g., acoustic waves).

For electromagnetic waves c in Eq. (3.74) is the speed of light. This equation is not relativistically correct, but the correction is small if $v \ll c$, which it is in all the atmospheric applications we know about. According to Eq. (3.74) there is no frequency shift of electromagnetic radiation when $\cos \vartheta = 0$. But according to relativity theory there is a shift (*transverse* Doppler shift), although appreciably smaller than the *longitudinal* Doppler shift.

We have to be careful when applying Eq. (3.74) to acoustic waves because, unlike electromagnetic waves, they propagate relative to a medium (e.g., air) which itself can be moving relative to transmitter and receiver.

For electromagnetic waves the relative frequency shift from Eq. (3.74) is

$$\frac{\nu' - \nu}{\nu} \approx \frac{v \cos \vartheta}{c}, \tag{3.75}$$

where $v \cos \vartheta$ is the component of the velocity of the receiver relative to the transmitter along the line joining them, positive if the two are approaching, negative if they are receding.

Now consider an air molecule illuminated by a monochromatic beam. The frequency of the beam in a coordinate system stationary with respect to the source is ν. But the molecule is, in general, moving relative to the source. As far as the molecule is concerned it is excited by, and hence scatters, radiation with frequency ν' given by

$$\nu' \approx \nu\left(1 + \frac{v_x}{c}\right), \tag{3.76}$$

where for conciseness we write $v_x = v \cos \vartheta$. But at the source an observer detects scattered radiation of frequency

$$\nu'' \approx \nu'\left(1 + \frac{v_x}{c}\right) \approx \nu\left(1 + \frac{v_x}{c}\right)^2 \approx \nu\left(1 + \frac{2v_x}{c}\right). \tag{3.77}$$

Molecular velocity components in any direction are distributed according to the Maxwell–Boltzmann distribution

$$p(v_x) = \sqrt{\frac{m}{2\pi k_B T}} \exp\left(-\frac{mv_x^2}{2k_B T}\right), \tag{3.78}$$

a member of the same family as the Maxwell–Boltzmann distribution for molecular kinetic energies (Sec. 1.2); m is the molecular mass, k_B Boltzmann's constant, and T absolute temperature. As with all continuous probability distributions the integral of p over any interval is the probability of a molecule having a velocity component in that interval. From Eq. (3.77) we obtain the *relative* frequency shift

$$\delta = \frac{\nu'' - \nu}{\nu} \approx \frac{2v_x}{c}. \tag{3.79}$$

Knowing the distribution function for v_x we can find that for δ:

$$p(\delta) = p(v_x)\frac{dv_x}{d\delta} = \sqrt{\frac{mc^2}{8\pi k_\mathrm{B} T}} \exp\left(-\frac{mc^2\delta^2}{8k_\mathrm{B} T}\right). \tag{3.80}$$

This probability distribution is peaked at and symmetric about $\delta = 0$. Values of δ for which the probability is one-half the peak are obtained from Eq. (3.80):

$$\delta = \pm 2\sqrt{\ln 2}\,\frac{v_\mathrm{m}}{c}, \tag{3.81}$$

where the *most probable molecular speed* is

$$v_\mathrm{m} = \sqrt{\frac{2k_\mathrm{B} T}{m}}. \tag{3.82}$$

We define the width Δ of the probability distribution for scattered frequencies (observed at the transmitter) as the range of relative frequency shifts for which the probability is at least one-half the peak:

$$\Delta = 4\sqrt{\ln 2}\,\frac{v_\mathrm{m}}{c}. \tag{3.83}$$

At normal terrestrial temperatures the most probable speed of air molecules is about $10^3\ \mathrm{m\,s}^{-1}$ whereas c is close to $3 \times 10^8\ \mathrm{m\,s}^{-1}$, and hence the relative Doppler shift, upward and downward, upon scattering of monochromatic radiation by moving molecules is of order 10^{-5}.

Doppler radar and lidar are used for measuring wind velocities (strictly, wind velocity components along the line from the transmitter-receiver to the point of observation). For example, particles carried by the wind scatter radiation Doppler shifted relative to the source. Wind speeds are of order $10\ \mathrm{m\,s}^{-1}$, so the relative Doppler shift is around 10^{-7}. This is not large but it can be measured. Indeed, frequency is the physical quantity that can be measured most precisely, about 1 part in 10^9.

Although we cannot see the consequences of the Doppler shift for light interacting with moving matter, we can hear the consequences of the Doppler shift of sound waves. The speed of sound in air is around $300\ \mathrm{m\,s}^{-1}$, and hence a *pure tone* from an object moving relative to an observer at $30\ \mathrm{m\,s}^{-1}$ will be Doppler shifted by about 10%, which can be detected by a human observer who is not tone deaf.

3.4.7   Interference of Waves with Different Directions

We showed that waves of different frequency do not interfere in the sense that the time-averaged power transmitted by them is the sum of the powers transmitted by each wave separately. What about waves with the same frequency but propagating in different directions?

You sometimes encounter assertions that waves propagating in different directions cannot or do not interfere. There are (at least) two interpretations of this assertion. One is unobjectionable. Suppose two beams are crossed. Outside the region of overlap each beam is exactly as it would be if it alone were propagating. Thus in this sense the two beams may be said to not interfere. This property of waves is not shared by two streams of automobiles. If they cross each other, collisions irreversibly change the automobiles. But this use of interfere is more akin to its meaning of bother, disturb, or molest: "Please don't interfere with my plans." In optics, interfere has a more restricted meaning, namely, waves interfere if the time-average power transmitted by their superposition is *not* the sum of the powers transmitted by each separately. Let us consider in this sense interference of two plane waves with the same frequency and amplitude but different directions. Superposition of two such waves yields

$$a \exp(i\mathbf{k}_1 \cdot \mathbf{x} - i\omega t) + a \exp(i\mathbf{k}_2 \cdot \mathbf{x} - i\omega t + i\Phi). \tag{3.84}$$

Given the phase difference

$$\Delta\varphi + \Phi = (\mathbf{k}_2 - \mathbf{k}_1) \cdot \mathbf{x} + \Phi \tag{3.85}$$

between these two waves, where Φ is a constant, what are we to make of assertions that waves with the same frequency propagating in different directions do not interfere?

To answer this choose $\mathbf{k}_1 = k\mathbf{e}_z$ and $\mathbf{k}_2 = \mathbf{k}_1 + \Delta\mathbf{k}$. These two directions determine a plane, which we take to be the yz-plane. Because the magnitude of both wavevectors is the same we have

$$\Delta\mathbf{k} = k\sin\vartheta\,\mathbf{e}_y + k(\cos\vartheta - 1)\mathbf{e}_z, \tag{3.86}$$

where ϑ is the angle between the two wavevectors. The phase shift $\Delta\varphi$ is therefore

$$\Delta\varphi = ky\sin\vartheta + kz(\cos\vartheta - 1). \tag{3.87}$$

Suppose that the detector is a disc centered at the origin and lying in the xy-plane. The power per unit area at each point of the detector is therefore [see Eq. (3.42)]

$$2a^2\{1 + \cos(\Delta\varphi + \Phi)\} = 2a^2\{1 + \cos\Delta\varphi\cos\Phi - \sin\Delta\varphi\sin\Phi\}, \tag{3.88}$$

where

$$\Delta\varphi = 2\pi y\sin\vartheta/\lambda, \tag{3.89}$$

and y lies between $-R$ and R, the radius of the detector. To simplify further, assume that $\vartheta \ll 1$ (i.e., the waves are almost in the same direction), which allows us to approximate $\sin\vartheta$ as ϑ. With this approximation Eq. (3.88) becomes

$$2a^2\{1 + \cos(2\pi y\vartheta/\lambda)\cos\Phi - \sin(2\pi y\vartheta/\lambda)\sin\Phi\}. \tag{3.90}$$

The cosine term oscillates between positive and negative values, its argument lying between $-2\pi R\vartheta/\lambda$ and $2\pi R\vartheta/\lambda$. The integral of the sine term over the circular disc vanishes. If the radius of the detector is such that the range of $2\pi y\vartheta/\lambda$ is much greater than 2π, that is, if

$$R \gg \frac{\lambda}{2\vartheta}, \tag{3.91}$$

then if Eq. (3.90) is integrated over the detector, all the positive oscillations of the cosine almost cancel all the negative oscillations, leaving the result that the total power received by the detector is $2a^2$, the sum of the powers of the two waves acting separately.

The condition Eq. (3.91) is likely to be satisfied at visible wavelengths. Suppose, for example, that the waves diverge by one degree. When superposed they will not yield an interference pattern if the radius of the detector is appreciably greater than 10 µm. But we can think of two uncontrived examples in which Eq. (3.91) is *not* satisfied. As we explain in the following subsection, the angular pattern of scattering by a particle is an interference pattern. A particle may be looked upon as a coherent array of dipoles excited to radiate waves by an incident beam. What is measured in a particular direction by a detector is the superposition of all these waves. Although they propagate in (slightly) different directions they still interfere. In typical laboratory experiments the detector may be at a distance of 0.5 m from a particle in the sample. Suppose that the linear size of a particle is 10 µm or less. The angular spread of the waves from all the dipoles making up the particle is therefore around 10^{-5} rad. This yields a value for $\lambda/2\vartheta$ of at least 1 cm, which is comparable with the linear dimensions of the detector. That is, for a detector at 0.5 m to resolve angular differences of about a degree the detector must be around 1 cm in diameter.

Now consider two identical particles. One may be looked upon as a detector of radiation from the other (and vice versa, of course). Suppose the particles are cloud droplets, which have typical dimensions of 10 µm. The average separation of cloud droplets is around 2000 µm, which corresponds to an angular spread of dipolar waves from one particle at the other particle of about 5×10^{-3} rad. At visible wavelengths $\lambda/2\vartheta$ is therefore around 50 µm. But the dimensions of the detector (the other particle) are smaller than this. Thus as far as this particle in its role as a detector is concerned, it receives waves from different directions that do indeed interfere. Waves from dipolar oscillators making up particles are not plane waves, closer to being spherical waves, which at sufficiently large distances can be considered to be locally planar (i.e., plane over a limited area). Nor are they all of equal amplitude. But these departures from the ideal do not change our general conclusion that waves in different directions can interfere.

3.4.8   Phase Shift on Scattering

Consider two identical dipoles (Fig. 3.5) illuminated by a monochromatic (scalar) plane wave, or by a source with a lateral coherence length much greater than the separation between them, with wavenumber $k = 2\pi/\lambda$. Assume that they are excited mostly by the incident wave, mutual excitation being negligible. Because the two dipoles are excited by the same incident wave, the waves scattered by them bear a fixed phase difference that depends on the separation between them, the wavelength of the illumination, and the direction of observation. Denote by e_i the direction of the incident wave, by e_{s1} and e_{s2} the directions of the scattered waves (directions toward the point of observation O), and by r_{12} the position of dipole 2 relative to dipole 1. Although the scattered waves are more or less spherical waves, we take the point of observation to be sufficiently far away that they can be considered to be locally plane waves. We consider scalar waves even though light is a vector wave because we are interested only in the phase difference. If the complex amplitude of the incident wave at 1 is a, at 2 it is $a\exp(ik e_i \cdot r_{12})$. Because the amplitude of each scattered wave is proportional to the

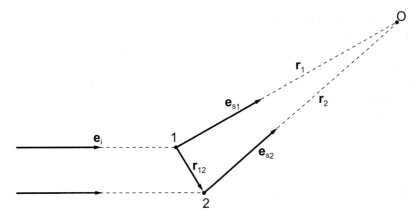

Figure 3.5: At O the total scattered field is the superposition of fields scattered by the two dipoles a fixed distance $\mathbf{r}_{12}$ apart. The phase difference between these two fields depends on the angle between the incident and scattered directions and the distance between the dipoles relative to the wavelength of the illumination.

amplitude of the wave that excites it, the two scattered waves at O are

$$\psi_1 = a\exp(ik\mathbf{e}_{s1}\cdot\mathbf{r}_1), \quad \psi_2 = a\exp(ik\mathbf{e}_i\cdot\mathbf{r}_{12} + ik\mathbf{e}_{s2}\cdot\mathbf{r}_1 - ik\mathbf{e}_{s2}\cdot\mathbf{r}_{12}), \tag{3.92}$$

where

$$\mathbf{r}_2 = \mathbf{r}_1 - \mathbf{r}_{12}. \tag{3.93}$$

Any factors common to the two waves are omitted. As the distance to the observation point O increases indefinitely, $\mathbf{e}_{s2} \rightarrow \mathbf{e}_{s1} = \mathbf{e}_s$, and the phase difference between the two waves approaches

$$\Delta\varphi = k(\mathbf{e}_i - \mathbf{e}_s)\cdot\mathbf{r}_{12} = \frac{2\pi}{\lambda}(\mathbf{e}_i - \mathbf{e}_s)\cdot\mathbf{r}_{12}. \tag{3.94}$$

From this simple equation flows an amazing amount of physical understanding.

The *forward direction* is the direction of the incident wave. According to Eq. (3.94), scattering in the forward direction ($\mathbf{e}_s = \mathbf{e}_i$) is always in phase ($\Delta\varphi = 0$) regardless of the separation of the two dipoles and the wavelength. This is the only scattering direction for which this is true, and hence the forward direction is singular, a point to which we return.

Any particle is a coherent array of N dipoles. Again, we assume that they are excited mostly by the incident wave. If the linear extent of this array is small compared with the wavelength, all the separate scattered waves are approximately in phase for all directions [see Eq. (3.42) with $\Delta\varphi = 0$ and $a_1 = a_2$], and hence the power scattered in any direction by N scatterers is N^2 times the power scattered by one. If the total volume of the particle is v and there are n dipoles per unit volume, the total number of dipoles is proportional to v. We therefore predict that the total power scattered by a particle small compared with the wavelength is proportional to v^2. As we show in Section 3.5, this turns out to be correct.

Interactions between dipoles are not negligible unless separated by sufficiently large distances. Although this condition is not satisfied by a small particle, scattering still increases as the square of its volume even when interactions are accounted for (by interactions is meant that the wave from each dipole appreciably excites its neighbors). The reason for this is that if a particle is sufficiently small, appreciable phase differences are not possible even with interactions. Hence the array of dipoles (particle) is approximately coherent and in-phase even though the dipoles excite each other.

Let us continue in this vein and see what general results we can obtain by physical reasoning based on Eq. (3.94). As the overall size of the particle increases, the phase differences between the waves scattered by its constituent dipoles also increase. When the particle is sufficiently small, all these waves are approximately in phase (constructive interference) for all scattering directions. But as the particle size increases, phase differences of order π or greater become possible for some directions. In particular, the phase difference between two or more dipolar waves can be such that in a particular direction they interfere destructively. Thus we conclude that although scattering by a particle increases as the square of its volume when the particle is small compared with the wavelength, this rapid increase cannot be sustained indefinitely. As particle size increases, scattering increases, but more slowly than volume squared. And again, this expectation is supported by detailed calculations (see Sec. 3.5.1). We can understand them, at least qualitatively, by simple phase difference arguments.

We show in Section 3.2 that scattering by a dipole (subject to restrictions) is approximately inversely proportional to the fourth power of the wavelength of the incident illumination. But this does not necessarily imply that scattering by a coherent array of such dipoles (a particle) also follows this wavelength dependence. As evidenced by Eq. (3.94), an additional wavelength dependence creeps in by way of the phase difference. Indeed, for sufficiently large particles, the net effect of the wavelength dependence of scattering by the individual dipoles nearly cancels the wavelength dependence originating from phase differences. In the following section we give a simple example to elaborate on this point.

Now let us turn to the directional dependence of scattering. Suppose that our two dipoles individually scatter the same in all directions. This does not then imply that scattering by the two together is the same in all directions. According to Eq. (3.94) the phase difference between the two scattered waves depends on the scattering direction. We can make this clearer by considering the special example of two dipoles on a line parallel to the incident wave ($\mathbf{r}_{12} = r\mathbf{e}_{\mathrm{i}}$), in which instance the phase shift (again, interactions are neglected) is

$$\Delta\varphi = \frac{2\pi r}{\lambda}(1 - \cos\vartheta), \tag{3.95}$$

where ϑ is the *scattering angle* (angle between incident and scattered waves). The power scattered in any direction by the two dipoles is determined by $\cos\Delta\varphi$. The quantity $1 - \cos\vartheta$ lies between 0 (forward direction) and 2 (backward direction), and hence the phase difference lies between 0 (forward) and $4\pi r/\lambda$ (backward). When the (backward) phase difference is an odd multiple of π, interference is destructive; when the phase difference is an even multiple of π, interference is constructive; and, of course, everything between these two extremes is possible. As r/λ increases so does the number of oscillations in the scattering diagram (scattering as a function of angle).

As a particle increases in size so does the average distance between its elements. We therefore expect, on the basis of the behavior of two dipoles, the scattering diagram for a particle to exhibit more maxima (minima) the greater its size relative to the wavelength. This expectation is borne out by detailed calculations (Sec. 3.5.1) as well as by measurements.

Because scattering is in-phase in the forward direction regardless of the wavelength and the separation between dipoles, we expect scattering in this direction to increase more rapidly with size than scattering in any other direction. Again, this expectation is borne out by calculations (Sec. 3.5.1) and measurements. A general result is that the larger the particle, the more that scattering by it is peaked in the forward direction.

The rate of change of the phase difference with respect to r

$$\frac{\partial \Delta \varphi}{\partial r} = \frac{2\pi}{\lambda}(1 - \cos \vartheta) \tag{3.96}$$

ranges from 0 in the forward direction to $4\pi/\lambda$ in the backward direction. The implication of this is that if we have a fixed number of dipoles (fixed particle volume) and we move them around (change the particle shape), scattering is least affected in the forward direction and most affected in the backward direction. Again, this is supported by calculations and measurements.

Much of the basic physics of scattering by particles is embodied in the simple phase difference given in Eq. (3.94). The rest is details, some of which we give in Section 3.5, but first we discuss scattering by air and by water.

3.4.9 Scattering by Air and Liquid Water Molecules

Air molecules near sea level are separated on average by about 3 nm ($\sim$10 molecular diameters). That is, the average volume allocated to each of N molecules in a volume V is V/N, the cube root of which is defined as the average separation. The separation r_{ij} between the ith and j^{th} molecules, however, is distributed statistically from some minimum (approximately a molecular diameter) to some maximum (approximately the cube root of V). Because the lateral coherence length of sunlight is around 50 μm, we have to add the waves, taking due account of phases, scattered by all the air molecules (about 10^9) in a volume with approximately this linear dimension, which is about 100 times the wavelengths of visible light.

Scattering by Spatially Uncorrelated Molecules

Suppose that N time-harmonic waves, all with the same amplitude a but different phases, are superposed. For sake of argument we take them to be plane waves, and omit the common factor $\exp(ikx - i\omega t)$. The sum of these waves is

$$\Psi = \sum_{j=1}^{N} a \exp(i\varphi_j). \tag{3.97}$$

Because these are time-harmonic waves, the time-averaged power transmitted by their sum is proportional to

$$\Psi\Psi^* = |\Psi|^2 = a^2 \sum_{j=1}^{N} \sum_{m=1}^{N} \exp\{i(\varphi_j-\varphi_m)\} = a^2N + a^2 \sum_{j=1}^{N} \sum_{m\neq j}^{N} \exp\{i(\varphi_j-\varphi_m)\}. \quad (3.98)$$

The sum in the rightmost side of this equation is the addition of phasors (vectors in the complex plane) of unit length but with different polar angles (phase differences $\Delta\varphi_{jk} = \varphi_j - \varphi_k$). We take Eq. (3.98) to be the total light scattered by the N air molecules in a volume equal to the coherence length cubed. The phase differences are given by Eq. (3.94), and hence Eq. (3.98) becomes

$$|\Psi|^2 = a^2N + a^2 \sum_{j=1}^{N} \sum_{m\neq j}^{N} \exp\{ik(\mathbf{e}_i - \mathbf{e}_s) \cdot \mathbf{r}_{jm}\}. \quad (3.99)$$

Air molecules move and their positions are almost completely uncorrelated. That is, a molecule can occupy almost any position in space regardless of the positions of other molecules because the volume accessible to a molecule is about 1000 times its volume. This in turn implies that all phase differences are equally probable, and hence the sum in Eq. (3.99) is approximately zero, which is readily evident if you draw a great many phasors with random directions emanating from a point, similar to the spokes in a bicycle wheel (laced carelessly). Because the positions of air molecules are uncorrelated, scattering by N (in a small volume) is N times scattering by one. As far as scattering by air molecules is concerned, it is *as if* phase does not exist. This is true for all scattering directions except the forward direction ($\mathbf{e}_s = \mathbf{e}_i$), for which scattering is in-phase regardless of the molecular separation relative to the wavelength. In fact, this in-phase scattering is the source of refraction (see Sec. 8.3 on atmospheric refraction and mirages). Historically, refraction has been treated separately from scattering, as if the two had nothing to do with each other. This is not true. Refraction is scattering; it can be looked upon as the coherent part of scattering. Scattered light not associated with refracted light is the incoherent part.

In 1899 Lord Rayleigh attributed the blue of the sky (see Sec. 8.1) to scattering by air molecules. Underlying his theoretical expression for the amount of light scattered by N air molecules was the assumption that the phases of the separate scattered waves are "entirely at random", and hence scattering by N air molecules is N times scattering by one. But he also recognized that "When the volume occupied by the molecules is no longer very small compared with the whole volume, the fact that two molecules cannot occupy the same space detracts from the random character of the distribution."

Scattering by air molecules is almost, but not quite, the same in all directions (isotropic), the reason for which is best left for the chapter on polarization (see Sec. 7.3; also Prob. 7.18). Angular scattering by air is also almost, but not quite, the same as that by a small (compared with the wavelength) sphere (see Fig. 3.13), small differences arising because the dominant air molecules are not spherically symmetric.

The Blue of the Sky: Scattering by Fluctuations or by Molecules?

You sometimes encounter the assertion that the blue of the sky (scattering by air molecules) is "really" a consequence of "scattering by fluctuations." This is piffle, reflecting ignorance of physics and its history, tantamount to denying the existence of molecules. The origins of this piffle go back almost 100 years to the work of Smoluchowski (1908) and Einstein (1910) who developed theories of scattering by dense media (e.g., liquids) by considering matter to be continuous but with spatially varying properties to which they applied thermodynamic arguments. Thermodynamics does not *explicitly* invoke molecules and so, of course, neither do the theories of Smoluchowski and Einstein, which, not surprisingly, contain thermodynamic variables such as temperature and isothermal compressibility. But this does not mean that Smoluchowski and Einstein believed that molecules are not the agents responsible for scattering. Einstein, in particular, recognized that his theory circumvented the difficulties of a molecular theory of scattering in fluids. He noted after his labors that "It is remarkable that our theory does not make *direct* use of the assumption of a discrete distribution of matter." But the fluctuation theories of Smoluchowski and Einstein have been distorted over the years into the fatuous notion that fluctuations, not molecules, do the scattering.

For an ideal gas, the theories of Smoluchowski and Einstein yield Rayleigh's result (which Einstein acknowledged) that scattering by N molecules is N times scattering by one, but their theories go further. In particular, they account for *critical opalescence*. At the critical point, the distinction between gas and liquid disappears and scattering greatly increases (according to theory it becomes infinite) as a fluid teeters between the liquid and gas phases. We give Einstein's scattering formula in Problem 5.16 and include a few problems that require this formula.

Many years later in 1945 Bruno Zimm tackled the problem of scattering in dense media by explicitly considering scattering by molecules. Zimm's own words demonstrate that he understood this scattering whereas the it-is-really-scattering-by-fluctuations folks do not: "the difficulty [of calculating interference between waves scattered by different molecules] was elegantly circumvented by Smoluchowski and Einstein, who considered the liquid as a continuous medium troubled by small statistical fluctuations in density. The extent of these fluctuations could be calculated from the macroscopic compressibility of the medium, and the intensity of the scattered light was obtained without discussing the individual molecules at all."

What Zimm did, in essence, was evaluate the sum in Eq. (3.99), approximating it by an integral and accounting for the correlations between molecular separations r_{ij} by using results from statistical mechanics. The first term in Eq. (3.99) is scattering by N isolated molecules, the second term a consequence of interference of waves from molecules correlated in position. Zimm's molecular theory of scattering reproduces the results of Einstein and Smoluchowski but goes a step further in that it does not yield infinite scattering at the critical point.

Now let us turn to an ordinary glass of water, absolutely free of contamination (setting aside the difficulty of preparing such water). Molecules in liquid water are separated by distances comparable with their diameter, and hence one molecule cannot move without pushing others aside. Because of this correlation in position, scattering by N water molecules is not simply N times scattering by one. Indeed, scattering per molecule in the liquid phase is appreciably less than in the gas phase. When clean water is illuminated by a monodirectional

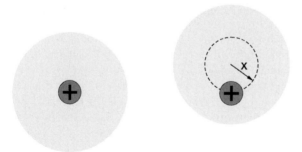

Figure 3.6: Simple model of an atom as a stationary positive nucleus (black sphere) surrounded by an equal but oppositely charged continuous (gray) electronic cloud. When the cloud is displaced a distance x, it experiences an electrostatic restoring force.

source of light, what is observed is described to good approximation by the laws of specular reflection and refraction. Although textbooks rarely say so, these are laws of scattering by coherent arrays. But there is always a residual incoherent component to the total scattering. When water is illuminated, most (but not all) of what is observed is accounted for by the laws of specular reflection and refraction. In particular, the reflected and refracted rays all lie in the plane determined by the normal to the surface of the water and the direction of the incident wave. Yet if we look in directions outside this plane we can observe scattered light, weak but measurable, and not necessarily the result of junk in the water. Water is intrinsically junky in that it is made up of discrete pieces (molecules), highly correlated pieces to be sure but not perfectly correlated.

Frequency Dependence of Scattering by Air Molecules

Light from the sky is a consequence of scattering by molecules essentially uncorrelated in position. The spectrum of this light is given approximately by Eq. (3.9), Rayleigh's scattering law. Air molecules are certainly small enough compared with the wavelengths of visible light that this condition for the validity of the law is satisfied. But there is another condition, evident from Eq. (3.8): the frequencies for which Eq. (3.9) is a good approximation must be much less than a resonant frequency of the scatterer. What is this frequency for an air molecule?

To answer this question we resorted to a crude model of an atom: a continuous (negative) charge distribution (electron cloud) surrounding a positively charged point nucleus (see Fig. 3.6). When the center of the electron cloud coincides with the nucleus, there is no net force, but when the cloud is displaced it experiences a restoring force. Let q be the charge on the nucleus. The charge density of the electron cloud is therefore $-3q/4\pi R^3$, where R is the atomic radius. When the cloud is displaced a distance x, it experiences an attractive force that pulls it back toward the center. This force is that between two point charges, the nucleus and one at the center with charge equal to the total (negative) charge within the sphere of radius x, which is $-qx^3/R^3$. From Coulomb's law we therefore have for the restoring force

$$F = \frac{q}{4\pi\varepsilon_0 x^2}\left(\frac{-qx^3}{R^3}\right) = -\frac{q^2 x}{4\pi\varepsilon_0 R^3} = Kx, \tag{3.100}$$

where ε_0 (the permittivity of free space) is a constant (8.85×10^{-12} in SI units). This equation should look familiar: it is the restoring force for a linear harmonic oscillator (Sec. 2.6). Thus the resonant frequency follows immediately from Eq. (3.76):

$$\omega_0 = \sqrt{\frac{K}{m}} = \sqrt{\frac{q^2}{4\pi\varepsilon_0 m R^3}}, \tag{3.101}$$

where we take m to be the mass of the electron cloud. If we take $R = 0.15$ nm for the atomic radius (strictly, this is the radius of molecular nitrogen), we obtain a resonant frequency that corresponds to a wavelength of about 0.1 µm, well into the ultraviolet.

Eq. (3.101) is an example of a *plasma frequency*. Plasma frequencies pop up all over the place, including in the ionosphere, for which plasma frequencies correspond to wavelengths of tens of meters. The plasma frequency is a cutoff frequency: below it a medium is reflecting, above it transmitting. This is why reflection of radio waves (for frequencies below the plasma frequency) by the ionosphere can be put to use for communication on Earth. To communicate beyond Earth would require waves above the plasma frequency. The late cosmologist, Fred Hoyle, wrote *The Black Cloud*, considered to be a classic of science fiction. Earth is menaced by an intelligent interstellar cloud that communicates with scientists on Earth by adjusting its plasma frequency.

3.5 Scattering by Particles

Although the form of Eq. (3.10) is the same regardless of the illuminated object, we use different symbols for molecular and particle scattering cross sections just as we did for absorption cross sections in order to emphasize that molecules and particles are different. Most of what we said in Section 2.8 about particle absorption cross sections also applies to particle scattering cross sections. But unlike absorption, scattering can be apportioned into directions. The *differential scattering cross section*, often written as $dC_{sca}/d\Omega$, is the contribution to the total scattering cross section C_{sca} from scattering into a unit solid angle (Sec. 4.1.1) in each direction, and hence

$$C_{sca} = \int_{4\pi} \frac{dC_{sca}}{d\Omega} \, d\Omega, \tag{3.102}$$

where integration is over all directions (4π steradians). The differential scattering cross section is *not* the derivative of C_{sca} with respect to the variable Ω, the notation being chosen merely to aid our memories.

The *scattering coefficient* β of a suspension of N identical particles per unit volume is defined similarly to its absorption coefficient:

$$\beta = N C_{sca}, \tag{3.103}$$

an expression readily generalized to a suspension of non-identical particles by adding their individual contributions to the total scattering coefficient. The sum of the absorption and scattering cross sections is called the *extinction cross section*:

$$C_{ext} = C_{abs} + C_{sca}, \tag{3.104}$$

and the sum of absorption and scattering coefficients is called the *extinction coefficient*. As with absorption cross sections, those for scattering and extinction are sometimes normalized by their geometrical (projected) cross-sectional areas G to yield dimensionless efficiencies or efficiency factors for scattering and extinction:

$$Q_{sca} = \frac{C_{sca}}{G}, \; Q_{ext} = \frac{C_{ext}}{G}, \tag{3.105}$$

neither of which has any physical significance (whereas Q_{abs} does) although they are computationally convenient. As we argued about absorption cross sections, volumetric scattering and extinction cross sections (for particles with well-defined volumes) are better measures of the efficiency of a particle at scattering and extinguishing radiation.

For a suspension of N identical particles Eq. (3.103) becomes [see Eq. (2.143)]

$$\beta = \frac{f C_{sca}}{v}, \tag{3.106}$$

where v is the volume of a single particle and f is the volume fraction of particles. For a suspension of nonidentical particles [see Eqs. (2.144)–(2.145)]

$$\beta = f \left\langle \frac{C_{sca}}{v} \right\rangle, \tag{3.107}$$

where f is the total volume fraction of all particles and the brackets indicate an ensemble average over all particle types. Underlying Eqs. (3.103) and (3.107) is the assumption that the consequences of interference are negligible (i.e., the suspension is an incoherent array). We know, however, that this cannot be true for arbitrary f. In the limit $f \to 1$ (if the distribution of size and shapes of particles is such that they can fill all space) the scattering coefficient should plummet (see Sec. 3.4.9 for the differences between scattering per molecule by air and by liquid water). But Eq. (3.107) predicts (incorrectly) a steadily increasing scattering coefficient. The same cautionary statements we made about absorption coefficients of suspensions are applicable to scattering coefficients.

In a transmission measurement the *sum* of absorption and scattering (i.e., extinction) is unavoidably measured. If multiple scattering is negligible a monodirectional beam is attenuated exponentially (Sec. 5.2.3):

$$F = F_0 \exp\{-N\langle C_{ext}\rangle x\}. \tag{3.108}$$

If we measure the ratio of transmitted to incident irradiances, knowing the path length and the number density of particles, we can, in principle, infer the (average) extinction cross section but we cannot determine the separate contributions of absorption and scattering to extinction. In any scattering medium, absorption is never totally absent; in any absorbing medium scattering is never totally absent. To separate extinction into its components requires either additional kinds of measurements, observations, or guidance from theory.

To get some understanding of how scattering cross sections of particles depend on their size and the wavelength of the illumination, we follow the same path as we did in trying to grasp absorption cross sections (Sec. 2.4). We consider reflection by a (negligibly absorbing) slab of uniform thickness h, area $A \gg \lambda^2$, and refractive index n, illuminated at normal incidence by a monodirectional, monochromatic beam with irradiance F_0. For simplicity we take the slab to be in air. The total power reflected by this slab is

$$W_r = F_0 A \tilde{R}, \tag{3.109}$$

where the coherent reflectivity (see Sec. 5.1.1) is

$$\tilde{R} = \frac{2R_\infty(1 - \cos\varphi)}{1 - 2R_\infty \cos\varphi + R_\infty^2}. \tag{3.110}$$

The reflectivity of the infinite (only one boundary) medium is

$$R_\infty = \left|\frac{n-1}{n+1}\right|^2, \tag{3.111}$$

and the phase difference as a consequence of two parallel boundaries a finite distance apart is

$$\varphi = \frac{4\pi n h}{\lambda}. \tag{3.112}$$

We use the term coherent reflectivity to emphasize that, like a particle, a slab is a coherent object.

The reflection cross section is defined as W_r/F_0, and hence the reflection cross section per unit volume $v = Ah$ is

$$C_{\text{ref}}/v = \frac{\tilde{R}}{h}. \tag{3.113}$$

For a sufficiently thin slab (relative to the wavelength), $\varphi \ll 1$, and Eq. (3.113) is approximately

$$C_{\text{ref}}/v \approx \frac{(n^2 - 1)^2 \pi^2 h}{\lambda^2}. \tag{3.114}$$

This expression resembles what we predicted in Sections 3.1 and 3.4.8 for scattering by a small particle (at wavelengths far from absorption bands), namely volumetric scattering proportional to size (volume) and to wavelength to a power (-4). A thin slab is not a dipole but can be looked upon as a sheet of dipoles, and hence the different (but similar) dependence of the volumetric reflection cross section on size and wavelength, increasing with increasing size and proportional to a power of the wavelength (-2). And note that this power is not exactly -2 because n also depends on wavelength. We note in passing that Eq. (3.114) is the source of the incorrect attribution of the blue sky (Sec. 8.1) to reflection by thin plates. Close but not close enough.

We can imagine the slab thickness to increase by piling one sheet of dipoles onto another. At first, the volumetric reflection cross section increases monotonically with each added sheet (Fig. 3.7) as long as all parts of the slab are excited in phase with all other parts. But once the total thickness reaches the point where appreciable ($\sim\pi/2$) phase differences among the

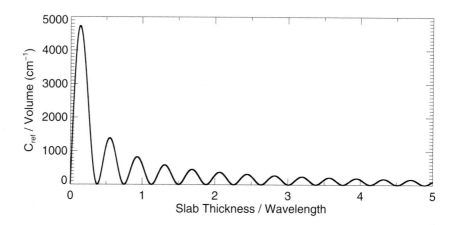

Figure 3.7: Reflection (scattering in the backward direction) cross section per unit volume of a transparent slab (with refractive index 1.33) illuminated at normal incidence as a function of slab thickness relative to the wavelength of the illumination (1 μm).

different parts are possible, the cross section decreases with increasing thickness, then oscillates with decreasing amplitude and decreasing distance between adjacent peaks. Moreover, the wavelength dependence of reflection ceases to resemble that for a dipole sheet. Because of interference, the sum of dipoles is not a dipole, which, in a nutshell, is what makes optics interesting.

Despite appearances to the contrary a plot of C_{ref}/v versus h for fixed frequency is *not* equivalent to a plot versus $1/\lambda$ (frequency) for fixed h. Frequency cannot be increased indefinitely and n remain constant. No such material exists, except a perfect vacuum, which also does not exist. Moreover, frequency cannot be varied arbitrarily without encountering non-negligible absorption (see Figure 3.8). If you pretend that absorption is negligible for more than a decade of wavelengths you probably have slipped into the world of fantasy.

3.5.1   Complex Refractive Index

The theory of scattering and absorption by a homogeneous particle contains a single material parameter, its *complex refractive index*. The simplest way to demonstrate that refractive indices can be written as complex numbers is to consider a plane, scalar wave propagating in the $+x$ direction in an optically homogeneous medium:

$$\psi = a \exp(ikx - i\omega t). \tag{3.115}$$

The phase speed v of this wave is ω/k. If c is the free-space speed of light we can rewrite this as

$$\psi = a \exp\{i\omega(nx/c - t)\}, \tag{3.116}$$

where

$$n = \frac{c}{v} \qquad (3.117)$$

is the refractive index of the medium in which the wave propagates. But this wave cannot be physically realistic because it is not attenuated if a is constant. So let's allow the wave to attenuate exponentially by writing its amplitude as

$$a \exp(-n''\omega x/c). \qquad (3.118)$$

Now the wave has the form

$$\psi = a \exp\{i\omega(Nx/c - t)\}, \qquad (3.119)$$

where the *complex refractive index* is

$$N = n' + in'' \qquad (3.120)$$

and we changed the symbol for its real part from n to n'. Alternatively, we can say that the wavenumber in Eq. (3.115) is complex and write it as

$$k = \frac{\omega}{c}N = k' + ik''. \qquad (3.121)$$

By now we should be comfortable with complex wavenumbers because we know that solutions to linear equations, such as the wave equations we have considered, can be expressed as complex functions (see Sec. 2.5). Because the time-averaged power transmitted by a wave in a medium with a complex refractive index is (except for a factor)

$$\psi\psi^* = |\psi|^2 = a^2 \exp\{-2n''\omega x/c\} = a^2 \exp\{-4\pi n''/\lambda\}, \qquad (3.122)$$

where $\lambda = 2\pi c/\omega$ is the free space wavelength, we immediately obtain the relationship between the imaginary part of the refractive index and the absorption coefficient denoted first by κ in Section 2.1, then changed to κ_b in Section 2.8 to distinguish between bulk (homogeneous) media and suspensions of particles:

$$\kappa_b = \frac{4\pi n''}{\lambda}. \qquad (3.123)$$

The symbol α is also used for the bulk absorption coefficient in Eq. (3.123).

The notation and conventions for complex refractive indices sometimes cause confusion and can lead to serious errors. For our time-harmonic convention, both n' and n'' are non-negative. And yet you often encounter complex refractive indices written as

$$N = n' - in'', \qquad (3.124)$$

where again both n' and n'' are non-negative. Which is correct, Eq. (3.120) or Eq. (3.124)? Both when used with the appropriate convention. If the convention is $\exp(-i\omega t)$, Eq. (3.120)

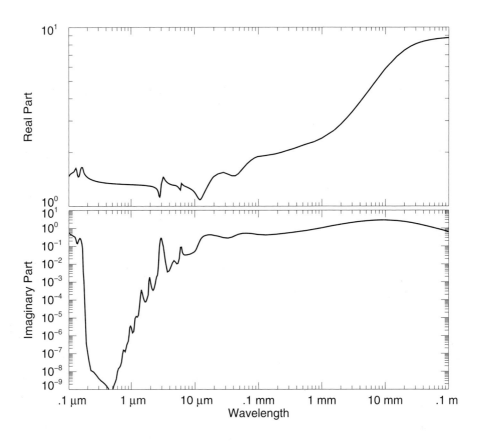

Figure 3.8: Optical constants (complex refractive index) of liquid water taken from the compilation by Querry *et al.* (1991) cited at the end of Chapter 2.

corresponds to waves that decrease in amplitude in the direction of propagation; if the convention is $\exp(i\omega t)$, Eq. (3.124) corresponds to such waves. If we could devise waves that increase in amplitude in the direction of propagation, without an external source feeding energy to them, the energy crisis would immediately evaporate. Conventions are neither right nor wrong. What is wrong is to mix conventions, a sure recipe for accidents (as is driving on the right-hand side of the road in Britain).

The complex refractive index often is written $N = n + ik$, with n and k called *optical constants* (especially by the folks who measure them). But just as the Lord Privy Seal is neither a lord nor a privy nor a seal, optical constants are neither optical (pertain to only visible or near-visible frequencies) nor constant. Indeed, they vary appreciably over frequencies from radio to ultraviolet (Fig. 3.8). Moreover, they are not independent, adjustable parameters that separately can take on any values. The real and imaginary parts are linked. If one varies, so must the other. For example, around 3 µm, a large variation in the imaginary part of the refractive index of water (absorption band) is accompanied by a variation in the real part (Fig. 3.8).

Real Refractive Index and Phase Speed

Few concepts are treated worse in textbooks, are encumbered by more myths and demonstrable nonsense, than the refractive index, complex or otherwise. For example, it has been stated countless times that the refractive index (by which is meant the real part) is greater in a denser medium. Yes, it is true that the refractive index of air is less than that of a thousand times denser water at most, maybe all (we don't know for sure), frequencies. But that's about as far as it goes. A delightfully humorous but edifying note about refractive indices and (mass) density was published half a century ago by the late R. Scott Barr in *American Journal of Physics* (which is supposed to be read by physics teachers and definitely must be read by authors of physics textbooks). The caption to Fig. 1 in Barr's note reads "Does index of refraction vary directly with density?" This figure shows refractive index versus density (at a given frequency in the visible) for a few dozen substances. The points are connected to answer the question in the caption: NO. Barr confesses that these points "were not selected entirely impartially", so shows another figure in which refractive index versus density is plotted (for a given temperature and wavelength) for 445 organic materials. As Barr notes, "this figure indicates that if a material has a high density it is more likely to have a high index than a low one, but that is about all that can be said." Actually, the picture is much bleaker than Barr paints. It is no trick at all to find two materials with different densities for which at one frequency the refractive index is greatest for the denser material but at another frequency is least. And now the clincher: there are plenty of dense, solid materials with *lower* refractive indices at some frequencies than that of such diaphanous substances as air. And these materials are not exotic or "anomalous", but close to hand. For example, at a wavenumber of about 530 cm^{-1}, the real part of the complex refractive index of magnesium oxide (made by burning magnesium wire in air) is 0.093. If magnesium oxide is too exotic for your tastes, try glass (silicon dioxide), with a real refractive index of 0.39 at 1250 cm^{-1}. And then there is ordinary table salt (sodium chloride), with a real refractive index of 0.14 at 200 cm^{-1}. Examples abound. We did not have to search high and low for them.

But having mentioned refractive indices near zero, we anticipate howls of outrage from those who have been taught that nothing can travel faster than light. It is indeed true that according to special relativity no *thing* can travel faster than light, but a surface of constant phase is not a thing. It is an imaginary surface moving with the phase speed v. Special relativity also dictates that *signals* cannot be propagated faster than c, but a plane, harmonic wave such as Eq. (3.33) cannot transmit a signal. You likely have an AM-FM radio. AM stands for *amplitude modulation*, FM for *frequency modulation*. Radio stations, despite proclaiming a single broadcast frequency, do not transmit absolutely monochromatic waves. Programs transmitted by such waves would be exceedingly dull: just a steady hum. To transmit a signal (information) requires modulating waves. Consider the simpler example of Morse code. A steady signal would convey no information. To convey information by Morse code requires turning something on for a brief instant, then off, varying the time intervals between off and on and the length of the time on. That is, the steady tone must be modulated. And so must speech in order to be understood. Try carrying on a conversation at a party by steadily grunting at a single frequency. You won't be invited again. Because the phase speed is neither the speed of a thing nor the speed with which signals are propagated, it is not constrained to be less than c, and hence the real part of the refractive index is not constrained to be greater than 1.

Students who stumble upon real refractive indices less than 1 may be placated by being told that the *group velocity* can't be greater than c. Alas, invoking the group velocity only makes matters worse. To understand why, we first need to explain the group velocity. Consider two waves of equal amplitude, propagating together, with wavenumbers $k \pm \Delta k/2$ and frequencies $\omega \pm \Delta \omega/2$, in a medium we take to be nonabsorbing ($N = n$ and k are real) for convenience. The sum of these two waves [after a bit of algebra, see Eq. (3.63)] is

$$\psi = 2a \exp\{ikx - i\omega t\} \cos\{\Delta k x/2 - \Delta \omega t/2\}. \tag{3.125}$$

This describes a plane harmonic wave with phase speed ω/k the amplitude of which is a wave with a much greater wavelength (much smaller wavenumber) and a correspondingly much smaller frequency. The speed of this long wavelength envelope modulating the much shorter wavelength wave is $\Delta \omega/\Delta k$, which leads to the definition of the *group velocity* (speed, really) v_g as the limit

$$\frac{1}{v_g} = \lim_{\Delta \omega \to 0} \frac{\Delta k}{\Delta \omega} = \frac{dk}{d\omega}. \tag{3.126}$$

From Eqs. (3.117) and (3.121) we have

$$k = \frac{\omega n}{c}, \quad \frac{1}{v} = \frac{k}{\omega} = \frac{n}{c}. \tag{3.127}$$

Combining Eqs. (3.126) and (3.127) yields

$$v_g = \frac{c}{n + \omega(dn/d\omega)}. \tag{3.128}$$

A medium in which n depends on ω is called *dispersive*, but because a truly non-dispersive medium does not exist (except possibly empty space), this term is not very distinctive. But the term dispersion does convey that a group of waves propagating together in a medium will not stick together. To disperse means "to become widely separated."

If $dn/d\omega = 0$, $v_g = v = c/n$, and a group of waves of different frequency and wavenumber would all have the same phase speed and hence would not disperse. We can rewrite Eq. (3.128) as

$$v_g = \frac{v}{1 + (\omega/n)(dn/d\omega)}. \tag{3.129}$$

The term *normal dispersion* is applied when $dn/d\omega > 0$, *anomalous dispersion* when $dn/d\omega < 0$, but there is nothing "anomalous" about anomalous dispersion. All materials exhibit anomalous dispersion at some frequencies, and when everything is anomalous, nothing is anomalous. In a frequency region of normal dispersion, $v_g < v$; in a region of anomalous dispersion, $v_g > v$. When we carefully examine Eq. (3.128) we arrive at a startling conclusion. Suppose that

$$n + \omega \frac{dn}{d\omega} < 0, \tag{3.130}$$

which can be satisfied only in a region of anomalous dispersion, and that

$$\left| n + \omega \frac{dn}{d\omega} \right| < 1. \tag{3.131}$$

If these two conditions are satisfied the group velocity is negative and its magnitude is greater than c: light breaks the speed limit while traveling the wrong way. This is why we said that invoking the group velocity when faced with phase speeds greater than c only makes matters worse. Are there any real materials that satisfy Eqs. (3.130) and (3.131)? Tons of them, as near as your breakfast table: the group velocity in table salt at a frequency of 200 cm^{-1} is about $-4c$.

Alleged violations of the speed limit c are a consequence of the careless, if not downright sloppy, ways in which the term speed of light is used. There are (at least) four different speeds of light: the phase speed, the group speed, the signal speed, and the speed of energy propagation. All four are different in a dispersive medium, but only one of them cannot be greater than c. As its name implies, the signal speed is the speed with which signals are propagated. Suppose that a source in a dispersive medium is suddenly turned on. After time t a detector a distance d from the source begins to register a signal. Special relativity says that the ratio d/t cannot be greater than c. As with Morse code, signals can be transmitted by light only by turning sources on and off.

Imaginary Refractive Index and Absorption by Particles

Although the real part of the complex refractive index is surrounded by misconceptions, when we turn to its imaginary part we encounter so much nonsense that our heads spin. If you think we exaggerate, consider the following story, which we did not make up (or could have). Several years ago we came across complex refractive indices inferred from measurements that at first glance seemed unassailable. Scattering of visible (laser) light versus scattering angle by single levitated spheres was fit by a nonlinear least-squares analysis to yield their complex refractive index. What could be more perfect? A single particle, known to be spherical and homogeneous, to which theory is unquestionably applicable. In the paper reporting the imaginary (in more than one sense) refractive indices so obtained, one of the materials was an oil with an imaginary index alleged to be 0.01. This oil came out of a bottle at least 10 cm in diameter. From Eq. (3.123) it follows that the corresponding absorption length of this oil at visible wavelengths is about 5 µm. Thus we would have to conclude that the oil was poured out of a bottle about 20,000 absorption lengths in diameter. To show that 0.01 is wrong by at least a factor of 10,000 all you have to do is pick up a bottle of the oil and note that you can see through it. What went wrong? In the first place, the experimenter did not know, did not understand, or was incapable of drawing any conclusions from Eq. (3.123), the fundamental equation relating n'' to a more directly measurable and understandable quantity. He also violated the first commandment of measurements: Thou shalt know the magnitude of what you intend to measure before you measure it. Without knowing this magnitude you cannot design a proper method for reliably measuring it. The method under scrutiny here lacked *sensitivity*. Within experimental uncertainty, the angular scattering pattern was the same for imaginary indices of 0.000001 and 0.1 and everything between. So the inferred, but hopelessly incorrect, imaginary index 0.01 was the experimental error for a correct value at least 10,000 times smaller.

Misconceptions about imaginary indices are the rule in the atmospheric science literature. They are treated as adjustable parameters, often without any understanding of their magnitudes or how they vary with frequency for different kinds of materials. It is common to find assertions that an imaginary index of 0.01 corresponds to "weak" absorption (at visible wavelengths), presumably because $0.01 \ll 1$. But as we have seen, an imaginary index of 0.01 corresponds to an absorption length of a few micrometers.

An aerosol is a suspension of particles in a gas and includes the gas. A great amount of pointless effort has been expended on determining the refractive index of the atmospheric aerosol even though it does not exist, just as the refractive index of a cow or the Amazon jungle does not exist. What we have here is the failure to distinguish between the properties of a material and of a body. A refractive index is a property of a (homogeneous) material, not of a heterogeneous body. There is no such entity as the complex refractive index of the atmospheric aerosol, an ever changing witch's brew containing particles of different and generally unknown composition, size, and shape. Individual particles may not even be composed of a single material but are aggregates of particles, like raisins in a loaf of bread coated with sesame seeds.

The complex refractive index of a homogeneous (on the scale of the wavelength) material is a more or less well-defined quantity and can in principle be measured, although in practice this may be difficult. Let's back up a step and outline how refractive indices are measured by experimenters who know what they are doing. Refractive indices cannot be measured directly but are inferred, usually from measurements of reflection and transmission by homogeneous samples (slabs) of the material of interest. These samples are optically smooth, of uniform and known thickness. The theory of reflection and transmission by such slabs as a function of refractive index is then used to invert measurements to obtain the desired refractive index. Sounds simple. But there are several catches. One is that the theory has to be applicable to the sample. Thus, if the theory is that for specular reflection by a uniformly thick, homogeneous sample, the sample bloody well better conform to this. Moreover, the measured quantities have to change when the refractive index changes. And if the refractive index only reproduces the measurements used to obtain it, it would be of little value. That is, if all we could do is infer a refractive index from measurements of reflectivity and transmissivity, then use it to calculate these same quantities (within experimental error), we'd just be chasing our tails. But properly measured optical constants have a wider validity. For example, they can be used to calculate reflection and transmission by slabs of *any* thickness, or scattering and absorption by homogeneous particles of *any* size and shape (for which a theory exists). This is not tail chasing.

Now consider particles in the atmosphere. In general, their size, shape, and composition are *not* known. They may not be homogeneous and their concentration (number density) varies from point to point, from time to time. Unlike samples prepared in the laboratory, experimenters have no control over the characteristics of atmospheric particles. Nevertheless, one can make various measurements of scattering by them. Problems arise, however, when one attempts to invert these measurements to obtain the (nonexistent) refractive index of the particles, a single complex number at a given wavelength that in some sense is a valid optical property. Between a measurement and a desired quantity to be inferred from it always lies a theory. What theory is used to invert measurements of scattering by atmospheric particles? Mie theory (see following section), of course, which is valid only for homogeneous spheres, possibly combined with radiative transfer theory for a plane-parallel, uniform medium. An inversion scheme based on this theory will dutifully give results, although as in the example of the oil droplets, they may be hopelessly wrong. All that one can say for sure is that the inferred quantity is consistent, within experimental error, with the theory: if you invert the measurements with Mie theory to obtain refractive indices, then use these refractive indices in Mie theory, you will recover (within error bars) the measurements. All you are doing here is chasing your tail.

Now contrast what folks have been trying to do for particles in the atmosphere with the oil droplet measurements discussed previously. The experimenter had absolute control over his sample: a single, homogeneous sphere, of known size and composition. And yet the error in his imaginary index inferred from measurements was at least a factor of 10,000. Think of how much more difficult it is to measure (remotely) the refractive index of particles, the size distribution, shape, composition, and concentration of which are unknown and over which the experimenter has no control.

Taken in the round, meteorologists don't know much about electromagnetic theory, optics, and solid state physics. But they do know a lot about fluid mechanics (or at least one branch of it). If we were to propose measuring the "effective" viscosity of the ocean–atmosphere system, meteorologists would double over in laughter. Would even the most naive modeler try to model the dynamics of the coupled ocean and atmosphere by treating this system as a single fluid with some kind of bogus effective viscosity? We think not. But these same folks swallow without hesitation the equally bogus notion of an effective refractive index of the atmospheric aerosol.

3.5.2   Scattering by an Isotropic, Homogeneous Sphere

An isotropic, homogeneous sphere is the simplest finite scatterer, the theory of scattering by which is attached to the name of Gustav Mie. So firm is this attachment that in defiance of logic and history every particle under the sun has been dubbed a "Mie scatterer", and Mie scattering has been promoted from a particular theory of limited applicability to the unearned rank of general scattering process.

Mie was not the first to solve the problem of scattering by an arbitrary sphere. It would be more correct to say that he was the last. He gave his solution in recognizably modern notation and also addressed a real problem: the colors of colloidal gold. For these reasons his name is attached to the sphere scattering problem even though he had illustrious predecessors, most notably Lorenz (not to be confused with Lorentz). This is an example in which eponymous recognition has gone to the last discoverer rather than to the first.

Mie scattering is not a physical process; *Mie theory* is one among many. Strictly speaking, it isn't even exact because it is based on continuum electromagnetic theory, itself approximate, and on illumination by a plane wave infinite in lateral extent.

Scattering by a sphere can be determined using various approximations and methods bearing little resemblance to Mie theory: Fraunhofer theory, geometrical optics, anomalous diffraction theory, coupled-dipole, T-matrix method, etc. Thus is a sphere a Mie scatterer or an anomalous diffraction scatterer or a coupled-dipole scatterer? The possibilities are endless.

When a physical process can be described by several different theories, it is inadvisable to attach the name of one of them to it.

There is no distinct boundary between so-called Mie and Rayleigh scatterers. Mie theory includes Rayleigh theory (for spheres), which is a limiting theory strictly applicable only as the size of the particle shrinks to zero. Even for spheres uncritically labeled "Rayleigh spheres", there are always deviations between the Rayleigh and Mie theories. By hobbling one's thinking with a supposedly sharp boundary between Rayleigh and Mie scattering, one risks throwing some interesting physics out the window. Whether a particle is a Mie or Rayleigh scatterer is not absolute. A particle may be graduated from Rayleigh to Mie status merely by a change of wavelength of the illumination. One often encounters statements about Mie scattering by cylinders, spheroids, coated spheres and other nonspherical or inhomogeneous particles. Judged historically, these statements are nonsense. Mie never considered any particles other than homogeneous spheres.

Logic would seem to demand that if a particle is a Mie scatterer, then Mie theory can be applied to scattering by it. This fallacious notion has caused and will continue to cause mischief, and is probably the best reason to cease referring to "Mie particles" or "Mie scatterers". Using Mie theory for scattering by particles other than spheres, especially near the backward direction, is risky.

More often than not, a better term than Mie or Rayleigh scattering is available. If the scatterers are molecules, molecular scattering is better than Rayleigh scattering (itself an imprecise term): the former term refers to an agent, the latter to a theory. Mie scatterer is just a needlessly aristocratic name for a humble sphere. Whenever Mie scatterer is replaced with sphere, the result is clearer. If qualifiers are needed, one can add small or large compared with the wavelength or comparable with the wavelength.

Briefly, the solution to the problem of scattering by an arbitrary homogeneous sphere illuminated by a plane wave can be obtained by expanding the incident, scattered, and internal electric and magnetic fields in series of vector spherical harmonics (general solutions to the equations of the electromagnetic field in spherical coordinates). The coefficients of these expansion functions are chosen so that the tangential components of the fields are continuous across the surface of the sphere. Thus this scattering problem is formally identical to reflection and refraction because of interfaces, although the sphere problem is considerably more complicated because the scattered and internal fields are not plane waves.

Observable quantities are expressed in terms of the complex scattering coefficients a_n and b_n in the expansions of the scattered electric and magnetic fields. For example, the cross sections are infinite series:

$$C_{\text{ext}} = \frac{2\pi}{k^2} \sum_{n=1}^{\infty} (2n + 1)\Re\{a_n + b_n\}, \tag{3.132}$$

$$C_{\text{sca}} = \frac{2\pi}{k^2} \sum_{n=1}^{\infty} (2n + 1)\{|a_n|^2 + |b_n|^2\}. \tag{3.133}$$

The scattering coefficients can be written

$$a_n = \frac{[D_n(mx)/m + n/x]\psi_n(x) - \psi_{n-1}(x)}{[D_n(mx)/m + n/x]\xi_n(x) - \xi_{n-1}(x)}, \tag{3.134}$$

$$b_n = \frac{[mD_n(mx) + n/x]\psi_n(x) - \psi_{n-1}(x)}{[mD_n(mx) + n/x]\xi_n(x) - \xi_{n-1}(x)}, \tag{3.135}$$

where ψ_n and ξ_n are Riccati–Bessel functions and the logarithmic derivative is

$$D_n(\rho) = \frac{d}{d\rho}\ln\psi_n(\rho). \tag{3.136}$$

The *size parameter* x is ka, where a is the radius of the sphere and k is the wavenumber of the incident radiation in the surrounding medium (assumed nonabsorbing); m is the complex refractive index (discussed in the previous subsection) of the sphere relative to the (real) refractive index of the surrounding medium. Equations (3.134)–(3.136) are one among many ways of writing the scattering coefficients, some of which are more suited to computations than others. Scattering in any direction for any state of polarization of the incident illumination is also determined by the scattering coefficients.

A good rule of thumb is that the number of terms required for convergence of the series in Eqs. (3.132) and (3.133) is approximately the size parameter $(2\pi a/\lambda)$. Raindrops are of order 1 mm, and hence their size parameter at visible wavelengths is of order 10,000. The details of rainbows do emerge from Mie theory but at the cost of summing 10,000 terms, which is why we often resort to approximations such as geometrical optics, which sheds light on some but not all features of rainbows (see Sec. 8.4.2). To describe all their features we have to resort to Mie theory, and even it isn't good enough because raindrops are not spheres (cloud droplets are), and their departures from sphericity have observable consequences. And this leads us to the dubious notion of an "equivalent sphere", the search for which rivals the quest of alchemists for recipes for transforming base metals into gold. Alas, just as there is no such recipe, there is no such thing as an equivalent sphere, one with all the same scattering and absorbing properties as a non-spherical particle. In the first place, such a particle is defined by what it is *not*, the only characteristic shared by all non-spherical particles. A non-cat is any animal that is not a cat, which leaves us with a menagerie housing everything from elephants and giraffes to porcupines and shrews. Because a collection of randomly oriented non-spherical particles has the same symmetry as a sphere, it is sometimes argued that the (ensemble averaged) scattering properties of the collection are the same as those of suitably chosen "equivalent spheres". Not true. The error here resides in confusing the symmetry of an ensemble of particles with that of a single particle. No matter what incantation is used for conjuring the properties of an equivalent sphere, differences between scattering and absorption by it and by a non-spherical particle always exist. Sometimes these differences are huge, sometimes not. Beware of equivalent spheres.

During the Great Depression mathematicians were put to work computing tables of trigonometric and other functions. The results of their labors now gather dust in libraries. Today, these tables could be generated more accurately in minutes on a pocket calculator. A similar fate has befallen Mie calculations. Before fast computers were inexpensive, tables of scattering functions for limited ranges of size parameter and refractive index were published. Today, these tables could be generated in minutes on a personal computer. Algorithms are more valuable and enduring than tables of computations, which are mostly useless except as checks for someone developing and testing algorithms. The primary tasks in Mie calculations are computing the functions in Eqs. (3.134) and (3.135) and summing series such as

Eqs. (3.132) and (3.133). Nowadays Mie codes abound. You can find them in books, on the Internet, and probably at upscale supermarket checkout stands.

Cross sections versus radius or wavelength convey physical information; efficiencies versus size parameter convey mathematical information. The size parameter is a variable with less physical content than its components, the whole being less than the sum of its parts. Moreover, cross sections versus size parameter (or its inverse) are not equivalent to cross sections versus wavelength. Except in dreamland, refractive indices vary with wavelength, and the Mie coefficients depend on x and m, wavelength being explicit in the first, implicit in the second.

Particles Much Smaller than the Wavelength

For sufficiently small x and $|m|\, x$, the volumetric extinction and scattering cross sections for spheres are approximately

$$C_{\text{ext}}/v \approx \frac{6\pi}{\lambda}\Im\left\{\frac{m^2-1}{m^2+2}\right\}, \tag{3.137}$$

$$C_{\text{sca}}/v \approx \frac{24\pi^3 v}{\lambda^4}\left|\frac{m^2-1}{m^2+2}\right|^2, \tag{3.138}$$

where v is the particle volume and λ is the wavelength in the (negligibly absorbing) material surrounding the sphere. Similar equations hold, in particular the volume dependence, for small, homogeneous particles of other shapes. Note the similarity of Eq. (3.138) to Eq. (3.114) for reflection by a thin slab. These equations are the source of a nameless paradox, which is disinterred from time to time, a corpse never allowed to rest in peace. If the sphere is nonabsorbing (m real), Eq. (3.137) yields a vanishing extinction cross section whereas Eq. (3.138) yields a non-vanishing scattering cross section, and yet extinction can never be less than scattering. Both equations were obtained from power series in the size parameter x. Equation (3.137) is the first term in the series. To be consistent, both cross sections must be expanded to the same order in x. When this is done, the paradox vanishes. In fact, it never existed.

Yet another pointless paradox arises from the curious definition of the *radar backscattering cross section* as 4π times the differential scattering cross section in the backward direction. For a small sphere this leads to a backscattering cross section 50% *greater* than its total scattering cross section, which at first glance certainly is cause for head scratching. The *radar reflectivity* is the sum of the radar backscattering cross sections of all the scatterers in a unit volume.

Figure 3.9 shows the scattering and absorption cross sections of a water droplet 20 µm in diameter over six wavelength decades. For wavelengths much greater than the diameter, scattering is a linear function (on a logarimthic plot) of wavelength with slope approximately -4, in accordance with Eq. (3.138). At these wavelengths, extinction is dominated by absorption, and so according to Eq. (3.137), absorption should decrease (again, on a logarithmic plot) linearly with slope -1. This is not quite what occurs (see Fig. 3.9) because the complex refractive index also varies with wavelength (Fig. 3.8) in this region.

Although Eq. (3.138) is the scattering cross section of a small particle, it still must contain enough molecules that it can be assigned a refractive index (a molecule cannot). Nevertheless, let's throw caution to the wind and extrapolate this equation to molecular sizes. If we use the refractive index of water and a particle diameter of 0.3 nm, we obtain a scattering cross section in the middle of the visible spectrum of about 0.7×10^{-19} µm^2. That for air (an average over all molecules but predominantly nitrogen) is about 4.6×10^{-19} µm^2. This isn't bad agreement considering that the scattering cross section of water vapor is less than that of air (nitrogen and oxygen), molecular diameters are not well defined, and that the scattering cross section depends on the sixth power of diameter.

Particles Much Larger than the Wavelength

Another paradox has a name: the *extinction paradox*. For a compact particle, such as a sphere, the extinction cross section has the limiting value

$$\lim_{a \to \infty} C_{\text{ext}} = 2G, \tag{3.139}$$

where a is a linear dimension of the particle (for a sphere its radius) and G its projected geometrical cross sectional area. The factor 2 in Eq. (3.139) has caused people to sweat: according to geometrical optics it should be 1. Equation (3.139) seems to imply that a large (compared with the wavelength) particle is too big for its britches by a factor of two. Geometrical optics, according to which a beam of light is imagined to be a bundle of rays, is reckoned to be a good approximation for objects much larger than the wavelength. Thus every ray that intersects a particle with geometrical area G should be either absorbed or deviated, whereas rays that lie outside this shadow region should pass unscathed. The catch here is that rays don't exist and no matter how large a finite particle is, it always exhibits some departures, possibly small, from geometrical optics, in this instance very close to the forward direction.

We noted previously (and show in Fig. 3.13) that scattering is more peaked in the forward direction the larger the particle. Theory counts scattered light as removed from a (monodirectional) beam no matter how small the scattering angle. To measure the full extinction cross section of an indefinitely large particle would require a detector with vanishingly small acceptance angle. But any real detector necessarily collects some of the near-forward scattered light, which reduces the extinction cross section from its theoretical maximum. When near-forward scattered light is included in the measurement, the limiting extinction cross section [Eq. (3.139)] drops to G, as expected on the basis of intuition molded by geometrical optics. The distinction here is between the real and the ideal. A real detector measures the extinction cross section

$$C_{\text{ext}} - \int_{\text{acc}} \frac{dC_{\text{sca}}}{d\Omega} \, d\Omega, \tag{3.140}$$

where C_{ext} is the ideal (theoretical) extinction cross section and integration is over the acceptance solid angle of the detector.

In Fig. 3.9 the two asymptotes, G and $2G$, are shown by dotted lines. At sufficiently short wavelengths, where extinction is dominated by scattering, the scattering cross section does indeed approach the asymptote $2G$. But note also that over a range of intermediate wavelengths the scattering and absorption cross sections are each approximately equal to G.

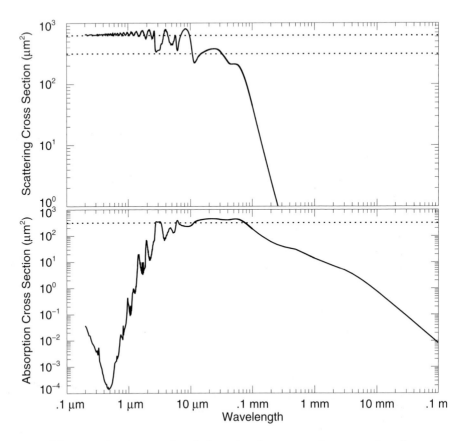

Figure 3.9: Scattering (top panel) and absorption (bottom panel) cross sections for a water droplet of diameter 20 µm. The two horizontal dotted lines in the top panel indicate the geometrical cross sectional area and twice this area; the horizontal dotted line in the bottom panel indicates the geometrical cross sectional area.

Those Elusive Small Droplets

We state in Section 1.4.2 that the question of what is *the* average size of a cloud droplet should be greeted with a horselaugh but only sketched the reason why. We can now expand further on this armed with the results of this section.

Cloud droplets are large compared with the wavelengths of visible radiation but small compared with those of microwaves and radar. At visible wavelengths droplet scattering cross sections are proportional to the square of droplet diameter. Extinction is dominated by scattering (Fig. 3.9) and extinction is nearly the asymptotic value [Eq. (3.139)]. At microwave wavelengths, however, extinction is dominated by absorption, and hence from Eq. (3.137), the absorption cross section is proportional to droplet diameter cubed. But at these wavelengths, the scattering cross section, and hence also the radar backscattering cross section, is propor-

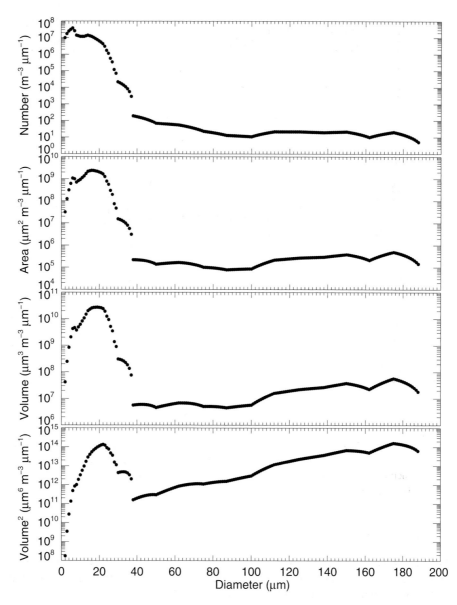

Figure 3.10: The top panel shows an *in situ* measurement of the droplet number size distribution in a stratus cloud. From this distribution the distribution functions for cross sectional area, volume, and volume squared follow.

tional to droplet diameter to the sixth power. These different dependences on size have profound consequences for inferring droplet properties remotely from radiation measurements.

Direct measurements made using an airplane flying through (water) stratus clouds near State College, Pennsylvania are displayed in Fig. 3.10, which shows the number distribution of droplets, from which area (diameter squared), volume (diameter cubed), and volume squared (diameter to the sixth power) distributions are calculated. The number density distribution (droplets per unit volume) peaks at about 6 μm. For diameters greater than about 40 μm, the number density is about six decades smaller than at the peak. Scattering of visible radiation peaks at about 16 μm, at which diameter the number density is more than a factor of two less than its peak value. The peak in the volume distribution is shifted even further, to about 20 μm. And although the peak in the volume squared distribution is shifted to only about 22 μm, droplets larger than about 40 μm contribute more to radar backscattering signals than the smaller droplets even though they are a million times more abundant.

Consider how depressing the message Fig. 3.10 conveys to anyone with hopes of inferring droplet size distributions remotely from radiation measurements. The number distribution is greatest for the smaller sizes, whereas properties on which any scheme for remotely inferring sizes might be based are distributed quite differently. For example, if the smallest droplets (< 6 μm) were to be removed entirely from the cloud, the volume, and even more so the volume-squared distributions would hardly miss them. What is needed in order to detect (remotely) the smaller droplets is some electromagnetic property that depends only on number density or, at worst, on diameter. But such a property does not seem to exist.

3.5.3 Some Observable Consequences of Scattering of Visible Radiation by Spherical Particles

The previous section is a panoramic view of scattering and absorption over many wavelength decades. In this section we focus on less than one-third of one of these decades with the aim of helping you to understand what you can see with your own eyes.

Size Dependence

Figure 3.11 shows the volumetric scattering cross section, in the middle of the visible spectrum, of water droplets in air and varying in diameter from about 10 nm to 1 mm. As predicted by simple coherence arguments (Sec. 3.4.8), volumetric scattering increases linearly (on a logarithmic scale) with increasing diameter until a droplet is approximately the size of the wavelength, reaches a peak, oscillates a bit, and decreases linearly (note the similarity to the volumetric reflection cross section in Fig. 3.7). This tremendous variation in volumetric scattering, about four decades, solely as a consequence of the state of aggregation of water molecules, has readily observable consequences. We can see through tens of kilometers of clear, moist air, but when some of the water vapor in it condenses into clouds only tens of meters thick they can obscure the sun. And yet we usually can see through rain shafts, which are born from clouds.

Wavelength Dependence

Now let's turn to the wavelength dependence over the visible spectrum of scattering by spheres of fixed size. Figure 3.12 shows scattering by oil droplets, assumed negligibly absorbing at visible wavelengths, with diameters much smaller than (0.05 μm), comparable with (0.8 μm), and much greater than (10 μm) the wavelengths of visible light. We took the (real) refractive index to be a constant 1.5, which does not introduce serious errors over this limited spectral

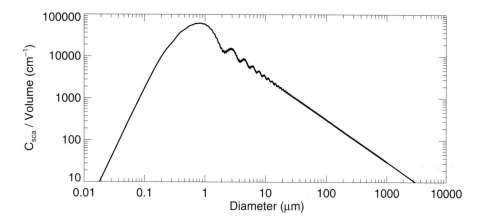

Figure 3.11: Volumetric scattering cross section, at a wavelength of 0.55 μm, for water droplets with diameters varying over six decades. Except for a proportionality factor, volumetric scattering is scattering per molecule.

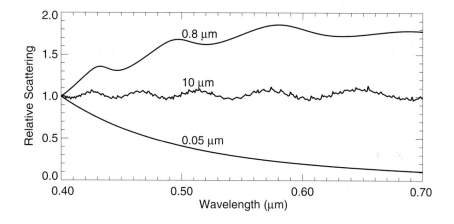

Figure 3.12: Spectral scattering relative to that at 0.4 μm for oil droplets ($n = 1.5$) of the three diameters shown on the curves.

range. As expected on the basis of the simple theory of dipolar scattering (Sec. 3.2), scattering by the smallest droplets follows Rayleigh's scattering law, whereas because of non-negligible phase differences, scattering by the largest droplets is nearly independent of wavelength. Note that the spectral dependence of scattering is the same for all droplets smaller than a certain size and larger than a certain size. But between these two extremes lies a comparatively narrow range of droplets that scatter light at the long wavelength end of the spectrum more than at the short end, the inverse of Rayleigh's law.

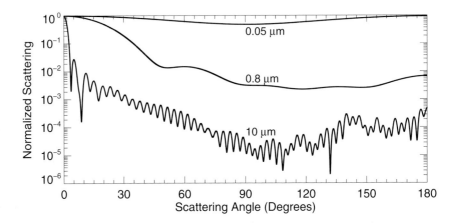

Figure 3.13: Angular dependence of scattering (normalized to 1 in the forward direction) of incident unpolarized light by water droplets of the three diameters shown on the curves. The wavelength of the incident illumination is 0.55 μm.

Sunlight or moonlight would be selectively (according to wavelength) transmitted by a thin suspension of the smallest droplets. The transmitted light would be reddened, which means that its spectrum would be shifted toward longer wavelengths relative to the incident spectrum, not necessarily that the transmitted light would be perceptually red, whereas the spectrum of light transmitted by the largest droplets would be essentially that of the incident light. Both observations are what we have come to consider the norm. We are not startled when we see a yellow, orange, or red sun or moon. But the spectrum of the light transmitted through a thin veil of oil droplets of intermediate size would be shifted toward shorter wavelengths, yielding a blue or green sun or moon, which would get our attention. This requires a cloud of droplets of just the right size and mostly this size, which is rare in the atmosphere, so rare that it occurs only "once in a blue moon." Astronomers call the second full moon in a month a blue moon, which has absolutely nothing to do with its color. The last reliable reports of blue and green suns and moons occurred in 1950, and were attributed to oily smoke (which is why we did calculations for oil droplets) from large forest fires in Canada.

Directional Dependence

Because scattering by two or more dipoles is coherent in phase in the forward direction, and only in this direction, we expect scattering by any particle to increase faster with increasing size in this direction than in any other direction. This expectation, based on simple physical reasoning, is supported by calculations (Fig. 3.13) of angular scattering of unpolarized visible light by water droplets with diameters much smaller than (0.05 μm), comparable with (0.8 μm), and much larger than (10 μm) visible wavelengths. The larger the droplet, the more that scattering by it is peaked in the forward direction. Also, the larger the droplet, the more maxima and minima in its scattering diagram, as expected on the basis of the simple interference arguments in Section 3.4.8.

Observable consequences of strong forward–backward asymmetry in scattering by droplets are not hard to come by, a spider's web festooned with drops like pearls on a string, for example. A web filament is a long, cylindrical scatterer, for which scattering of sunlight is highly concentrated in one direction. Unless you happen to be looking in that direction, you won't see webs, which is why you run into them on walks through woods. But if a web has collected fog droplets, you can readily see it, especially if you are looking toward the sun. And then the web almost disappears if you view it with the sun at your back. While driving across the Great Plains of the United States we have watched the greatly changing brightness of spray from center-pivot irrigation as our viewing angle changed.

The wiggles and bumps in Figs. 3.11, 3.12, and especially 3.13 are what might be called *obvious* manifestations of interference. Absorption dampens and may completely iron out such wiggles and bumps, which should hardly come as a surprise: interference filters are not made of strongly absorbing materials. But there is just as much interference in a metallic particle with a smooth scattering diagram as in a negligibly absorbing particle. The scattering pattern for both kinds of particles results from interference between all the waves scattered by their constituent molecules. We have become accustomed to associating interference with wiggles and bumps, but their absence does not necessarily signal no interference.

References and Suggestions for Further Reading

For a biography of Maxwell see Ivan Tolstoy, 1981: *James Clerk Maxwell: A Biography*. University of Chicago. This is a model of scientific biography comprehensible to a non-scientific audience. It is short, well written, and without a single equation conveys the contributions of a scientist whose fame rests mostly on equations. For a much more detailed and technical history of electromagnetism (including electromagnetic radiation) see Edmund Whittaker, 1987: *A History of the Theories of Aether and Electricity. I. The Classical Theories, II. The Modern Theories*. Tomash/American Institute of Physics.

The microscopic (scattering) theory of reflection and refraction is discussed in Max Born and Emil Wolf, 1965: *Principles of Optics*. 3rd rev. ed. Pergamon, pp. 98–108.

A derivation of the radiative reaction is given in John David Jackson, 1975: *Classical Electrodynamics*, 2nd ed., John Wiley & Sons, Ch. 17.

Our treatment of the vibrating string was heavily influenced by Dudley H. Towne, 1967: *Wave Phenomena*, Addison-Wesley, one of the best intermediate-level physics textbooks ever published.

For an expository article on coherence see A. J. Forrester, 1956: On coherence properties of light waves. *American Journal of Physics*, Vol. 24, pp. 192–6.

We criticize, on physical grounds, the usual textbook treatment of the two-slit interference experiment. Tony Rothman, 2003: *Everything's Relative and Other Fables from Science and*

Technology, John Wiley & Sons, Ch. 2 criticizes it on historical grounds. He casts serious doubts on whether Thomas Young ever did the experiment so often attributed to him.

The simple theory of the diffraction grating is discussed in, for example, Francis A. Jenkins and Harvey E. White, 1976: *Fundamentals of Optics*, 4th ed. McGraw-Hill, Ch. 17. Several papers on grating anomalies, including those by Rayleigh and Wood, are reprinted in Daniel Maystre, Ed., 1992: *Selected Papers on Diffraction Gratings*. SPIE Optical Engineering Press, Sec. 5.

For more on laser speckle see J. C. Dainty, Ed., 1984: *Laser Speckle and Related Phenomena*, 2nd ed., Springer, and M. Francon, 1979: *Laser Speckle and Applications in Optics*, Academic.

For more on optical heterodyning, with atmospheric applications, see James C. Owens, 1969: Optical Doppler measurement of microscale winds. *Proceedings of the IEEE*, Vol. 57, pp. 530–7; Robert T. Menzies, 1976: Laser heterodyne detection techniques, in *Laser Monitoring of the Atmosphere*, E. D. Hinkley, Ed., Springer, pp. 297–353; Rob Frehlich, 1996: Coherent Doppler lidar measurements of winds, in *Trends in Optics: Research, Development and Applications*, Anna Consortini, Ed., Academic, pp. 351–70.

Our treatment of the Doppler effect (Sec. 3.4.6) was inspired by Thomas P. Gill, 1965: *The Doppler Effect: An Introduction to the Theory of the Effect*, Logos Press. p. vii: "One might think of a Doppler effect arising when letters are posted from successive railway stations on a long journey." Gill is one of the few authors to note that the Doppler effect has nothing fundamental to do with waves, although he adds that "it must be at once admitted that all practical applications of any importance are concerned with periodic processes."

For a simple demonstration of the Doppler effect, a discussion of its history, and some of its applications, see Craig F. Bohren, 1991: *What Light Through Yonder Window Breaks?*, John Wiley & Sons, Ch. 14.

For a good derivation of the relativistic Doppler effect, including the longitudinal Doppler effect, see Thomas M. Helliwell, 1966: *Introduction to Special Relativity*, Allyn and Bacon, pp. 116–22. This is an excellent little book on relativity. Especially recommended is Appendix B (pp. 148–50), which lists the advantages and disadvantages of a velocity-dependent mass. Contrary to what still seems to be widely believed, a velocity-dependent mass is not an essential part of relativity. To use it or not is a matter of taste, not necessity.

Because of television it is fairly well known that Doppler radar is used to observe weather systems (see, e.g., Louis J. Battan, 1979: *Radar Observation of the Atmosphere*, University of Chicago; Richard J. Doviak and Dusan S. Zrnic, 1993: *Doppler Radar and Weather Observations*, 2nd ed., Academic). Perhaps less well known is an application closer to home: observation of the human (and animal) heart. Both the senior author and one of his dogs, Zas, have had echo cardiograms in which Doppler-shifted ultrasound (high frequency sound waves) is used to measure blood flow in the heart. And both have slight mitral valve leaks, detected by means of the Dopper effect. As the veterinary cardiologist said upon examining Zas's echo cardiogram, "The apple doesn't fall far from the tree." For an elementary treatment of the use of Doppler ultrasound in medicine see Frederick W. Kremkau, 1998: *Diagnostic Ultrasound: Principles and Instruments*, 5th ed., W. B. Saunders.

An English translation of Einstein's 1910 paper, Theory of the opalescence of homogeneous liquids and mixtures of liquids in the vicinity of the critical state, is in Jerome Alexander, Ed., 1926: *Colloid Chemistry*, Vol. I, Chemical Catalog Co., pp. 323–39. This same collection contains an excellent overview (pp. 340–52) of light scattering by gases, liquids, and solids, W. H. Martin's ,The scattering of light in one-phase systems. Although Martin's article is now a bit dated, it is still worth reading because of its clarity and the insights it provides. Martin claims to have been the first to state (in 1913) "based on direct experimental evidence that any dust-free medium scatters light."

What Einstein did without molecules was done with them by Bruno H. Zimm, 1945: Molecular theory of the scattering of light in fluids. *Journal of Chemical Physics*, Vol. 13, pp. 141–5.

For many more details about absorption and scattering by particles see Hendrik C. van de Hulst, 1981: *Light Scattering by Small Particles*. Dover; Diran Deirmendjian, 1969: *Electromagnetic Scattering on Polydispersions*. Elsevier; Milton Kerker, 1969: *The Scattering of Light and Other Electromagnetic Radiation*. Academic; Craig F. Bohren and Donald R. Huffman, 1983: *Absorption and Scattering of Light by Small Particles*. Wiley-Interscience; H. Moysés Nussenzeig, 1992: *Diffraction Effects in Semiclassical Scattering*. Cambridge University Press; Walter T. Grandy, Jr., 2000: *Scattering of Waves from Large Spheres*. Cambridge University Press; Michael I. Mishchenko, Larry D. Travis, and Andrew A. Lacis, 2002: *Scattering, Absorption, and Emission of Light by Small Particles*. Cambridge University Press.

For a brief overview of scattering by particles see Craig F. Bohren, 1995: Scattering by particles in *Handbook of Optics*, Vol. 1, 2nd ed., Michael Bass, Eric W. Van Stryland, David R. Williams, William L. Wolfe, Eds. McGraw-Hill, pp. 6.1–6.21.

For a biographical sketch of Mie see Pedro Lilienfeld, 1991: Gustav Mie: the person. *Applied Optics*, Vol. 30, pp. 4696–8. Pedro was granted a unique opportunity to have a bit of word fun. At the time of publication his affiliation was MIE, Inc. (Measuring Instruments for the Environment), and is so denoted in his paper.

For brevity we call the theory of scattering of a plane wave by a homogeneous sphere Mie theory, but you also will find it called Mie–Debye theory, Lorenz–Mie theory, or even Lorenz–Mie–Debye theory. Once again Stiglitz's law of eponymy (Sec. 2.1) holds. Mie was almost the *last* to solve this scattering problem (Debye's paper was published a year later) but his solution is in recognizably modern notation and he applied it to a real problem, the colors of metals in the colloidal state. For an overview of the contributions of Ludvig Lorenz see Helge Kragh, 1991: Ludvig Lorenz and nineteenth century optical theory: the work of a great Danish scientist. *Applied Optics*, Vol. 30, pp. 4688–95. And for a comprehensive historical survey see Nelson A. Logan, 1965: Survey of some early studies of the scattering of plane waves by a sphere. *Proceedings IEEE*, Vol. 53, pp. 773–85.

A history of light scattering, From Leonardo to the graser: Light scattering in perspective, was published serially by John D. Hey in *South African Journal of Science*: 1983, Vol. 79, pp. 11–27, 310–24; 1985, Vol. 81, pp. 77–91, 601–13; Vol. 82, pp. 356–60.

Many of the seminal papers on light scattering are reprinted in the compilations by Milton Kerker, Ed., 1988: *Selected Papers on Light Scattering* (2 Vols.), SPIE Optical Engineering Press, 1988 and Philip L. Marston, Ed., 1993: *Selected Papers on Geometrical Aspects of Light Scattering*. SPIE Optical Engineering Press.

For some of the most recent work on scattering by nonspherical particles see Michael I. Mischchenko, Joop W. Hovenier, and Larry D. Travis, Eds., 2000: *Light Scattering by Nonspherical Particles: Theory, Measurements, and Applications*. Academic.

The diagrams in Figs. 3.11, 3.12, and 3.13 are for spheres, which are *compact* particles, and also negligibly absorbing at the wavelengths of interest. A compact particle is one that more or less fills a continuous region of space (without macroscopic gaps). But not all particles are compact and hence one should be wary about generalizations based on the scattering properties of compact particles. Soot aggregates (produced in fires) are examples of particles that are not compact. Such particles have been called fractal particles, but fractal-like would be more appropriate given that a true fractal is self-similar at *all* scales, whereas soot aggregates and other fractal-like particles are not. For a good review see Chris M. Sorensen, 2001: Light scattering by fractal aggregates: a review. *Aerosol Science and Technology*, Vol. 35, pp. 648–87.

A must read is E. Scott Barr, 1955: Concerning index of refraction and density. *American Journal of Physics*, Vol. 23, 623–4.

It is depressing to realize that as long ago as 1899 it was known that there is a distinction between signal and group velocity, and by 1907 that the signal velocity cannot be greater than *c*. By 1914 two papers had appeared, back to back, the first by Arnold Sommerfeld, the second by Léon Brillouin, in what was arguably the leading physics journal of the day, in which wave propagation in a dispersive medium was worked out in elaborate detail. English translations of both papers are given as Chapters 2 and 3 in Léon Brillouin's *Wave Propagation and Group Velocity*, 1960, Academic. These chapters require knowledge of the theory of functions of a complex variable, but the introductory chapter does not make great mathematical demands of readers and the introductory section of the second chapter is mostly qualitative. Figure 19 in Chapter 3 deserves careful study.

For a superb compendium of optical constants see Edward D. Palik, Ed., 1998: *Handbook of Optical Constants of Solids* (3 Vols.). Academic. Each volume is divided roughly into two equal sections. The first section describes methods for measuring optical constants and their physical interpretation. The second section is a compilation of measured optical constants for various materials. Perusal of these volumes makes it abundantly clear that determining optical constants even for well-defined, homogeneous samples of known geometry is difficult, and that much of what is in textbooks about refractive indices is wrong or at best misleading.

A refractive index is an average property. For an average to mean something it has to be taken over a sufficiently large number of values, which is why a single (complex) number describes to good approximation the response of a molecular medium to excitation by radiation of a given frequency. Such a medium (e.g., pure water) is composed of a great many individual scatterers (i.e., molecules) in a volume of one cubic wavelength. A corollary of this is that a medium composed of many particles in a cubic wavelength also can be assigned a refractive index. To the exciting radiation, the particles are large molecules. Determining the effective refractive index of mixtures of molecules and its extension to composite particulate media is nearly 200 years old, the first such problem being that of determining the refractive index of air, a mixture of gases. To good approximation, the refractive index of a mixture of ideal gases is the volume-weighted average of the refractive indices of its components. But this simple mixing rule does not hold as well for liquids or, worse yet, mixtures of particles and over the years many attempts have been made to formulate mixing rules with a greater range of validity. For more on this see Akhlesh Lakhtakia, Ed., 1996: *Selected Papers on Linear Optical Composite Materials*, SPIE Optical Engineering Press. This compendium contains many of the classical papers on the subject of homogenizing composite media.

Spectral measurements of a blue sun observed at Edinburgh in 1950 are reported by Robert Wilson, 1951: The blue sun of 1950 September. *Monthly Notices of the Royal Astronomical Society*, Vol. 111, pp. 478–89, reprinted in the compendium *Selected Papers on Scattering in the Atmosphere* cited at the end of Chapter 8.

Problems

3.1. If multiple scattering is negligible, a beam of radiation is attenuated exponentially by scattering just as it is by absorption (see Sec. 5.2.3). Thus the irradiance of a monodirectional beam transmitted through a scattering medium (with negligible absorption) is $F = F_o \exp(-\beta z)$, where β is the scattering coefficient. As in Section 2.3, we can determine the negative of the flux divergence: $\beta F_o \exp(-\beta z)$. But this leads to a perplexing result: the negative flux divergence and hence the heating rate appear to be nonzero even in a medium with no absorption; moreover, this heating rate (at $z = 0$) increases with increasing scattering coefficient. Explain.

3.2. In Section 3.3 we derive a wave equation for the (small) displacements of a string under tension. The motion of the string is not damped. Generalize this one-dimensional wave equation to a string with damping. Use the simplest damping law you can think of. After you have obtained the wave equation, find harmonic solutions (sinusoidal waves). Interpret your solutions. Find the power transmitted by the waves. Interpret your result. This problem is much easier if you use complex solutions to the wave equation.

3.3. Several years ago we received a letter from a former student. He had seen a series of dark bands in the sky, which he interpreted as an "interference pattern". To the extent that we understood the student's reasoning, it seemed that he attributed this pattern to interference of the light scattered by ice particles in a cirrus cloud. Suppose that you had received this letter. What would have been your response?

3.4. Why does the wave function $A \exp(ikx - i\omega t)$ *not* strictly describe a laser beam?

HINT: This is not a difficult problem. You do not need to know anything about lasers but you should have at least seen one. You may assume that light from a laser is exactly monochromatic and exactly monodirectional. This is not a problem about lasers but rather about reality, idealizations of reality, and the differences between them.

3.5. The following appeared in a paper (Sean Twomey, 1980: Direct visual photometric technique for estimating absorption in collected aerosol samples. *Applied Optics*, Vol. 19, pp. 1740–1) in a scientific journal several years ago:

"There is at the present time considerable interest in absorption of solar radiation by aerosol particles in the atmosphere, and it has been speculated that such absorption may have climatic significance.

Ground-based measurements of solar radiation are fundamentally incapable of giving such absorption directly and without assumptions, since . . ."

Please supply the missing arguments in support of the author's assertion in the second paragraph (italicized). Before you write a long dissertation, reflect for a moment that the author of this paper needed only 17 words with which to make his point forcefully and clearly.

3.6. Estimate an upper limit for the imaginary part of the refractive index of ordinary window glass in the middle of the visible spectrum. Then do the same for tinted glass.

3.7. Sir William Henry Bragg wrote two delightful works of popular science, one entitled *The Universe of Light*, the other *The World of Sound*. Discuss the difference in the titles.

3.8. The frequency ν of incident light scattered by a moving scatterer is Doppler shifted to a frequency ν'. But the energy of a photon is $h\nu$, which means that the energy of a photon is changed upon scattering. Doesn't this violate conservation of energy? Suppose that a (nonabsorbing) gas is illuminated by light. What happens to the energy of the gas as a consequence of the Doppler shift?

HINT: It might help to review Section 1.1.

3.9. These days it has become fairly common in the United States to be driving along a highway and come upon a sign at the side of the road flashing YOUR SPEED IS X MPH. Is this speed equal to, greater than, or less than your actual speed? Explain your answer. Do not invoke "experimental error" or "friction" or some other such nonsense.

3.10. In H. G. Wells's novel, *The Invisible Man*, the title character makes himself invisible by drinking a potion that makes his refractive index equal to that of air at visible wavelengths. This then enables him to do all kinds of dastardly things that he would not have been able to do if he were visible. Setting aside the fact that drinking something that would transform the refractive index of all human cells to 1 is absurd, everything else in this novel is scientifically correct with the exception of one subtle but fatal flaw. What is it? The Invisible Man functions just like ordinary human beings.

3.11. We state in Section 1.1 that frequency is more fundamental than wavelength but did not explain why. Supply the missing explanation.

3.12. Suppose that a blackbody with temperature T is moving relative to a stationary observer with line-of-sight velocity v. Derive an expression for the difference between the frequency of blackbody peak irradiance measured by the stationary observer and that measured by an observer moving with the blackbody. Don't bother to use the relativistically correct expression for the Doppler shift. The approximate expression, Eq. (3.74), is accurate (within about 10%) for speeds less than about $0.5c$.

HINT: You can use results in Section 1.2 or simply write down the answer. What does this result tell you about the velocity required such that the perceived color of a (visible) blackbody (e.g., a star) is appreciably different for the two different observers? This problem has historical significance. Doppler's original work was aimed at explaining the different observed colors of stars by their different speeds. You might revisit this problem after reading Section 4.3.

3.13. We state in Section 3.4.6 that the distribution function Eq. (3.78) for molecular velocity components along a particular direction is a member of the same family as the Maxwell–Boltzmann distribution for molecular kinetic energies. Show that this distribution function, Eq. (1.8), follows from Eq. (3.78).

HINT: For a gas in equilibrium the probability distributions for velocity components in three orthogonal directions must be the same and independent. Transform from the probability distribution in rectangular Cartesian coordinates (in velocity space) to the probability distribution in spherical polar coordinates, which then leads to the probability distribution for kinetic energies by the theorem for transforming variables of integration.

3.14. A refractive index is an average quantity that describes, to good approximation (often but not always), the electromagnetic response of systems composed of huge numbers of scatterers (e.g., atoms and molecules). This is possible only because one cubic wavelength of the media to which refractive indices usually are applied (solids and liquids even at ultraviolet wavelengths) contain a great many scatterers, and hence an average over the relevant volume means something. But to a wave of sufficiently long wavelength, a cloud droplet (or a snow grain) is a molecule insofar as its scattering properties are concerned (i.e., a dipole scatterer). For what wavelengths is it therefore not absurd to speak of the refractive index of an ordinary cloud? What about for a snowpack (snow on the ground)?

3.15. We found the following example of the Doppler effect "in every day life" on a web site devoted to explaining the principles of Doppler echocardiography: "an observer stationed on a highway overpass readily notices that the pitch of the sound made from the engine of a passing automobile changes from high to low as the car approaches and then travels into the distance. The engine is emitting the same sound as it passes beneath, but the observer notices a change in pitch dependent on the speed of the auto and its direction." Sounds good but we don't believe it. Why?

3.16. Equation (3.111) for normal-incidence reflection was given for a negligibly absorbing medium (real refractive index). But this equation is valid even if the refractive index is complex. Under what general conditions (magnitudes of the real and imaginary parts of the complex refractive index) is this reflectivity close to 1?

3.17. A particle is a collection of molecules, and hence scattering by a particle is the sum of scattering by all its constituent molecules. Suppose that there were no such thing as interference of light. What would the scattering diagram (scattered irradiance as a function of direction) of a particle look like?

3.18. Is the scattering cross section at visible wavelengths of a small (compared with the wavelength) water droplet equal to, greater than, or less than the scattering cross section of a bubble of the same size in water? The source of illumination is the same for both scatterers. Try to answer this by physical reasoning before attempting a calculation. If there is a difference, try to give a physical explanation.

3.19. Fog lamps on automobiles often are yellow. Perhaps there are sound reasons for this. Be that as it may, various reasons have been given. Here is one from a website that answers questions about science. "It is important for fog lights to be one [color] (rather than white) because the different wavelengths (colors) of visible light scatter off the fog droplets differently. This phenomenon is known as 'dispersion', because the different colors of light in an image will separate from each other... If you illuminate the road with only one wavelength (color) of light, the images of the objects you will see still become somewhat blurry because of the scattering of the light by the fog, but at least you won't have the extra problems from dispersion. So, if we want to use just one wavelength of light, which wavelength should we use? It turns out that light with short wavelengths scatters more than light with long wavelengths. So a long wavelength is best." Discuss this explanation in light of what is in this chapter and what you know about fog (as the term is used by meteorologists and most people). We revisit this problem in Chapter 4 (see Prob. 4.64).

3.20. A plane harmonic wave transmitted in an optically homogeneous medium is the superposition of an incident (exciting) wave in free space and all the forward-scattered waves it excites. Thus the phase of the transmitted wave is, in general, shifted from that of the incident wave. Show that the (real) refractive index is a phase-shift parameter. This is a simple problem to reveal another – and in our opinion better – interpretation of the refractive index than the ratio of the speed of light in a vacuum to the phase speed in a medium. Phase speeds have no physical significance whereas phase shifts do. Moreover, once we look upon the refractive index as a phase shift parameter, this should put an end to hand-wringing over supposedly forbidden ($> c$) "speeds of light" because we are not encumbered by expectations that transmitted waves must always be retarded or advanced.

3.21. In his contribution to *Colloid Chemistry* (cited in the references) W. H. Martin (p. 549) asserts that "From the standpoint of the diffraction theory a perfectly regular arrangement of molecules within a crystal would result in no scattering". By "scattering" here he means lateral scattering. Examine this assertion by determining scattering by a one-dimensional crystal: N identical dipolar scatterers equally spaced along a line. The direction of the incident wave is parallel to this line. Follow the same approach as that in Section 3.4.9. Equation (1.39) is useful. The expression you obtain for the scattered energy might look familiar. Consider the limiting cases of a separation between adjacent scatterers much smaller than and comparable with the wavelength. Carefully examine the strict validity of every assumption and approximation you make, especially when the separation is much smaller than the wavelength. This is an open-ended problem that you can explore almost endlessly. Its purpose is to make you think about the perils of making absolute statements ("no scattering") on the basis of approximate theories, especially if the magnitude of the physical quantity to be predicted is small in some sense.

3.22. In Section 3.4.6 we show that the frequency of radiation shifts because of interaction with moving matter. But this means that photon energies also shift, which seems to suggest that energy is not conserved. Show that it is. For simplicity, take the moving matter to be a disc of area A, much larger than the wavelength, moving with a speed v much smaller than c. The disc is illuminated at normal incidence and is nonabsorbing with reflectivity R and transmissivity T.

3.23. Consider two stretched strings, both infinitely long, with different mass densities, and joined at $x = 0$. The tension in both strings is, of course, the same. A time-harmonic wave propagating in the $+x$ direction and incident on this interface between two different strings gives rise to a reflected wave in the string $-\infty < x < 0$ and a transmitted wave in the string $0 < x < \infty$. At $x = 0$ the string displacement y and its slope must be continuous. Determine the amplitudes of the reflected and transmitted waves (relative to that of the incident wave). Show that energy flux (time-averaged) is conserved. This problem is essentially identical to reflection and transmission of electromagnetic waves. Note the similarity between your result and the Fresnel coefficients in Section 7.2.

3.24. This problem is related to Problem 3.23. Consider a finite stretched string of length h, attached at one end ($x = 0$) to an infinite string with a different mass density and at the other end ($x = h$) to another infinite string with the same mass density as the string $-\infty < x < 0$; an incident harmonic wave in this region propagates in the $+x$ direction. Find the ratio of the reflected ($x < 0$) energy flux (time-averaged) to the incident flux (reflectivity) and the ratio of the transmitted ($x > h$) energy flux to the incident flux (transmissivity). Compare your result with Eq. (5.24). Show that energy is conserved (reflectivity + transmissivity = 1).

HINT: Two sets of boundary conditions now have to be satisfied (at $x = 0$ and $x = h$).

3.25. We show in Section 2.8 that the absorption cross section per unit volume of a large (compared with the wavelength), weakly absorbing particle is approximately equal to the bulk absorption coefficient of the particle material. Show that this same approximation holds (to within about a factor of 2) at the other extreme, a particle small compared with the wavelength, provided that the real part of the refractive index lies between about 1 and 2 and the imaginary part is sufficiently small. Take the particle to be a sphere in air. This result is a useful rule of thumb if used with due caution.

3.26. How does the scattering cross section of atmospheric molecules compare with their geometrical cross section? Keep in mind that the diameter of a molecule is not a precisely defined quantity.

3.27. According to Eqs. (3.134) and (3.135) the extinction and scattering cross sections for a small sphere (in air, say) become infinite if the complex refractive index has a particular (finite) value. What is that value? Can you find materials that, at some wavelengths, have approximately this value?

3.28. Derive the Doppler line shape [Eq. (2.135)] beginning with Eqs. (3.79) and (3.80). Keep in mind that the Doppler shift for absorption is half that for scattering.

3.29. The scattering cross section of a particle is often greater in air than in a negligibly absorbing medium with a higher refractive index (e.g., water). But is this *always* true? Can you find any exceptions?

HINTS: First try to answer this question in a general way for an arbitrary sphere, which for simplicity take to be negligibly absorbing. Keep in mind that the wavelength for the size parameter is that in the medium surrounding the particle. The chain rule for differentiation is useful. To obtain a specific example, consider a sphere very small compared with the wavelength.

3.30. Give the simplest example you can think of to show that the scattering cross section of a particle has no necessary relation to its mass.

3.31. Consider two dipolar antennas each driven to radiate by their own independent power sources. Absorption is negligible. The antennas are mutually coherent and their power sources are fixed. Move the antennas closer and closer together until the radiation from them is in phase. Now the total radiated power should be twice what it was when the antennas were far apart (relative to the wavelength). But this seems to violate conservation of energy because the power sources are fixed. Explain.

3.32. Many years ago one of the authors used to give lectures as part of a short course on atmospheric aerosols. One of his topics was optical constants. He would pull out of his pocket a chunk of obsidian, a black glass of volcanic origin, wave it in the air and, with lots of dramatic flourishes, proclaim it to be the blackest substance on the planet, "blacker than the inside of a bruised crow", and assert that it therefore *must* have a huge imaginary index. Then he'd ask his audience to estimate its value. These were seasoned engineers and scientists, not innocent students. As expected (and hoped for) almost everyone would fall into the trap and blurt out idiotic responses: ten, one hundred, one thousand, a million. One of those who estimated a million earned his living (predictably) from remotely measuring the imaginary index of the atmospheric aerosol. But a very few clever folks didn't fall into the trap and were able to come up with a reasonable estimate. How?

3.33. We note in Section 1.1 that photons carry linear momentum, and hence a beam of radiation can exert a force (radiation pressure) on objects (see Prob. 1.32). The force exerted by a monodirectional beam on a particle is proportional to the beam's momentum flux, which is proportional to its irradiance. This force is the product of the momentum flux and the radiation pressure cross section. A rigorous derivation of this cross section is difficult, even for a sphere, but by physical grounds you should be able to write down the correct expression for an arbitrary particle.

We also note in Section 1.1 that photons can carry angular momentum, a consequence of which is that a circularly polarized beam (Sec. 7.1.3) can exert a *radiation torque* on a particle. This torque is the product of the angular momentum flux (proportional to irradiance) and the radiation torque cross section. A rigorous derivation of this cross section for a sphere is even more formidable than a derivation of the radiation pressure cross section. See Philip L. Marston and James H. Crichton, 1984: Radiation torque on a sphere caused by a circularly-polarized electromagnetic wave, *Physical Review A*, Vol. 30, pp. 2508–16. The senior author of this book was the second reviewer of this paper. The first reviewer had rejected it on the grounds that the result was obvious. No, the result was simple and plausible, which is not the same as obvious.

4 Radiometry and Photometry: What you Get and What you See

What the eye or photographic film or instruments such as photomultiplier tubes *get* is incident radiant energy over a time interval (radiant power). But film and instruments cannot be said to *see* anything. Only the eye can see, and seeing is mostly the processing of radiant energy by the organs of seeing – retina, optic nerve, and, most important, the brain – not the simple formation of images on the retina. It is sometimes said that the eye is just like a camera. It would be more accurate to say that the eye is *not* just like a camera except in the most trivial sense, namely, they both contain arrangements of lenses to form images. Yet even the most complicated camera, a mere chunk of metal and glass, cannot begin to perform the feats routinely done with ease by the human and animal eye–brain combination. Our brains create a visual world beginning with light as the raw material.

Wavelength is not a synonym for color nor is radiant power a synonym for brightness. But we begin with these strictly physical properties of light (radiometry), then proceed to the sensations it produces in the human observer (photometry).

4.1 The General Radiation Field

Light is a superposition of electromagnetic waves, intertwined electric fields $\mathbf{E}$ and magnetic fields $\mathbf{H}$. Because these fields are vectors, so are electromagnetic waves. They satisfy vector wave equations similar to the scalar wave equation derived in Section 3.3 for the vibrating string. We usually are most interested in the rate at which radiant energy is transported by electromagnetic waves. The electric and magnetic fields determine this transport rate by way of the *Poynting vector*

$$\mathbf{S} = \mathbf{E} \times \mathbf{H}, \tag{4.1}$$

where $\mathbf{E}$ and $\mathbf{H}$ are the fields at a point. If this point is on a surface (real or mathematical), the rate at which radiant energy is transported across unit area of that surface is

$$\mathbf{S} \cdot \mathbf{n}, \tag{4.2}$$

where $\mathbf{n}$ is a unit vector normal to the surface. The derivation of Eq. (4.1), boiled down to its essence, is fundamentally no different from, although more complicated than, the derivation in Section 3.3 of the energy flux vector for a vibrating string. We determined the time rate of change of kinetic and potential energy of a finite length of string, then noted that this was

Fundamentals of Atmospheric Radiation: An Introduction with 400 Problems. Craig F. Bohren and Eugene E. Clothiaux
Copyright © 2006 Wiley-VCH Verlag GmbH & Co. KGaA, Weinheim
ISBN: 3-527-40503-8

equal to the difference in energy fluxes at its end points. Similarly, Eq. (4.1) is obtained by determining the time rate of change of electric and magnetic energy within a bounded volume and noting that this is equal to the integral of the Poynting vector over the bounding surface. The energy flux vector for the string, Eq. (3.27), is the product of two functions. Similarly, the energy flux vector for the electromagnetic field, Eq. (4.1), is the (vector) product of two fields.

The scalar quantity Eq. (4.2), with the dimensions of power per unit area, is what we are after, but to get it we need two vector fields ($\mathbf{E}$ and $\mathbf{H}$) in order to determine a third ($\mathbf{S}$). This is possible in principle, but, except in very restricted circumstances, is essentially impossible in practice. We don't live in a world of simple electromagnetic waves. Just look around the room in which you are reading these words. You receive light from all directions, differing in amount and in its spectrum for each direction, a complex mosaic that changes when you move. How on earth could anyone ever sum the electric and magnetic fields originating from so many sources: walls, floor, ceiling, furniture, this book itself? Fortunately, we often don't have to determine these vector fields but can go straight to the desired scalar quantity, radiant energy transport. We can circumvent the electromagnetic field because most sources of visible and near-visible radiation are incoherent: there is no fixed phase relation between radiation from the walls and from the floor, from one part of a wall, and from another. Thus we can add the radiant energy from each source and ignore phase differences because they wash out when integrated over space or time. Another reason we can circumvent electromagnetic fields is that the wavelengths of the radiation of interest usually are much smaller than the objects with which the radiation interacts. If our radiation environment consisted of waves with wavelengths of order meters or more we might be in trouble. As we show in Section 3.4.2, the lateral coherence length of a source increases with its wavelength.

Radiometry is based on approximating electromagnetic radiation as a gas of photons. Like gas molecules, photons may be distributed in energy and in direction, and the properties of a photon gas may vary from point to point and from moment to moment. But photons do *not* interfere (see Sec. 3.4): they don't have phases. Or perhaps it would be more correct to say that phase differences wash out when we take averages over space or time, and so it is *as if* they had no phases. Ignoring phases results in an enormous simplification. Indeed, it makes radiometry (the measurement of radiation) possible. But all simplification comes with a price. We have to be alert to possible discrepancies between theory and observations. A theory in which phase is tossed out the window cannot possibly account for phenomena in which interference plays an important role.

The fundamental radiometric quantity is *radiance*, which to understand requires a thorough grasp of solid angle, which we turn to next. Solid angle is almost entirely absent from electromagnetic theory but plays a central role in radiometry.

4.1.1   Solid Angle

Any direction in a plane can be specified by a vector. Suppose that a vector $\mathbf{r}$ in the plane is rotated about a point O into another direction. The tip of the rotated vector traces out an arc of length s. The angle between these two directions, denoted by ϑ, is *defined* as the ratio s/r, and lies between 0 and 2π *radians* (the circumference of a circle of radius r is $2\pi r$). But another interpretation of angle is that it is the *measure* (size) of the set of *all* directions from O to points on the arc. Length is the measure of the set of all points lying on a line between

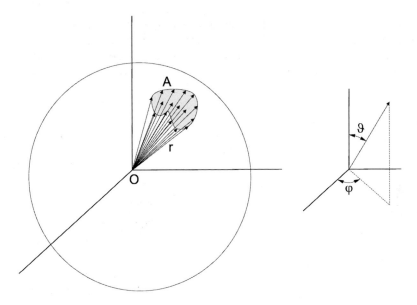

Figure 4.1: The solid angle of the infinite set of directions originating at O and ending on a surface with area A on a sphere of radius r is A/r^2.

two points. Area is the measure of the set of all points within a closed curve on a surface. And volume is the measure of the set of all points in space within a closed surface. These measures provide a way of comparing the size of one set of points with that of another. Far from being abstract, they are the means by which prices are assigned to rope, parcels of land, and gasoline (the gallon and liter are volumetric measures). Similarly, angle provides a way of comparing sets of directions in the plane. But directions are not confined to two-dimensional space. What is the measure of directions in three-dimensional space?

Consider a spherical surface of radius r on which a closed curve is inscribed (Fig. 4.1); the area of that part of the surface within this curve is A. Every vector from the origin O to a point on A specifies a direction in space. The measure of the set of all these directions is its *solid angle*

$$\Omega = \frac{A}{r^2}, \tag{4.3}$$

which lies between 0 and 4π *steradians* (abbreviated as sr). In principle, Ω can be determined by evaluating a surface integral:

$$\Omega = \frac{1}{r^2} \iint r^2 \sin\vartheta \, d\vartheta \, d\varphi = \iint \sin\vartheta \, d\vartheta \, d\varphi, \tag{4.4}$$

where ϑ and φ are spherical polar coordinates (co-latitude and azimuth, respectively) and the limits of integration are determined by A. Let's use Eq. (4.4) to determine the solid angle subtended by the sun at the surface of Earth, by which we mean the solid angle of the set of all directions from a point on Earth to the sun. We need this quantity later. The sun is azimuthally symmetric and its angular width, ϑ_s, is about $0.5°$ ($\pi/360$ rad). The solid angle subtended by the sun is therefore

$$\Omega_s = \int_0^{2\pi} \int_0^{\vartheta_s/2} \sin\vartheta \, d\vartheta \, d\varphi = 2\pi \left(1 - \cos\frac{\vartheta_s}{2}\right) \approx 6 \times 10^{-5} \text{ sr.} \tag{4.5}$$

Because $\vartheta_s \ll 1$, we can expand the cosine in Eq. (4.5) and truncate after the first two terms to obtain

$$\Omega_s \approx \pi\vartheta_s^2/4 \text{ sr.} \tag{4.6}$$

We could have obtained Eq. (4.6) by dividing the area of a small (planar) disc of radius $r\vartheta_s/2$ by r^2.

Any direction in space can be specified by a unit vector $\boldsymbol{\Omega}$, and so we sometimes write Eq. (4.4) symbolically as

$$\Omega = \int_\Omega d\boldsymbol{\Omega}. \tag{4.7}$$

This does *not*, however, denote integration over the variable $\boldsymbol{\Omega}$ just as a volume integral written as

$$\int_V dV \tag{4.8}$$

does not denote integration over the variable V. Equation (4.7) is simply a more compact way of writing Eq. (4.4).

4.1.2   Radiance

Before tackling radiance we need one more result. Consider a monodirectional, monochromatic, uniform beam of light. To obtain a measure of the amount of radiant energy transported by the beam we imagine a surface A to be placed in the beam with the normal to the surface parallel to it (Fig. 4.2). We can in principle determine how many photons in unit time N_o cross A ; N_o multiplied by the photon energy is the amount of radiant power (energy per unit time) crossing A. Divide this quantity by A to obtain the radiant power crossing unit area. This quantity is not solely a property of the beam (the radiation field): it also depends on the orientation of A. Tilt A so that its normal makes an angle ϑ with the direction of the beam (chosen so that $\cos\vartheta$ is positive). Now we measure N photons crossing A per unit time. This is related to the previous number by

$$N_o \cos\vartheta = N. \tag{4.9}$$

It follows from this equation that

$$\frac{N_o}{A} = \frac{N}{A\cos\vartheta}, \tag{4.10}$$

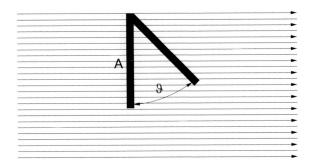

Figure 4.2: The rate at which radiant energy is transported across an area A in a monodirectional radiation field depends on the orientation of A.

and hence the quantity

$$\frac{N}{A \cos \vartheta},$$

(4.11)

where N is the rate at which photons cross A for any ϑ, is a property solely of the radiation field. Geometrically, $A \cos \vartheta$ is the area of the surface projected onto a plane perpendicular to the direction of the beam.

We use the term radiation *field* to describe the beam. For our purposes a field is any physical quantity that varies in space and time, usually continuously except possibly across surfaces. Field quantities often satisfy differential equations.

The (scalar) radiation field (not to be confused with the underlying vector electromagnetic field) is specified by the *radiance L*, a non-negative distribution function much like the distribution functions discussed in Section 1.2. As we show in Section 6.1.2, L satisfies an integro-differential equation, another reason for saying that L specifies a radiation field. Like all distribution functions, L is defined by its integral properties, and in general depends on position, direction, frequency, and time, so we sometimes write it as $L(\mathbf{x}, \mathbf{\Omega}, \omega, t)$ to explicitly indicate these dependencies. At any point in space consider a planar surface of area A, a set of directions with solid angle Ω, a set of frequencies between ω_1 and ω_2, and a time interval between t_1 and t_2. The total amount of radiant energy confined to this set of frequencies and directions, and crossing this surface in the specified time interval is given by

$$\int_{t_1}^{t_2} \int_{\omega_1}^{\omega_2} \int_A \int_\Omega L(\mathbf{x}, \mathbf{\Omega}, \omega, t) \cos \Theta \, d\Omega \, dA \, d\omega \, dt,$$

(4.12)

where Θ is the angle between the normal to the surface and the direction $\mathbf{\Omega}$. The cosine factor is introduced so that L is a property solely of the radiation field, not of the orientation of A [see Eq. (4.11)]. The dimensions of L are power per unit area, per unit solid angle, per unit frequency. The radiance defined by Eq. (4.12) is sometimes called the spectral or monochromatic radiance, and its dependence on frequency or, equivalently, wavelength sometimes indicated by a subscript: L_ω, L_ν, L_λ . The total or integrated radiance is the integral of L over a range of frequencies. Unless specified otherwise, by radiance we mean spectral radiance.

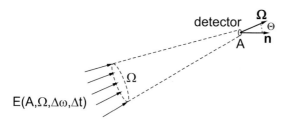

Figure 4.3: Radiant energy $E(A, \Omega, \Delta\omega, \Delta t)$, confined to a solid angle Ω around the direction Ω, is incident on a detector with area A. The unit vector $\mathbf{n}$ is normal to A.

Before writing Eq. (4.12) in a different way, we review the mean-value theorem of integral calculus, which we invoked in Section 1.2. According to this theorem (for a one-dimensional integral), between any two values x_1 and x_2, there is some intermediate value, call it $\overline{x}$, such that

$$\int_{x_1}^{x_2} f(x)\, dx = f(\overline{x})(x_2 - x_1), \tag{4.13}$$

where f is any continuous and bounded function. The mean-value theorem doesn't tell us how to find $\overline{x}$, only that it exists. The geometrical interpretation of this theorem is straightforward. The integral in Eq. (4.13) is the area of the region bounded by the continuous curve $y = f(x)$, the x-axis between x_1 and x_2, and lines perpendicular to the x-axis of length $y_1 = f(x_1)$ and $y_2 = f(x_2)$. According to the mean-value theorem there is some value $\overline{y}$ on this curve such that the area is $\overline{y}(x_2 - x_1)$.

The mean-value theorem also holds for multiple integrals such as Eq. (4.12), which therefore is equal to

$$\overline{L \cos\Theta}\; A\Omega\, \Delta\omega\, \Delta t, \tag{4.14}$$

where $\overline{L}$ and $\overline{\cos\Theta}$ indicate some value of L and $\cos\Theta$ over the domain of integration, $\Delta\omega = \omega_2 - \omega_1$, and $\Delta t = t_2 - t_1$. Equation (4.14) provides a means by which we can (in principle) measure the radiance at a point and in a particular direction. Place a detector with area A at the point where L is to be measured (Fig. 4.3). The detector is collimated in that it receives radiation only over a set of directions with solid angle Ω, around some direction Ω, and is equipped with a filter that passes only radiation in some frequency interval $\Delta\omega$. Measure the total radiant energy $E(A, \Omega, \Delta\omega, \Delta t)$ received by this detector over some time interval Δt. Divide this energy by $A\cos\Theta\Omega\Delta\omega\Delta t$ to obtain an estimate for L. (A can be oriented relative to Ω so that $\cos\Theta$ has the limiting value 1 as Ω shrinks.) Form the quotient

$$\frac{E(A, \Omega, \Delta\omega, \Delta t)}{A\cos\Theta\Omega\Delta\omega\Delta t} \tag{4.15}$$

for ever-decreasing values of A, Ω, $\Delta\omega$, and Δt until it no longer changes, then stop. At this point a fractional change in either A, Ω, $\Delta\omega$, or Δt leads to the same fractional change in $E(A, \Omega, \Delta\omega, \Delta t)$. The radiation field is now uniform over the geometric quantities A and Ω

and the intervals Δt and $\Delta \omega$. The quotient so obtained is L at space point **x**, in the direction Ω, for frequency ω, at time t.

We cannot let A become indefinitely small because at some point $(A < \lambda^2)$ the concept of a continuous radiance becomes invalid. But this is hardly cause for concern because we run into this kind of limitation all the time. For example, the density ρ at a point in a fluid often is defined as

$$\rho = \lim_{V \to 0} \frac{M}{V}, \tag{4.16}$$

where V is a volume containing the point and M the mass of fluid within this volume. But the limit of V here is not literally 0. When V shrinks to molecular dimensions (or smaller), the quotient in Eq. (4.16) undergoes wild fluctuations depending on whether V contains molecules or not. So the limit in Eq. (4.16) is interpreted to mean that we shrink V until the quotient no longer changes, then stop and call that quotient the density at a point. But a better definition of density, in our view, is that it is a distribution function. Its integral over any arbitrary (within limits) volume is the mass enclosed by that volume.

The only way to truly grasp radiance, or indeed any physical concept, is to become familiar with its properties, to observe how it behaves in as many contexts as possible. Defining radiance is only a first small step toward understanding it. One essential property of radiance is that it is *additive*: if several incoherent sources contribute to the radiance at a particular point and in a particular direction, the total radiance is the sum of the radiances from each source as if it were acting alone. Another property of radiance is its *invariance*, which we turn to next.

4.1.3   Invariance of Radiance

If absorption and scattering by the medium in which radiation propagates is negligible, radiance is invariant along a particular direction. By this is meant the following. Go to any point in the radiation field and determine the radiance there in the direction Ω. Now proceed along this direction. At any point on this path the radiance in the direction Ω is the same. The proof is as follows.

At any point insert a planar surface with area A, sufficiently small that L in the direction Ω is the same over every point of A, oriented so that its normal is parallel to Ω. The quantity LA is therefore the amount of radiation crossing A per unit solid angle around the direction Ω. A surface A'' at a distance r from A receives an amount of radiant energy

$$LA\Omega' = LA\frac{A''}{r^2}, \tag{4.17}$$

where Ω' is the solid angle subtended by A'' at A (both of which are $\ll r^2$). This is shown schematically in Fig. 4.4. If A'' is sufficiently small, the power intercepted by it (per unit area) is uniform and equal to

$$\frac{LA}{r^2}. \tag{4.18}$$

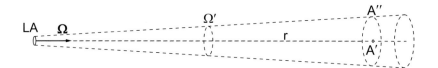

Figure 4.4: Invariance of radiance. The top figure shows the amount of radiation intercepted by A'' a distance r from A where the radiance is L. A smaller area A' within A'' is shown in the bottom figure. The scale is changed between top and bottom. The radiant energy intercepted by A' is LAA'/r^2, which equals $L'A'A/r^2$ in the absence of attenuation.

The power intercepted by a smaller area A' within A'' is therefore

$$P' = \frac{LA}{r^2} A'. \tag{4.19}$$

A is represented as much larger than A' in the bottom half of Fig. 4.4 because of the difficulty of representing all distances and areas to the same scale. At A' the radiance in the direction Ω is L'. In the absence of attenuation all of P' must have originated from A and only from A, and hence

$$L' = \frac{P'}{A'\Omega} = \frac{LA}{r^2\Omega} = L, \tag{4.20}$$

where $\Omega = A/r^2$ is the solid angle subtended by A at A'. Equation (4.20) is the invariance principle for radiance, which also follows from the equation of radiative transfer (see Sec. 6.1.2) when absorption and scattering are negligible. This principle may come as a surprise in light of the oft-repeated mantra that radiation decreases as the inverse square of distance. Some radiometric quantities under some circumstances obey this law, but radiance does not. And radiance is usually what we observe, as we show in the following section.

4.1.4   Imaging Devices and Radiance

An imaging device, such as the human eye, maps radiances onto a surface. To show this, consider a uniformly luminous disc with radius h at a distance $s(\gg h)$ from a simple lens with area A_{lens} (Fig. 4.5). This object disc, with area $A = \pi h^2$, is centered on the optic axis of the lens, and the normal to the disc is parallel to the optic axis. The lens forms an inverted image of the disc at a distance s' from the lens. The radius h' of the image disc is given by

$$\frac{h'}{s'} = \frac{h}{s}, \tag{4.21}$$

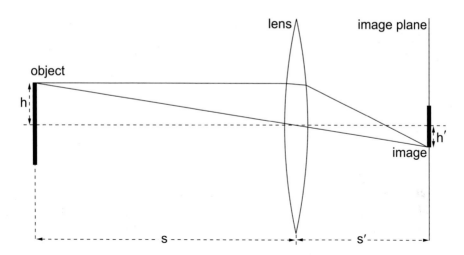

Figure 4.5: The size of an image in the image plane of a simple lens depends on the distance s between object and lens and its focal length s'.

and hence the area A' of the image disc relative to that of the object disc is

$$\frac{A'}{A} = \frac{s'^2}{s^2}.$$

(4.22)

The solid angle subtended by the lens at the object is approximately

$$\Omega_{\text{lens}} = \frac{A_{\text{lens}}}{s^2}$$

(4.23)

if $s \gg \sqrt{A_{\text{lens}}}$. Thus the radiant power from the disc intercepted by the lens is

$$L A \Omega_{\text{lens}} = L A \frac{A_{\text{lens}}}{s^2},$$

(4.24)

where L is the radiance from the object disc (in the direction of the optic axis). Implicit in Eq. (4.24) is the assumption that $\cos \Theta \approx 1$ for the angle Θ between the normal to the object disc and any ray from the object disc to the lens. If reflection and absorption by the lens are small relative to transmission, almost all the radiant energy [Eq. (4.24)] illuminates the image disc. Thus the amount of radiant power per unit area of the image disc is

$$L \frac{A_{\text{lens}}}{s^2} \frac{A}{A'},$$

(4.25)

which combined with Eq. (4.22) yields

$$L \frac{A_{\text{lens}}}{s'^2} = L \Omega'_{\text{lens}}$$

(4.26)

for the radiant power per unit area of the image disc. At each point of the image plane, the radiant power per unit area depends only on the radiance at the object point and a fixed quantity, the solid angle Ω'_{lens} subtended by the lens at the image plane. This comes about because of two opposing factors. As the object is moved farther from the lens, the radiant power it intercepts decreases as the inverse square of distance s [Eq. (4.24)]. But the area of the image also decreases in the same way [Eq. (4.22)], and hence the power per unit area of the image does not change.

Suppose that the lens is the human eye and the image plane is the retina. We can imagine any scene to be made up of small discs with different radiances, high radiances corresponding to what we call bright parts of the scene, and low radiances corresponding to dark parts. The radiances here are those in the direction toward the eye of an observer. Each disc in the scene is mapped onto its image on the retina. The size of the image disc decreases with distance but not the amount of radiant power per unit area of each image disc, which is proportional to the radiance of the object disc.

You can verify this for yourself by observing a white disc at ever-increasing distances on a clear day. Although the (angular) size of the disc decreases with distance, you perceive it to be just as bright.

But now we have to add some psychological and physical caveats. Although the size of retinal images decreases with distance, the human observer perceives the size of familiar objects to be more or less the same regardless of distance (within limits). This is called *size constancy*, a mechanism built into our brains and of which we are not directly aware. If we were to interpret our visual surroundings according to the changing sizes of retinal images, we might go mad. Familiar objects would appear to shrink as they recede from us. But this is not what we perceive, which you can verify by walking away from a familiar object such as a dog and noting that its perceived size does not markedly change. You do not gasp in amazement as you move away from a Doberman Pinscher and it shrinks to a Miniature Pinscher, then back to a Doberman as you return.

Another caveat is that the perceived brightness of an object does depend on its angular size. Again, you can verify this by cutting out circular discs of different sizes from the same sheet of white paper and pasting them onto the same large, uniform backdrop (say a large sheet of gray paper) illuminated uniformly by bright sunshine. Observe the discs at different distances and note which seems brighter. Objectively the radiance from all discs is the same if they are composed of the same material, against the same backdrop, and illuminated by the same light, but subjectively they are different. The objective pattern of radiant energy on the human retina is just the beginning of visual perception. The retina is connected by way of the optic nerve to the brain, which processes retinal images to create a visual world. Radiance variations are the bricks and mortar out of which the brain builds this world.

Now the physical caveat. As the distance s increases the solid angle subtended by the disc (source) at the lens decreases to the extent that the lateral coherence length (see Sec. 3.4.2) is greater than the diameter of the lens, which then produces an interference pattern. And as soon as interference comes into play, the assumptions underlying Eq. (4.26) break down. For example, Eq. (4.21) for the size of retinal images is no longer valid. From Eq. (3.50) it follows that we begin to run into the limitations of theory when

$$\frac{s}{h} > \frac{d_{\text{lens}}}{\lambda}, \tag{4.27}$$

where d_{lens} is the diameter of the lens. This is why Eq. (4.26) is not valid for stars, which are essentially point sources in that their details cannot be resolved with telescopes. The image size of a star is not determined by Eq. (4.21), which predicts (incorrectly) that this size steadily decreases with increasing distance s (for fixed h). When s/h is such that Eq. (4.27) is satisfied, the size of the image becomes greater than that predicted by Eq. (4.26), less distinct, and its size constant. The amount of radiant power received by the lens, however, is inversely proportional to distance squared, and because the image size is constant, the amount of radiant power per unit area of the image decreases similarly.

A distinction also must be made between an imaging device and a flat-plate detector. Such a detector receives light from scenes in which radiance may vary markedly with direction. The power recorded by the detector is proportional to an average radiance because each point on the plate receives light from many directions. If we want the distribution of radiance, we would have to collimate the detector to exclude light except that from a small set of directions. The purpose of lenses in imaging devices (eye, camera, slide projector) is to ensure that each point of the detector (retina, film, screen) receives light from only one point of the scene to be imaged. As we note in Section 1.4.7, the essence of imaging is one-to-one mapping.

4.1.5   A Simple Lens Cannot Increase Radiance

Many of us as children have used a magnifying glass to burn a piece of paper or to incinerate some unfortunate ant or even to start a fire. Indeed, the focal point of a lens is called a *caustic*, and something that is caustic burns: caustic soda burns skin, a caustic remark burns your ears. This common experience with lenses has unfortunately engendered the misconception that lenses can increase radiance.

Consider a piece of white paper illuminated by sunlight on a clear day. If we neglect attenuation by the atmosphere, the solar radiance at the paper is the same as that at the surface of the sun. From the definition of radiance, and given that the solid angle subtended by the sun at the surface of Earth is small, the radiant power incident on unit area of the paper is $L_s\Omega_s$, where L_s is the radiance of the sun. For simplicity we take the direction of the sun to be normal to the paper. Now suppose we interpose a lens between the paper and the sun at a distance f, the focal length of the lens, above the paper. Thus the sun is imaged onto the paper. If A_s is the (projected) area of the sun, the area of its image A_i follows from Eq. (4.22):

$$A_i = \frac{f^2}{r^2} A_s = f^2 \Omega_s, \tag{4.28}$$

where r is the Earth–sun distance.

The radiant power intercepted by the lens is $L_s\Omega_s A_{lens}$. We neglect reflection by the lens, which reduces this power by a few percent. The radiant power per unit area of the image of the sun on the paper is therefore

$$\frac{L_s\Omega_s A_{lens}}{A_i} = \frac{L_s\Omega_s A_{lens}}{\Omega_s f^2} = L_s\Omega_{lens}, \tag{4.29}$$

where Ω_{lens} is the solid angle subtended by the lens at its focal point. With no lens in place, each unit area of the paper receives $L_s\Omega_s$. With a lens in place, a unit area of the image of the sun (*not* the region surrounding this image) on the paper receives $L_s\Omega_{lens}$. In the limited image region, therefore, the radiant power (per unit area) increases if $\Omega_{lens} > \Omega_s$. But the radiance in this region is that of the sun, which follows from Eq. (4.29), the radiant power per unit area of the sun's image. To convert this to radiance, divide by the solid angle of the lens, which yields L_s.

To make this clearer, imagine that we can shrink ourselves to the size of an ant and position ourselves on the image of the sun on the paper. What would we experience? First, we would be uncomfortably hot because the radiant power per unit area increases. But we would see the sun just as bright as without the lens. That is, the radiance of the sun would not change, although there would be an observable difference in the appearance of the sun: its angular width would be greater seen through the lens. The function of magnifying lenses is to increase *angular* sizes. Thus an ant at the focal point would see a sun of larger angular size. How much larger depends on the magnification of the lens ($2\times$, $5\times$, $10 \times \ldots$).

The origins of the misconception about lenses increasing radiance are not difficult to discover. A uniform surface illuminated by sunlight is uniformly bright. Take a lens and focus an image of the sun onto such a surface. The result is a bright spot (the image of the sun) surrounded by a much larger, darker area. All a lens can do is redistribute solar radiation, and hence if the radiant power per unit area of the image increases, the radiant power per unit area surrounding the image must decrease accordingly so that total radiant power is conserved. You can demonstrate this for yourself with a magnifying lens. Image the sun onto a uniform surface. If you are indoors, an incandescent lamp will do. What you'll observe is a bright spot surrounded by a darker area. It is this increase in *contrast* that is mistaken for an increase in the incident radiance. Humans respond to relative, not absolute differences. The radiance of gray paper illuminated by sunlight is likely to be greater than the radiance of white paper illuminated indoors by an incandescent lamp, and yet we perceive the white paper to be brighter than the gray because the white paper is brighter than its surroundings.

4.1.6   Radiance Changes Upon (Specular) Reflection and Refraction

Denote by L_i the radiance in a particular direction Ω_i incident on an optically smooth interface between two optically homogeneous media. This direction is specified by the angle ϑ_i between Ω_i and the normal to the surface. Let A be an area on this surface, sufficiently small that the radiance over A is approximately constant. The radiant power incident on A is

$$L_i A \cos \vartheta_i \Omega_i, \tag{4.30}$$

where Ω_i is the solid angle of a set of directions centered around the direction of incidence and sufficiently small that the radiance is approximately constant. Let L_r be the radiance in the direction of reflection, which makes an angle $\vartheta_r = \vartheta_i$ with the normal. If the reflectivity is 100%, the reflected power is

$$L_r A \cos \vartheta_r \Omega_r, \tag{4.31}$$

where Ω_r is the solid angle of that set of reflected directions corresponding to all the incident directions in Ω_i. These two solid angles are equal, which follows from the law of specular

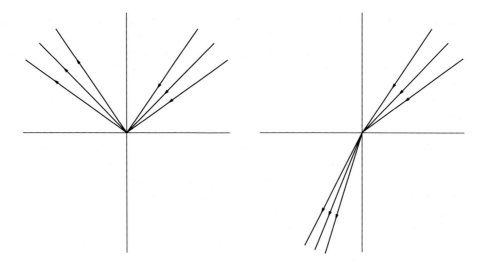

Figure 4.6: In specular reflection (left), incident rays in a set of directions with a given solid angle are mapped by reflection into a set of directions with the same solid angle. But in specular refraction (right), the solid angle of the set of refracted rays is less than that of the incident rays if the refractive index of the illuminated medium is greater than that of the medium containing the source.

reflection. That is, a set of incident directions is mapped into an equal set of reflected directions (Fig. 4.6). Conservation of radiant energy requires that Eqs. (4.30) and (4.31) be equal:

$$L_r = L_i. \tag{4.32}$$

This equation can be verified by looking at an object (a piece of white paper, say) and its reflection (image) in a mirror simultaneously. The image will be seen to be almost as bright as the object. We say "almost" because ordinary mirrors have reflectivities of not much more than 90% at visible wavelengths. If the reflectivity is denoted by $R(\vartheta_i)$, Eq. (4.32) becomes

$$L_r = R(\vartheta_i)L_i. \tag{4.33}$$

Equation (4.33) can be verified, at least qualitatively, as follows. The reflectivity of *any* surface approaches 100% as ϑ_i approaches 90° (glancing incidence). Take a black container – we used a black frying pan – and fill it with water to its brim. Hold a white object – we used a pill bottle – just above the surface of the water and observe the reflection of the object from different directions. If you crouch down so that your line of sight almost grazes the surface, you'll see a reflected image almost as bright as the object. Because the reflectivity of water at near-normal incidence at visible wavelengths is about 0.02 (see Fig. 7.6), the image is much less bright if you look down at the reflection.

The 50-fold reduction of radiance at near-normal incidence upon specular reflection by water is appreciable but not nearly so great as the reduction upon *diffuse* reflection (e.g., by snow and clouds), which we discuss in Section 4.2.

At visible wavelengths, the reflectivity of ice is almost the same as that of liquid water. You often can see from the windows of an airplane the consequences of specular reflection of sunlight by ice crystals. Although you might think of a mirror as being a more or less continuous surface, this is not necessarily so. A mirror can be distributed in space (smash a mirror and sprinkle the shards on the floor and you'll see what we mean). An example of this in nature is a collection of small ice crystals (e.g., plates) falling in air with their tip angles (the angle between the normal to a plate and the vertical) distributed around some small average angle. Perfect alignment would require all crystals to have a tip angle of zero. These crystals reflect an image of the sun, called a *subsun* (because it lies below the sun in angle), different from the sun in two respects. The radiance of the image is about 50 times less than that of the sun; the image is not perfectly circular because of the distribution of tip angles. Subsuns are one of the most frequently seen UFOs, although they are neither "unidentified" nor "flying" nor "objects". They are distorted images of the sun, perfect candidates for UFOs because they are much brighter than clouds, elliptical, and no matter how fast an airplane flies, always keep pace with it (as long as the supply of ice crystals lasts).

Now consider transmitted (refracted) radiance L_t. For simplicity we take the incident light to be propagating in air. The refractive index of the (negligibly absorbing) medium on which the light is incident is n. If we assume that the transmissivity of the medium is 1 (no reflection), energy conservation again requires that

$$L_i A \cos \vartheta_i \Omega_i = L_t A \cos \vartheta_t \Omega_t, \tag{4.34}$$

where ϑ_t is the angle of refraction and Ω_t is the solid angle of the set of transmitted directions corresponding to the set of incident directions. The two solid angles follow from Eq. (4.4):

$$\Omega_i = \sin \vartheta_i \Delta\vartheta_i \Delta\varphi_i, \ \Omega_t = \sin \vartheta_t \Delta\vartheta_t \Delta\varphi_t. \tag{4.35}$$

From the law of refraction we have

$$\sin \vartheta_i = n \sin \vartheta_t. \tag{4.36}$$

Square both sides of Eq. (4.36) and take the derivative of the result with respect to ϑ_t:

$$2 \sin \vartheta_i \cos \vartheta_i = 2n^2 \sin \vartheta_t \cos \vartheta_t \frac{d\vartheta_t}{d\vartheta_i}. \tag{4.37}$$

Now approximate the derivative in this equation as the ratio of differences to obtain

$$\sin \vartheta_i \cos \vartheta_i \Delta\vartheta_i \approx n^2 \sin \vartheta_t \cos \vartheta_t \Delta\vartheta_t. \tag{4.38}$$

This result, Eqs. (4.34) and (4.35), and azimuthal symmetry ($\Delta\varphi_i = \Delta\varphi_t$) yield

$$n^2 L_i = L_t. \tag{4.39}$$

Thus for $n > 1$ the transmitted radiance is *greater* than the incident radiance (assuming no reflection). The reason for this is that upon refraction a set of incident directions is mapped into a smaller set of refracted directions (Fig. 4.6). Radiance is power per unit solid angle (and per

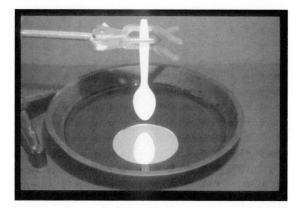

Figure 4.7: A white spoon suspended above a black pan filled with water and with a mirror lying on the bottom of the pan. The image of the spoon reflected by the mirror, and twice refracted, is almost as bright as the spoon. If you look carefully you can see the faint image of the spoon specularly reflected by the water.

unit area), and so a decrease in solid angle, all else being equal, results in an increase in radiance. If the rays in Fig. 4.6 are reversed, what is meant by incident and transmitted is reversed and the transmitted radiance is *less* than the incident radiance. Consider a 100%-reflecting mirror under water, which, as shown in Section 2.1, is negligibly absorbing at visible wavelengths over distances of tens of centimeters. What is the radiance seen by someone above the water looking at the mirror under water? If transmission from air to water increases radiance, reflection by the mirror doesn't change it, and transmission from water to air decreases it, the net effect should be zero. The radiance transmitted into the water is

$$L_{t1} = n^2 L_i. \tag{4.40}$$

Reflection does not change the radiance

$$L_r = L_{t1} = n^2 L_i. \tag{4.41}$$

L_i is the radiance incident at the air–water interface, and hence L_r is the radiance at the water–air interface. Thus the radiance transmitted from water to air is

$$L_{t2} = \frac{L_r}{n^2} = L_i, \tag{4.42}$$

as expected.

A simple demonstration of the reversibility implied by Eq. (4.42) can be obtained with a mirror under water in a black container. Suspend a white object (as a light source) above the surface of the water, and observe this object together with its transmitted-reflected-transmitted image (Fig. 4.7).

The assumptions of no reflection and $n \neq 1$ are inconsistent because the latter implies a nonzero reflectivity. But this often is not a serious error provided that we confine ourselves to near-normal angles of incidence. For example, the transmissivity of water (from air) at visible and near-visible wavelengths is about 0.98 (Fig. 7.6) for angles within about $40°$ of normal incidence, whereas n^2 is about 1.77.

Because Eq. (4.39) can be written

$$L_i = \frac{L_t}{n^2}, \tag{4.43}$$

one sometimes finds the assertion that radiance is not invariant in a particular direction (in the absence of attenuation) but rather radiance divided by refractive index squared. This needs to be qualified by the caveat that departures of the transmissivity from unity are neglected.

4.1.7   Luminance and Brightness

Up to this point we implicitly assumed that human observers respond to radiance, which is not incorrect but is incomplete. We respond to *integrated* radiance. Moreover, light of each wavelength over the visible spectrum does not evoke the same sensation of brightness even for equal radiance. The peak *luminous efficiency* of the human eye occurs in the middle of the visible spectrum (550 nm) and drops to nearly zero at 400 nm on the short-wavelength side of the peak and at 700 nm on the long-wavelength side. We show the luminous efficiency in Section 1.2 in our discussion of distribution functions, but for convenience reproduce it in Fig. 4.8. Strictly, this figure shows the luminous efficiency for *photopic* vision, when the eye is adapted to what loosely may be called normal light levels (e.g., daylight) in contrast with *scotopic* vision, when the eye is adapted to low light levels (e.g., moonlight). In photopic vision brightness matches are determined mostly by the cones in the retina, whereas in scotopic vision they are determined mostly by the rods; *mesopic* lies between these two extremes. But it is difficult to specify the exact conditions under which each type of vision predominates. The scotopic luminous efficiency curve has about the same shape as the photopic luminous efficiency but its peak is shifted about 50 nm toward the blue, giving rise to the *Purkinje effect*. This is manifested by changes in brightness of red and blue objects under different levels of illumination. If they appear equally bright in daylight, the blue object may be perceived to be brighter after the sun has set. Figure 4.8 is not that for a particular human observer but rather an average for many observers.

The eye integrates the spectral radiance over the visible spectrum weighted by the luminous efficiency. This integrated quantity, called the *luminance*, is

$$K \int LV \, d\lambda, \tag{4.44}$$

where V is the (dimensionless) luminous efficiency of the human eye (also called luminosity, relative luminance, etc.). The constant K is 683 lumens per watt (lm W^{-1}), and hence luminance has the units of lumens per square meter per steradian, called the *nit*, an appropriate term (although not likely chosen to be humorous) given the level of nitpicking in photometry. The *candela* is the lumen per steradian, and hence another unit for luminance is candela per square meter (cd m^{-2}). The integral in Eq. (4.44) is over wavelength but could just as well be over frequency. Luminance, unlike radiance, takes into account the spectral response of

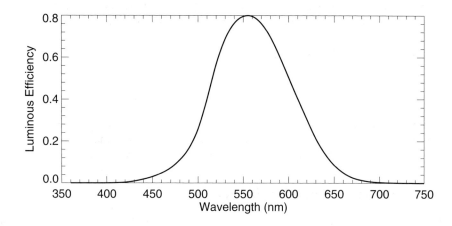

Figure 4.8: Luminous efficiency for the (average) human observer under photopic conditions.

a particular detector, the human eye (by which we mean the retina connected to the brain). Luminance is a *photometric* quantity; its counterpart, radiance, is a *radiometric* quantity. The qualifier photometric signals that the response of the human eye is taken into account. Luminance is a *psychophysical* quantity. The physical part comes from L in Eq. (4.44); the psychological part comes from V, which is obtained, in essence, by asking human observers about what they see.

The observable consequences of Fig. 4.8 and Eq. (4.44) were driven home to us when a student came to class wearing a white jacket with red and green fluorescent stripes on its sleeves. At a glance it appeared that these stripes might be brighter (higher luminance) than the rest of the jacket. We had a photometer near to hand, and so were able to make the necessary measurements. Indeed, the stripes were brighter. These measurements fly in the face of most of our everyday experiences with colored objects. Dye a white T-shirt, for example. The dye preferentially absorbs light of different visible wavelengths, which reduces the reflected radiance relative to that of the white shirt under the same illumination. Thus the luminance of the dyed shirt decreases, and is perceived to be darker than its white cousin. This is what usually happens. But there is a catch. Suppose that some light on the short-wavelength side of the peak of the luminous efficiency curve is shifted to longer wavelengths instead of being absorbed. The amount of reflected radiant energy is not increased but its spectrum is shifted to a region of higher luminous efficiency, and hence a higher luminance. This can happen with *fluorescent* objects, especially ones with high fluorescent yield (one incident photon gives rise to about one photon but of a different frequency).

This experience caused us to wonder about the function of the fluorescent orange vests worn by highway workers and hunters. The spectrum of such a vest (in sunlight) is shown in Fig. 4.9 together with that of an adjacent piece of white paper and of an orange (fruit), the paragon of orangeness. The vest is perceived to be orange, but its luminance is about 20% greater than that of the orange. The kinds of orange objects found in nature (in the woods during deer hunting season, for example) are appreciably darker than fluorescent vests, which

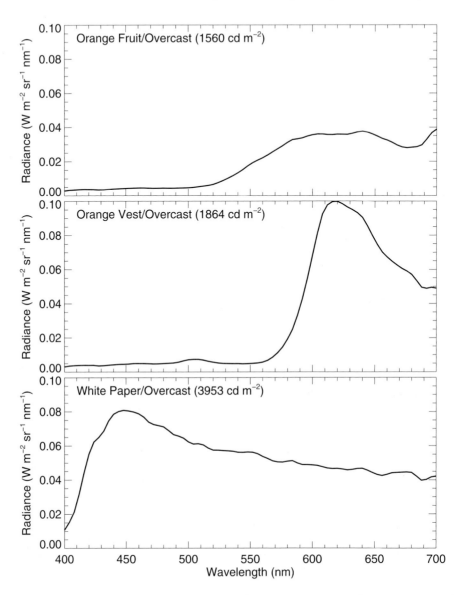

Figure 4.9: Measured spectra and luminances of an orange, an orange fluorescent vest, and white paper, all illuminated by an overcast sky.

by their unusually high luminances are seen to be unnatural, or perhaps extra-natural would be a better term. A deer hunter in the woods does not want to blend in with the surroundings but rather to stand out. A white object satisfies this criterion, but, alas, so does the rear end of a whitetail deer. An orange fluorescent vest or cap is bright and not likely to be confused with less bright orange vegetation.

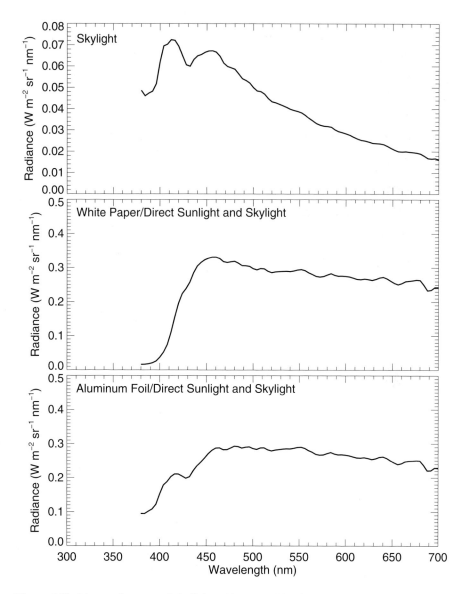

Figure 4.10: Measured spectra of skylight, white paper illuminated by daylight (direct sunlight and skylight), and aluminum foil illuminated by daylight. Note that the dip in the skylight spectrum at around 430 nm is absent in the spectrum of the white paper but present in the spectrum of the aluminum foil. The paper fluoresces whereas the foil does not.

Not only are fluorescent orange vests unnaturally orange, some white paper is unnaturally white, which we discovered by accident. To reduce direct sunlight, we measured reflection by white paper illuminated by sunlight and skylight (daylight); skylight spectra could be obtained without the white paper intermediary. The overhead skylight spectrum in a clean environment is approximately the solar spectrum modulated by a smooth function, the inverse-fourth-power law (Fig. 8.1). Thus peaks and valleys in the sunlight and skylight spectra must originate in the solar spectrum. To our surprise, however, features around 400–450 nm in skylight spectra disappeared in light reflected by white paper (Fig. 4.10). To discover why we measured daylight reflected by crumpled aluminum foil, which restored the skylight spectral features. Clearly, white paper can distort spectra. To dig deeper we used an ultraviolet source with a narrow ($\approx$30 nm) peak around 375 nm. Light reflected by crumpled aluminum foil faithfully follows this source spectrum, whereas that reflected by white paper bears almost no resemblance to it because of fluorescence (Fig. 4.11). Incident light at short wavelengths is transformed into light in wavelength regions where the luminous efficiency of the human eye is greater. Fluorescent whiteners are added to laundry detergents to make clothes "whiter than white" and to white paper to make it brighter. To obtain a reflector that doesn't distort spectra requires inorganic (negligibly absorbing) powders of various oxides and carbonates.

The eye is neither a radiance detector nor a luminance detector but rather a *brightness* detector. Radiance has units (e.g., $W\,m^{-2}\,sr^{-1}$) and can be measured in princip. Luminance also has units and can be obtained from radiance by integration. But brightness, which has no units, is the "term most commonly used for the attribute of sensation by which an observer is aware of differences of luminance." What this means is that the eye can tell (within limits) when the luminances of two objects are equal but cannot tell their absolute luminances. The luminance of an object we would call gray is more than a hundred times greater when illuminated by direct sunlight than when illuminated by indoor lighting, and yet we still call it gray because we compare it to objects in its surroundings, not with some absolute standard tucked away in our heads.

4.1.8   A Few Words about Terminology and Units

By conscious design, not accident, we gradually introduced radiometric and photometric terms, as few as possible and only as needed. The tragedy of radiometry (and photometry) is that an inherently interesting subject – the best scientific instrument we carry with us everywhere is our eyes, and we are constantly immersed in a world of ever-changing brightness and color – has been made dreadfully boring by wallowing in a mire of terminology and units. This harsh view is shared by others, expressed with biting wit by R. C. Hilborn, who defined radiometry as an acronym for revulsive, archaic, diabolical, invidious, odious, mystifying, exotic terminology regenerating yawns. Terms are multiplied without end. Units in photometry border on the fantastic (foot-candle, talbot, nit, troland, candela, lux, lumen, stilb, foot-Lambert, nox, skot, and so on *ad nauseam*). Radiometry comes across as the science of terminology, its seeming objective being to multiply distinctions endlessly and thereby coin as many terms as possible. To make matters worse, the symbols are ghastly, quantities that are not derivatives yet written as derivatives, and strange-looking ones at that. Mass density rarely, if ever, is written as the derivative of mass with respect to volume [see Eq. (4.16)], but this sort of thing is done all the time in radiometry even though radiometric quantities are distribution functions fundamentally no different from mass density.

Other than in this paragraph we do not use the term *intensity* in this book. Intensity is an Alice-in-Wonderland term that means whatever you want it to mean. It is not uncommon

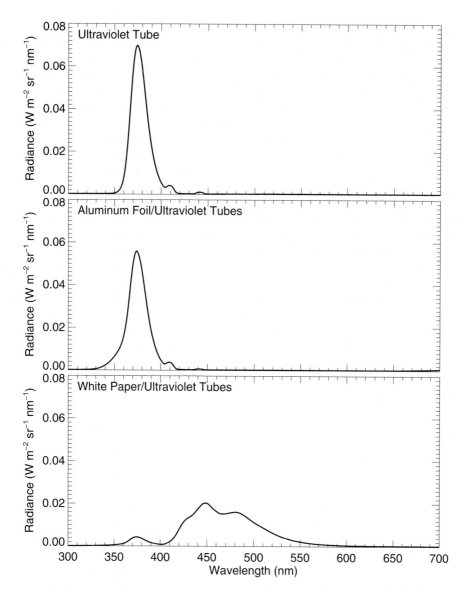

Figure 4.11: Measured spectra of an ultraviolet tube, tube light reflected by aluminum foil, and tube light reflected by white paper. The spectrum reflected by the foil faithfully follows that of the source, whereas because of fluorescence by the paper, its reflected spectrum bears little relation to the source.

to find it used for two (or more) different radiometric quantities in the same paper or book. A humorous yet serious attack on the misuse of intensity was published by J. C. Palmer in a journal devoted to measurement. The first sentence of his abstract hints at what follows: "The current misuse of the term *intensity* in physics is deplored." Palmer notes that he has seen intensity used in five different ways in optics and, like us, has encountered authors who change their meaning as the mood strikes them. After a tongue-in-cheek search for even more terms, Palmer halts abruptly and shouts (in boldface type) that "intensity is an SI base quantity." That is, (luminous) intensity "takes its place alongside the other six SI base quantities: length (metre), mass (kilogram), time (second), electric current (amperes), thermodynamic temperature (kelvin), and amount of substance (mole)." Intensity used in its proper sense is radiant (or luminous) power per steradian. Palmer's parting shot is, "The message is clear; those who use *intensity* in a context **other** than W/sr are either uninformed or just plain careless and sloppy. I can't comprehend a 'special reason' to redefine an SI base quantity, can you?" Using intensity to mean radiance is akin to using length to mean what to the rest of the world is area.

We do not especially like the term radiance, if for no other reason than that one has to be sober in order to pronounce it clearly enough for it to be distinguished from irradiance (Sec. 4.2). But we accept this term because it has been recommended by the Optical Society of America (OSA) for more than 50 years and endorsed by other international organizations devoted to standardization. OSA is the largest optical sciences society in the world, and the bulk of papers on optical science (including atmospheric optics) has been and continues to be published in its journals, the editors of which are supposed to frown on manuscripts in which intensity is used for radiance (or for whatever else strikes the mood of their authors).

The recommendation by OSA's Committee on Colorimetry that brightness be used to mean "that attribute of sensation by which an observer is aware of differences of luminance" goes back more than 50 years, although brightness seems to be used by astronomers for what optical scientists would call radiance. But brightness also has been used as a synonym for luminance. To distinguish between the two, the term photometric brightness has been suggested for luminance.

We have found that for our purposes two radiometric quantities (radiance and irradiance) and two photometric quantities (luminance and brightness), possibly a third, illuminance (the photometric counterpart of irradiance), are all that we need. We do not use intensity because we rarely encounter a genuine need for it and because it has ceased to have any meaning given that it has so many. Irradiance is often called flux, but this ought to be flux density.

Rather than scrap all the units in photometry they could be recycled as characters in a kind of Lord of the Rings fantasy. Once upon a time, in the land of the Nits, dwelt a king named Troland with his beautiful daughter Candela. And so on.

4.2   Irradiance

Although irradiances made brief appearances in previous chapters, the radiation fields were restricted to monodirectional beams. Now we remove that restriction. Unlike radiance, irradiance is not solely a property of the radiation field but depends, in general, on the orientation of the detector. Consider a plane surface in a radiation field. We call one side of this plane the upper side, the other the lower side. Radiation in all directions crosses this plane, so we restrict ourselves to a hemisphere of directions from lower to upper. Construct a coordinate system centered on a small area A of the plane; ϑ and φ are the spherical polar coordinates in this system. The radiance in each direction Ω_i contributes (approximately) an amount

$$L(\Omega_i) A \cos \vartheta_i \Omega_i \tag{4.45}$$

to the total radiant power incident on A, where ϑ_i is the angle between Ω_i and the normal to A (directed in the sense lower to upper) and Ω_i is the solid angle of a small set of directions around Ω_i. Thus the total radiant power crossing unit area is (approximately)

$$\sum_i L(\Omega_i) \cos \vartheta_i \Omega_i, \tag{4.46}$$

where the sum is over all upward directions. In the limit of smaller and smaller solid angles, this sum becomes an integral

$$F_\uparrow = \int_{2\pi} L(\Omega) \cos \vartheta \, d\Omega, \tag{4.47}$$

where 2π indicates integration over a hemisphere of directions. As shown in Section 4.1.1, integration over solid angle is the same as integration over a unit sphere:

$$F_\uparrow = \int_0^{2\pi} \int_0^{\pi/2} L(\vartheta, \varphi) \cos \vartheta \sin \vartheta \, d\vartheta \, d\varphi. \tag{4.48}$$

We may call $F_\uparrow$ the upward irradiance, but keep in mind that up and down are arbitrary designations for two opposite directions. For our purposes radiation does not know that gravity exists.

We have to be careful in specifying the downward irradiance as a positive quantity. In Eq. (4.48) the limits of integration of ϑ are such that $\cos \vartheta$ is positive. For ϑ from $\pi/2$ to π (lower hemisphere) $\cos \vartheta$ is negative. This is fixed by interchanging the limits of integration so that the integral giving the downward irradiance is positive:

$$F_\downarrow = \int_0^{2\pi} \int_\pi^{\pi/2} L(\vartheta, \varphi) \cos \vartheta \sin \vartheta \, d\vartheta \, d\varphi. \tag{4.49}$$

Two properties of irradiance must be kept in mind. First, the upward and downward irradiances depend, in general, on the orientation of the reference plane dividing the set of all directions into upward and downward hemispheres. Second, although radiance determines irradiance, the converse, in general, is not true: if we know the irradiance, we cannot uniquely determine the corresponding radiance. The one exception is an *isotropic* radiation field. A completely isotropic radiation field is one for which radiance does not depend on direction. A radiation field also can be isotropic over the upward or downward hemispheres. Suppose that the upward radiance is independent of direction. From Eq. (4.48) it therefore follows that

$$F_\uparrow = L \int_0^{2\pi} \int_0^{\pi/2} \cos \vartheta \sin \vartheta \, d\vartheta \, d\varphi = \pi L. \tag{4.50}$$

We can write this another way:

$$L = \frac{F_\uparrow}{\pi}. \tag{4.51}$$

At first glance Eq. (4.51) may seem to be missing a factor 2. After all, a hemisphere subtends 2π sr. The apparently missing factor is accounted for by the cosine of the angle between the normal to the surface and Ω. Another way to look at this is to recognize that the area of a hemisphere of unit radius projected onto a plane is π. Even in an isotropic radiation field the rate at which radiation crosses unit area per unit solid angle still depends on direction because of the cosine factor [Eq. (4.45)].

In Section 1.2 the Planck function P_e [Eq. (1.11) or (1.20)] was taken as that for irradiance to avoid a digression on radiance, which also would have required a preceding digression on solid angle. Within an opaque container at constant temperature the (equilibrium) radiance P_e/π, sometimes denoted as B, is isotropic.

4.2.1   Diffuse Reflection

Suppose that a diffuse reflector such as snow is illuminated solely by direct sunlight (we neglect illumination by skylight). The direction of the sunlight makes an angle Θ_s with the normal to the surface of the reflector. The downward irradiance is

$$F_\downarrow = \int L_s \cos \vartheta \, d\Omega, \qquad (4.52)$$

where L_s is the radiance of the sun. Because the solid angle the sun subtends is small, this integral is approximately

$$F_\downarrow \approx L_s \cos \Theta_s \Omega_s. \qquad (4.53)$$

The upward irradiance is a consequence of reflection of the downward irradiance (emission by snow is negligible at visible and near-visible frequencies):

$$F_\uparrow = R_d F_\downarrow = R_d L_s \cos \Theta_s \Omega_s, \qquad (4.54)$$

where the reflectivity (also called albedo) R_d of the diffuse reflector is that for reflection into all directions (in the upward hemisphere), although the contributions to R_d are, in general, different in different directions (per unit solid angle). Also, R_d depends, in general, on Θ_s. Assume that the upward radiance is isotropic. This is never true – there are no perfectly diffuse reflectors – but sometimes is a good approximation for a large range of directions (but not an entire hemisphere). With this assumption of isotropy the radiance of the diffuse reflector is

$$L_d = L_s R_d \cos \Theta_s \frac{\Omega_s}{\pi}. \qquad (4.55)$$

Even with the sun directly overhead illuminating clean, fine-grained, highly reflecting snow (see Fig. 5.16), its radiance is almost 10^5 times smaller than the sun's radiance, which is high only in a small set of directions. The radiance of snow is much less, but over an entire hemisphere of directions.

Now consider a flat-plate detector. Such a detector responds to the net amount of radiation it receives. The radiation illuminating the detector is sunlight (again, we neglect the contribution from the sky, which we could exclude with a suitable collimator). As in the previous example, the downward irradiance is given by Eq. (4.53) and the upward irradiance by

Eq. (4.54), where R_d is the reflectivity of the detector. The detector responds to the difference between these two irradiances, the rate at which radiant energy is absorbed by the detector:

$$F_{\downarrow} - F_{\uparrow} = L_s \cos \Theta_s \Omega_s (1 - R_d).\tag{4.56}$$

The response of the detector is proportional to this quantity. If we neglect the dependence of L_s on solar zenith angle Θ_s because of atmospheric attenuation, the response of the detector follows a cosine law if R_d is independent of Θ_s. Such an idealized detector is sometimes called a perfectly diffuse or *Lambertian* detector. Real detectors are never Lambertian, and so a correction has to be made for their departure from the ideal.

Equipped with an understanding of the difference between radiance and irradiance, we can return to the problem of the sun and a lens. We now recognize Eq. (4.29) as specifying the incident (downward) irradiance on the paper at the image of the sun. Assume that this paper is 100% reflecting (it is not) and is oriented so that sunlight normally illuminates it. Assume further that the paper is Lambertian (also not strictly true). With these assumptions, the radiance of the image of the sun on the piece of paper is

$$L_s \frac{\Omega_{\text{lens}}}{\pi}.\tag{4.57}$$

From this equation we might conclude that the radiance of the sun's image could be made greater than that of the sun if the solid angle of the lens subtended at its focal point were greater than π. This is a difficult order to fill. It requires a lens with a large collecting area and a short focal length. Unfortunately, the focal length of a lens is proportional to its radius of curvature, so small lenses (with small radii of curvature) have short focal lengths and large lenses (with large radii of curvature) have long focal lengths. Thus it would be difficult to make a lens that subtends an angle greater than π at its focal length.

We also now can revisit the problem of the radiance of an object under water, this time an isotropic diffuse reflector (diffuser) rather than a specular reflector (mirror). The diffuser is parallel to the air–water interface. First take the diffuser to be in air, illuminated by a beam with radiance L_i and solid angle Ω_i incident at an angle ϑ_i. The radiance of the diffuser in air is

$$L_a = \frac{L_i \Omega_i \cos \vartheta_i}{\pi},\tag{4.58}$$

where for simplicity we take the reflectivity to be 1. According to Eq. (4.39), the diffuser under water is illuminated by a beam with radiance $L_i n^2$. The radiance of the diffuser in the water is

$$L_r = \frac{L_i n^2 \Omega_t \cos \vartheta_t}{\pi}.\tag{4.59}$$

An implicit assumption underlying this expression is that the reflectivity of the diffuser is the same in air as in water. The radiance L_r [Eq. (4.59)] is transmitted to an observer (in air), who detects radiance $L_w = L_r/n^2$. All these results can be combined to yield

$$L_w = \frac{L_r}{n^2} = \frac{L_i \Omega_t \cos \vartheta_t}{\pi} = \frac{L_i \Omega_i \cos \vartheta_i}{\pi n^2} = \frac{L_a}{n^2},\tag{4.60}$$

where we also used

$$\Omega_i \cos \vartheta_i = n^2 \Omega_t \cos \vartheta_t, \tag{4.61}$$

which follows from Eqs. (4.35) and (4.38).

Subject to all the assumptions we made, explicit and implicit, the visible radiance of a diffuser under water, unlike that of a mirror, is about 0.56 times its radiance above water under the same illumination. You can observe this by immersing a diffuser (white plastic spoons serve well) partly in water. There will be a distinct brightness change at the interface between the part of the diffuser in air and the part in water. But the precise geometry of the diffuser and even the nature of the water container will affect what is observed. For best results (greatest contrast), the container should be black so that light reflected by it does not illuminate the diffuser. And if the object is not flat, it should be oriented so that light reflected by one part of it does not illuminate other parts. Figure 4.12 shows spoons in different orientations partly immersed in water in black and white pans. Equation (4.60) should be looked upon as specifying the *maximum* reduction in radiance. Moreover, although the reflectivity of an ordinary silvered mirror does not change appreciably from air to water, that of glass does. And, as we show in Section 5.3.1, the reflectivity of a porous material (such as sand) decreases upon wetting. But the radiance decrease described by Eq. (4.60) is solely a consequence of the geometrical properties of radiance and irradiance.

You can see the same kind of brightness change when you partly immerse your hand in a basin filled with water. The part of your hand below the water is darker than that above, and the difference is more striking if the basin is stainless steel or dark porcelain rather than white porcelain.

Scientific progress undoubtedly has been retarded by the switch from taking baths to taking showers. Nothing of great moment has been discovered in the shower, whereas the bathtub is a natural laboratory. Archimedes did not burst from his shower shouting Eureka. If you take leisurely baths instead of showers (the fast food of bathing) you can't help but notice a darkening of your outstretched thighs at the interface between air and water.

To underscore the difference between radiance and irradiance consider the downward irradiance above and below the surface of water in a black container (so that no radiation is reflected by it). Assume that the downward radiance above the water is isotropic. With this assumption the downward irradiance is

$$\int_0^{2\pi} \int_0^{\pi/2} L_i \cos \vartheta_i \sin \vartheta_i \, d\vartheta_i \, d\varphi_i = \pi L_i. \tag{4.62}$$

If we neglect the departure of the transmissivity from 1, the downward radiance in the water is $L_t = L_i n^2$. This downward radiance is constant over a range of directions but not over a complete hemisphere. As we saw in Section 4.1.6, the hemisphere of downward incident directions is not mapped into a complete hemisphere of downward transmitted directions. That is, there are no rays in the water making an angle greater than ϑ_c with the normal to the surface, where this *critical angle* is given by

$$\sin \vartheta_c = \frac{1}{n}. \tag{4.63}$$

This follows from the law of refraction for an angle of incidence $\pi/2$.

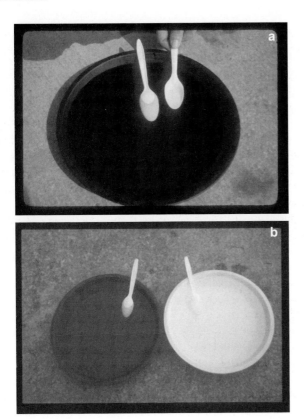

Figure 4.12: a) White plastic spoons oriented differently and partly immersed in water in the same black pan. Note the marked change of brightness of the spoons across the air–water boundary, especially for the spoon with its bowl oriented convex. b) White plastic spoons partly immersed in water in white and black pans. Both spoons are oriented with their bowls convex, and yet the brightness change across the air–water boundary is much greater for the spoon in the black pan.

The downward irradiance in the water is therefore

$$\int_0^{2\pi}\int_0^{\vartheta_c} L_t \cos\vartheta_t \sin\vartheta_t \, d\vartheta_t \, d\varphi = 2\pi L_i n^2 \int_0^{\vartheta_c} \frac{1}{2}\frac{d}{d\vartheta_t}(\sin^2\vartheta_t)\, d\vartheta_t = \pi L_i, \quad (4.64)$$

which is equal to the downward irradiance above the water. Thus irradiance is conserved in going from air to water whereas radiance is not. Again, we emphasize that this result is based on the assumption that reflection because of the air–water interface is negligible. This is not a bad assumption given that at visible wavelengths the ratio of reflected to (isotropic) incident irradiance is about 6%.

4.2.2　Flux Divergence

Consider a closed (imaginary) surface S in a radiation field. The outward unit normal at each point is $\mathbf{n}$. Radiation crosses this surface from outside to inside the region enclosed by S and vice versa. At any point the net rate at which radiation crosses unit area of the surface is

$$-\int_{4\pi} \mathbf{n} \cdot \mathbf{\Omega} L \, d\Omega. \tag{4.65}$$

The minus sign comes about because radiation transported into the region enclosed by S is counted as positive and radiation transported out is counted as negative. We may define the vector irradiance (or vector flux) at a point as

$$\mathbf{F} = \int_{4\pi} \mathbf{\Omega} L \, d\Omega. \tag{4.66}$$

With this definition, the total net radiation transported into the region enclosed by S is

$$-\int_{S} \mathbf{F} \cdot \mathbf{n} \, dS. \tag{4.67}$$

From the divergence theorem we have

$$\int_{S} \mathbf{F} \cdot \mathbf{n} \, dS = \int_{V} \nabla \cdot \mathbf{F} \, dV, \tag{4.68}$$

where V is the volume enclosed by S. Thus the rate of radiant energy deposition per unit volume is the negative of the vector flux divergence

$$-\nabla \cdot \mathbf{F}. \tag{4.69}$$

This is a generalization of the result we obtained previously for a monodirectional beam propagating in a purely absorbing medium. Equation (4.69) is not restricted to such a beam or to such a medium.

From Eq. (4.66) we have

$$\nabla \cdot \mathbf{F} = \int_{4\pi} \nabla \cdot (\mathbf{\Omega} L) \, d\Omega = \int_{4\pi} \mathbf{\Omega} \cdot \nabla L \, d\Omega. \tag{4.70}$$

A necessary condition for deposition (transformation) of radiant energy is that the gradient of the radiance cannot vanish for all directions. But this is not sufficient. The absorption coefficient must also be nonzero, which makes physical sense and is proved in Section 6.1.2.

4.3　Color

Color used as a synonym for wavelength (or frequency) is a common crutch of popular science writers, who seem to think that wavelength cannot be swallowed unless sugar-coated with color, thereby impeding understanding of both color and wavelength as well as depriving readers of something marvelous. Colors are produced in our brains. Sometimes these

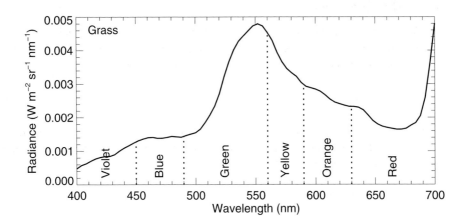

Figure 4.13: Measured spectrum of grass illuminated by daylight. The visible spectrum is divided into six regions to which color names are given.

colors originate from external stimulation, sometimes not; sometimes the stimulation is light, sometimes not. We may dream in color despite the absence of light on our retinas. We may see colors by taking psychotropic drugs or by being hit in the head. Even when light is the stimulus that produces color, there is not a one to one correspondence between a perceived color and the spectrum of the light that produces it. The histologist Santiago Ramón y Cajal, arguably Spain's most famous scientist, succinctly and beautifully captured the essence of human vision. "By virtue of a marvelous alchemy, begun in the retina and completed in the central nervous system, that which in the surrounding ether is simple undulatory motion, is converted in the brain into something completely new and purely subjective: sensations, perceptions, visual memories, associations of images, ideas, and wills."

Physicists are thoroughly steeped in spectra, especially line spectra because of their central role in the evolution of quantum mechanics. But much of what passes for knowledge about color among physicists is demonstrably wrong. Take, for example, the following assertion by a doctor of physics who is also a doctor of medicine, although the two doctorates seem to reside in different parts of his brain: "A leaf looks green because it is absorbing all the colors of the spectrum except green, which it is reflecting." This is not a carefully chosen exotic blooper but rather common currency. Consider this assertion about a green leaf in light of a measured visible spectrum of grass (Fig. 4.13). Although we label six bands in this spectrum with color names, they are somewhat arbitrary in that there are no sharp boundaries between different perceived colors. Note that some yellows and oranges are more intense than some greens. Moreover, light of *all* wavelengths is in light from green grass. And this is true of most colored objects in our surroundings. An apple is not red because, as we have heard many times, it reflects only red light. Red is merely the perceptually dominant color. We are incapable of looking at a source of visible light and assessing how much red light, green light, and so on it is composed of. We are not spectrometers for electromagnetic radiation although we are to a degree for acoustic radiation. We can distinguish the separate instruments in an

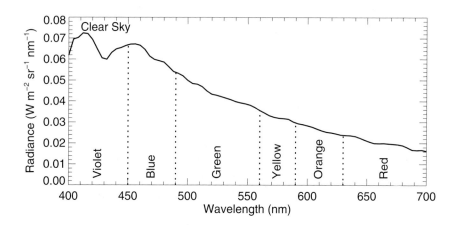

Figure 4.14: Measured skylight spectrum (near zenith). The visible spectrum is divided into six regions to which color names are given. Although the spectrum peaks in the violet, skylight is perceived to be blue.

orchestra, from the high-frequency piccolos to the low-frequency bassoons, and sense their relative contributions to the composite sound of all instruments playing simultaneously.

But, you might protest, the spectrum in Fig. 4.13 *peaks* in the green. So it does, which is largely irrelevant, as evidenced by Fig. 4.14, a measured spectrum of a clear, blue zenith sky. Although we give the color name blue to the sky, its spectrum peaks in the violet and, as with the spectrum of green grass, contains light of all wavelengths. Indeed, there is more light that is *not* blue in skylight than light in the band labeled blue. An even more striking example is provided by the display of a computer adjusted to show yellow (Fig. 4.15). Light in the band labeled yellow is overshadowed by that in the bands labeled green and orange. The sensation we call yellow can indeed be produced by monochromatic sources with wavelengths in the band 560 to 590 nm but also by mixing red and green beams containing *no* light of these wavelengths. What about that paragon of yellowness, the banana? Surely it must be a source of only yellow light. Alas, measurements show otherwise. The spectrum of a banana (Fig. 4.16) contains as much green and orange and red as yellow, if by these colors is meant light confined to certain bands of wavelengths.

None of these measured spectra are contrived even though they handily refute assertions that have been made many times about color, much of the confusion about which stems from a failure to distinguish between wavelength, color, and color names.

Countless students have been introduced to that mythical character Roy G. Biv as a way of learning the names of the seven colors supposedly identified by Newton. What did Newton really say? In his *Opticks*, Prop. II, Theor. II, he notes that "the Spectrum... did... appear tinged with this Series of Colors, violet, indigo, blue, green, yellow, orange, red, *together with all their intermediate Degrees in a continual Succession perpetually varying. So that there appeared as many Degrees of Colours, as there were sorts of Rays differing in Refrangibility*" [emphasis added]. The widespread notion that there are seven and only seven colors in nature

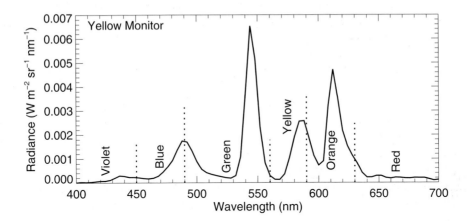

Figure 4.15: Measured spectrum of computer display yellow pixels. The visible spectrum is divided into six regions to which color names are given. Although the amount of radiation in the band labeled yellow is overshadowed by radiation in the bands labeled green and orange, the display is still perceived to be yellow.

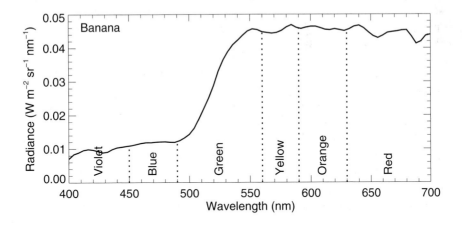

Figure 4.16: Measured spectrum of a banana illuminated by daylight. The visible spectrum is divided into six regions to which color names are given. A yellow banana is far from being a reflector of only yellow light.

is a consequence of the failure to read what Newton wrote and to come to grips with the nature of color. Newton did not believe that there are only seven colors, he merely gave seven color names. Why did he stop at seven? This is the magic number par excellence: seven pillars of wisdom, days of the week, deadly sins, wonders of the world, intervals of the diatonic scale, and so on. In a famous essay about Newton, John Maynard Keynes noted that "Newton was not the first of the age of reason. He was the last of the magicians..." Even today, three hundred years after Newton, many people still believe that numbers have magical powers, some being lucky, others unlucky. Yet we supposedly live in a scientific age. Newton lived in an age when magic, including number magic, still held sway over people's minds.

Every language has a finite number of color words, ranging from three to eleven, by which we mean ones used in everyday speech. We do not count the exotic words confected by paint manufacturers or fashion designers. Depending on our native language and our experience we use a small number of color names, most of them learned as children. And we can determine when two objects are perceived to have the same color, the basis of colorimetry, to which we turn next.

4.3.1   Colorimetry: The CIE Chromaticity Diagram

Colorimetry, the measurement of color, is based on color matching by human observers. All we can do is match two colors, and different observers are not likely to agree when colors exactly match. Suppose that we look at a large black screen with two identical circular holes in it. On the other side is a white screen that can be illuminated by two different sources. A partition is arranged so that we see only light from one source through one hole, light from the other source through the other hole. We wait long enough for our eyes to be fully adapted to photopic illumination, then adjust the sources until we judge the two discs of light, side by side, to be indistinguishable.

One source, the *sample*, has a narrow spectrum centered on some visible wavelength. We can adjust this source so that its radiant power is the same for every wavelength. The other source is a composite obtained by superposing (adding) light from three lamps (*primaries*), which we call red, green, and blue because these are the color sensations each separately evokes. We can independently adjust the power of each primary. For samples corresponding to each visible wavelength, we adjust the three primaries until we get a color and brightness match, then record the luminances of the primaries. For each visible wavelength we obtain three luminances, the *tristimulus* values, denoted as $\overline{r}$, $\overline{g}$, and $\overline{b}$. Each set of tristimulus values yields light indistinguishable from the sample. We do this for all visible wavelengths, thereby obtaining three *tristimulus curves*. But there is a catch. We cannot beg, borrow, or steal three real lamps that enable us to match all visible wavelengths. Despite the blather about three primary colors, they do not exist. The only way to match some sample wavelengths is to take one of the three lamps and add its light to that of the sample, the combination of which then can match the remaining two lamps. This results in some tristimulus values that are negative because one of the three lamps has been removed from the trio. This is deemed unacceptable, or at least inconvenient. But this inconvenience is removed once we recognize that the only essential aspect of color matching is that it requires *three* numbers, the absolute values of which are irrelevant. Thus any linear transformation

$$\overline{x} = a_{11}\overline{r} + a_{12}\overline{g} + a_{13}\overline{b}, \tag{4.71}$$

$$\overline{y} = a_{21}\overline{r} + a_{22}\overline{g} + a_{23}\overline{b}, \tag{4.72}$$

$$\overline{z} = a_{31}\overline{r} + a_{32}\overline{g} + a_{33}\overline{b}, \tag{4.73}$$

yields an acceptable set of tristimulus values, and we can choose the coefficients a_{ij} such that $\overline{x}$, $\overline{y}$, and $\overline{z}$ are never negative. The $\overline{g}$ curve obtained with human observers is remarkably sim-

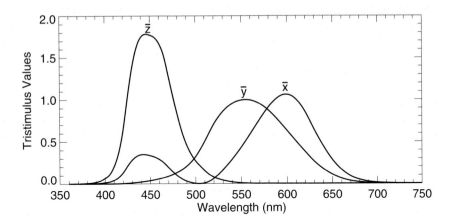

Figure 4.17: CIE 1931 tristimulus functions. The area under each curve is the same; $\bar{y}$ is the luminous efficiency (Fig. 4.8). These functions are obtained by transformations of data obtained from color matching experiments with human observers.

ilar to the curve of luminous efficiency (Fig. 4.8), so we choose the coefficients in Eq. (4.72) such that $\bar{y}$ is *exactly* the luminous efficiency. Then we choose the coefficients in Eqs. (4.71) and (4.73) such that the area under all three tristimulus curves is the same. The result, shown in Fig. 4.17, is the set of three curves for the CIE (2°) 1931 *standard observer*. CIE is the abbreviation for Commission Internationale de l'Eclairage (International Commission on Illumination), 2° is the field of view for which the color matching measurements were made, and 1931 is the year in which these values were adopted. The standard observer is not kept in a cage at the National Institute of Standards and Technology in Boulder, Colorado, to be trotted out on ceremonial occasions, but is rather a composite of many observers with normal color vision.

From the tristimulus values $\bar{x}$, $\bar{y}$, $\bar{z}$ for each visible wavelength follow the tristimulus values X, Y, Z for *any* source of visible light with (spectral) radiance L:

$$X = k \int L\bar{x}\, d\lambda, \tag{4.74}$$

$$Y = k \int L\bar{y}\, d\lambda, \tag{4.75}$$

$$Z = k \int L\bar{z}\, d\lambda, \tag{4.76}$$

where k is a constant we choose to suit our fancy. Note that because $\bar{y}$ is the luminous efficiency, Y is proportional to the luminance of the source [see Eq. (4.44)], which could be a *primary* source (e.g., lamp, direct sunlight) or a *secondary* source (light excited by a primary source).

What exactly do X, Y, Z signify? Two light sources with the same tristimulus values for standard conditions (field of view, absolute luminance, surroundings, adaptation, etc.) will be indistinguishable to the standard observer (who does not exist). In general, any set X, Y, Z can be obtained in an *infinite* number of ways, that is, by an infinite set of spectra. A spectrum determines a set of tristimulus values, but not the converse. This undeniable, experimentally verifiable fact, called *metamerism*, shatters fatuous notions about color being simply a synonym for wavelength.

Even though the standard observer is fictitious, real observers with normal color vision will agree that two light sources with the same tristimulus values are nearly indistinguishable. But the tristimulus values do not provide a complete description of the visual appearance of colored objects. The tristimulus values in Eqs. (4.74)–(4.76) are based on radiances, which depend on direction, whereas they could just as well be based on irradiances. Two colored objects may have the same (irradiance) tristimulus values and yet still appear different because of their different textures (and hence different radiances in different directions). For example, a glossy and a matte surface with identical tristimulus values usually would be perceived to be different. And then there is *simultaneous color contrast*: two colored objects identical in all respects may be perceived to have different colors if their surroundings are different. Simultaneous color contrast is often subtle but can be demonstrated dramatically. In a room illuminated only by red light, we asked several people to describe the color of a sheet of orange construction paper on a black backdrop. The unanimous response was red. But when we placed a sheet of red paper beside the orange sheet, everyone described it as orange. The spectrum of the light from the sheet had not changed, only its surroundings.

Because one of the tristimulus values of a source is proportional to its luminance, the other two must specify its chromatic characteristics. This in turn implies that color (as opposed to brightness) has *two* qualities, which could be dominant wavelength and purity. According to the Optical Society of America's definition in *The Science of Color* (p. 42) "...*dominant wavelength*... is the wavelength that appears to be dominant in the light... *purity* may be said to be the degree to which the dominant wavelength appears to predominate in the light." Although easy to grasp, dominant wavelength and purity are not used much by color scientists. Two associated terms are *hue* and *saturation*. Again we turn to *The Science of Color* (p .101) for definitions: "Hue is the attribute most commonly associated with the wavelength or dominant wavelength of the stimulus and designated by such terms as red, yellow, green, blue. Saturation is the degree to which a chromatic color sensation differs from an achromatic color sensation of the same brightness. For instance, a typical pink is a red of low saturation and high brightness, whereas the light from sodium vapor is a yellow of high saturation".

We can make the qualitative concepts of dominant wavelength and purity quantitative by means of the tristimulus values. Because Y is the luminance, we could factor it out by dividing X and Z by Y. Although this is not what is done, dividing X and Y by the sum of all three tristimulus values to obtain the *chromaticity coordinates*

$$x = \frac{X}{X+Y+Z}, \tag{4.77}$$

$$y = \frac{Y}{X+Y+Z}, \tag{4.78}$$

is equivalent. The set of all possible points x and y constitutes a two-dimensional, bounded color space, the *chromaticity diagram*, a geometrical representation of the gamut of colors perceived by someone with normal color vision. The chromaticity diagram is that region

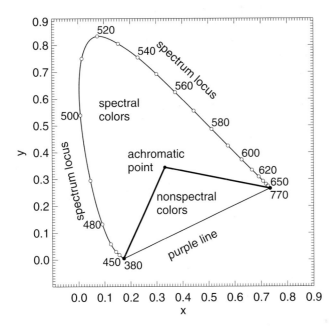

Figure 4.18: The CIE 1931 chromaticity diagram is a color space showing the gamut of colors perceived by the standard human observer. Regions of this color space are labeled by their salient characteristics: spectrum locus (set of all pure colors), spectral colors (those to which a dominant wavelength can be assigned), nonspectral colors (those to which a dominant wavelength cannot be assigned), the purple line (mixtures of violet and red), and the achromatic (or white) point.

in Fig. 4.18 bounded by the curved *spectrum locus*, the set of points corresponding to light sources of a single wavelength (a few wavelengths are labeled for reference), and the *purple line*. Purple is not a spectral color in that it cannot be obtained with a monochromatic source, whereas red, for example, can. Purple is inherently a mixture of violet and red light. The *achromatic* or *white point* is specified by the coordinates of a source we agree to call white. As we show in the following section, there is no such thing as absolute white, and hence the position of the achromatic point is not absolute.

Color space can be divided into two regions, one representing *spectral colors*, the other *nonspectral colors*. The spectral region is that part of the chromaticity diagram defined by all possible straight lines through the achromatic point and connecting two points on the spectrum locus. These two points correspond to *complementary colors*, in the sense that adding them in the proper proportion yields the sensation white. For example, a pure green with wavelength around 492 nm and a pure orange with wavelength around 605 nm are complementary colors: add two monochromatic sources of these wavelengths in the proper proportion and the result is white light (Fig. 4.19). The nonspectral region is that part of the chromaticity diagram defined

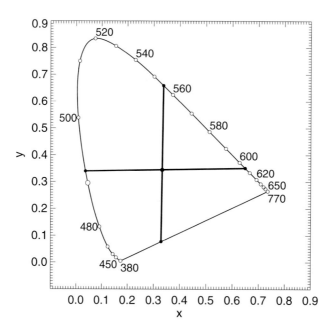

Figure 4.19: The CIE 1931 chromaticity diagram showing complementary wavelengths 492 nm and 605 nm. A suitable mixture of two sources of these wavelengths yields the sensation white. Although points on the purple line do not correspond to specific wavelengths, the complementary wavelength of such a point is defined by that of the wavelength (555 nm in the example shown) that must be subtracted from white light to yield the color on the purple line.

by all possible straight lines through the achromatic point and connecting a point on the purple line with a point on the spectrum locus. Because monochromatic sources of purple light do not exist, we cannot add a pure purple to its complementary pure color (the intersection point on the spectrum locus) to obtain white. But we can *subtract* from white light the complementary color, a green with wavelength 555 nm for the purple shown in Fig. 4.19.

Suppose that a source is represented by a point on the chromaticity diagram. From the achromatic point draw a straight line that intersects the source point and extend this line to the spectrum locus. The wavelength at the intersection point is the dominant wavelength. If a monochromatic source of this wavelength were to be added to white light in the proper proportion, the result would be indistinguishable in color from the source. The term *purity* is used somewhat carelessly (we confess to having committed this sin) without qualification, but there are two purities, one with a simple geometrical definition, the other with a simple photometric definition. The *excitation purity* is the length of the line from the achromatic point to the source point relative to the distance from the achromatic point to the spectrum locus point (dominant wavelength), usually expressed as a percentage. An achromatic source has an excitation purity of 0%, a monochromatic source an excitation purity of 100%. This

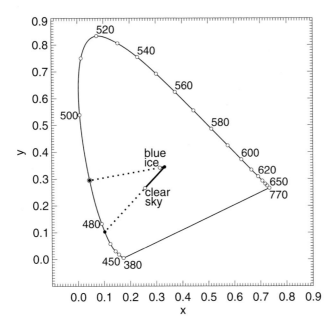

Figure 4.20: CIE 1931 chromaticity diagram with measured colorimetric points for a clear sky and blue ice. The dominant wavelength for the blue ice is 490 nm; that for the blue sky is 477 nm.

purity is a measure of how close the source point is to the spectrum locus, and its value is evident from a glance at the CIE diagram. But it lacks a simple photometric definition. The *chromatic purity* is not evident from the CIE diagram but is defined in what might be called the natural way: if Y_0 is the luminance of an achromatic source and Y_1 that of a monochromatic source such that the sum of these two sources matches the luminance Y and color of the source of interest, its chromatic purity is Y_1/Y. The two purities coincide at the end points (0 and 100%) but elsewhere are different.

Figure 4.20 shows chromaticity points obtained from spectral measurements of blue ice (frozen waterfall) and a clear sky near zenith (see Fig. 5.17 for the spectrum of the blue ice). The excitation purity of the sky is about 33%, its chromatic purity about 13%, and its dominant wavelength about 477 nm, solidly in the blue. The excitation purity of the blue ice is considerably lower, about 6%, its chromatic purity about 5%, and its dominant wavelength about 490 nm, what we might call blue–green. And indeed these calculations are in accord with what can be observed in winter on a clear day. Bubbly ice (a frozen waterfall in a road cut, for example) is a much less vivid blue than the zenith sky, and the ice has a noticeably more greenish cast than the sky.

The CIE diagram helps to support the assertion in Section 1.4.4 about the restricted validity of color temperature. Figure 4.21 shows the curve of chromaticity points for blackbodies with a large range of absolute temperatures. Despite this range, blackbodies can't come close

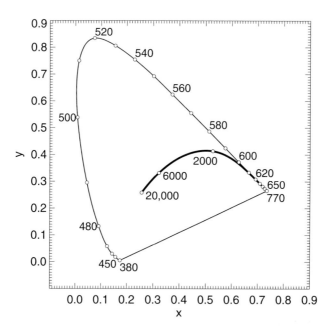

Figure 4.21: CIE 1931 chromaticity diagram showing the locus of colorimetric points for black-bodies with temperatures in the range 1000–20,000 K.

to matching the entire gamut of colors. At temperatures less than about 1000 K, the light from a blackbody is a red of high purity, whence the term red hot. The melting point of carbon is about 3800 K, that of tungsten about 100 K lower, which gives upper limits on the color temperature of heated carbon or tungsten filaments. The boiling point of tungsten is about 6300 K, approximately the color temperature of sunlight. As the absolute temperature increases without limit (don't try this at home) the color of a blackbody eventually becomes blue although never of high purity. A step beyond color temperature is the *correlated color temperature* of a source, the temperature of a blackbody with chromaticity point closest to that of the source. But even correlated color temperature becomes nearly meaningless for points far from the blackbody curve.

Figure 4.22 shows chromaticity points for more than a dozen sources: white light such as daylight reflected by magnesium oxide powder, snow, cloud, paper, fruit, a green dish, dirt and a fluorescent vest; zenith and horizon skylight; black light; the display of a computer set for various colors. The chromaticity points for all the white sources are different but clustered in the same neighborhood. And note that a red apple is far from being a pure red. Even the colors on the display, set to the highest purity, are not on the spectrum locus. The purest color is that for the orange fluorescent vest. The pattern that emerges here is clear: the colors in our everyday lives, even those purposely designed to be vivid, are usually far from being pure. Our visual world is mostly pastel.

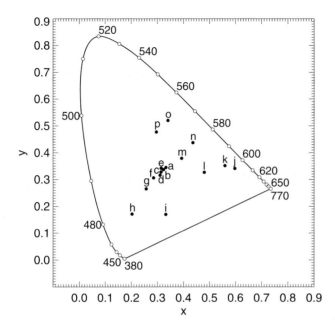

Figure 4.22: CIE 1931 chromaticity diagram with colorimetric points for several different sources: (a) MgO; (b) snow; (c) cloud; (d) thesis paper; (e) blue ice; (f) clear horizon sky; (g) clear zenith sky; (h) blue computer display pixels; (i) black light; (j) orange vest; (k) red computer display pixels; (l) apple; (m) dirt; (n) banana; (o) green dish; (p) green computer display pixels. Note that none of these points are on or even close to the spectrum locus.

Although the CIE 1931 diagram has long been the workhorse of colorimetry, it is being replaced gradually by the CIE 1976 UCS diagram, where UCS stands for *uniform chromaticity scale*. This diagram represents an attempt to make equal displacements correspond to equal perceptual differences. The coordinates (u', v') in this system are obtained from the tristimulus values:

$$u' = \frac{4X}{X + 15Y + 3Z},\tag{4.79}$$

$$v' = \frac{9Y}{X + 15Y + 3Z}.\tag{4.80}$$

Because of the one-to-one relation between (x, y) and (u', v'), the CIE 1976 UCS diagram contains no new information but just displays old information in a new way. Figure 4.23 shows the CIE 1976 UCS diagram with a curve of blackbody chromaticity points.

Because chromaticity diagrams are infinite sets of points, we might be tempted to conclude that the number of colors is infinite, which is what Newton implied in his statement about colors "perpetually varying." Although wavelength may be said to vary perpetually, color cannot because no human observer can detect differences between sources represented by arbitrarily

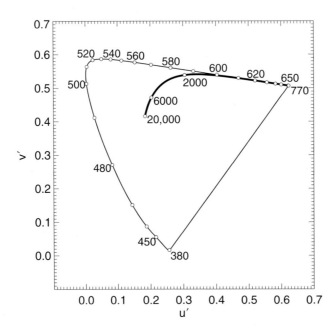

Figure 4.23: CIE 1976 UCS chromaticity diagram with blackbody points as in Fig. 4.21. The 1976 diagram contains no more information than the 1931 diagram but displays it differently.

close points on a chromaticity diagram. Around any point on a chromaticity diagram is a neighborhood of points representing sources indistinguishable to almost all observers. These neighborhoods are the *MacAdam ellipses*, elliptical subsets of color space specified by a fixed precision in color matching. Because of the finite area of these ellipses, the number of perceptually different colors is finite. Ellipses in different regions of color space do not have the same area or shape, and UCS diagrams are attempts to distort color space such that all ellipses are transformed into equal-area circles, an elusive goal not yet or ever likely to be reached. If to the two colorimetric coordinates is added luminance, we obtain ellipsoids in a three-dimensional space that specify the inherent fuzziness of the human observer's ability to distinguish between sources of different color and brightness.

There are an infinite number of visible wavelengths, a large but indeterminate number of perceptually different colors corresponding to fuzzy regions in color space, and a small range of colors words in everyday use in all languages.

4.3.2   The Nonexistence of Absolute White

To end, we return to the seemingly heretical statement in the previous section about the nonexistence of absolute white. White light often is defined as a source with all visible wavelengths in equal proportions, that is, a flat spectrum. We call this physicist's white. It doesn't exist in nature nor can you readily buy a source of such light, although with effort it might be created with carefully chosen filters. More important, this definition is largely irrelevant to the perception of white. This is most easily demonstrated by comparing different spectra of indisputably white light sources. Figure 4.24 shows four such spectra: snow illuminated by daylight, and white paper illuminated by an incandescent bulb, a fluorescent bulb, and a fluorescent tube. Despite the great differences in these spectra, to the human observer they all are perceived as white. Note in particular how much the spectrum of the fluorescent bulb departs from white light defined as a flat spectrum.

If you look at a sheet of white paper illuminated by an incandescent lamp, then take the paper into a room illuminated by a fluorescent lamp, then take the paper outdoors, you still see it as white. It doesn't change color although its spectrum changes. A spectrophotometer does not have a brain, whereas you do, and without your conscious effort it continually maintains color constancy so that under any illumination white paper is perceived to be white. Color constancy was strikingly demonstrated to us on a visit to the National Institute of Standards and Technology in Gaithersburg, Maryland. Scientists there had made a video of a scene illuminated by a source the color temperature of which could be varied continuously over a large range. As we watched this scene with the color temperature of the ambient light rapidly changing by thousands of degrees we barely noticed any differences. But when we compared (by splitting the screen) the scene illuminated by a source at 2000 K with the same scene illuminated by a source at 20,000 K, the difference was striking. The scene at 2000 K was noticeably redder than that at 20,000 K.

Only when you compare two sources of white light side by side can you detect any differences. The spectrum of an incandescent bulb is considerably richer in red light than is the spectrum of, say, snow (Fig. 4.24). To observe this look at the reflection (to reduce the luminance) of an incandescent bulb in a window that looks out onto snow. Here the bulb is noticeably yellowish-orange.

A striking demonstration can be obtained in a room illuminated solely by red light. Despite the red illumination of everything in the room, familiar objects in it known to be white are perceived to be more or less white because they are the brightest objects.

Although the human observer does not perceive any differences between white light sources with markedly different spectra, a camera does because it is not equipped with a brain. Thus if you use outdoor film for indoor photography (or vice versa), your camera cannot adapt, and the result may be photos of your friends and family with toothy, greenish grins. To avoid unnatural colors you must use daylight film for outdoor photography and tungsten film for indoor photography or use daylight film indoors with a blue filter on the camera. Digital cameras face the same problem of reproducing colors as faithfully as possible regardless of the ambient illumination, and are equipped with circuits and algorithms to achieve this end. And photos taken with digital cameras can be processed afterwards to obtain more faithful reproductions of what the human observer sees. The digital camera represents an attempt to more closely duplicate with silicon chips what the eye–brain system does.

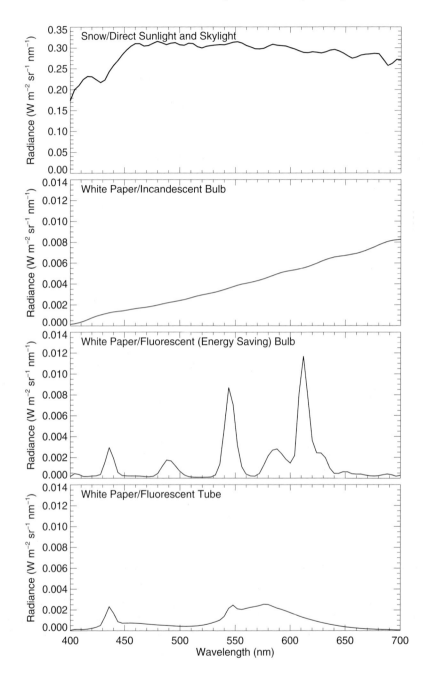

Figure 4.24: Measured spectra of snow illuminated by daylight and white paper illuminated by an incandescent bulb and by two different fluorescent bulbs. All these spectra are objectively quite different but subjectively the same: white.

References and Suggestions for Further Reading

In denoting $\mathbf{E}$ as the electric field (also called electric vector, electric field intensity, electric field strength) and $\mathbf{H}$ as the magnetic field (also called magnetic vector, magnetic field intensity, magnetic field strength), we follow the current convention despite its historical and physical incorrectness. The fundamental fields, the movers and shakers of charges, are $\mathbf{E}$ and $\mathbf{B}$, as evidenced by the Lorentz equation for the force on a charge q with velocity $\mathbf{v}$: $\mathbf{F} = q(\mathbf{E} + \mathbf{v} \times \mathbf{B})$. $\mathbf{B}$ is now often called the magnetic induction, whereas this term should be used for $\mathbf{H}$. In free space $\mathbf{B}$ is proportional to $\mathbf{H}$ by way of a universal (scalar) constant. In material media, however, there is no such general relation, but rather many (assumed) particular relations. For more on this see W. G. Dixon, 1978: *Special Relativity: The Foundation of Macroscopic Physics*, Cambridge University Press, p. 200, also Werner S. Weiglhofer and Akhlesh Lakhtakia, Eds., 2003: *Introduction to Complex Mediums for Optics and Electromagnetics*, SPIE Press, pp. 31 & 348. One author gets around this notational confusion by referring simply to the "B-field and H-field" (Günther Scharf, 1994: *From Electrostatics to Optics: A Concise Electrodynamics Course*, Springer, p. 157).

For a good general reference on radiometry and photometry see Robert W. Boyd, 1983: *Radiation and the Detection of Optical Radiation*, John Wiley & Sons.

For a nonmathematical discussion of solid angle, radiance, irradiance, luminance, brightness and examples of the subjectivity of brightness see Craig F. Bohren, 1991: *What Light Through Yonder Window Breaks?*, John Wiley & Sons, Ch. 15.

Poynting's theorem is derived in treatises on electromagnetic theory, for example, Julius Adams Stratton, 1941: *Electromagnetic Theory*, McGraw-Hill, pp. 131-7; John David Jackson, 1975: *Classical Electrodynamics*, 2nd ed., John Wiley & Sons, pp. 236–7. Poynting's original paper is reprinted in *Collected Scientific Papers by John Henry Poynting*, Cambridge University Press (1920).

We owe the insight that solid angle is a measure to Rudolph W. Preisendorfer, 1965: *Radiative Transfer on Discrete Spaces*, Pergamon, p. 23.

For an illuminating schematic diagram showing how mass density [Eq. (4.17)] depends on volume see Ludwig Prandtl and Oskar G. Tietjens, 1957: *Fundamentals of Hydro- and Aeromechanics*. Dover, Fig. 1, p. 9. A similar diagram is in George Keith Batchelor, 1967: *An Introduction to Fluid Mechanics*. Cambridge University Press. Fig. 1.2.1, p. 5. This figure is fundamental to understanding all densities, that is, limits of quotients, whether explicitly called densities or not.

For more on how the human perceptual system imposes constancy of size, shape, color, and brightness see treatises on perception, for example Herschel W. Leibowitz, 1965: *Visual Perception*, MacMillan; Tom Cornsweet, 1970: *Visual Perception*, Harcourt Brace Jovanovich; Robert Sekuler and Randolph Blake, 1985: *Perception*. Alfred A. Knopf.

For a discussion and photograph of the subsun see Robert Greenler, 1980: *Rainbows, Halos, and Glories*. Cambridge University Press, pp. 73-4, and Alistair B. Fraser, David K. Lynch, and Stanley D. Gedzelman, 1994: Subsuns, Bottlinger's ring and elliptical halos, *Applied Optics*, Vol. 33, pp. 4580-6.

The invariance principle $L/n^2 = $ const. [Eq. (4.43)] is in a review article by E. A Milne, 1930: Thermodynamics of the stars. *Handbuch der Astrophysik*, Vol. 3, Part I, p. 74 (reprinted in the collection edited by Donald H. Menzel, 1966: *Selected Papers on the Transfer of Radiation*. Dover) but goes back even earlier to Max Planck's *Theory of Heat Radiation* (p. 35), the second German edition of which was published in 1913, its 1914 English translation reprinted by Dover in 1959.

For a discussion of the luminous efficiency curve and the Purkinje effect see Yves Le Grand, 1957: *Light, Colour and Vision*, John Wiley & Sons, Chs. 4 & 6. You can open this book almost at random and find something about vision worth reading.

References relevant to fluorescence are given at the end of Chapter 1.

R. C. Hilborn's amusing, but sadly true, quip about radiometry is in *American Journal of Physics* (Vol. 52, 1984, p. 668).

For a humorous but biting criticism of the sloppy use of intensity see J. M. Palmer, 1993: Getting intense on intensity. *Metrologia*, Vol. 30, 371–2.

The quotation in the first paragraph in Section 4.3 is our translation of Santiago Ramón y Cajal, 1943: *El Mundo Visto a los Ochenta Años*, 5th ed. Espas-Calpe Argentina, pp. 22–23.

To find out what Newton really wrote about colors (or anything) we suggest the radical step of reading his own words. Newton's *Optiks* is readily available as a Dover edition published in 1952.

Keynes's essay, Newton, The man is reprinted in Robert Karplus, 1970: *Physics and Man*, W. A. Benjamin, pp. 22–9.

For an ethnolinguistic view of color words see Brent Berlin and Paul Kay, 1969: *Basic Color Terms: Their Universality and Evolution*, University of California Press. For recent statistical evidence in support of the hypothesis in this book of "a total universal inventory of exactly 11 basic color categories...from which the 11 or fewer basic color terms of any given language are always drawn" see Paul Kay and Terry Regler, 2003: Resolving the question of color naming universals. *Proceedings of the National Academy of Sciences*, Vol. 100, pp. 9085–9. A subsequent paper Terry Regler, Paul Kay, and Richard S. Cook, 2005: Focal colors are universal after all. *Proceedings of the National Academy of Sciences*, Vol. 102, pp. 8386–91 presents evidence that the most likely set of terms are the six corresponding to English white, black, red, green, yellow, and blue.

Books written by committees are rarely worth reading. A notable exception is *The Science of Color* (Optical Society of America, 1963), written by the Committee on Colorimetry of the Optical Society of America.

For a figure (p. 50) showing the CIE diagram with color names assigned to regions in this diagram see Fred W. Billmeyer, Jr. and Max Saltzman, 1981: *Principles of Color Technology*, 2nd ed. John Wiley & Sons.

An encyclopedic treatise on colorimetry, chock full of data, is Günter Wyszecki and W. S. Stiles, 1982: *Color Science: Concepts and Methods, Quantitative Data and Formulae*, 2nd ed. John Wiley & Sons.

For a detailed discussion of the MacAdam ellipses see David L. MacAdam, 1985: *Color Measurement: Theme and Variations*, 2nd ed. Springer.

For a recent review of colorimetry, which begins with the assertion that "Since perceived color is a property of the human eye and brain, and not a property of physics...colorimetry is inextricably linked" to understanding the biology of vision, see Andrew Stockman, 2004: Colorimetry in *The Optics Encyclopedia*, Vol. 1, Thomas G. Brown, Katherine Creath, Herwig Kogelink, Michael A. Kriss, Joanna Schmit, Marvin J. Weber, Eds. Wiley-VCH, pp. 207-26.

Problems

4.1. What is the solid angle at O of the sphere in the following diagram as a function of the distances d_c and r_s?

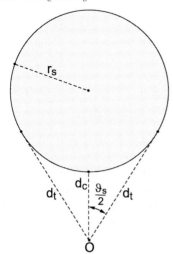

4.2. The solid angle approximation of Eq. (4.6) is equivalent to $\pi\{d_t\vartheta_s/2\}^2/d_t^2$ in the notation of Problem 4.1. Three other approximations are $\pi r_s^2/(d_c + r_s)^2$, $\pi\{d_t\sin(\vartheta_s/2)\}^2/d_t^2$, and $\pi\{d_t \sin(\vartheta_s/2)\}^2/\{d_t \cos(\vartheta_s/2)\}^2$. All of these approximate expressions produce similar (and accurate) values if $d_c \gg r_s$. If d_c is not much greater than r_s which approximate expression is most accurate? To explore this question find d_c (in terms of r_s) that leads to a 10% error in $\pi r_s^2/(d_c + r_s)^2$ relative to the exact solid angle. For this value of d_c rank the performance of the four approximate expressions. Using a figure similar to the one in Problem 4.1, give geometrical interpretations of the four approximations and use them to provide plausible explanations for the rankings obtained.

4.3. In the derivation of Eq. (4.5) we implicitly assume that Earth is a point. By how much does the solid angle subtended by the sun at Earth vary when we account for its finiteness? Consider only daylight locations on Earth. Take the radius of the sun to be 6.6×10^8 m, that of Earth to be 6.4×10^6 m, and the distance between the centers of Earth and sun to be 1.5×10^{11} m.

4.4. If you are inside a hollow sphere what is its solid angle? What is the solid angle of the sphere if you are just outside its surface? What is the solid angle of an infinite plane if you are 1 m from it? What if you are 10^{11} m from it? If you are inside a hollow cube what is its solid angle? What about just outside the cube and located at the center of one of its faces? Just outside the cube and located at the center of one of its edges? And what about just outside the cube at one of its corners?

4.5. A disc of radius r_s is a source of uniform and isotropic radiance L_s. A target disc of radius r_t is at a distance d_{st} from the source disc. The centers of these two discs lie on a common axis perpendicular to both with $r_s, r_t \ll d_{st}$. Show that the radiant power incident on the target disc is $L_s(2\pi^2 r_s^2)(1 - \cos\vartheta_{st})$, where $\vartheta_{st} = \sin^{-1}\{r_t/\sqrt{r_t^2 + d_{st}^2}\}$, and that a good approximation to the radiant power is $L_s\pi^2 r_s^2 r_t^2/d_{st}^2$.

4.6. In the accompanying diagrams the isotropic radiance from disc A_s is L_s. Derive expressions for the radiant power incident on disc A_t in terms of A_s, A_t, h and d assuming that h is much greater than the dimensions of A_t and A_s.

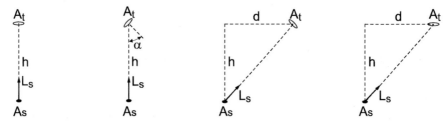

4.7. A disc is a uniform source of isotropic radiance. What is the total power through a target disc? The two discs are parallel to each other, and a common axis goes through their centers. The distance h between them is small compared with the target disc radius r_t but large compared with the source disc radius r_s. Is the irradiance over the target disc uniform? Why or why not? Now change this problem. Find the irradiance at the center of the target disc for arbitrary source disc radius and distance between the two discs. Check your solution against limiting cases, which is always good advice when you have solved any problem.

4.8. Assume the sun to be an isotropic emitter with uniform radiance L_s. Show that the expression for the irradiance at A_t (with $\vartheta_e = 0°$) is $L_s\pi r_s^2/d_{se}^2$. There are three ways to solve this problem. First, use the irradiance at the surface of the sun and spherical symmetry;

radiance is not needed. In the second approach use only Eq. (4.49) and appropriately adjusted limits of integration to separate those angles for which L_s is zero and not zero. Finally, solve this problem the hard way. This approach requires both setting up a double integral with the proper limits over that part of the sun that emits to Earth and using the angle ϑ'_s and the distance d'_{se}. This integral can be evaluated analytically for L_s uniform and isotropic, but would have to be evaluated numerically if the variation of L_s over the sun were taken into account. If evaluating the exact integral proves to be too difficult, try evaluating it approximately for $r_s/d_{se} \ll 1$. How would the solar irradiance at A_t change if it were located at latitude ϑ_e in the diagram?

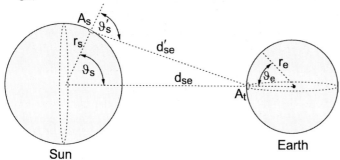

4.9. We are so accustomed to writing the solar irradiance as $L_s\Omega_s$, where L_s is the (assumed uniform and isotropic) radiance of the sun and Ω_s its solid angle subtended at Earth, that it is easy to forget that this expression is approximate if we use the exact expression, Eq. (4.5), for the solid angle. How much error do we make when we use the exact solid angle to determine the solar irradiance?

4.10. To our eyes the clear night sky is black even with a full moon. But film can respond to light our eyes cannot by increasing the exposure time. Thus one should be able to photograph a blue moonlit sky by attaching a camera to a tripod, pointing the camera at a patch of moonlit sky, opening the shutter and leaving it open. The only question is, How long? To answer this question, we pointed a camera at a patch of blue sunlit sky and noted that, for the film and aperture (f-stop) chosen, the exposure time selected by the camera's light meter was 1/250 s. Estimate the exposure time in moonlight (for the same film and aperture). Assume a full moon and take the reflectivity of the moon for visible light to be 10%.

4.11. Several years ago the rules for hunting in Pennsylvania were changed. Previously, hunters were not allowed to fire a rifle within 100 yards ($\approx$100 m) of a residence; now the minimum distance is 150 yards ($\approx$150 m). Assume that there are open fields near a house and that a single hunter is standing at the minimum distance. How has the new regulation changed the probability that a resident of the house would be hit accidentally by a bullet fired by this hunter? Needless to say, the hunter is not trying to shoot the resident. If trees surround the house, a bullet might hit a branch before reaching the house. Assume that the probability a bullet travels 100 yards without hitting a branch is 0.5. What is the probability that it travels 150 yards without hitting a branch? State all your assumptions. NOTE: This problem is not about hunting.

4.12. Suppose that you have been given the task of measuring the irradiance from a snow-pack illuminated by direct sunlight. Take the detector to be a circular disc of diameter D. This detector necessarily casts a shadow on the snow, and the shadow region does not contribute much to the irradiance measured by the detector. Thus the presence of the detector introduces an error in the measurement of the upward irradiance. Estimate the height h above the snow-pack that you must place the detector so that this shadow error is a few percent. Take the sun to be directly overhead and a parallel source of light. Don't worry about minor details. Give the quickest and dirtiest – but still correct – answer you can come up with.

4.13. We once received a message from a woman in Michigan who wondered "why water in lakes and streams appears much darker, often black in the winter". What is a possible explanation for what she observed?

HINT: This explanation makes no use of any detailed knowledge about lakes and streams in Michigan.

4.14. The following was taken from an advertisement for an "ultra bright laser pointer": "The 650 nm (nanometer, wavelength) laser light is five times brighter than 670 nm laser because the 650 nm is closer to the 633 nm of red light." Discuss.

4.15. A *pyranometer* measures the time-varying downward solar irradiance during daylight hours. On days with broken clouds the irradiance sometimes is observed to be greater than it would be if the sky were clear. Explain. What is the upper limit of the factor by which the measured irradiance can be greater when broken clouds are present than when the sky is clear, all else being equal? For simplicity, you may take the sun to be directly overhead.

4.16. Why does the temperature of an electric stove coil not increase without limit after it is turned on? Estimate the distance (relative to the size of the coil) from the coil at which the rate you are radiatively heated by it is comparable with the rate you are radiatively heated by the surroundings. You may assume a coil temperature of 1500 K. Make any reasonable assumptions. Elaborate derivations and detailed calculations are neither necessary nor desirable.

4.17. Some people still believe that seasonal temperature variations in midlatitudes are a consequence of the eccentricity of the Earth's orbit around the sun (i.e., the distance between the Earth and sun varies over the year by about 3%). What simple but convincing quantitative arguments can you make to show that variations in the distance between Earth and sun are not the cause of the seasons? What simple but convincing quantitative arguments can you make to show that the true cause of seasonal temperature variations is the obliquity of the Earth's axis (i.e., the rotational axis is inclined by about 23 degrees to the plane of the Earth–sun orbit)?

HINT: The easiest way to do this problem is to base your arguments on surface temperatures on the moon, which has no atmosphere or oceans to complicate matters.

4.18. We note in Section 1.6 that Earth, from the point of view of an observer on the moon, is an infrared sun with an effective blackbody temperature of around 255 K. What is the corresponding terrestrial irradiance at the moon? The distance from Earth to moon is about 60 Earth radii.

4.19. The albedo of Earth was known before there were artificial satellites (the moon is a *natural* satellite). Given that the luminance of the full moon is 9300 times that of the new moon, estimate Earth's albedo. This is revisited in Problem 5.23.

Determining Earth's albedo by measuring the *earthshine* (faint moonlight resulting from sunlight reflected by Earth) has a long and interesting history and is a topic of considerable current interest. For more on this see André Danjon, 1954: Albedo, color, and polarization of the Earth, in *The Earth as a Planet*, Gerard P. Kuiper, Ed., University of Chicago Press, pp. 726–38; Donald R. Huffman, Charles Weidman, and Sean Twomey, 1990: Repetition of Danjon earthshine measurements for determination of long term trends in the Earth's albedo. *Colloque André Danjon*, N. Capitaine and S. Debarbat, Eds., Observatoire de Paris, Journees 1990, Systemes de Reference Spatio-Temporels, Paris, 28-29-30 Mai, pp. 111–6; P. R., Goode, J, Qiu, V. Yurchyshyn, J. Hickey, M-C. Chu, E. Kolbe, C. T. Brown, and S. E., Koonin, 2001: Earthshine observations of the Earth's reflectance. *Geophysical Research Letters*, Vol. 28, pp. 1671–4.

4.20. Estimate the ratio of the total visible radiance from black paper (with reflectivity 10%, say) illuminated by bright sunlight to the radiance of white paper (with reflectivity 90%, say) illuminated by a 100 W light bulb at the kind of distances you would be from a reading lamp. What does this ratio tell you about the meaning of the terms black and white?

4.21. Estimate how much the solar irradiance (over a small area) can be increased with an ordinary magnifying glass (e.g., the kind used to read maps). All you need is a magnifying glass and ruler to answer this. Then estimate the maximum temperature (again, over a small area) to which an object might be raised by placing it at the focal point of a magnifying glass exposed to direct sunlight. Be sure that your answer makes sense.

4.22. Estimate by how much subsuns are brighter than clouds.

4.23. Using only your eyes investigate the degree to which an ordinary blackboard or white paper is a diffuse (isotropic) reflector of visible radiation.

4.24. Is the solid angle subtended by the sun at Earth the same as the solid angle subtended by Earth at the sun?

4.25. A colleague forwarded us the following email message (spelling and grammatical errors have not been corrected) that he, and many other people, received:

> "Everyone should mark their calendars this month for the 'Last Lunar Harrah' of the Millennium: This year will be the first full moon to occur on the winter solstice, Dec. 22…Since a full moon on the winter solstice occurred in conjunction with a lunar perigee (point in the moon's orbit that is closest to Earth), the moon will appear about 14% larger than it does at apogee (the point in its elliptical orbit that is farthest from the Earth). Since the Earth is also several million miles closer to the sun at this time of the year than in the summer, sunlight striking the moon is about 7% stronger, making it brighter. Also, this will be the closest perigee of the Moon of the year since the moon's orbit is constantly deforming. If the weather is clear and there is a snow cover where you live…it is believed that even car headlights will be superfluous.
>
> In layman's terms it will be a super bright full moon, much more than the usual AND it hasn't happened this way for 133 years! Our ancestors 133 years ago saw this. Our descendants 100 or so years from now will see this again."

Quantitatively evaluate the assertion that the moon will be "super bright" to the extent that driving in moonlight without headlights would be possible.

HINT: Determine by how much the lunar irradiance at Earth's surface increases.

4.26. What is the total amount of light scattered by a particle with scattering cross section C_{sca} at a height h above an infinite plane surface at every point of which the isotropic radiance is L? Now suppose that this surface is a diffuse reflector with reflectivity 1 illuminated normally by sunlight. What is the ratio of the total scattering of light from the surface to total scattering of direct sunlight?

HINT: The first part of this problem is done most easily in a cylindrical polar coordinate system with the z-axis passing through the particle and normal to the surface.

4.27. What would be the range of wavelengths (frequencies) over which the human eye is sensitive to electromagnetic radiation if the number of octaves were the same as that for the range of frequencies over which the human ear is sensitive to acoustic radiation (sound)? Take the center of this range of wavelengths to be such that the number of octaves above the middle of the visible spectrum equals the number below. To answer this question you will have to learn the range of audible sound frequencies and what is meant by an octave (how many octaves on a piano?).

4.28. Can you think of reasons why the human eye would not have evolved to respond to the range of wavelengths calculated in the previous problem?

HINT: Figure 2.25 may help.

4.29. In *Optical Treatise on the Gradation of Light*, published in 1760 (an earlier version appeared in 1729), Pierre Bouguer asserted that "at a depth of about 311 ft in sea water, the light of the sun becomes equal to that of the full moon seen at the surface of the earth." Verify Bouguer's assertion. Make any reasonable assumptions. All that is wanted is a rough check. Was Bouguer at least approximately correct (more than 240 years ago) or was he hopelessly wrong? Figure 5.12 should be helpful.

4.30. A man once told us that at a summer dog show in Arizona he was advised by a judge to never wet dogs to cool them. Water drops on the fur act as little lenses that greatly increase the rate of solar heating of the dog, and hence the dog's temperature would rise. What do you think of this advice? How would you give different advice in a way that could be understood by someone with little scientific training? Can you think of any convincing experiments or demonstrations to support your advice?

4.31. Show that combining any two visible sources S_1 and S_2 in any proportion yields colorimetric coordinates (x, y) that lie along the straight line between the respective colorimetric coordinates (x_1, y_1) and (x_2, y_2) of the sources.

HINT: For this and related problems the safest approach is to begin with X, Y, and Z.

4.32. Use the results of Problem 4.31 to show that the perceived color of any source of light, except a strictly monochromatic source (i.e., one with chromaticity coordinates lying on the spectrum locus), can be obtained in an infinite number of ways.

4.33. Use the results of Problem 4.31 to show that any spectral color can be obtained by adding white light to a monochromatic source.

4.34. Use the results of Problem 4.31 to show that any nonspectral color can be obtained by subtracting from white light a monochromatic source with wavelength equal to that of the complementary color of the nonspectral color.

4.35. Derive an expression for the excitation purity as a function of the colorimetric coordinates of the achromatic point, of the spectrum of interest, and the dominant wavelength. Derive a similar expression for the chromatic purity. Finally, determine the relation between excitation and chromatic purity. At what points are the two purities the same? Where is the difference between them greatest?

4.36. Can the perceived color we call blue be obtained by superposing sources of light that contain *no* blue light in the sense of light with wavelengths between 450 and 490 nm? If so, how?

HINT: Use the CIE diagram.

4.37. We stated in Section 4.3 that the light from three different (real) lamps cannot be added so as to match all visible wavelengths. This is equivalent to saying that three lamps cannot yield all points of color space. What about adding three laser beams of different wavelengths? If three won't do the job, what about four? If not four, what about five, and so on?

HINT: Use the CIE chromaticity diagram. Laser beams are represented by points on the spectrum locus. An analytical proof is not needed (perhaps not possible) here. You can answer these questions with crude sketches.

4.38. This problem is related to the previous one. For what shape of color space (chromaticity diagram) could all visible wavelengths be matched by combining three laser beams?

4.39. First do Problem 4.37, then answer the following questions. Would it be possible using various (finite) combinations of inks or dyes to obtain the full gamut of colors in color space? Is truly full-color photographic film possible (i.e., film that exactly reproduces the colors seen by humans)? What about truly full-color television?

4.40. Derive an expression for transforming from (x, y) color space to (u', v') color space. Is the excitation purity, defined as a ratio of distances on the CIE 1931 chromaticity diagram, the same as an excitation purity defined in a similar way for the CIE 1976 UCS chromaticity diagram?

4.41. G. I. Taylor is well known to meteorologists because of his many contributions to meteorology. What is less well known is that his first piece of research, done while he was a graduate student, was, in effect, photography photon by photon for very long times. That is, Taylor adjusted the irradiance of his light source so that only one photon at a time occupied the volume of his apparatus. For sake of argument, assume that this volume was that of a box 10 cm on a side (the approximate volume of a camera) with the source being one side of the box. Estimate the irradiance of Taylor's light source such that only one photon at a time occupies this volume.

HINT: See Problem 1.1 in Chapter 1.

4.42. Estimate the brightness temperature of the full moon at, say, the middle of the visible spectrum. You may take the visible reflectivity of the moon to be 10%. Estimate the color temperature of the moon. If these two temperatures are different, explain. Describe the color of a blackbody at the brightness temperature you obtain for the full moon. Explain any difference between this color and that of the moon.

4.43. You can buy at grocery stores Food Colors & Egg Dyes, four small (29 ml) vials containing different dyes (red, yellow, green, and blue). The color associated with each vial is indicated by the color of its cap. But the vial with the yellow cap presents us with a puzzle. The dye in this vial is red, and yet a few drops added to a glass of water turns the water yellow. Explain.

4.44. This problem is related to the previous one. A drop or two of the yellow dye added to a glass of water colors the water yellow. And a drop or two of the blue dye added to a glass of water colors the water blue. But if you look at the glass of yellow water through the glass of blue water (or vice versa), you see what is unquestionably green. Explain. This has a practical application: plastic re-sealable zipper bags with seals composed of two parts, a grooved yellow strip, and its blue mate. When the two are snapped together tightly the result is a green strip.

4.45. *Fahrenheit 451*, by the science-fiction writer Ray Bradbury, is an anti-utopian novel about a future world in which firemen, instead of putting out fires, burn books. The title comes from the *ignition temperature* of paper, the temperature to which it must be raised in order to burst into flames. What is the *minimum* (net) irradiance of direct sunlight illuminating paper in order for it to reach this temperature? Show that even allowing for a high reflectivity of paper for solar radiation, attenuation by the atmosphere, reflection and absorption by a magnifying lens, such a lens can focus solar radiation to yield irradiances well in excess of this minimum. To do this problem you'll have to get your hands on a magnifying lens. And once you have it, do some paper-burning experiments.

4.46. No one should measure any quantity without knowing beforehand its expected range of values. Estimate the maximum luminance of a diffusely-reflecting object illuminated by daylight.

HINT: Figure 1.5 should be helpful. All that is wanted here is a rough estimate.

4.47. This problem is related to the previous one. Estimate the maximum luminance of the (full) moon. You may take the visible reflectivity of the moon to be 10%. Then estimate the maximum luminance of a snowpack in moonlight. What do these values and that obtained in the previous problem tell you about the dynamic range of the human eye to different magnitudes of luminance?

4.48. This problem is related to the previous one. Estimate the luminance of a snowpack illuminated by a crescent moon (just before the new moon). To do this problem requires a moon calendar showing phases of the moon.

4.49. According to electromagnetic theory, which underlies all of optics, the fundamental measureable quantity (other than the separate electric and magnetic fields) is the Poynting vector, which is essentially an irradiance. Yet according to radiometry, the fundamental quantity is radiance, from which irradiance can be obtained by integration. This seems contradictory. Explain.

4.50. In order to check our result for Problem 4.47, we measured the luminance of the (nearly) full moon on a fairly clear night. But a problem we faced is that the field of view of our instrument is one degree whereas the moon subtends half a degree. How did we account for this?

HINT: The previous problem should be of help.

4.51. What is the relation between the spectral (or total) energy density u of an isotropic radiation field (e.g., the radiation field inside a cavity in equilibrium) and the spectral (or total) radiance?

4.52. This problem is related to the previous one. We stated without proof in Problem 1.40, that the pressure of a photon gas (for cavity radiation) is one-third its (total) energy density. Prove this. You will need the result from Problem 1.23.

HINT: Determine the total pressure on a specularly reflecting wall with 100% reflectivity.

4.53. This problem requires working the previous two. What must be the total radiance of equilibrium radiation (within a cavity) such that its pressure is one atmosphere? How does this radiance compare with that of the sun?

4.54. The *Bond albedo* for a spherical body is defined as the ratio of the total amount of reflected radiation to the total amount of incident (monodirectional) radiation. Derive an expression for the Bond albedo for a spherical body with a reflectivity R that depends on the angle of incidence. Assume azimuthal symmetry and uniformity of the body (reflectivity does not depend on position). Show that this albedo is the same as that for an isotropic source illuminating the body at any point.

4.55. We showed at the end of Section 4.2.1 that total irradiance is conserved across an air–water interface. In so doing we assumed that the transmissivity from air to water is close to 1 (reflectivity negligible). The justification for this was that the reflected (isotropic) irradiance is about 6% of the incident (isotropic) irradiance. Now let's turn this problem around. Suppose that a source of isotropic radiation is in the water. Is total irradiance still conserved across the interface using logic similar to Eqs. (4.62)–(4.64)? If not, what needs to be done to ensure that it is conserved?

HINT: Divide the angles of incidence into those less than the critical angle and those greater, and make a simple approximation.

4.56. A factor is missing from each of the terms in Eq. (1.65), but this makes no difference because it is common to all of them. What is this factor and why do you think that we omitted it?

4.57. Find the radiance from an infinite, diffusely reflecting plane when illuminated by a uniform and isotropic source disc at a distance z from the plane and parallel to it. The dimensions of the area A of the disc are $\ll z$. Express your answer as a function of the angle between the normal to the source disc and the line connecting the center of the source disc to a point on the plane.

4.58. Consider an infinite planar source, isotropic and uniform, above which and parallel to it is an infinite planar isotropic reflector with reflectivity R. The two planes are displaced relative to each other by a distance d along the z-axis. The source plane occupies the region in the xy-plane $-\infty < x < \infty$, $0 < y < \infty$, whereas the reflector occupies the region $-\infty < x < \infty$, $-\infty < y < 0$. Ignoring multiple reflections between the two planes find the reflected radiance at every point of the reflector. Your answer can be expressed succinctly as a function of an angle and checked using the known result for a source plane $-\infty < x < \infty$, $-\infty < y < \infty$.

4.59. To avoid wallowing in terminology, the bane of radiometry and photometry, we do not define *hemispherical emissivity* in Chapter 1, although we do note that emissivity depends on direction, in general. Hemispherical emissivity is the ratio of emitted irradiance to the Planck irradiance, whereas directional emissivity is the ratio of emitted radiance to the Planck radiance. Show how the hemispherical emissivity can be obtained from the directional emissivity. Assume azimuthal symmetry, that is, emission depends only on the angle between the direction of emission and the normal to the emitting surface. First guess what the relation might be, then obtain it carefully. If there is a discrepancy between your guess and the result of a careful analysis, ask yourself why.

4.60. On page 92 of *Mathematics: The Loss of Certainty* (1980, Oxford University Press) Morris Kline adduces the following example that "the notion of equality cannot be applied automatically to experience": "...colors a and b may seem to be the same as do colors b and c but a and c can be distinguished." Explain. We note that Kline wrote many excellent books on mathematics worth reading.

4.61. On page 131 of Dudley Towne's *Wave Phenomena* (cited at the end of Ch. 3) we find the following problem: "White light is normally incident from air upon the surface of a medium of high dispersion, the index of refraction for blue light being greater than that for red. Describe the color of the reflected light." Answer this question.

HINT: You will need to bring together ideas from Chapter 3, this chapter, and possibly even refractive index data for real materials. We suspect that our answer to Towne's question is not that expected by him. This is a simple question with a complex answer.

4.62. Consider a white disc of radius a. The disc is a uniform source of isotropic radiation of radiance L_0. Scattering and absorption by the medium between the disc and a point of observation is negligible. As the disc is moved away from this point, in what sense does the radiance of the disc change? This should become obvious if you sketch the radiance as a function of all relevant variables. How does your answer depend on the point of observation?

HINT: This is not a difficult problem but it does require recognizing that radiance is a function. Begin this problem by first understanding the second part of Problem 4.7and then move the point in the second part of Problem 4.7 away from the center of the target disc.

4.63. Consider a cylindrical tube of radius a, the inside walls of which are black. A uniform, isotropic source with radiance L_0 is placed at one end of the tube ($z = 0$). What is the radiance at any depth z in the tube? Is the radiance function at a given z the same at all radial points $0 \le r \le a$? What about the irradiance? Obtain an approximate expression for the (average) irradiance as a function of z for $z \gg a$. Try to devise a simple experiment to test your result, at least qualitatively.

4.64. We considered one explanation for the yellow of fog lamps in Problem 3.19. Here is another: the human eye is more sensitive to yellow light. Discuss. In particular, suppose that we were to place a yellow filter over a white headlight. Would the luminance change and, if so, in what direction? Under what simple condition could you be reasonably certain that the luminance of a yellow headlight was greater than that of a white headlight? Why did we say "reasonably" rather than absolutely? For more on fog lamps see J. H. Nelson, 1938: Optics of headlights. *Journal of Scientific Instruments*, Vol. 15, pp. 317–22. You might want to revisit this problem one more time after reading Sections 8.2 and 8.4.1.

4.65. A pinhole camera is, in essence, an opaque screen with a tiny hole in it. The hole serves the same purpose as a lens in that it results in a one-to-one transformation between object and image. The hole must be small for a sharp image. How does the radiance of the sun's image seen on the image plane (assumed to be a diffuse reflector) of the pinhole camera depend on the distance from the pinhole to this plane and other relevant physical variables? What does this result tell you about the limitations of a pinhole camera? You can do a simple experiment to verify your result. Poke a hole in a sheet of aluminum foil with a pin, image the sun onto a sheet of paper, and vary its distance from the hole.

4.66. The ASA number on film indicates how "fast" it is: the greater the ASA number, the shorter the exposure time (for a given light source and aperture size). Typically, one might use ASA 100 or 200 for outdoor photography. What ASA would be required for a pinhole camera to be able to duplicate what an ordinary camera (with a lens) does? You'll have to make some rough measurements to answer this question.

4.67. A lightning flash is an intense but very brief source of radiation. Approximate a vertical flash as a cylindrical source of radius R and length H, where $H \gg R$. Assume that emission by the flash is uniform in space over H and uniform in time over a time interval Δt. During this time N_{p} photons in the frequency interval $\Delta \nu$ centered on ν ($\Delta \nu \ll \nu$) are emitted isotropically. What is the irradiance of the flash? What is the corresponding radiance? Suppose that an irradiance detector is at a distance D ($R \ll D \ll H$) from the center of the flash, in a horizontal plane that intersects the flash at its midpoint. In clear air (scattering and absorption negligible), what is the time interval over which the detector will receive radiation from the flash? What is the average irradiance measured by the detector during this time interval? Describe qualitatively how this time interval and average irradiance change if the flash occurs in a cloud, in which the detector is also embedded, for which $\beta D \gg 1$, where β is the scattering coefficient (absorption assumed negligible). Also describe how the directionality of the radiation field at the detector changes because of the cloud.

4.68. Suppose we assume that there exists a spectrum we call "physicist's white", namely a perfectly flat spectrum over the visible. We still have to specify what we mean by flat, that is, for what variable (frequency, wavelength, etc.) is the spectrum flat? Suppose that it is flat as a function of frequency. What is the corresponding wavelength spectrum? Suppose that the spectrum is flat as a function of wavelength. What is the corresponding frequency spectrum?

5 Multiple Scattering: Elementary

This is the first of two chapters on multiple scattering. We could define a multiple-scattering medium as one for which the separate parts are excited to radiate not only by an external source of radiation but also by the radiation from each other. According to this unqualified definition, however, almost nothing, with the possible exception of a single (structureless) electron, is *not* a multiple-scattering medium. Multiple scattering is always a matter of degree, sometimes negligible, sometimes not. We pointed out in Section 3.4.9 that the light reflected by a glass of water is a consequence of scattering by a huge array of individual scatterers (molecules), separated by distances small compared with the wavelength and highly correlated in position. It follows that there is just as much multiple scattering in a water puddle as in a water cloud, both illuminated by sunlight. This is lost sight of – indeed, sometimes denied – because the usual theory of (specular) reflection by, say, a puddle, is based on continuum theory in which multiple scattering is not explicit. We call such a medium a *coherent* multiple-scattering medium. But from now on, unless stated otherwise, we confine ourselves to *incoherent* multiple-scattering media, ones for which, because of the arrangement of their separate parts, we are spared the drudgery of having to keep track of phases of scattered waves. For such media it is *as if* light were not a wave, which results in considerable simplification.

To understand what one sees in the natural world (e.g., the variation in color and brightness of the clear sky, the brightness and darkness of clouds, etc.) requires understanding multiple scattering, a topic conspicuous by its absence in optics textbooks. Because of this we sneak up on our subject rather than rush to embrace it, proceeding from a pile of plates to a two-stream theory, then, in the following chapter, N-stream theory, and finally the integro-differential equation of transfer. We first crawl, then walk, then run, and finally jump hurdles.

5.1 Multiple Scattering by a Pile of Parallel Plates

The simplest approach to a multiple-scattering medium, which demonstrates many of its characteristics, is by way of a pile of identical parallel plates, uniformly thick and infinite in lateral extent. We take the incident illumination to be perpendicular to the pile, although our analysis in no essential way depends on this. A single plate has a reflectivity R_1 and a transmissivity T_1. What is the reflectivity R_2 and transmissivity T_2 of two such plates (Fig. 5.1)? We assume that we can ignore the consequences of interference, which is valid if the separation between plates is appreciably larger than the wavelengths of the illumination and the irradiances are not strictly monochromatic but integrated over a range of wavelengths. We could approach this problem as we did in Section 1.4.1 where we determined the radiation field between two identical emitting plates by summing an infinite series. But here we take a different approach

Fundamentals of Atmospheric Radiation: An Introduction with 400 Problems. Craig F. Bohren and Eugene E. Clothiaux
Copyright © 2006 Wiley-VCH Verlag GmbH & Co. KGaA, Weinheim
ISBN: 3-527-40503-8

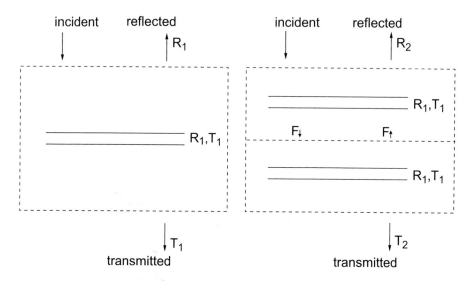

Figure 5.1: Given the reflectivity R_1 and transmissivity T_1 of an infinite (in lateral extent), uniform plate, the corresponding quantities for two such parallel plates, sufficiently far apart (relative to the wavelength of the incident illumination) that interference is negligible, follow by solving for the irradiances $F_{\downarrow}$ and $F_{\uparrow}$ between them.

to show that we can solve the same problem by different methods. The downward irradiance between the two plates is denoted by $F_{\downarrow}$, the upward irradiance by $F_{\uparrow}$; the incident irradiance has magnitude 1 (the units are irrelevant). Downward and upward (the terms downwelling and upwelling are also used) are just convenient labels for two opposite directions and have nothing fundamental to do with the direction of a gravitational field. For our purposes, light is not affected by gravity. These two irradiances satisfy

$$F_{\downarrow} = T_1 + F_{\uparrow} R_1, \tag{5.1}$$

$$F_{\uparrow} = F_{\downarrow} R_1. \tag{5.2}$$

These two equations in the two unknown irradiances can be solved to obtain

$$F_{\downarrow} = \frac{T_1}{1 - R_1^2}, \tag{5.3}$$

$$F_{\uparrow} = \frac{R_1 T_1}{1 - R_1^2}. \tag{5.4}$$

The reflectivity and transmissivity of the two plates depend on these two irradiances (for unit incident irradiance):

$$R_2 = R_1 + F_{\uparrow} T_1, \tag{5.5}$$

$$T_2 = F_{\downarrow} T_1. \tag{5.6}$$

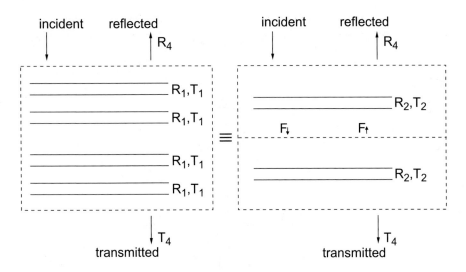

Figure 5.2: A pile of four identical parallel plates is equivalent to a pile of two composite plates, the reflectivity and transmissivity of which are determined by those of a single plate (Fig. 5.1). Thus the reflectivity and transmissivity of $2N$ identical plates readily follow from those of N plates.

Combining Eqs. (5.3)–(5.6) yields

$$R_2 = R_1 + \frac{T_1^2 R_1}{1 - R_1^2}, \tag{5.7}$$

$$T_2 = \frac{T_1^2}{1 - R_1^2}. \tag{5.8}$$

So far our results are general. Now we add the assumption that the plates are nonabsorbing ($R_1 + T_1 = 1$). With this assumption Eqs. (5.7) and (5.8) become

$$R_2 = \frac{2R_1}{1 + R_1}, \tag{5.9}$$

$$T_2 = \frac{1 - R_1}{1 + R_1}. \tag{5.10}$$

As required by conservation of radiant energy, $R_2 + T_2 = 1$.

We can look upon the two plates as a single composite plate with reflectivity given by Eq. (5.9), transmissivity by Eq. (5.10). Thus we have a simple way of obtaining the reflectivity R_4 and transmissivity T_4 of four identical plates (Fig. 5.2): in Eqs. (5.9) and (5.10) substitute R_2 for R_1. The result for the reflectivity, after a bit of rearrangement, is

$$R_4 = \frac{4R_1}{1 + 3R_1}. \tag{5.11}$$

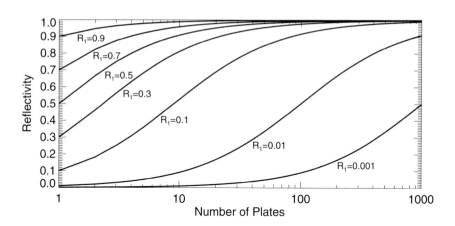

Figure 5.3: Reflectivity of a pile of identical, uniform, parallel plates, each with reflectivity R_1, versus number of plates. Because they are taken to be nonabsorbing, the reflectivity of the pile asymptotically approaches 1.

Now we can consider the four identical plates to be a single composite plate, and substitute R_4 in Eq. (5.9) to obtain the reflectivity of eight plates, then 16, 32, 64 and so on *ad infinitum*. Although we assumed no absorption, this does not affect our general conclusion that if we can determine the reflectivity and transmissivity of only two identical plates, we can determine these quantities for any multiple of two.

The form of Eq. (5.11) suggests that

$$R_N = \frac{N R_1}{1 + (N-1)R_1}, N = 1, 2, 4, 8 \ldots \tag{5.12}$$

Thus for any nonzero R_1, R_N approaches 1 as N approaches infinity (Fig. 5.3). Even though the reflectivity of a single plate may be small, a sufficiently large number of plates is highly reflecting. You can demonstrate this with a pile of plastic bags or overhead transparencies or plastic food wrap. For example, put one transparency on a black background (black paper or cloth) and compare it with a pile of several transparencies. As you add more and more transparencies the pile will become brighter and brighter but eventually reach a point of diminishing returns in which the addition of another transparency yields no perceptible increase. Or just take a long sheet of plastic food wrap and fold it in half, then in half again, and again. What you'll see is the curve in Fig. 5.3 growing before your eyes.

If N is large compared with 1, Eq. (5.12) is approximately

$$R_N \approx \frac{N R_1}{1 + N R_1}. \tag{5.13}$$

From this equation it follows that by measuring R_N we cannot separately determine N and R_1, only their product. The cumulative effect of many weakly reflecting plates is identical with that of a few highly reflecting plates. Measuring T_N would be of no help because it is

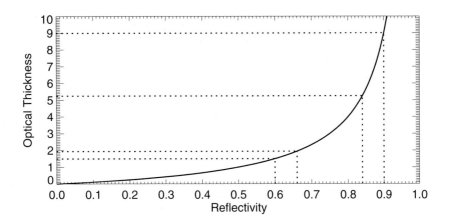

Figure 5.4: A fixed uncertainty (error) in reflectivity does not correspond to a fixed error in inferred optical thickness of a pile of nonabsorbing parallel plates. The greater the reflectivity, the greater the error in optical thickness. The error approaches infinity as the reflectivity approaches 1.

$1 - R_N$. Let's give the quantity NR_1 a symbol, τ, and a name, *optical thickness*, so that Eq. (5.13) can be written

$$R_N \approx \frac{\tau}{1+\tau}. \tag{5.14}$$

Measuring one quantity (or quantities) to obtain another quantity (or quantities) is an example of an *inversion*. Here we invert the measured reflectivity to obtain the desired optical thickness. But we already have encountered one limitation (among several) of inversions: what we want may be unobtainable. If we want N and R_1 we are out of luck: we have to content ourselves with their product. Another limitation is that an inversion interposes a theory between measurement of one physical quantity (e.g., reflectivity) and inference of another (e.g., optical thickness). Thus for an inversion to have a hope of yielding accurate results the theory must be applicable to the system of interest, in this instance a pile of identical parallel plates for which interference can be neglected. To understand a third limitation of inversions, we invert Eq. (5.14) to obtain

$$\tau \approx \frac{R_N}{1 - R_N}. \tag{5.15}$$

If we were to measure R_N in a perfect world, one in which measurements are free of errors, we could determine optical thickness by substituting reflectivity in Eq. (5.15). But perfection is not of this world, and depending on where measurements lie on the τ-R_N curve, a small error in R_N may correspond to a huge error in τ. Note in Fig. 5.4 that for R_N near 1, even a small error in this quantity corresponds to a huge error (or uncertainty) in τ. The slope of this

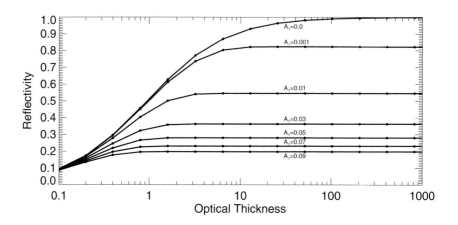

Figure 5.5: Reflectivity versus optical thickness for a pile of identical parallel plates, each with absorptivity A_1.

curve

$$\frac{d\tau}{dR_N} \approx \frac{1}{(1-R_N)^2} \tag{5.16}$$

is infinite at $R_N = 1$. The consequences of this can be made clearer by approximating the derivative in Eq. (5.16) by the ratio of differences and combining it with Eq. (5.15):

$$\frac{\Delta\tau}{\tau} \approx \frac{\Delta R_N}{R_N} \frac{1}{1-R_N}. \tag{5.17}$$

If we interpret $\Delta\tau/\tau$ as the fractional error in optical thickness for a given fractional error in reflectivity, it follows from this equation that for reflectivities near 1 even a small error in reflectivity corresponds to a huge error in optical thickness. We have run head on into the limitations of an inversion scheme.

Now let us return to the general problem of N plates that may be absorbing (more physically realistic than no absorption). We already solved this problem by the *doubling method* embodied in Eqs. (5.7) and (5.8):

$$R_{2N} = R_N\left[1 + \frac{T_N^2}{1-R_N^2}\right], \tag{5.18}$$

$$T_{2N} = \frac{T_N^2}{1-R_N^2}, N = 1,2,4,\ldots \tag{5.19}$$

Figure 5.5 shows how the reflectivity increases with optical thickness for $R_1 = 0.05$ and different values of A_1, the absorptivity of a single plate. On physical grounds it follows that as N approaches infinity, the transmissivity of the pile of plates approaches zero. If you need a proof of this, consider that for fixed R_1, T_1 for an absorbing plate is always less than for

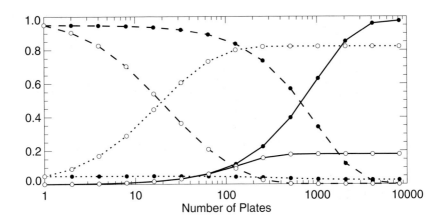

Figure 5.6: Absorptivity (solid curves), reflectivity (dotted curves), and transmissivity (dashed curves) of a pile of separated plates (open circles) and the same number of plates fused together (solid circles). The absorptivity of a single plate is 0.001 and the reflectivity because of a single interface of a plate is 0.025; normally incident radiation.

a nonabsorbing plate. This implies that R_2 and T_2 for an absorbing plate are less than the corresponding quantities when the plates are nonabsorbing. And similarly for any number of plates. We already showed that for nonabsorbing plates, the reflectivity approaches 1, and hence the transmissivity approaches 0 as the number of plates increases without limit. The transmissivity for a pile of absorbing plates, must approach the same limit. Define A_∞ as the limit of $1 - R_N - T_N$ as N approaches infinity, which is $1 - R_\infty$. Thus on a plot of reflectivity versus optical thickness (NR_1) for different values of A_1, the difference between 1 and the asymptotic value for the reflectivity R_∞ is A_∞ (Fig. 5.5).

If the absorptivity of a single plate is zero, the absorptivity of any number of plates is also zero, and the asymptotic reflectivity is 1. But if the absorptivity of a single plate is as small as 0.001, the asymptotic reflectivity drops by almost 20%. Further increases in the absorptivity of a single plate cause the asymptotic reflectivity to drop even more, but not proportionately. That is, the magnitude of dR_∞/dA_1 (which is negative) is greatest at $A_1 = 0$.

How the reflecting, absorbing, and transmitting properties of a given amount of material are affected by how the material is dispersed is shown in Fig. 5.6, where these properties for piles of separated plates are compared with those for the same number of plates fused together. The plate separation is sufficiently large that interference can be neglected. The reflectivity of the separated plates greatly outstrips that for the fused plates, but at the cost of a lower transmissivity and absorptivity. Both transmissivities approach zero as the number of plates increases, but transmissivity of the separated plates is always lower, often much lower.

When absorption is not zero, measurement of reflection and transmission yields two distinct quantities. But, alas, we now have three quantities to be determined: number of plates and any two of the set $\{R_1, A_1, T_1\}$. So again inverting measurements cannot give us all that we might want.

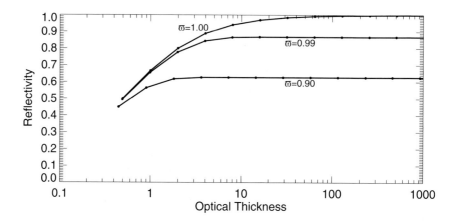

Figure 5.7: Reflectivity versus optical thickness for a pile of identical parallel plates for various values of the single-scattering albedo ϖ (the probability that an incident photon is not absorbed by a single plate) and asymmetry parameter $g = 0$.

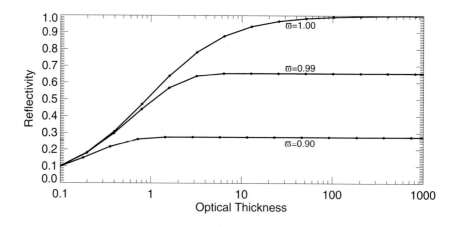

Figure 5.8: Same as Fig. 5.7 but with $g = 0.8$.

We can recast this in a different form. No new wine, just old wine in a new bottle. The quantity

$$\varpi = R_1 + T_1 = 1 - A_1 \tag{5.20}$$

is the probability that incident light is *not* absorbed by a single plate. We call this quantity the *single-scattering albedo* (the reason for this strange term will become apparent in following sections); its values lie between 0 and 1. Incident light not absorbed by a plate is either reflected, its direction changed by $180°$, or transmitted, its direction not changed at all. This leads us to define the *asymmetry parameter* g as the average of the cosine of the angle by which the incident light is changed:

$$g = \frac{T_1}{T_1 + R_1} \cos(0) + \frac{R_1}{T_1 + R_1} \cos(\pi) = \frac{T_1 - R_1}{T_1 + R_1}. \tag{5.21}$$

The asymmetry parameter lies between 1 (no reflection) and -1 (no transmission). We can solve Eqs. (5.20) and (5.21) for R_1 and T_1 in terms of ϖ and g:

$$R_1 = \varpi \frac{1 - g}{2}, \; T_1 = \varpi \frac{1 + g}{2}. \tag{5.22}$$

The three independent parameters τ, ϖ, g are equivalent to the three independent parameters N and any two of the set $\{R_1, T_1, A_1\}$. Figures 5.7 and 5.8 show reflectivity versus optical thickness for various values of g and ϖ. Note that the consequence of increasing g (for fixed ϖ) is to reduce the reflectivity. The greater the value of g, the greater the transmissivity of a single plate and of a pile of plates.

5.1.1 Why We Sometimes Can Ignore Interference and Sometimes Not

We began the previous section by asserting that we can ignore the consequences of interference (phase differences) in determining reflection and transmission by a pile of plates if they are separated by distances large compared with the wavelength and if the reflectivities and transmissivities are averages over bands of wavelengths. Because this is such an important yet often misunderstood point we digress briefly to support our assertion.

Consider a single transparent (negligible absorption) plate of uniform thickness h and refractive index n illuminated at normal incidence by light of wavelength λ. If we ignore interference, the reflectivity of this plate is

$$R = \frac{2R_\infty}{1 + R_\infty}, \tag{5.23}$$

where R_∞ is the reflectivity of a single interface of the plate, also the reflectivity of an infinitely thick slab of the same material. The similarity of this equation to Eq. (5.9) is not an accident. A single plate has two identical interfaces, so is formally equivalent to two identical plates. Both equations were derived under the same assumption, namely that we can add irradiances of different beams of radiation without having to concern ourselves with phase differences. Note that the thickness of the plate is neither explicit nor implicit in Eq. (5.23), and wavelength is only implicit through the possible dependence of R_∞ on λ.

Now consider the reflectivity of this same plate but taking into account interference (the wave nature of light):

$$\tilde{R} = \frac{2R_\infty(1 - \cos\phi)}{1 - 2R_\infty \cos\phi + R_\infty^2}, \tag{5.24}$$

where $\phi = 4\pi nh/\lambda$. This equation can be derived by adding waves, taking account of phase differences, multiply reflected because of the two interfaces of the plate (see Prob. 7.55). We find the total reflected electromagnetic wave, which has amplitude and phase, then square it to obtain reflected irradiance. We call Eq. (5.24) the *coherent reflectivity* to emphasize that it was derived taking interference into account and to distinguish it from the *incoherent reflectivity* [Eq. (5.23)]. These two equations are quite different. For example, the minimum value of $\tilde{R}$ is zero, its maximum value is $4R_\infty/(1 + R_\infty)^2$. Both equations apply to the same plate. Which is correct? Well, they *both* are correct, although in a strict sense Eq. (5.24) is more correct. To make sense out of this seemingly paradoxical statement consider the average of Eq. (5.24):

$$\langle \tilde{R} \rangle = \frac{1}{2\pi} \int_0^{2\pi} \frac{2R_\infty(1 - \cos\phi)}{1 - 2R_\infty \cos\phi + R_\infty^2} \, d\phi. \tag{5.25}$$

This integral can be looked upon as either an average over one period (2π), any integral number of periods, or any non-integral number appreciably greater than 1. With a fair amount of labor, using tables of integrals or attacking Eq. (5.25) with hammer and tongs, we obtain

$$\langle \tilde{R} \rangle = \frac{2R_\infty}{1 + R_\infty} = R. \tag{5.26}$$

Note that because $\phi = 4\pi n h\nu/c$, where c is the free-space speed of light and ν is the frequency, an average over ϕ is essentially an average over a band of frequencies (if the band is sufficiently narrow that n does not vary appreciably). For a narrow range of frequencies $\Delta\nu$ the corresponding range of phases ϕ is

$$\Delta\phi = 4\pi \frac{nh}{\lambda} \frac{\Delta\nu}{\nu}, \tag{5.27}$$

which is many multiples of 2π only if $nh/\lambda \gg 1$. If this condition is satisfied, the average of $\tilde{R}$ over this frequency interval is R to good approximation. But if the plate is thinner than or comparable with the wavelength, the average of $\tilde{R}$ over the frequency interval is not likely to be R except by accident (see Fig. 5.9).

We could have obtained Eq. (5.23) by first deriving Eq. (5.24). This would have required us to obtain $\tilde{R}$ from electromagnetic theory (a theory of waves with amplitudes and phases), which is not especially difficult but entails considerably more labor than deriving R. Then we would have had to evaluate the integral Eq. (5.25), which is difficult (but not impossible). This laborious procedure would have been like shooting an ant with a machine gun if we could have gone directly to Eq. (5.23) from the outset because the plate of interest is many wavelengths thick. We always are allowed to do more work than necessary, but life is short so we look for shortcuts. And keep in mind this was the easiest problem we could come up with to illustrate our point. A more complicated problem might not have been soluble exactly (i.e., within the framework of electromagnetic theory) or its solution might have been extremely tedious.

We hope that now you have a better understanding of why we could ignore interference when deriving the reflectivity of a pile of plates subject to the requirement that the separation between them is large compared with the wavelength. Note, however, that the plates themselves need not be thick compared with the wavelength. We could use Eq. (5.24) for the individual plates, then find the reflectivity of a pile of them using Eq. (5.12) if the separation between plates is large compared with the wavelength and if by reflectivity we mean a simple average over a range of frequencies.

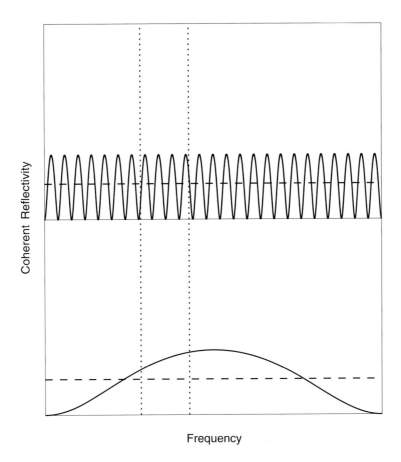

Frequency

Figure 5.9: Coherent reflectivity of a single, uniformly thick, nonabsorbing plate versus frequency (solid line). The top curve is for a plate with thickness times refractive index about 25 times the wavelength of the incident illumination; the bottom curve is for a plate with thickness times refractive index of about one wavelength. The horizontal dashed lines are the incoherent reflectivity obtained by ignoring interference. Vertical dotted lines show the range of frequencies over which the coherent reflectivity is averaged. For the plate much thicker than the wavelength, this average is essentially the incoherent reflectivity, whereas this is not true for the plate with thickness comparable with the wavelength.

5.1.2   Radiative Transfer in Plane-Parallel Media

A pile of parallel plates is a *discrete* medium consisting of alternating regions with abrupt, regular, and fixed boundaries between them. Of course all media are discrete at some scale: air is composed of discrete molecules moving about in empty space; clouds are composed of discrete droplets suspended in air. But the fleeting and amorphous discreteness of air and clouds is on a much smaller scale than that of a pile of plates. Moreover, atmospheres have no well-defined upper boundaries; like old soldiers, they just fade away.

In our analysis of a pile of plates we did not determine the radiation field everywhere, only the irradiances reflected and transmitted by the entire pile. All mathematical models are idealizations of reality, but if we hope to understand radiation fields at each point in an atmosphere or within clouds as well as above and below them, we need a model closer than a pile of plates to the reality of air and clouds. So we construct (on paper or in our minds) hypothetical *continuous* media the properties of which vary in only one direction. These media extend to infinity in directions in a plane, and hence are designated as *plane-parallel*. If there are boundaries within them we assume that they are abrupt (i.e., scattering and absorption properties change discontinuously at boundaries). No such media exist. An atmosphere, as its name implies, is spherical. Real clouds are finite, their properties vary in all directions, and they have jagged boundaries. Nevertheless, as long as we are careful to recognize the limitations of theories of radiative transfer in plane-parallel media (i.e., unreal media) they will help us understand radiative transfer in real media.

We assume that temperatures and wavelengths of interest are such that emission within media is negligible. All emission (i.e., external sources of radiation such as sunlight) is assumed to occur outside media of interest.

We also ignore the finite time it takes radiation to propagate, which is why time is absent from the equations of transfer in this book. Where time may enter is in the time-dependence of external sources (i.e., boundary conditions). If the boundary illumination changes with time, the consequences are assumed to be propagated instantaneously throughout the medium because of the huge value of the speed of light relative to distances and times of interest. For example, illumination of a cloud by sunlight may vary over times of order 10^2 s, whereas sunlight propagates thousands of meters in a cloud or air in times of order 10^{-5} s. What this means is that as sunlight changes, the radiation field everywhere in a cloud or in air adjusts almost instantaneously to these changes.

An example of when time lags may not be negligible is a cloud illuminated by a short pulse of radiation. A source (e.g., laser) is turned on suddenly, then turned off suddenly after a short time interval t. But the time interval during which scattered photons are received by a detector located at the source is greater than t because each photon has traveled a different path length. Moreover, the shape (in time) of the return pulse relative to that of the transmitted pulse depends on the characteristics in the clouds (see Sec. 6.4.2).

We assume further, as we did with the pile of plates, that the consequences of interference are negligible, which means that we can add radiant power from different sources. This is an inherent contradiction in radiative transfer theory. In deriving the equations of transfer in this book we take limits as distances shrink to zero, which implies that all quantities are mathematically continuous. But if media were truly continuous we could not apply to them a theory based on adding radiant power. We would have to add waves taking account of phase differences, then determine the radiant power. An example is a glass of water, which comes as close to a continuous medium as anything in nature. The water molecules are separated by distances small compared with the wavelengths of sunlight and their positions are strongly correlated (i.e., one molecule can't move without other molecules getting out of its way). Such a medium composed of a huge number of closely packed, correlated molecules is a multiple-scattering medium, just as a cloud of water droplets is. The difference between them lies only in the theories applicable to them. The radiative transfer theory in this book applied to a glass of water does not yield the laws of specular reflection and refraction, not even approximately. To determine scattering by this coherent array of molecules we would have to find the waves scattered by all of them, add all these waves taking account of phase differences, then square the resultant to obtain the power. We therefore assume continuity for mathematical purposes when we derive the equation of transfer but apply it to discrete media because only for such media can we ignore phase differences. We can get away with this for air because even though its molecules are separated by distances small compared with the wavelengths of sunlight, they are in constant motion and their positions are essentially uncorrelated (see Sec. 3.4.9). And we can get away with this for clouds because the constantly changing separation between droplets is large compared with the wavelengths of solar and terrestrial radiation and also with their lateral coherence lengths (see Sec. 3.4.2).

Because we assumed incoherent scattering, the photon language is the natural one for discussing the radiative transfer theory considered here. We look upon photons as discrete blobs of energy without phases, so that the energy transported by two photons traveling in the same direction is the sum of the energy transported by each one separately. Our radiative transfer theory is a theory of multiple scattering of photons rather than of waves.

5.1.3 Mean Free Path

What happens to photons as we imagine them launched from a particular point (call it the origin of the x-axis) in a medium? Photons obey statistical laws. We cannot determine what happens to a particular photon (photons are indistinguishable so the concept of a "particular photon" is meaningless) but we can determine what happens in a statistical sense to an ensemble of many photons. For example, what is the probability that a photon propagating along the x-axis beginning at $x = 0$ is absorbed between x and $x + \Delta x$? This is given by the integral of the probability distribution function $p(x)$

$$\int_x^{x+\Delta x} p(x)\, dx. \tag{5.28}$$

From Eq. (2.7) for exponential attenuation by absorption it follows that the probability a photon is *not* absorbed over a distance x from the origin is $\exp\{-\kappa x\}$, the probability it is *not* absorbed over a distance $x + \Delta x$ is $\exp\{-\kappa(x+\Delta x)\}$, and hence the probability it *is* absorbed in the interval Δx is the difference

$$\int_x^{x+\Delta x} p(x)\, dx = \exp\{-\kappa x\} - \exp\{-\kappa(x + \Delta x)\}. \tag{5.29}$$

If $p(x)$ is continuous and bounded then according to the mean value theorem of integral calculus the integral in Eq. (5.29) is

$$p(\overline{x})\Delta x = \exp\{-\kappa x\} - \exp\{-\kappa(x + \Delta x)\}, \tag{5.30}$$

where $\overline{x}$ lies between x and $x + \Delta x$. Divide both sides of this equation by Δx and take the limit as $\Delta x \to 0$:

$$p(x) = -\frac{d}{dx}\exp(-\kappa x) = \kappa \exp(-\kappa x). \tag{5.31}$$

With this probability distribution, the integral of which over all x is 1, we can find the mean distance a photon propagates before being absorbed:

$$\langle x \rangle = \int_0^\infty x\kappa \exp(-\kappa x)\,dx = \frac{1}{\kappa} = \ell_{\mathrm{a}}. \tag{5.32}$$

Thus the physical interpretation of κ is that its inverse is the mean free path for absorption ℓ_{a}. Attenuation by scattering also is exponential provided that a photon is removed from a beam of radiation if it is scattered once in any direction whatsoever. If we denote by β the scattering coefficient in the expression $\exp(-\beta x)$, the mean free path for scattering ℓ_{s} is $1/\beta$, and the total mean free path is

$$\ell_{\mathrm{t}} = \frac{1}{\kappa + \beta}, \tag{5.33}$$

where

$$\frac{1}{\ell_{\mathrm{t}}} = \frac{1}{\ell_{\mathrm{a}}} + \frac{1}{\ell_{\mathrm{s}}}. \tag{5.34}$$

We need one more result for the analysis that follows. What is the probability that given that a photon reaches x without being absorbed, it is absorbed in Δx? What we are after here is a *conditional probability*, the probability of absorption in Δx given that a photon has reached x, which is related to, but not the same as Eq. (5.30). The probability that a photon propagates a distance x from the origin without absorption is $\exp(-\kappa x)$. The probability that a photon is absorbed in Δx is given by Eq. (5.30). So this probability divided by the previous one is the conditional probability we are after:

$$1 - \exp(-\kappa \Delta x). \tag{5.35}$$

This quantity is the probability that if a photon finds itself at x it is absorbed as it propagates a further distance Δx. If $\kappa \Delta x \ll 1$, this probability is approximately $\kappa \Delta x$. And similarly, the probability of scattering is $\beta \Delta x$ provided that $\beta \Delta x \ll 1$.

Although the absorption coefficient κ and scattering coefficient β are fundamental quantities in the theory of radiative transfer, determining their values lies outside this theory. We either have to obtain them from measurements or appeal to a theory that comes to grips with the discreteness of molecules and particles.

5.2   Two-Stream Theory of Radiative Transfer

The preceding sections were fairly general, applicable to all theories of radiative transfer. To proceed further we have to make some specific assumptions, and the simplest one is that the radiation field consists of irradiances F in two and only two directions (streams), denoted as upward and downward. This is an idealization given that strictly monodirectional irradiances do not exist; even a laser beam has a finite angular spread. Scattering can therefore occur in only these two directions: a photon directed downward can be scattered only downward or upward, and similarly for a photon directed upward. We also ignore the polarization state of the

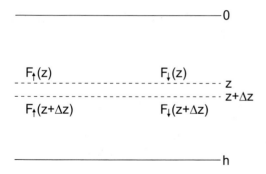

Figure 5.10: Downward ($\downarrow$) and upward ($\uparrow$) irradiances F are different at z and $z + \Delta z$ because of absorption and scattering within Δz. Note that the positive z-axis is downward.

radiation, which here makes no difference because the polarization state of light scattered by a spherically symmetric medium is not changed upon scattering in the forward and backward directions (see Sec. 7.4).

Conservation of downward radiant energy in a thin (relative to the mean free path) layer between z and $z + \Delta z$ (Fig. 5.10) yields

$$F_{\downarrow}(z + \Delta z) = F_{\downarrow}(z) - \kappa \Delta z F_{\downarrow}(z) - \beta \Delta z p_{\downarrow\uparrow} F_{\downarrow}(z) + \beta \Delta z p_{\uparrow\downarrow} F_{\uparrow}(z + \Delta z). \quad (5.36)$$

This is the mathematical form of the following statement. At the bottom of the layer the downward radiation is that incident at the top *decreased* by absorption ($\kappa \Delta z$ is the probability of absorption) and by scattering upward in Δz, but *increased* because upward radiation incident at the bottom of the layer is scattered downward in Δz. We can ignore the probability that a photon is scattered more than once in Δz (which depends on higher powers of $\beta \Delta z$) if we take $\beta \Delta z \ll 1$. The quantity $p_{\downarrow\uparrow}$ is the (conditional) probability that given that a downward photon is scattered, it is scattered in the upward direction, and similarly for $p_{\uparrow\downarrow}$. Divide both sides of Eq. (5.36) by Δz and take the limit as $\Delta z \to 0$:

$$\frac{dF_{\downarrow}}{dz} = -\kappa F_{\downarrow} - \beta p_{\downarrow\uparrow} F_{\downarrow} + \beta p_{\uparrow\downarrow} F_{\uparrow}. \quad (5.37)$$

A similar radiant energy conservation argument applied to the upward irradiance yields

$$\frac{dF_{\uparrow}}{dz} = \kappa F_{\uparrow} + \beta p_{\uparrow\downarrow} F_{\uparrow} - \beta p_{\downarrow\uparrow} F_{\downarrow}. \quad (5.38)$$

The sign reversal is a consequence of attenuation of upward radiation in the direction of decreasing z.

It is the presence of the third term on the right sides of Eqs. (5.37) and (5.38) that makes radiative transfer non-trivial. Were it not for this term, the solutions to both equations would be simple exponentials. What complicates matters is that downward radiation is a source for upward radiation and vice versa: the upward and downward irradiances are coupled.

Because photons can be scattered only upward or downward the sum of probabilities must satisfy

$$p_{\downarrow\uparrow} + p_{\downarrow\downarrow} = p_{\uparrow\downarrow} + p_{\uparrow\uparrow} = 1. \quad (5.39)$$

By using Eq. (5.39) we can rewrite Eqs. (5.37) and (5.38) as

$$\frac{dF_\downarrow}{dz} = -(\kappa + \beta)F_\downarrow + \beta(p_{\downarrow\downarrow}F_\downarrow + p_{\uparrow\downarrow}F_\uparrow),$$

(5.40)

$$\frac{dF_\uparrow}{dz} = (\kappa + \beta)F_\uparrow - \beta(p_{\uparrow\uparrow}F_\uparrow + p_{\downarrow\uparrow}F_\downarrow).$$

(5.41)

These equations are easier to interpret. The first term on the right side expresses all the ways radiation can be removed from a particular direction (expenses), the second term all the ways radiation can be added to that direction (income). If you can balance a checkbook you can understand these equations.

Now we further assume that the medium is *isotropic*: $p_{\downarrow\uparrow} = p_{\uparrow\downarrow}$, $p_{\downarrow\downarrow} = p_{\uparrow\uparrow}$. Such a medium is rotationally symmetric: rotate it by any amount and you can't tell that it has been rotated. A suspension of spherically symmetric scatterers is an isotropic medium, as is a suspension of randomly oriented asymmetric scatterers. An example of an anisotropic medium is a suspension of oriented, asymmetric scatterers.

We have to take some care with the term isotropic because it is used in different ways, sometimes in the same breath. An *isotropic radiation field* here is defined by $F_\uparrow = F_\downarrow$; *isotropic scattering* by $p_{\uparrow\downarrow} = p_{\uparrow\uparrow}$. Isotropic scattering does not necessarily imply an isotropic radiation field and conversely. To make matters more confusing, there are no isotropic scatterers of electromagnetic waves (acoustic waves are another story) in nature in the sense that they scatter equally in *all* directions (see Sec. 7.3). You can't beg, borrow, or steal such scatterers, but it is possible to find ones that scatter equally in two opposite hemispheres of directions (e.g., scattering of sunlight by air molecules).

As with the pile of plates, we define the asymmetry parameter g as the mean cosine of the scattering angle, which has only two values, 1 and -1:

$$g = p_{\downarrow\downarrow}(1) + p_{\downarrow\uparrow}(-1).$$

(5.42)

With this definition and the assumption of an isotropic medium the various probabilities in Eqs. (5.40) ánd (5.41) can be expressed solely in terms of g:

$$p_{\downarrow\uparrow} = p_{\uparrow\downarrow} = \frac{1-g}{2}, \quad p_{\downarrow\downarrow} = p_{\uparrow\uparrow} = \frac{1+g}{2}.$$

(5.43)

It often is convenient to transform from physical depth z to *optical depth* τ:

$$\tau = \int_0^z (\kappa + \beta)\, dz = \tau_a + \tau_s.$$

(5.44)

We have to take care when we encounter this term because total optical depth τ is the sum of absorption optical depth τ_a and scattering optical depth τ_s. Authors often write optical depth as τ and leave it to readers to guess which one of the three possibilities is meant. If κ and β are independent of z it follows from Eq. (5.33) that τ is physical depth measured in units of total mean free path.

By using Eqs. (5.43) and (5.44) we can write Eqs. (5.40) and (5.41) as

$$\frac{dF_\downarrow}{d\tau} = -F_\downarrow + \varpi\left\{\frac{1+g}{2}F_\downarrow + \frac{1-g}{2}F_\uparrow\right\},$$

(5.45)

$$\frac{dF_\uparrow}{d\tau} = F_\uparrow - \varpi\left\{\frac{1+g}{2}F_\uparrow + \frac{1-g}{2}F_\downarrow\right\}, \tag{5.46}$$

where the *single-scattering albedo* ϖ is $\beta/(\beta+\kappa)$. More compact versions of these equations can be obtained by adding and subtracting them:

$$\frac{d}{d\tau}(F_\downarrow - F_\uparrow) = -(1-\varpi)(F_\downarrow + F_\uparrow), \tag{5.47}$$

$$\frac{d}{d\tau}(F_\downarrow + F_\uparrow) = -(1-\varpi g)(F_\downarrow - F_\uparrow). \tag{5.48}$$

In general, optical depth τ, single-scattering albedo ϖ, and asymmetry parameter g depend on the frequency of the radiation. The latter two quantities also may vary with physical depth z or, equivalently, optical depth; ϖ lies between 0 and 1, g between -1 and 1, although the end points of these two intervals never occur in reality.

5.2.1   Conservative Scattering

A single-scattering albedo of 1 is sometimes referred to as *conservative scattering*, which is not a comment on the political affiliation of photons but rather signals that radiant energy is conserved; nonzero absorption implies that radiant energy is converted into other forms although, as always, total energy is conserved. As noted at the end of the previous paragraph $\varpi = 1$ does not exist in nature, but the assumption that it does sometimes leads to no serious errors and simplifies solutions to the equations of transfer. Indeed, with this assumption and the assumption that g is independent of τ, the difference of irradiances in Eq. (5.47) is constant, and their sum in Eq. (5.48) is a linear function of τ, which yields

$$F_\downarrow = B + C(1 - \tau^*), \quad F_\uparrow = B - C(1 + \tau^*), \tag{5.49}$$

where B and C are constants determined by conditions at the upper ($\tau = 0$) and lower ($\tau = \overline{\tau}$) boundaries of the medium. Although written without a subscript, here optical depth is scattering optical depth. The quantity $\tau^* = \tau(1 - g)$ is the *scaled optical depth*. Scaled here means scaled to isotropic scattering. Two media, one composed of isotropic scatterers ($g = 0$), the other of anisotropic scatterers ($g \neq 0$), are equivalent if the optical depth of the latter is the optical depth of the former divided by $1 - g$. The *optical thickness* of the entire medium (from $z = 0$ to $z = h$) is

$$\overline{\tau} = \int_0^h \beta\, dz. \tag{5.50}$$

Optical thickness and optical depth often are used interchangeably, but we try to be careful to reserve optical thickness for the optical depth of an entire medium.

5.2.2   Conservative Scattering: Equilibrium Solution

If the medium is infinite ($h \to \infty$), for the two irradiances in Eq. (5.49) to be finite requires $C = 0$, which in turn implies that the upward and downward irradiances are equal and independent of τ. Note that here the radiation field is isotropic even though scattering need not be (g is arbitrary). Infinite media don't exist, so let's consider a medium illuminated by downward irradiance F_0 at its upper boundary, and with a mirror (reflectivity 1) at its lower boundary. The upper boundary condition is $F_\downarrow = F_0$, the lower boundary condition is $F_\uparrow = F_\downarrow$. Again the solution is $C = 0$, and the two irradiances are uniform and equal to the downward irradiance at the upper boundary. A uniform, isotropic radiation field is called the *equilibrium solution*. To understand why, consider the medium to be sandwiched between two mirrors (with reflectivity 1). Again, the solution to Eq. (5.49) is $C = 0$, but B is arbitrary. The interpretation of this result is that if photons are injected into the medium they can neither escape from nor be absorbed by it, and hence the radiation field eventually becomes uniform and isotropic. Strict equilibrium occurs when all gradients are ironed out.

5.2.3   Conservative Scattering: Reflection and Transmission

Although the solutions in the previous section are illuminating (forgive the pun), they are not especially realistic. So let's consider a more realistic problem in which photons that leak out the bottom boundary never return (i.e., $F_\uparrow = 0$ at $\overline{\tau}$), either because there is nothing to return them or they are absorbed before doing so; the downward irradiance at the top boundary is F_0. With these boundary conditions the solution to Eq. (5.49) yields for the reflectivity R and transmissivity T

$$R = \frac{F_\uparrow(0)}{F_0} = \frac{\overline{\tau}(1-g)/2}{1+\overline{\tau}(1-g)/2} = \frac{\overline{\tau}^*/2}{1+\overline{\tau}^*/2}, \tag{5.51}$$

$$T = \frac{F_\downarrow(\overline{\tau})}{F_0} = \frac{1}{1+\overline{\tau}(1-g)/2} = \frac{1}{1+\overline{\tau}^*/2}. \tag{5.52}$$

As required by radiant energy conservation, $R + T = 1$. But wait! We've seen Eq. (5.51) before. It has the same form as Eq. (5.14), the reflectivity of a pile of plates, the only difference being that we called $\tau = NR_1 = N(1-g)/2$ [see Eq. (5.21)] the optical thickness. We did so because calling this quantity one-half the scaled optical thickness might have been perplexing. But now that you are inoculated against perplexity we can rewrite history and say that in Eq. (5.14) we really meant to write $\overline{\tau}$ for N. Does it make sense that N, the number of plates, is analogous to optical thickness? We think it does. Although N can take on only discrete values $(1, 2, 3 \ldots)$ whereas $\overline{\tau}$ is continuous, $\overline{\tau} = 1$ corresponds (roughly) to a probability of 1 for a photon to be scattered at least once, $\overline{\tau} = 2$ corresponds to a probability of 1 for a photon to be scattered at least twice, and so on. But this is analogous to what happens when we add plates to a pile. Every plate added (in discrete steps) increases the probability of a photon being scattered. With 2 plates a photon is likely to be scattered (reflected) at least twice, with 3 plates, at least 3 times, and so on. With the wisdom of hindsight, and a cup of physical intuition, we could have solved the equation of transfer (without even being aware of its existence) to obtain Eqs. (5.51) and (5.52) knowing the solution to the pile-of-plates problem and understanding the physical meaning of optical thickness. Moreover, now we don't have to plot Eq. (5.51): we already did so (Figs. 5.7 and 5.8).

From Eq. (5.51) it follows that reflectivity asymptotically approaches 1 as optical thickness approaches infinity. But it is not necessary for the medium to be infinite in order for it to be *optically thick*, by which is meant indistinguishable from an infinite medium. For example, the average contrast threshold of the human eye is 2%, which means that humans cannot detect the difference between two light sources with luminances different by less than 2%. If we adopt this criterion, a scaled optical thickness $\bar{\tau}(1 - g)$ of around 100 corresponds to a reflectivity of 0.98, which is within 2% of the reflectivity of an infinite medium.

With the help of Eq. (5.43) we can write Eq. (5.52) in a form that is easier to interpret:

$$T = \frac{1}{1 + \bar{\tau}p_{\downarrow\uparrow}}. \tag{5.53}$$

This equation tells us that attenuation of incident light is a consequence only of downward photons scattered upward. Downward photons scattered downward continue to contribute to the downward irradiance.

With the assumption that $\bar{\tau}^*/2 \ll 1$, Eq. (5.52) can be expanded in a power series and truncated after the second term:

$$T = 1 - \bar{\tau}^*/2. \tag{5.54}$$

But this is also the first two terms in the expansion of the exponential function:

$$T \approx \exp(-\bar{\tau}^*/2). \tag{5.55}$$

Exponential attenuation by scattering would prevail only if multiple scattering were negligible, that is, if photons scattered out of the forward direction never returned to that direction. Contrary to what the term "multiple" might lead you to think, attenuation for which multiple scattering is taken into account is always *less* than attenuation for which single scattering is assumed (Fig. 5.11). No equations or figures are needed to prove this. On the basis of single-scattering arguments, a photon scattered out of its original direction never returns to that direction, whereas multiple scattering gives the photon additional chances to be scattered back into that direction.

Clouds often are said to be white (strictly, the spectrum of light reflected by them is not appreciably different from that of the light illuminating them) because scattering of visible light by cloud droplets is independent of wavelength. This leads to the notion (which takes on various guises) that when a multiple-scattering medium is observed to be white, this infallibly signals the presence of "big" (compared with the wavelength) particles. Here we have a failure to distinguish between a necessary and a sufficient condition. It is indeed true that cloud particles (water droplets, ice particles) are sufficiently large compared with the wavelengths of visible light that they scatter more or less independently of wavelength (Fig. 3.12). This is a sufficient condition for the cloud to be white upon illumination by white light, but it is not necessary, which you can demonstrate for yourself. Paint the insides of two aluminum pie pans black. Fill them with water. To one add just a few drops of milk, to the other so much milk that the bottom of the pan is not visible. Observe these two pans side by side in bright sunshine. The dilute (optically thin) suspension has a bluish cast, whereas the more concentrated (optically thick) suspension is white. And yet the particles in both pans are the same. This is yet another example in which single-scattering arguments applied to a multiple-scattering medium lead to erroneous conclusions. The individual particles in milk are sufficiently small that they scatter more at the short-wavelength end of the visible spectrum

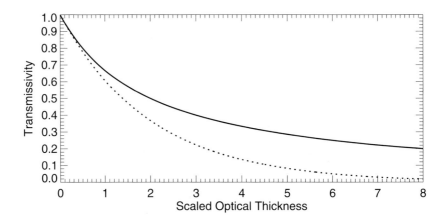

Figure 5.11: Because of multiple scattering, transmission by a plane-parallel medium decreases less rapidly (solid curve) with scaled optical thickness than exponentially (dashed curve).

than at the long-wavelength end. But in an optically thick medium of many such particles the cumulative effect of multiple scattering is to wash out this spectral dependence, which follows from differentiating Eq. (5.51) to obtain the wavelength dependence of reflectivity:

$$\frac{dR}{d\lambda} = \frac{dR}{d\bar{\tau}^*}\frac{d\bar{\tau}^*}{d\lambda} = \frac{1}{2(1+\bar{\tau}^*/2)^2}\frac{d\bar{\tau}^*}{d\lambda}. \tag{5.56}$$

The spectral dependence of $\bar{\tau}^*$ is a consequence of that of the scattering coefficient β and of g (usually much weaker). Consider the two limits, optically thin and optically thick:

$$\frac{dR}{d\lambda} \approx \frac{d\bar{\tau}^*}{d\lambda} \ (\bar{\tau}^*/2 \ll 1), \tag{5.57}$$

$$\frac{dR}{d\lambda} \approx 0 \ (\bar{\tau}^*/2 \gg 1). \tag{5.58}$$

These are markedly different spectral dependences and yet the scatterers are the same. What is different is only their amount. For the optically thin medium [Eq. (5.57)], the spectral dependence of reflectivity is essentially that of the individual scatterers; for the optically thick medium [Eq. (5.58)], reflectivity is essentially independent of wavelength.

At least two caveats should accompany pronouncements about the whiteness of clouds. You may have noticed pastel colors, called *iridescence*, in thin clouds or at the edges of thick ones when looking toward the sun. Or you may even have seen colored rings a few degrees across, called the *corona*, around the sun or moon through thin, even imperceptible, clouds (see Sec. 8.4.1). The key word here is "thin". Although *total* scattering (in all directions) of visible light by cloud droplets is to good approximation independent of wavelength, *angular* scattering is not: light is scattered more in some directions than in others depending on wavelength. We can express this within the context of our simple two-stream theory by saying

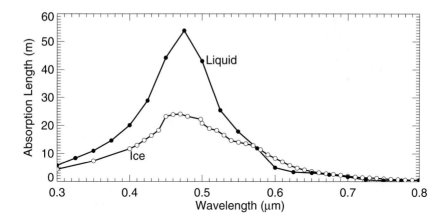

Figure 5.12: Absorption length (inverse absorption coefficient) of pure ice and liquid water over the visible spectrum. The data for liquid water were taken from Querry *et al.* (1991), those for ice from Warren (1984).

that g depends on wavelength. This dependence is weak but may be observable in thin clouds. Here is yet another example of how multiple scattering qualitatively changes what is observed.

Although clouds seen by *reflected* light are usually white (again, assuming incident white light), light *transmitted* by very thick clouds may have a bluish cast. This is not evident from Eq. (5.52) nor can it be because it was derived under the assumption that absorption is negligible. And it often is – but not always. Pure water, including ice, has an absorption minimum in the blue–green part of the visible spectrum and rises sharply toward the red (Fig. 5.12). But absorption by water at all visible wavelengths is weak in the sense that visible light has to be transmitted many meters through water before being perceptibly attenuated by absorption (which is why we don't think of a glass of water as a glass of blue dye). Although the total liquid water paths of thick clouds are of order a centimeter or less, multiple scattering in very thick clouds can greatly magnify the effective transmission distance. We try to make this point clearer in Section 5.3 on multiple scattering in absorbing media.

5.2.4   Conservative Scattering: Diffuse Radiation

Now we turn our attention to the radiation field within and transmitted by a multiple-scattering medium such as a cloud illuminated at its upper boundary by irradiance F_0. The irradiance of the light that has *not* been scattered, upward or downward, is an exponential function of optical depth into the medium:

$$F_\downarrow^u = F_0 \exp(-\tau), \tag{5.59}$$

where the superscript "u" indicates unscattered. By definition the *diffuse* downward irradiance $D_\downarrow$ is light that *has* been scattered, and hence is the difference between the total downward irradiance and the unscattered downward irradiance:

$$D_\downarrow = F_\downarrow - F_\downarrow^u. \tag{5.60}$$

The upward irradiance is necessarily diffuse because it is light that must have been scattered. For the sake of harmonious notation, we denote $F_\uparrow$ as $D_\uparrow$. With the assumption that radiation transmitted through the lower boundary of the medium ($\tau = \bar\tau$) is not returned [$D_\uparrow(\bar\tau) = 0$], the solution to Eq. (5.49), combined with Eq. (5.60), yields the two diffuse irradiances at any optical depth τ within the medium:

$$\frac{D_\downarrow}{F_0} = \frac{1 + (\bar\tau - \tau)(1 - g)/2}{1 + \bar\tau(1 - g)/2} - \exp(-\tau), \tag{5.61}$$

$$\frac{D_\uparrow}{F_0} = \frac{(\bar\tau - \tau)(1 - g)/2}{1 + \bar\tau(1 - g)/2}. \tag{5.62}$$

If we assume that the medium is optically thick [$\bar\tau(1 - g)/2 \gg 1$], the downward irradiance, Eq. (5.61), at $\tau = 2/(1 - g)$ is approximately

$$\frac{D_\downarrow}{F_0} = 1 - \exp\{-2/(1 - g)\}. \tag{5.63}$$

The argument of the exponential ranges from -1 to $-\infty$, so for optical depths greater than $2/(1 - g)$ it is always a good approximation to neglect the exponential term in Eq. (5.61):

$$\frac{D_\downarrow}{F_0} = \frac{1 + (\bar\tau - \tau)(1 - g)/2}{1 + \bar\tau(1 - g)/2}. \tag{5.64}$$

At optical depths into the medium greater than about $2/(1 - g)$ the total irradiance is dominated by the diffuse irradiance. From Eqs. (5.62) and (5.64) the two diffuse irradiances are approximately equal if

$$(\bar\tau - \tau)(1 - g)/2 > 1 \text{ or } \tau < \bar\tau - 2/(1 - g). \tag{5.65}$$

At (optical) distances less than $2/(1 - g)$ from the lower boundary, the black underlying medium [i.e., $D_\uparrow(\bar\tau) = 0$] makes its presence felt.

Thus within the region lying at least $2/(1 - g)$ from the upper and lower boundaries, the diffuse radiation field is approximately isotropic. This makes sense if we recognize that over the optical distance $\tau p_{\uparrow\downarrow} = \tau p_{\downarrow\uparrow} = \tau(1 - g)/2 = 1$, a photon is nearly certain to have been turned around. And recall that the radiation field was isotropic for a medium sandwiched between two mirrors (equilibrium solution). The medium to which Eqs. (5.61) and (5.62) apply is also sandwiched between two mirrors, but they are *distributed* over an optical thickness (approximately) $2/(1 - g)$ rather than being *localized* at the upper and lower boundaries. So with a bit of physical intuition and an understanding of $\tau(1 - g)/2$ we could have bypassed Eqs. (5.61) and (5.62) and gone directly to the final result, namely an isotropic radiation field within the core of a multiple-scattering medium.

You can observe Eqs. (5.61) and (5.62) unfolding before your eyes as you descend in an airplane into a thick cloud. At first you can tell (optically) up from down. The light

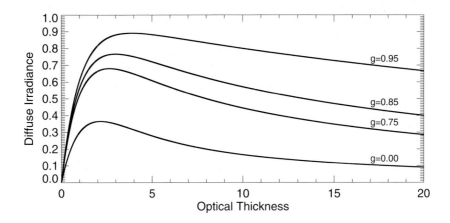

Figure 5.13: Diffuse irradiance beneath a plane-parallel, nonabsorbing medium as a function of its optical thickness. The higher values of the asymmetry parameter g are typical for clouds at visible and near-visible wavelengths; $g = 0$ is for molecular scattering.

above is brighter than that below. But as you descend farther you enter a world in which everything looks the same in all directions. Indeed, a former student of ours carried a small radiometer with him on an airplane flight and verified for himself that within a cloud the radiation field is more or less isotropic. Suspended in this featureless world, you may become a bit anxious, then breathe a sigh of relief when you finally notice darkness below, signaling that your airplane is about to leave the cloud.

The diffuse downward irradiance [Eq. (5.61)] *beneath* a cloud

$$\frac{D_\downarrow(\bar{\tau})}{F_0} = \frac{1}{1 + \bar{\tau}(1-g)/2} - \exp(-\bar{\tau}) \tag{5.66}$$

rises steeply from 0 for an optical thickness $\bar{\tau} = 0$ to a maximum at approximately $\bar{\tau} = (1/g)\ln[2/(1-g)]$ $(g > 0.5)$, then gradually decreases for increasing optical thickness (Fig. 5.13). As with the isotropic radiation field within a cloud, you can readily observe Eq. (5.66) being played out in nature – and you don't even have to board an airplane. The overhead sky (except, of course, directly toward the sun) is not nearly as bright as many clouds. This is evident when you observe patches of dark blue sky in the transitory openings in cloud decks (Fig. 1.9). Clouds increase the diffuse downward radiation – but only up to a point. With ever-increasing optical thickness clouds eventually reduce the downward radiation to less than what it would be from a clear sky. Clouds are both givers and takers of light. If not too thick, they can make the sky brighter than the clear sky, but very thick clouds result in a dark, gloomy day.

We sometimes encounter assertions that reflection by the ground can color the bottoms of clouds (e.g., the bottoms of clouds are green because of reflection by green grass). We can investigate these assertions within the framework of the two-stream theory by solving Eq. (5.49) subject to the lower boundary condition $F_\uparrow(\bar{\tau}) = R_s F_\downarrow(\bar{\tau})$, where R_s is the reflectivity of the underlying surface:

$$D_\downarrow(\overline{\tau}) \approx F_\downarrow(\overline{\tau}) = \frac{F_0(1 - R)}{1 - RR_s}. \tag{5.67}$$

We used Eq. (5.51) for the reflectivity R over a non-reflecting surface to cast our solution in a simple form, easy to interpret: $1 - R$ is the transmissivity T of the cloud, $1/(1 - RR_s)$ is a consequence of multiple reflections between the cloud and the underlying ground. If we assume that $RR_s \ll 1$, we can expand the denominator in Eq. (5.67) and truncate the series after the second term:

$$D_\downarrow(\overline{\tau}) \approx F_0(1 - R)(1 + RR_s). \tag{5.68}$$

The spectrum of the diffuse downward radiation beneath the cloud is therefore not that of the ground $[F_0(1 - R)R_s]$ but rather diluted, so to speak, as a consequence of multiplication by R and addition of 1. From this result it is difficult to fathom how reflection by colored ground could do more than tinge the light from the bottoms of more or less continuous clouds. Broken clouds are another matter (see Prob. 5.38).

Now consider an extreme case: $R_s = 1$ over some narrow band of wavelengths and 0 at other wavelengths. Over this band the downward irradiance [Eq. (5.67)] is F_0, whereas outside this band is $F_0(1 - R)$. So indeed there could be a spike in the spectrum over the narrow band of wavelengths if $R \approx 1$. This result required invoking reflection by ground that does not generally exist in nature at visible wavelengths. But we must be careful here not to extrapolate results from the visible to other wavelengths (e.g., the near-infrared) where surface reflectivity may be much greater. Alexander Marshak and his colleagues have demonstrated that the increase in the reflectivity of vegetated surfaces from visible (0.65 µm) to near-infrared wavelengths (0.80 µm) is sufficient to give appreciable difference in the downward radiances from cloud base at these two wavelengths (see Prob. 5.35). Nevertheless, we conclude that if the bottoms of continuous clouds are markedly colored at visible wavelengths, we ought to look to causes other than reflection by the ground. One such cause, (selective) absorption, is briefly mentioned at the end of the previous section. Now we elaborate on this.

5.3   Multiple Scattering in an Absorbing Medium

Up to this point we have assumed that absorption is negligible and taken κ to be identically zero. But an absolutely nonabsorbing medium does not exist. What about free space? It, too, does not exist. Even the vast reaches of interstellar space are populated sparsely with molecules and particles. If you know how to make an absolute vacuum, run, do not walk, to the Patent Office. Moreover absorption is rarely if ever absolutely negligible: it may be negligible for some purposes but not others. So the term "nonabsorbing" should be read as shorthand for "negligibly absorbing for our particular purposes." In this section we do not neglect absorption.

If we assume that ϖ and g are independent of τ and differentiate Eq. (5.47) with respect to τ, then substitute Eq. (5.48) in the resulting equation, and vice versa, we obtain two second-order differential equations of the form

$$\frac{d^2 F}{d\tau^2} = K^2 F, \tag{5.69}$$

where

$$K = \sqrt{(1 - \varpi)(1 - g\varpi)} \tag{5.70}$$

and F is either the sum or difference of $F_\uparrow$ and $F_\downarrow$. As a rule, differentiating differential equations to obtain ones of higher order is jumping out of the frying pan into the fire. But here the result was an equation [Eq. (5.69)] the solutions to which are simple exponential functions $A \exp(\pm K\tau)$, which can be verified by differentiating them twice; A is a constant of integration determined by boundary conditions. For simplicity, take the medium to be infinite, which means that for the irradiances to be finite we have to exclude solutions with positive exponent. Subject to the boundary condition that the downward irradiance at $\tau = 0$ is F_0 and the requirement that the sum and differences of the two irradiances be linked by Eq. (5.47) or, equivalently, by Eq. (5.48), the solution to Eq. (5.69) for an infinite medium is

$$F_\downarrow = F_0 \exp(-K\tau), \ F_\uparrow = F_0 R_\infty \exp(-K\tau), \tag{5.71}$$

where the reflectivity $R_\infty = F_\uparrow(0)/F_0$ is

$$R_\infty = \frac{\sqrt{1 - \varpi g} - \sqrt{1 - \varpi}}{\sqrt{1 - \varpi g} + \sqrt{1 - \varpi}}. \tag{5.72}$$

As a check on the correctness of this solution, note that $R_\infty = 0$ when $g = 1$, which is what we expect on physical grounds: if scattering is entirely downward incident photons cannot contribute to the reflected (upward) irradiance.

Unlike in a nonabsorbing (infinite) medium, the radiation field in an absorbing medium is not isotropic, the upward irradiance being less than the downward irradiance, although the radiation field is approximately isotropic if R_∞ is close to 1. Where the really striking differences occur, however, is in attenuation. According to Eq. (5.71) both irradiances are attenuated exponentially with (optical) depth. For ease of physical interpretation it will be more convenient to transform to physical depth z. If we take κ, β, and g to be independent of z we can write

$$K\tau = \sqrt{\kappa\{\kappa + \beta(1 - g)\}} \ z = \alpha z, \tag{5.73}$$

where we call α the *attenuation coefficient*. Note that K is dimensionless whereas α has the dimensions of inverse length. A necessary and sufficient condition for attenuation is that κ be nonzero, and hence attenuation is a consequence of absorption but with a twist: scattering amplifies this absorption. If $\beta(1 - g) = 0$, $\alpha = \kappa$, as we expect. But suppose that $\beta(1 - g)$ is not zero, indeed, suppose that it is much larger than κ. In this instance α can be considerably larger than κ:

$$\alpha = \kappa\sqrt{1 + \frac{\beta(1 - g)}{\kappa}} \approx \kappa\sqrt{\frac{\beta(1 - g)}{\kappa}} \approx \kappa\sqrt{\frac{1 - g}{1 - \varpi}}. \tag{5.74}$$

Another way of looking at this is that attenuation at physical depth $\overline{z}$ in a medium for which $\beta(1-g) \gg \kappa$ is the same at depth $z > \overline{z}$ in a medium for which $\alpha = \kappa$:

$$\overline{z} \approx z\sqrt{\frac{1-\varpi}{1-g}}. \tag{5.75}$$

Multiple scattering increases photon path lengths, and the longer the path, the greater the chance of absorption.

Perhaps the most striking difference between a nonabsorbing and an absorbing medium is in the evolution of the transmitted spectrum with depth. We showed previously that the radiation field in an infinite nonabsorbing medium is isotropic and constant with depth. From Eq. (5.52) the spectrum of the radiation transmitted by a finite nonabsorbing medium of physical thickness z is (for sufficiently large optical thicknesses)

$$F_{\downarrow}(z) \approx \frac{2F_0}{\beta(1-g)z}, \tag{5.76}$$

if β is independent of z, whereas the spectrum of the downward irradiance at depth z in the infinite absorbing medium is

$$F_{\downarrow}(z) \approx F_0 \exp(-\alpha z). \tag{5.77}$$

Although the *magnitude* of the irradiance in Eq. (5.76) decreases with depth, the *shape* of the spectrum is invariant: F_0 modulated by the wavelength variation of $1/\{\beta(1-g)\}$ at all depths. But in the absorbing medium the shape of the transmitted spectrum can change markedly with depth. For simplicity take F_0 to be constant. It follows from Eq. (5.77) that the ratio of the irradiance at the wavelength for which attenuation is a minimum (over the range of wavelengths of interest) to the irradiance at any other wavelength is

$$\exp\{(\alpha - \alpha_{\min})z\}, \tag{5.78}$$

where $\alpha_{\min}$ is the minimum value of the attenuation coefficient. Because $\alpha > \alpha_{\min}$ the limit of this ratio as z becomes indefinitely large is infinity. Thus with increasing depth the transmitted spectrum is dominated more and more by light at and around the wavelength of minimum attenuation (of course, at large depths the magnitude of the irradiance may be such that the light is imperceptible). This makes sense given that the light surviving to a depth z is that which has *not* been absorbed. The wavelength dependence of attenuation can be a consequence of the wavelength dependence of κ, of β, or of g, of any two of these quantities, or of all three [Eq. (5.74)], although in the examples we consider in following sections the wavelength dependence of α is essentially that of κ.

Now let us contrast the spectra of reflected and transmitted light. In the spectrum of transmitted light one restriction on the medium is that κ is not zero. With the additional restriction that $\kappa \ll \beta(1-g)$ (implying $\beta \gg \kappa$ and ϖ close to 1), the reflectivity R_{∞} [Eq. (5.72)] is approximately

$$R_{\infty} \approx 1 - 2\sqrt{\frac{\kappa}{\beta(1-g)}}. \tag{5.79}$$

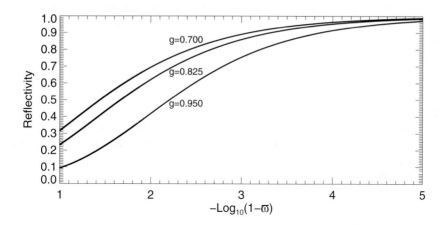

Figure 5.14: Reflectivity of an absorbing, plane-parallel medium of infinite depth for various values of the asymmetry parameter g. The horizontal axis is the negative logarithm of the ratio of absorption to absorption plus scattering.

It follows that R_∞ is close to 1, and hence the spectrum of the reflected light is essentially that of the incident light. This is much different from our result in the previous paragraph that the spectrum of light transmitted sufficiently deep in the medium can be markedly different from the incident spectrum. Why the difference?

To resolve this we turn to Eq. (5.51) for the reflectivity of a finite, nonabsorbing medium of physical thickness h, according to which the reflectivity is within 10% of its asymptotic value (1) when $\beta(1 - g)h$ is about 20. This in turns implies that the reflectivity of the infinite absorbing medium [with $\kappa \ll \beta(1-g)$] is a consequence mostly of scattering within a distance h of the upper boundary, where

$$h = \frac{20}{\beta(1 - g)}.$$ (5.80)

How much attenuation occurs in this layer? According to Eqs. (5.71) and (5.74) attenuation over a distance h is determined by the negative exponential of

$$\sqrt{\kappa\beta(1 - g)}\, h = 20\sqrt{\frac{\kappa}{\beta(1 - g)}}.$$ (5.81)

Thus if $\kappa/\beta(1 - g)$ is sufficiently small, so is the argument [Eq. (5.81)] of the exponential function, and hence attenuation is negligible. As far as reflection is concerned most of the action occurs in a layer near the surface sufficiently thin that few incident photons traversing this layer are absorbed, whereas the deeper they penetrate into the medium the fewer escape absorption.

Before turning to particular observations you can make for yourselves that will breathe life and meaning into these mathematical results we consider some general conclusions that follow from Eq. (5.72). The relative change of R_∞ with respect to ϖ ($g \neq 1$)

$$\frac{1}{R_\infty}\frac{\partial R_\infty}{\partial \varpi} = \frac{1}{\varpi\sqrt{1-\varpi}\sqrt{1-\varpi g}} \tag{5.82}$$

at $\varpi = 1$ ($R_\infty = 1$) is infinite. Thus the addition of a small amount of absorption (relative to scattering) to a nonabsorbing medium results in a relatively much greater decrease in its reflectivity: the purer you are, the more easily your corruption is evident for all to see. Equation (5.79) suggests that we should plot R_∞ versus the logarithm of $1 - \varpi$ (Fig. 5.14). We may interpret $1 - \varpi$ as the probability that a photon is absorbed in a single interaction with a molecule or particle, whereas $1 - R_\infty$ is the probability of absorption in many such interactions. From Fig. 5.14 it follows that even when the former is quite small, the latter can be many times greater. Once again we see that, as with the pile of absorbing plates (Sec. 5.1), a little bit of absorption goes a long way in a multiple-scattering medium.

5.3.1   Clouds, Snow, Paint, Frozen Waterfalls, Wet Sand, and Broken Beer Bottles

First, let's apply some of the results in the preceding section to clouds. For simplicity consider a uniform plane-parallel cloud of thickness h_c composed of N water droplets per unit volume, all with the same radius a, per unit volume of cloud. The total volume of water in this cloud per unit (horizontal) cross-sectional area is

$$Nvh_c = fh_c = h_w, \tag{5.83}$$

where v is the volume of a single cloud droplet, $f = Nv$ is the volume fraction of the cloud occupied by liquid water, and h_w is the liquid water path of the cloud expressed as the depth of water that would result if the cloud were compressed into a continuous slab of water. A large value for h_w is of order centimeters, and even this puny amount requires clouds thousands of meters thick. Thus the volume fraction f of water in clouds is quite small, of order 10^{-6}. Despite their apparent solidity, clouds are mostly air. It follows from Fig. 5.12 that a slab of liquid water (or ice) a few centimeters thick is insufficient to markedly attenuate visible light of any wavelength. But, as we noted previously, multiple scattering can greatly amplify photon paths. To find out how much we have to estimate the magnitude of the product of the attenuation coefficient α [Eq. (5.74)] and h_c. Because the single-scattering albedo for cloud droplets is very close to 1 over the visible spectrum (Fig. 5.15) we have

$$\alpha h_c \approx \kappa h_c \sqrt{\frac{\beta(1-g)}{\kappa}}. \tag{5.84}$$

The absorption and scattering coefficients are

$$\kappa = NC_{\text{abs}}, \quad \beta = NC_{\text{sca}}, \tag{5.85}$$

where C_{abs} is the absorption cross section of a droplet and C_{sca} its scattering cross section (Secs. 2.8 and 3.5). Cloud droplets are large compared with the wavelengths of visible light

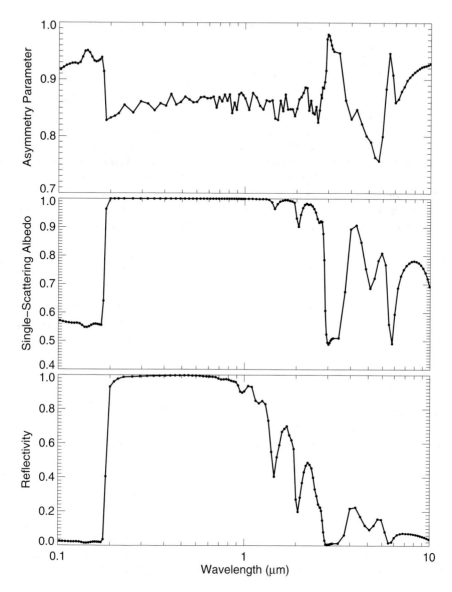

Figure 5.15: Asymmetry parameter and single-scattering albedo, from UV to IR, of a water droplet 10 μm in diameter. The reflectivity, from Eq. (5.72), is for an infinite medium composed of water droplets this size.

and weakly absorbing in that the product of droplet diameter and the absorption coefficient of water, κ_{w}, is much less than 1. Because of this the absorption cross section is approximately $\kappa_{\mathrm{w}} v$, where v is the droplet volume (Sec. 2.8). This expression is not exact but is good enough for our purposes. The scattering cross section is approximately $2\pi a^2$. With these approximations and a bit of algebra Eq. (5.84) can be written

$$\alpha h_c \approx f h_c \kappa_w \sqrt{\frac{3(1-g)}{\kappa_w d}} = h_w \kappa_w \sqrt{\frac{3(1-g)}{\kappa_w d}}, \tag{5.86}$$

where d is the droplet diameter. Although we took all droplets to be the same size (not true for most clouds, especially those with appreciable liquid water paths), the form of Eq. (5.86) is unchanged even if the droplets are distributed in size. All we need do is interpret d as the ratio $\langle d^3 \rangle / \langle d^2 \rangle$, where $\langle \rangle$ indicates an average over the distribution. Thus the cloud is equivalent (as far as attenuation is concerned) to a slab of water of thickness

$$h_w \sqrt{\frac{3(1-g)}{\kappa_w d}}. \tag{5.87}$$

Over the visible spectrum κ_w varies but is of order 0.1 m^{-1} (Fig. 5.12), whereas d is of order $10 \, \mu\text{m}$ (10^{-5} m) and $3(1-g)$ for cloud droplets is around 0.5. Thus the equivalent thickness of the cloud is amplified by nearly a factor of 1000, and a cloud with liquid water path 1 cm gives the same attenuation as a slab of water about 10 m thick. This much water *can* appreciably attenuate visible light. Because absorption by pure liquid water has a minimum in the blue–green part of the visible spectrum (Fig. 5.12), the bottoms of very thick clouds can (and do) have a bluish cast even though the spectrum of visible incident sunlight is nearly unchanged upon reflection. Although these conclusions are based on the assumption that clouds are composed of liquid water, the same is true for clouds composed of ice particles or of mixtures of both because the visible absorption coefficients for ice and liquid water are nearly the same (Fig. 5.12). But let's turn now to a medium that is composed of ice particles, snow, by which we mean a snowpack, snow on the ground rather than in the air.

Snow and Paint

Christmas cards notwithstanding, snowflakes do not long retain their intricate dendritic shapes, if they ever had them, after deposition but instead metamorphose into more or less rounded grains. Applying the radiative transfer theory in this book to snow has been criticized or, at the very least, has evoked astonishment that it agrees with measurements, because the grains in snow are touching. This is largely a red herring. For radiative transfer theory to be valid for an ensemble of particles requires that some characteristic length be *large compared with the wavelength*, as we saw in the discussion of Fig. 5.9. To expand on this point we first note that a particulate media, and here the particles could be molecules, is optically homogeneous if it contains many particles in a cubic wavelength. For N identical particles per unit volume this condition is

$$N\lambda^3 = Nv \frac{\lambda^3}{v} = f \left(\frac{\lambda}{d}\right)^3 \gg 1, \tag{5.88}$$

where v is the volume of a particle and d its cube root. This condition is satisfied for air. Snow grains are of order 1 mm; the volume fraction f for snow varies but a high value would be around 0.5. Thus the condition Eq. (5.88) is not satisfied for snow at visible wavelengths, and so it is not optically homogeneous.

But what is the condition under which scattered power is additive and hence the scattering coefficient increases linearly with N? This is difficult to answer with certainty, although on the basis of physical reasoning we expect that we can add scattered powers when some characteristic linear dimension is large compared with the wavelength. According to the kinetic theory of gases, the collision mean free path ℓ of a molecule is $1/N\sigma_c$, where N is the number of molecules per unit volume and σ_c is the collision cross section of a molecule. Note that this mean free path is formally identical to the mean free photon paths discussed in Section 5.1.3. The molecular mean free path is the average distance a molecule travels before encountering another molecule. Although molecules are in motion, this is largely irrelevant. We can imagine all the molecules to be frozen in place. On average, ℓ is the distance one of these molecules can be translated before encountering another molecule. We can extend this to an ensemble of particles, which may be stationary, and take

$$\ell = \frac{1}{N\sigma_c} = \frac{v}{Nv\sigma_c} = \frac{d}{f} > \lambda \tag{5.89}$$

as the condition for the scattering coefficient to be proportional to N. The collision cross section is approximately the geometrical cross section d^2. Equation (5.89) is approximately satisfied for air, the scattering coefficient for which is indeed proportional to N (see Sec. 3.4.9). And at visible wavelengths the condition Eq. (5.89) is certainly satisfied for snow.

But if snow grains were comparable with or smaller than the wavelengths of visible light, radiative transfer theory might not give such good results. Indeed, this is what happens in paint films. High-quality white paint is a suspension of titanium dioxide particles. The chapter entitled "Titanium" in Primo Levi's *Periodic Table* begins with a "tall man" painting a closet white. A woman watching him asks, "Why is it so white?" After thinking for a while he replies, "Because it is titanium." Titanium dioxide is inert, absorption by its particles is negligible at visible wavelengths, and its high refractive index yields considerable scattering per particle. This is what efficient opacifiers (i.e., whiteners) are intended to do. Titanium dioxide particles are used in all kinds of manufactured white objects: plastic cutlery, deck chairs – the list goes on and on. The titanium dioxide particles in paint are comparable with the wavelengths of visible light, about 0.2 μm, a size chosen to maximize the scattering cross section per unit particle volume.

According to Eq. (5.51) for the reflectivity of a nonabsorbing medium of thickness h, the reflectivity of a paint film daubed onto a black surface should steadily increase with particle volume fraction. To make this clear we rewrite Eq. (5.51) as

$$R = \frac{fh(1-g)(C_{\text{sca}}/v)}{2 + fh(1-g)(C_{\text{sca}}/v)}, \tag{5.90}$$

where the scaled scattering cross section per unit volume $(1-g)C_{\text{sca}}/v$ is fixed for a given particle. This equation shows why one wants to use particles that maximize $(1-g)C_{\text{sca}}/v$ for a fixed fh, the amount of material, per unit area, of the film, which is what you pay for. According to this equation R should increase steadily with increasing volume fraction (i.e., particle concentration). And this is what happens – at first. But as the concentration increases

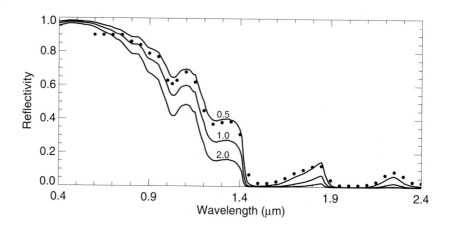

Figure 5.16: Reflectivity of snow calculated from the two-stream theory for three grain diameters (mm). The solid circles are measurements for snow of unspecified grain size made by O'Brien and Munis (1975).

a point is reached where the reflectivity begins to *decrease*. The paint folks refer to this as "crowding", but there is just as much crowding in snowpacks as in paint films. The difference between them lies not in their volume fractions, which are similar, but in their particle sizes. For ice grains in snow this is many times the wavelengths of visible light, whereas for paint particles it is comparable with or even appreciably less.

Now that we understand why radiative transfer theory *should* work for snow, let's show that it *does* by comparing theory with measurements and observations. From Eq. (5.79) and the same arguments that led to Eq. (5.86), the reflectivity of snow is approximately

$$R_\infty \approx 1 - 2\sqrt{\frac{\kappa_i d}{3(1-g)}}, \tag{5.91}$$

where κ_i is the absorption coefficient of ice and d is the grain diameter. Now we cheat (slightly) and "reverse approximate" Eq. (5.91):

$$R_\infty \approx \exp\left(-2\sqrt{\frac{\kappa_i d}{3(1-g)}}\right). \tag{5.92}$$

This equation is essentially the same as the previous one for small values of the argument but has the advantage of never giving negative reflectivities, whereas Eq. (5.91) pushed too far will. But when the reflectivity is sufficiently small we don't care what its exact value is, and hence Eq. (5.92), although incorrect, is good enough for our purposes. Figure 5.16 shows R_∞ calculated using Eq. (5.92) over the solar spectrum from 0.4 μm to 2.5 μm for various grain sizes. At visible wavelengths R_∞ is close to 1 and nearly independent of wavelength. That is, snow is white (no surprise here), which we can see with our own eyes. What we can't see is the steep plunge of reflectivity beyond about 1.0 μm where absorption by ice greatly increases. As predicted by Eq. (5.92) the wiggles and bumps in the reflection spectrum of snow faithfully follow those in the absorption spectrum of ice (Fig. 2.2). Calculated reflection is consistent with measurements except that theory usually predicts higher values over the visible than those measured. This is likely a consequence of trace amounts of contaminants such as soot (a catchall name for the carbonaceous products of combustion). At visible wavelengths absorption by soot is around a million times greater than that by ice, so soot in snow in parts per million can lower its visible reflectivity. In the infrared, however, ice is sufficiently absorbing that the addition of a bit of soot to snow does not markedly change its reflectivity.

   Figure 5.16 supports an assertion about snow that at first glance seems the product of a disturbed mind: snow is both the whitest natural substance on earth and the blackest. At visible wavelengths the reflectivity of clean, fine-grained snow is as high as that of anything you'll find in nature, whereas well into the infrared its reflectivity plunges to near zero. We are easily led astray by extrapolating outside the narrow range of visible wavelengths to which we are sensitive.

Frozen Waterfalls

Although snow is white, bubbly ice may not be. The scatterers in snow are ice grains suspended in air whereas the scatterers in bubbly ice are air grains suspended in (absorbing) ice. We can approximate the absorption coefficient of bubbly ice as $\kappa_i(1 - f)$, where f is the volume fraction of bubbles. By the same arguments that yielded Eqs. (5.91) and (5.92) the reflectivity of bubbly ice is

$$R_\infty = \exp\left(-2\sqrt{\frac{\kappa_i d(1-f)}{3f(1-g)}}\right), \tag{5.93}$$

where d is the bubble diameter. This equation, compared with Eq. (5.92), tells us that bubbly ice is equivalent to snow with an effective grain diameter

$$d_{\text{eff}} = \frac{(1-f)d}{f}, \tag{5.94}$$

which makes physical sense. If f is low, say 0.01, the effective grains in bubbly ice are 100 times larger (for equal d) than the ice grains in snow. Because of the spectral dependence of κ_i (Fig. 5.12), light reflected by bubbly ice can have a bluish or blue–green cast. You can readily observe this if you live in a part of the world with real winters. Frozen waterfalls (icefalls) on road cuts are often perceptibly blue (Fig. 5.17). Moreover, this blueness is made more striking if the icefalls are partly covered by snow. You can see, side by side, the visual consequences of Eqs. (5.92) and (5.93): darker, bluish bubbly ice flanked by brighter, white snow.

Snow Holes and Crevasses

Although snow is usually white seen by reflection, the light transmitted into snow and other natural ice bodies is sometimes blue, the purity of which increases with increasing depth. The deep blueness of holes in snow, crevasses, and ice caves has been commented on by

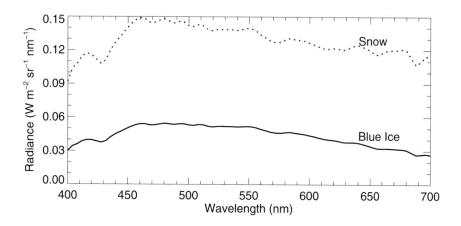

Figure 5.17: The measured radiance of blue ice is appreciably less than that of nearby snow. Although not immediately obvious from these measurements, the ice spectrum is slightly richer in shorter wavelength radiation. The ratio of blue (472 nm) to red (632 nm) radiances for the snow is about 1.2, whereas that for the ice is about 1.4; this difference is readily perceptible.

many writers. In 1923 Raman published a paper in which he attributed the blueness of ice to molecular scattering, that is, scattering described by Rayleigh's law, and such is Raman's prestige that this explanation, which is not difficult to demolish, is still widespread today.

Raman's argument, in essence, is that the scattering coefficient β of snow has the inverse fourth power of wavelength dependence of molecular scattering and that absorption plays no role. But the solution to Eq. (5.49) for an infinite (i.e., optically thick), nonabsorbing medium is a uniform, isotropic irradiance equal to the incident irradiance everywhere in the medium. Snow is often optically thick in the sense that not much need accumulate before its reflectivity does not change with a further increase in thickness. As a result, the reflectivity of the underlying surface is irrelevant. But suppose that the snow is not optically thick. Now Eq. (5.51) yields the reflected irradiance for an optically thin medium which is proportional to β [strictly, to $\beta(1-g)$, although for molecular scattering $g = 0$]. Also, Eq. (5.76) predicts that for a thick medium, the downward (transmitted) irradiance is *inversely* proportional to β and the shape of the transmitted spectrum does not change with increasing physical depth. Thus if the scattering coefficient of snow were the molecular scattering coefficient, the light reflected by a thin layer of snow should have the blueness of the sky and the light transmitted by a thick layer should be reddish but with no change in purity with depth. This is not observed. A word of caution is in order before we proceed. Snow in shadow sometimes is perceived to be blue when the dominant source of illumination is the blue sky. You can amuse yourselves by poring over art books or strolling through an art museum and carefully examining snowscapes. Artists usually paint shadows on snow blue, but of a purity not observed in nature, presumably to trumpet that they know shadows on snow can be blue.

The observation that the light in snow holes becomes an increasingly purer blue with increasing depth indicates that absorption must play a role, as evident from Eq. (5.71), which shows an exponentially changing spectrum of both the upward and downward irradiances. For Raman's theory to have even a hope of being correct, the attenuation coefficient [Eq. (5.73)] must be dominated by β (with $g = 0$), and hence is given by

$$\alpha \approx \sqrt{\kappa\beta}. \tag{5.95}$$

But if β is the molecular scattering coefficient (and κ is independent of wavelength), attenuation is greatest at the short wavelength end of the visible spectrum and least at the long wavelength end. We therefore predict that the light transmitted into snow and other natural ice bodies should be a red the purity of which increases with increasing depth, which is contrary to what is observed.

Where did Raman go wrong in his explanation of blue ice resulting from molecular scattering? It is indeed true that the water molecules in ice – indeed, the molecules in anything – do the scattering, but the nature of this scattering depends on the distance over which they are correlated. Scattering by a single molecule is indeed proportional to inverse fourth power of wavelength, but this does not mean that an ice grain composed of many such molecules scatters this way. A grain is a coherent object: every molecule in it is correlated with the position of every other molecule over a characteristic distance, the size of the grain. The total light scattered by the grain is the superposition of all the waves scattered by its constituent molecules taking due account of the phase differences between them, which in turn depend on the size of the grain relative to the wavelength. This additional wavelength dependence of the scattering by a coherent object, if sufficiently large, is what effaces the wavelength dependence of the scattering by its individual molecules. Stated another way, scattering is a consequence of heterogeneity, be it on the molecular scale or larger. And in snow or bubbly ice scattering as a consequence of heterogeneity on the scale of grains or bubbles swamps scattering as a consequence of heterogeneity on the molecular scale.

This provides a beautiful example of how a medium can be both coherent and incoherent. A single ice grain (or cloud droplet) is a coherent scatterer in the sense that to determine the light scattered by such a scatterer one *must* take into account the phase differences between the waves scattered by the individual molecules. But a snowpack (or a cloud) is an incoherent scatterer in the sense that to determine the light scattered by many grains (or droplets) one *may* ignore the phase differences between the waves scattered by the individual grains.

Wet Sand

A common observation is that when sand and many other porous media are wet with water or other liquids they become darker. Why? To answer this let's first pretend that the sand is nonabsorbing and overlies a black substrate. From Eq. (5.51) the relative change in R with respect to g is

$$\frac{1}{R}\frac{\partial R}{\partial g} = \frac{-1}{(1-g)[1 + \overline{\tau}(1-g)/2]} \tag{5.96}$$

and hence R always decreases with increasing g. In the limit $g \to 1$, Eq. (5.96) approaches infinity. Thus the biggest change in reflectivity occurs when g is largest. Sand grains are much larger than the wavelengths of visible light. Scattering by such particles is highly peaked in the forward direction. But although total (all directions) scattering is independent of the medium in which the sand grains are suspended the angular distribution of this scattering (i.e., g) is not. In particular, the more closely the surrounding medium optically matches the particles, the more scattering is peaked in the forward direction (g closer to 1). Thus we expect g for sand grains in water (wet sand) to be greater than for sand grains in air (dry sand), which from Eq. (5.96) implies a lower reflectivity. As g increases photons are more likely to penetrate to the black substrate and be absorbed there. But how do we know it isn't absorption by the water in a multiple-scattering medium that causes the reflectivity to decrease? Wet sand (composed of nonabsorbing grains in absorbing water) is similar to bubbly ice, and so we can apply Eq. (5.93) to such sand, where d is the size of the sand grains and κ_i is the absorption coefficient of liquid water (at visible wavelengths similar to that of ice). Sand grains are comparable in size to snow grains and the volume fraction f of grains in snow (as opposed to bubbles in ice) is of order 0.5, which yields only a small increase in effective size [Eq. (5.94)]. So if dry sand (optically thick) has a high reflectivity, that of the same sand when wet should not change much because of absorption by water.

Real sand is not strictly nonabsorbing. We once bought some clean, fine-grained aquarium sand at a pet shop. To our eyes it had a pastel reddish-orange cast. When we examined it under a microscope we discovered that it was composed mostly of large, transparent quartz grains seasoned with much smaller reddish-orange grains (presumably iron oxide). Scattering therefore was dominated by the quartz grains, absorption by the iron oxide grains. From Eq. (5.72) the relative change of R_∞ with g for an optically thick absorbing medium ($\varpi \neq 1$) is

$$\frac{1}{R_\infty}\frac{\partial R_\infty}{\partial g} = \frac{-\sqrt{1-\varpi}}{(1-g)\sqrt{1-\varpi g}}. \tag{5.97}$$

As with a nonabsorbing (finite) medium the greatest relative decrease in R_∞ occurs for g close to 1 (for fixed ϖ). The only difference between a finite nonabsorbing medium overlying a black surface and an optically thick (effectively infinite) absorbing medium is that in the former absorption occurs outside the medium whereas in the latter it occurs inside. For both media a decreased reflectivity upon wetting is a consequence of (single) scattering peaking more sharply in the forward direction. The experimental test of this is to wet sand with various transparent liquids of different refractive index ("transparent" means that you can see through a bottle of the liquid). The more closely the liquid matches the grains that dominate scattering, the darker the sand.

Broken Beer Bottles

And now we finally arrive at broken beer bottles. Many beer bottles are colored, some brown, some green. Indeed, it may very well be that only colored bottles contain beer worth drinking, but we leave that for readers to determine experimentally. Suppose that we recklessly, even violently, toss lots of green beer bottles into a dumpster, breaking most of them, with the result that the dumpster is piled with glass shards. This pile will still be seen as green. Now take a hammer and smash the shards more and more. We eventually end up with ground glass, but it is white, not green. What happened? Suffice it to say that pulverizing beer bottles doesn't change the chemical composition of the glass but it does decrease the size of the glass particles.

Suppose that d ($\gg \lambda$) is a characteristic linear dimension of a glass particle (grain) and that κ_g, the absorption coefficient of bulk glass (not of glass particles), is such that $\kappa_g d \ll 1$. You can verify that this condition is satisfied by observing that a sliver of glass shows no color. With these restrictions the absorption cross section of a glass grain is proportional to its volume (d^3) and κ_g, and its scattering cross section is much larger and proportional to d^2. It follows from Eqs. (5.72) and (5.79), also Fig. 5.14, that the reflectivity of an optically thick medium decreases with decreasing $1 - \varpi$, which for glass grains is

$$1 - \varpi = \frac{\kappa}{\kappa + \beta} \approx \frac{\kappa}{\beta} \approx C \frac{\kappa_g d^3}{d^2} = C \kappa_g d, \tag{5.98}$$

where C is all the constants of proportionality swept together. Absorption by a grain decreases more rapidly with decreasing size than does scattering, a consequence of which is that R_∞ increases with decreasing size.

5.4   Emission

Up to this point we have assumed that emission within a medium is negligible whereas scattering is not. Now we do the reverse: ignore scattering but allow for emission. The medium of interest could be the clear troposphere or a cloud. Around the wavelengths ($\sim 10\,\mu\text{m}$) at which the Planck function has a maximum at typical terrestrial temperatures (~ 300 K), scattering by air is negligible compared with absorption. This follows from extrapolating the scattering optical thickness of the atmosphere as a function of wavelength (Fig. 8.2) into the infrared. Because scattering by air molecules is approximately inversely proportional to wavelength to the fourth power, the scattering optical thickness of the atmosphere at 10 μm is 10^4 times *smaller* than at 1 μm. And the single-scattering albedo for a typical cloud droplet plunges from visible to infrared wavelengths (Fig. 5.15).

Before proceeding we need Kirchhoff's law *within* a medium. All the assumptions in Section 1.4 underlying the derivation of this law apply here. Consider a thin slab of thickness Δz and area A (Δz much less than the dimensions of A) in an opaque cavity at temperature T. We assume that Δz is sufficiently small that the emissivity of the slab is proportional to Δz: $\varepsilon = \chi \Delta z$. That is, the emissivity is linear in the amount of matter as long as the matter has little chance of absorbing the radiation it emits. This is the first term in a Taylor series expansion of the emissivity in powers of Δz. At equilibrium, emission by the slab (neglecting the edges) at a given frequency is balanced by absorption of incident radiation:

$$2 P_e \chi \Delta z A = 2 P_e \kappa \Delta z A, \tag{5.99}$$

where P_e is the Planck function (Sec. 1.2). The factor 2 arises because the slab emits in and is illuminated from two directions. From this equation it follows that

$$\chi = \kappa, \tag{5.100}$$

which is closer to the original Kirchhoff's law than the form obtained in Section 1.4. His form was that the ratio of what he called the emissive power (our χP_e) to what he called the absorptive power (our κ) is a universal function of wavelength and temperature, which much later was shown to be the Planck function. Kirchhoff's terminology was poorly chosen because the dimensions of these two quantities are not the same and κ is solely a material property whereas χP_e is not. Planck called χP_e the coefficient of emission and κ the coefficient of absorption, terminology as objectionable as Kirchhoff's and for the same reasons. Equation (5.100) is what is needed to determine upward and downward emission by a thin slab within a medium:

$$\kappa \Delta z P_e. \tag{5.101}$$

First consider the downward irradiance, the governing equation for which is Eq. (5.40) with the scattering terms on the right side removed and an emission term [from Eq. (5.101)] added:

$$\frac{F_\downarrow}{dz} = -\kappa F_\downarrow + \kappa P_e\{T(z)\}, \tag{5.102}$$

where $T(z)$ is the absolute temperature profile in the medium. As before we transform from physical depth z to (absorption) optical depth τ [Eq. (5.44)], although we omit the subscript a:

$$\frac{dF_\downarrow}{d\tau} + F_\downarrow = P_e\{T(\tau)\}. \tag{5.103}$$

At the top of the medium ($\tau = 0$) the downward irradiance is zero (no external source of radiation). Multiply both sides of Eq. (5.103) by $\exp(\tau)$:

$$e^\tau \frac{dF_\downarrow}{d\tau} + e^\tau F_\downarrow = \frac{d}{d\tau} e^\tau F_\downarrow = e^\tau P_e\{T(\tau)\} \tag{5.104}$$

and integrate the result from the top of the medium to its bottom ($\bar{\tau}$):

$$F_\downarrow(\bar{\tau}) = \int_0^{\bar{\tau}} \exp\{-(\bar{\tau} - \tau)\} P_e\{T(\tau)\} \, d\tau. \tag{5.105}$$

The interpretation of this equation is so straightforward that we could have obtained it without solving any differential equation. At each optical depth τ an amount of radiation per unit optical depth $P_e\{T(\tau)\}$ is emitted downward. The fraction of this emitted radiation that reaches the bottom of the medium is $\exp\{-(\bar{\tau} - \tau)\}$, where $\bar{\tau} - \tau$ is the optical distance from the point of emission within the medium to its bottom. The sum (integral) over all points is the total radiation at the bottom.

As a check on this result consider a medium with a uniform temperature. This is not a bad approximation for the troposphere or clouds when we realize that variations in *absolute* temperature within them are only about 10–20%. With this assumption Eq. (5.105) is readily integrated:

$$F_\downarrow(\bar{\tau}) = \{1 - \exp(-\bar{\tau})\} P_e(T), \tag{5.106}$$

which implies that the emissivity of the medium is

$$\varepsilon = 1 - \exp(-\bar{\tau}). \tag{5.107}$$

The sum of the reflectivity, transmissivity, and absorptivity must be 1. Because scattering was assumed to be zero, the reflectivity is 0. The transmissivity is $\exp(-\overline{\tau})$, and hence the absorptivity is $1 - \exp(-\overline{\tau})$. But by Kirchhoff's law, absorptivity equals emissivity, which leads to Eq. (5.107) by the route that led to Eq. (2.24). If at a particular wavelength $\overline{\tau} \gg 1$, then $\varepsilon \approx 1$, and at that wavelength the medium can be said to be black.

We could set up another differential equation for the upward irradiance, but this is hardly necessary because we can go straight to its solution just as we could have gone straight to Eq. (5.105), bypassing Eqs. (5.102)–(5.104):

$$F_\uparrow(0) = F_\uparrow(\overline{\tau})\exp(-\overline{\tau}) + \int_0^{\overline{\tau}} \exp(-\tau)P_e\{T(\tau)\}\,d\tau. \qquad (5.108)$$

The upward irradiance at the top of the medium is the sum of transmitted upward irradiance emitted at the bottom boundary plus all the unattenuated radiation emitted upward at each point of the medium.

Equations (5.105) and (5.108) are at the heart of remote sensing of atmosphere temperature profiles, either from the ground or from a satellite. According to these equations the radiation at the top or bottom of the atmosphere depends on the temperature profile, which is buried within an integral. The challenge is to dig it out by making spectral radiation measurements.

References and Suggestions for Further Reading

The doubling equation for reflectivity [Eq. (5.12)] is getting on in years. It can be found in an 1862 paper by Sir George Gabriel Stokes, On the intensity of the light reflected from or transmitted through a pile of plates, in *Proceedings of the Royal Society*. A more readily obtainable source for this paper than the original journal is Stokes's *Mathematical and Physical Papers*, Vol. IV, pp. 143–56 (Cambridge University Press, 1904). In a footnote Stokes states that in an 1856 paper "I find that the formulae for the particular case of perfect transparency have already been given by M. Neumann. His demonstration does not appear to have been published." To this is added, by the editor of Stokes's papers, Sir Joseph Larmor, that the "efficiency of a pile of plates was also considered by Fresnel himself; his solution, which neglects opacity, was published posthumously" (in 1868). Fresnel died in 1827.

Although the doubling method in Section 5.1 is applied to (implicitly) optically homogeneous plates, it is just as valid for optically inhomogeneous (plane-parallel) media (see Prob. 5.33). Indeed, this method has been applied to media described by the equation of radiative transfer in Section 6.1.2; for a good review see James E. Hansen and Larry D. Travis, 1974: Light scattering in planetary atmospheres. *Space Science Reviews*, Vol. 16, pp. 527–610. This method has been extended to media containing spatially non-uniform sources of radiation (e.g., thermal radiation); see Warren J. Wiscombe, 1976: Extension of the doubling method to inhomogeneous sources. *Journal of Quantitative Spectroscopy and Radiative Transfer*, Vol. 16, pp. 477–89.

Equation (5.24) is derived in, for example, O. S. Heavens, 1991: *Optical Properties of Thin Solid Films*, Dover, pp. 55–9, Max Born and Emil Wolf, 1965: *Principles of Optics*, 3rd rev. ed., Pergamon, pp. 61–2. This equation can be derived without explicitly solving the equations of the electromagnetic field (see Prob. 7.41).

It should become evident in the following chapter that an infinite number of two-stream equations could result from approximating the integro-differential equation of radiative transfer for a plane-parallel medium. The two-stream theory of Section 5.2, derived by physical arguments rather than by approximating the equation of transfer, is discussed in an expository paper on multiple scattering by Craig F. Bohren, 1987: Multiple scattering of light and some of its observable consequences. *American Journal of Physics*, Vol. 55, pp. 524–33, reprinted in the collection edited by Craig F. Bohren, 1989: *Scattering in the Atmosphere*, SPIE Optical Engineering Press. The granddaddy of two-stream theories is that by Arthur Schuster, 1905: Radiation through a foggy atmosphere. *Astrophysical Journal*, Vol. 21, pp. 1–22, reprinted in the collection edited by Donald H. Menzel, 1966: *Selected Papers on Transfer of Radiation*, Dover.

The absorption length for liquid water in Fig. 5.12 is from Table 1 in Marvin R. Querry, David M. Wieliczka, and David J. Segelstein, 1991: Water (H_2O), pp. 1059–77 in Edward D. Palik, Ed., 1991: *Handbook of Optical Constants of Solids*, Vol. II, Academic Press. That for ice is from the compilation by Stephen G. Warren, 1984: Optical constants of ice from the ultraviolet to the microwave. *Applied Optics*, Vol. 23, pp. 1206–25.

The reflectivity measurements for snow in Fig. 5.16 are from Harold W. O'Brien and Richard H. Munis, 1975: Red and near-infrared spectral reflectance of snow. CRREL Research Report 332, Cold Regions Research and Engineering Laboratory, Hanover, New Hampshire.

The method alluded to in Section 5.2.4 for determining cloud properties, using ground-based measurements, by exploiting the quite different reflection properties of vegetation at visible and near-infrared wavelengths is discussed in Alexander Marshak, Yuri Knyazikhin, Anthony B. Davis, Warren J. Wiscombe, and Peter Pilewskie, 2000: Cloud–vegetation interaction: use of Normalized Difference Cloud Index for estimation of cloud optical thickness. *Geophysical Research Letters*, Vol. 27, pp. 1695–98; Alexander Marshak, Yuri Knyazikhin, Keith D. Evans, Warren J. Wiscombe, 2004: The "RED versus NIR" plane to retrieve broken-cloud optical depth from ground-based measurements. *Journal of the Atmospheric Sciences*, Vol. 61, pp. 1911–25.

For a comprehensive review of theories of multiple scattering by snow (on the ground) and comparison with measurements see Stephen G. Warren, 1982: Optical properties of snow. *Reviews of Geophysics and Space Physics*, Vol. 20, pp. 67–89.

For a discussion of the colors of natural ice bodies, accompanied by color photographs of icefalls and holes in snow, see Craig F. Bohren,1983: Colors of snow, frozen waterfalls, and icebergs. *Journal of the Optical Society of America*, Vol. 73, pp. 1646–52.

The blue of natural ice bodies discussed in this chapter should not be confused with the blueness of shadows on snow. For more on this see Michael E. Churma, 1994: Blue shadows: physical, physiological, and psychological causes. *Applied Optics*, Vol. 33, pp. 4719–22.

C. V. Raman's incorrect explanation of the colors of glaciers is in his 1923 paper Thermal opalescence in crystals and the colour of ice in glaciers, *Nature*, Vol. 111, pp. 13–14.

For detailed calculations of the optimum diameter of titanium dioxide particles (0.22 μm), based on the two-stream theory of this chapter, see Bruce R. Palmer, Penelope Stamatakis, Craig F. Bohren, and Gary G. Salzman, 1989: A multiple-scattering model for opacifying particles in polymer films. *Journal of Coatings Technology*, Vol. 61, pp. 41–7.

Primo Levi's story about titanium is in *The Periodic Table* (1984), Schoken Books, p. 166.

For theory and measurements of the darkening of sand when wet by various liquids see Sean A. Twomey, Craig F. Bohren, and John L. Mergenthaler, 1986: Reflectance and albedo differences between wet and dry surfaces. *Applied Optics*, Vol. 25, pp. 431–37. For a non-mathematical account of the brightness differences between dry and wet sand, also sand of different grain sizes, see Craig F. Bohren, 1987: *Clouds in a Glass of Beer*, John Wiley & Sons, Ch. 15. For a different interpretation of why some things are darker when wet see John Lekner and Michael C. Dorf, 1988: Why some things are darker when wet, *Applied Optics*, Vol. 27, pp. 1278–80.

Related to the question raised in Problem 5.15 is one that never seems to go away: Can you get a sunburn through glass? For an experimental answer to this question see Richard Bartels and Fred Loxsom, 1995: Can you get a sunburn through glass? Theory and an experiment. *The Physics Teacher*, Vol. 33, pp. 466–70.

Problems

5.1. Prove Eq. (5.12).

HINT: You can do this proof by induction. Assume that the truth of Eq. (5.12) for arbitrary N implies the truth of it for $2N$ using Eq. (5.9), the general rule for finding the reflectivity of $2N$ plates given that for N plates. We know that Eq. (5.12) is true for $N = 2$ and $N = 4$.

5.2. In Section 5.1.3 we determine the mean free path for absorption (scattering). What is the root-mean-square free path? Determine the nth-root-mean free path, that is, $\langle x^n \rangle^{1/n}$. What is the median free path? What is the most probable free path?

5.3. To convince yourself that the denominator in Eq. (5.67) is a consequence of multiple reflections between cloud and ground, derive this equation by summing the infinite series of such reflections. Treat the cloud as a slab with reflectivity R and transmissivity T overlying a surface with reflectivity R_s. You can check this solution by solving the equations of radiative transfer subject to suitable boundary conditions.

5.4. A magazine advertisement for Bermuda once caught our eye. The ad showed a couple in bathing suits sitting together romantically on a beach. A distinct line running parallel to the water separated bright white sand from darker pink sand (at water's edge), and the couple was sitting on the pink sand. The caption read "Does pink sand feel softer?" Don't bother answering this question but instead explain the abrupt change in color (and brightness) of the beach.

HINT: It might help to learn a bit about coral.

5.5. Estimate the wavelength of incident radiation such that a snowpack can be considered optically homogeneous. Take the snow volume fraction f to be 0.3, a typical value.

5.6. We once were asked the following question by a professor at another university: "The bottoms of thick clouds often are dark or gray. Yet scattering by cloud droplets is strongly peaked in the forward direction. Doesn't this imply that the bottoms of clouds should be bright?" Answer this question.

5.7. We once received an anguished message from a scientist interested in transmission of visible light by suspensions of particles. He stated that "as the beam marches through the sample it is assumed to be extinguished (scattering + absorption) exponentially as $\exp(-\alpha_{\mathrm{ext}}h)$," where h is the sample thickness and $\alpha_{\mathrm{ext}} = fC_{\mathrm{ext}}/v$, f is the volume fraction of particles in the suspension, C_{ext} is the extinction cross section, and v is the volume of a single particle.

This scientist stated that using values of extinction for a (nonabsorbing) sphere of diameter 1.05 µm, he computed a value for α_{ext} of 0.443 mm^{-1} ($f = 10^{-4}$). He further noted that "for our path length of 23 mm, the incident beam [according to calculations] is attenuated to about 3.8×10^{-5} of its original value (i.e., almost totally extinguished). This attenuation is far too great. We know [from laboratory observations] that a 10^{-4} suspension is very transparent [i.e., visible light is transmitted by it]. What are we doing wrong?"

Respond to this message. If necessary, draw upon examples in the atmosphere to help this person.

5.8. Within the simple two-stream model of multiple scattering, it is not possible to determine how the reflectivity of a cloud depends on the angle of incidence of the illumination (angle between the normal to the cloud and the incident beam). Nevertheless, on the basis of physical intuition acquired from the two-stream model you should be able to guess how the reflectivity of a cloud depends on this angle. Sketch this dependence and briefly explain your reasoning.

5.9. This problem is related to the previous one. Using a simple extension of the two-stream theory of reflection by clouds, and guided by physical intuition, you should be able do more than just make a crude sketch of cloud reflectivity versus solar zenith angle. You should be able to determine the slope of this curve and its limiting value.

5.10. Find the most general expression for the rate (per unit volume) at which radiant energy (of a given frequency) is deposited (transformed) within a medium according to the two-stream theory.

HINT: It might help to contemplate Fig. 5.10 and review the section on flux divergence (2.3).

5.11. Given the result of Problem 5.10, find the rate of transformation of radiant energy at each point of an infinite, absorbing medium according to the two-stream theory. The medium is illuminated from above. As a way of checking your result integrate this local energy transformation rate over the entire medium. You should obtain an expression you could have written down immediately without doing any integration.

5.12. This problem is related to the previous two. Pure lake or seawater (water containing no solid particle or biological organisms) is heated by absorption of solar radiation through a depth of tens of meters (see Figs. 2.2 and 5.12). Suppose that particles are added to the water. What happens to the radiative heating rate just below the surface (assuming no change in the incident solar radiation)? You may assume that the particles are nonabsorbing at solar wavelengths and that the reflectivity of the water is zero. Your first reaction to this question might be that because reflection by the water is increased by the particles, the heating rate decreases. Although this is true *globally* (for the medium as a whole) show either by simple mathematical or physical arguments (or some combination of both) that the *local* heating rate may not necessarily decrease. How might the particles change the temperature and its gradient in the water? Devise a simple experiment for testing your conclusions.

5.13. If the atmosphere were isothermal but not uniform in composition (i.e., the concentration of infrared active gases varies with height), would emission by the atmosphere to space be the same as, greater than or less than emission by the atmosphere to the ground? Here emission to space does not include radiation from the ground transmitted by the atmosphere. You may assume that the atmosphere emits only up or down.

5.14. On a sunny day we once noticed small clouds scattered over the horizon in all directions. The sun was perhaps $30-40°$ above the horizon. Clouds toward the sun were bright, but those away from the sun were darker. Moreover, it was possible to find pairs of clouds of apparent similar size, although in different directions, one of the pair dark, the other bright. We also observed that among the dark clouds (those opposite the sun) the brighter ones were also the largest. Explain. A diagram here will be helpful.

5.15. You occasionally come across the assertion that it is easier to get a sunburn on a cloudy day than on a clear day. The explanation sometimes offered for this is that clouds "transmit more ultraviolet radiation" (by "more" here presumably is meant more than on a clear day). Discuss this explanation. If you don't believe it, and can back this up with physical arguments, put forward an alternative explanation (under the assumption that the assertion is true).

5.16. Einstein's formula (see references at the end of Ch. 3) for the scattering coefficient (which he called an absorption coefficient because it determines the *apparent* absorption of a beam of light) of a nonabsorbing, one-phase, homogeneous medium can be written

$$\beta_{mol} = \frac{k_B T}{6\pi} \frac{(n^2 - 1)^2 (n^2 + 2)^2}{9} \kappa_T \left(\frac{2\pi}{\lambda}\right)^4,$$

where n is the refractive index of the medium and κ_T its isothermal compressibility. This is the scattering coefficient of a medium as a consequence of its irreducible graininess at the molecular level. With the use of this formula, critically discuss the notion that the blueness of the deep sea is a consequence solely of molecular scattering. The isothermal compressibility of water is about $5 \times 10^{-10} Pa^{-1}$.

5.17. Suppose that you are faced with the task of measuring the imaginary part of the refractive index of pure water at, say, visible wavelengths by measuring (relative) transmission of a beam. What is the irreducible error in the imaginary index? Is it positive or negative? By irreducible here is meant that there is *no* error in the transmission measurement. Suppose that you needed to know the imaginary index to within 10%. What lower limit does this then set on the actual value of the imaginary index? That is, what is the value of the actual imaginary index below which it is impossible to attain the desired accuracy by measuring transmission? HINT: This problem is related to the previous one.

5.18. Show that total emission to the surface in the radial (vertical) direction from a molecular species with an absorption coefficient that decreases exponentially with scale height (e-folding distance) H is the same as that for a uniform, finite layer of thickness H and absorption coefficient equal to the surface value. Take the (slab) atmosphere to be isothermal.

5.19. We assert in Section 2.2 that the increasing infrared ($\sim$10 µm) brightness temperature of the clear atmosphere with increasing zenith angle is a consequence mostly of an increasing path length (emissivity). Yet the average temperature along a path in the atmosphere does increase with increasing zenith angle because temperature usually decreases with height. Show that this is not the major contribution to the variation of brightness temperature with zenith angle. That is, estimate the maximum correction to emission as a result of assuming that the atmosphere is isothermal with a temperature equal to the surface value. Before tackling this problem first decide on physical grounds the direction for which the error is greatest, then do your analysis for this direction. You may assume that temperature decreases linearly with height at a rate of $6.5\,°\mathrm{C/km}$.

HINT: Do the previous problem and Problem 2.12 first.

5.20. Suppose that scatterers are added to an absorbing ($\varpi = 0$) plane-parallel medium. What happens to its emissivity? How does your answer depend on the asymmetry parameter? Detailed analysis and calculations are not necessary to answer these questions.

5.21. First we derived two-stream equations of transfer, Eqs. (5.45) and (5.46), under the assumption of no emission. Then we derived an equation for the downward irradiance, Eq. (5.103), under the assumption of no scattering. Without much effort you should be able to write down two-stream equations for a medium for which both emission and scattering must be taken into account.

5.22. Solve the equations derived in the previous problem to obtain the emissivity of a uniform, isothermal, optically thick plane-parallel medium. This problem requires a fair amount of labor, so before tackling it make some plausible estimates, based on physical reasoning, of the functional dependence of emissivity on single-scattering albedo and asymmetry parameter.

5.23. More than half a century ago, long before artificial satellites circled Earth, its albedo for solar radiation was measured using visual observations of a natural satellite, the moon (see Prob. 4.10). The albedo for solar radiation, which is mostly a consequence of clouds, especially thick ones, is the ratio of solar radiation incident on Earth to the solar radiation reflected by Earth. These measurements gave a value for the average albedo of around 0.38 (38%). Many years later, measurements from artificial satellites yielded the now accepted value of around 0.30 (30%). Why the discrepancy?

HINT: The answer is not friction or "experimental error" or "better instruments." You may assume that the uncertainties in albedos measured by the two different methods are smaller than the difference between albedos obtained by these two methods.

5.24. On examinations students have hazarded the guess that the difference between the albedos in the previous problem is a consequence of "atmospheric attenuation." Show that this guess is not correct.

5.25. Estimate the optical thickness of a cloud such that its transmitted (diffuse) radiance is equal to the attenuated (by the cloud) radiance of direct sunlight. This is an estimate of how optically thick a cloud must be such that the disc of the sun cannot be seen through it.

HINT: Assume that the cloud is sufficiently thick that the downward radiance emerging from the cloud is isotropic.

5.26. Estimate the least optical thickness of a cloud layer such that it is as bright (as seen from below cloud) as the clear (zenith) sky. Then estimate the physical thickness of such a cloud.

All that is wanted here is a rough estimate: meters, tens of meters, hundreds of meters... ? Estimate the cloud optical thickness beyond which the light emerging from the bottom of a cloud is *less* bright than the clear sky. And again estimate the physical thickness of such a cloud. Ignore absorption and make any reasonable simplifying assumptions.

5.27. You are flying over a vast snow-covered expanse above clouds and notice that the clouds are not as bright as the snow. This may seem puzzling. Explain.

5.28. In the preceding problem, the snow was brighter than the clouds. But suppose that you observe the clouds to be brighter than the snow. Under what viewing conditions and cloud optical thicknesses might this be possible?

5.29. According to the *Guinness Book of World Records*, aerogel is the world's lightest (lowest density) solid. Composed of tiny pores in a silicon dioxide (glass) matrix, it is 99.8% air. We once used aerogel in an examination. We placed two sheets of paper, one black, one white, side by side on a desk, with a slab of aerogel (perhaps a few centimeters thick) straddling the sheets. The aerogel over the black sheet had a definite blue cast, whereas that over the white sheet was yellow. Explain.

5.30. In Section 1.3 we noted that the emissivity of a single *particle* can be greater than 1 at some frequencies. What about the emissivity of a suspension of many such particles (take the lateral dimensions of the suspension to be much larger than the wavelength)?

5.31. We cited in the references at the end of Chapter 2 a method for determining the thermodynamic phase (ice or water) of cloud particles by making use of differences between near-infrared absorption by ice and water (see accompanying figure). A criticism of this method when first presented was that a change in cloud reflectivity could be a consequence of a change in particle size. And this is a valid criticism given that the product of absorption coefficient and particle size [e.g., Eq. (5.91)] determines the amount to which reflectivity is reduced below unity. What experimental evidence, in addition to different magnitudes of reflectivity, would be necessary to rebut this criticism?

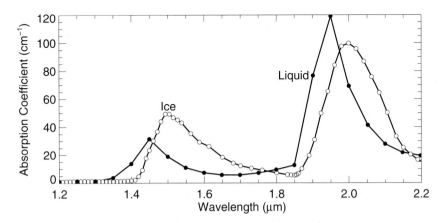

5.32. Suppose that all the droplets in a cloud coalesce to form rain. By how much does the scattering optical thickness of the cloud change at visible wavelengths? The typical size of a cloud droplet is of order 10 μm, that of a raindrop is of order 1 mm.

5.33. The reflectivity [Eq. (5.51)] for a plane-parallel medium of arbitrary optical thickness was obtained by solving the two-stream equations of transfer. But it could be obtained even if you had never seen these equations by using the doubling formula for a pile of plates [Eq. (5.12)] beginning with the reflectivity for an optically thin plane-parallel medium. Do so.

HINTS: Divide the medium of interest into N layers of equal optical thickness. By making N arbitrarily large the optical thickness of a layer can be made arbitrarily small. The reflectivity of a single thin layer is readily obtained from single-scattering theory.

5.34. Show that, in principle, one can determine the visible reflectivity of clouds by making measurements from the ground of downward radiation at two different wavelengths, one in the visible and one in the near-infrared, provided that the reflectivity of the ground is quite different at these two wavelengths whereas that of the cloud is not. For simplicity, assume that the reflectivity of the ground is 0 in the visible and 1 in the infrared.

5.35. Solve Eq. (5.69) for a finite medium with K given by Eq. (5.70) for downward irradiance F_0 at $\tau = 0$ and upward irradiance $F_\uparrow = R_g F_\downarrow$ at $\tau = \bar\tau$, where the reflectivity of the ground is R_g.

5.36. Interference colors often are seen in light reflected by *thin* films (e.g., oil slicks illuminated by sunlight). Why must they be "thin"? And what exactly is meant by "thin"? After all, Eq. (5.24) is for a film of *arbitrary* thickness.

HINT: Analysis is simplified if you take $R_\infty \ll 1$ and ignore the wavelength dependence of refractive index.

5.37. At the end of Section 5.2.4 we assert that "it is difficult to fathom how reflection by colored ground could do more than tinge the light from the bottoms of more or less continuous clouds. Broken clouds are another matter." Can you imagine a cloud cover and ground radiance spectrum (physically realistic even if improbable) such that the bottom of the cloud (at least part of it) would have the color of the ground?

HINT: See Problem 4.58.

5.38. We state in Section 3.4.9 that scattering (of visible radiation) per molecule by air is much greater than that by liquid water. By how much (at standard pressure and temperature)?

HINT: Problem 5.16 is useful. You also need a bit of elementary thermodynamics and some refractive index data.

5.39. We state in Section 3.4.9 that scattering per molecule by air is much greater than that by liquid (and ask you to estimate by how much in Problem 5.38). We have encountered the following explanation for why scattering per molecule in condensed matter (such as liquid water) is *less* than that in a gas. For each molecule in liquid water one is almost certain to find another molecule such that the waves scattered by the two (in the same direction, of course) are exactly out of phase, and so the waves interfere destructively. Do this for every pair of molecules and you end up with essentially no (laterally) scattered light. Do you accept this explanation? Why or why not?

5.40. For what kind of medium does the Einstein equation in Problem 5.16 predict identically zero lateral scattering? Does such a medium exist?

5.41. We state in Section 3.4.9 that the Einstein theory of scattering (see Prob. 5.16) predicts infinite scattering at the critical point. Why?

HINT: To answer this question requires some familiarity with thermodynamic phase diagrams.

5.42. Consider a thin suspension of particles with scattering cross section nearly independent of wavelength but absorption cross section strongly dependent on wavelength and not negligible compared with the scattering cross section (this is neither unphysical nor unusual). What is the reflection spectrum of this suspension if multiple scattering is identically zero? If you have difficulty answering this question, retreat to the simpler problem of a pile of plates.

5.43. Problem 5.35 required you to obtain an expression, using the two-stream theory, for the reflectivity of a finite medium. Use that expression to test the correctness of your answer to Problem 5.42. That is, consider the reflectivity in the limit of small optical thickness for a medium with a scattering coefficient that does not depend on wavelength but an absorption coefficient that does.

5.44. In Section 5.3.2 we showed that attenuation in a multiple-scattering medium is not necessarily exponential. This is true for irradiance and for plane-parallel illumination, by which is meant that the illumination is uniform over an infinite plane. As a result, if we use a detector with line of sight perpendicular to the plane-parallel medium to measure the irradiance, its location in the plane perpendicular to its line of sight does not affect the irradiance it measures. But what about for a narrow, collimated beam illuminating the plane-parallel medium and a detector with a small aperture and a narrow field of view and pointed directly at the beam?

5.45. Show that the optical thickness of a medium composed of (incoherent) scatterers does not change if the medium is uniformly compressed (but not to the point where the scatterers fuse together) or expanded.

5.46. It is often said that Mie theory (Sec. 3.5.1) is "exact". This cannot be literally true because this theory is based on the (false) assumption that matter is homogeneous at all scales (continuum electromagnetic theory). Estimate how inexact Mie theory is by considering the ratio of intrinsic scattering by a given volume of water to scattering predicted by continuum theory for that same volume of water in the form of a droplet in air. Take the droplet diameter to be typical of that for cloud droplets and the wavelength in the middle of the visible spectrum. You will need the Einstein equation in Problem 5.16. How does your result help to understand Raman's incorrect explanation of the blueness of natural ice bodies (Sec. 5.3.1)?

5.47. This problem is related to the previous one, in which we considered a droplet much larger than the wavelength. Now consider the other extreme: a droplet much smaller than the wavelength. For what size droplet will scattering on the molecular scale be comparable with that on the macroscopic scale? Try to guess the answer (on physical grounds) before doing the calculations. You'll need results from Section 3.5.1.

5.48. Derive the expression Eq. (5.23) for the incoherent reflectivity of a nonabsorbing plate by adding up multiple reflections (see Sec. 1.4.1). When is it obvious that this equation cannot be correct? Go one step further and derive the incoherent transmissivity T of an absorbing plate of thickness h and absorption coefficient κ. What condition would have to be satisfied in order to have a hope of inferring the absorption coefficient from the slope of $-\ln T$ versus h and R_∞ from the intercept of this curve? Even if this condition were satisfied, what are some of the pitfalls one would face in trying to use this curve to infer R_∞ and κ by measuring transmission by plates of the same material but different thicknesses?

5.49. Soot is a catchall name for the carbonaceous products of combustion. It has no definite chemical composition but is highly absorbing. Typically, soot particles are much smaller than the wavelengths of visible light. The (visible) reflectivity of clean, fine-grained snow is close to 1. How much soot added to a snowpack would reduce its reflectivity sufficiently that the human observer could notice a difference between clean and dirty thick snowpacks (observed side by side)? Express your answer as the ratio of soot volume to ice volume. The real part of the refractive index of substances called soot is around 2, its imaginary part around 1, and both don't vary much over the visible spectrum. The exact values are meaningless because soot is not an exact substance. Make as many simple approximations as you can in order to quickly estimate how much soot is needed (parts per thousand, million, billion?) to visibly change a snowpack.

5.50. To use the doubling method for a scattering and absorbing medium of arbitrary optical thickness requires the reflectivity and transmissivity of a very optically thin slice of that medium. You should be able to obtain them (in the two-stream approximation) by strictly physical arguments. You can check your result by considering the solution to Problem 5.35 (for $R_g = 0$) in the limit of small optical thickness. Be careful to take this limit consistently. You need a correct power series expansion of the reflectivity and transmissivity to first order in optical thickness. If you are ambitious you can write a doubling program to use the reflectivity and transmissivity you obtain to calculate that of an arbitrary medium and compare the computed (approximate) results with the exact (two-stream) theory. Note in particular how the error changes with increasing optical thickness. Also, try to reduce the error by reducing the thickness of the initial slab.

5.51. There are three ways of determining reflection and transmission by an arbitrary finite slab (within the two-stream theory) overlying a plane with an arbitrary reflectivity R_g. Solve the appropriate differential equations [Eqs. (5.47) and (5.48)] subject to suitable upper and lower boundary conditions. Or begin with the solution to reflection and transmission by the slab over a non-reflecting plane and then either follow the same approach that yielded Eqs. (5.7) and (5.8) or sum the same kind of infinite series (infinite multiple reflections) as in Section 1.4.1. Try all three methods. Which is easier? Which gives you the most physical insight? One of the tests of the correctness of equations for reflectivity and transmissivity of a slab (over a black surface) is to examine whether their sum is 1 for no absorption. Can this same test be applied if the slab overlies a surface with arbitrary reflectivity? Why or why not? If the usual test fails, come up with one that will not.

5.52. Carbon black is a suspension of very small carbon particles. Absorption by these particles is much greater than scattering, and you can take their absorption cross section to be inversely proportional to wavelength. What is the color of an infinitely thick carbon black film? Suppose the film is over a substrate with reflectivity 1 at visible wavelengths. How do the color and luminance of the film vary as its thickness varies from 0 to a sufficiently large value that it can be considered infinitely thick? You can answer these questions using the two-stream theory or by physical reasoning.

5.53. Derive the reflectivity and transmissivity of an absorbing slab of uniform thickness h and absorption coefficient κ, infinite in lateral extent and normally illuminated. The slab is optically smooth and optically homogeneous. R is the reflectivity and T the transmissivity at the slab *interface*. Ignore interference. Do this problem two ways: by summing infinite series of rays reflected and transmitted by the *slab* (and not just by its interface with the medium in which it is embedded); and by solving the two-stream equation of radiative transfer with $\beta = 0$ and carefully choosing the boundary conditions. Keep in mind that the two-stream solution applies only to interior points of the medium ($0 < z < h$).

5.54. For the slab in Problem 5.53, show by physical arguments that $R + T = 1$. The absorptivity of the slab $\overline{A}$ is $1 - (\overline{R} + \overline{T})$. Take κh to be fixed. If $\overline{A}_0$ is the absorptivity for a non-reflecting slab ($R = 0$), show that the absorptivity $\overline{A}$ for any $R > 0$ is less than $\overline{A}_0$.

5.55. In Chapter 2 we derived an approximate expression for the volumetric absorption cross section of a uniformly thick slab, normally illuminated, for "weak absorption" [Eq. (2.148)] and "strong absorption" [Eq. (2.149)]. We neglected reflection by the slab. Use the results of Problem 5.54 to remove this restriction. Does reflection increase or decrease the volumetric absorption cross section?

5.56. Clouds do not have optically smooth surfaces (at least not at visible and near-visible wavelengths). But there are some scattering media that do, for example, bubbly ice and water containing suspended particles. Derive the reflectivity of a scattering and absorbing medium the illuminated surface of which is optically smooth (i.e., specularly reflecting and transmitting). For simplicity take the medium to be infinitely thick. You can do this problem two ways. The hard way is to solve the two-stream equations subject to the boundary conditions in Problem 5.53 then interpret the result physically. The easy way is to use physical arguments to solve the problem, then write down the solution. Does the absorptivity of the infinite slab increase or decrease because of the addition of scatterers?

5.57. This problem is related to Problems 5.53–5.56, but is more of a small term project than a problem to be dashed off in a few hours. Mie theory is fundamentally incapable of treating absorption by spheres that are themselves cloudy. And yet such spheres exist: milk drops, for example. Perhaps we can obtain some insights by considering a simpler problem, absorption by a cloudy, uniform slab, much larger than the wavelength, and illuminated at normal incidence. Is there any combination of reflectivity R, absorption coefficient κ, scattering coefficient β, and thickness h such that the absorptivity of a cloudy slab is greater than that of a clear slab, all else being equal. For simplicity, take the asymmetry parameter of the scatterers in the slab to be zero. Also assume that the absorption coefficient with and without scatterers is the same. The problem of absorption by a cloudy sphere was considered by Sean A. Twomey, 1987: Influence of internal scattering on the optical properties of particles and drops in the near infrared. *Applied Optics*, Vol. 26, pp. 1342–47. Twomey concluded that "absorption by a weakly absorbing particle can be increased substantially beyond that in the absence of internal scattering." What do you conclude on the basis of calculations for a slab?

5.58. Reflections between parallel plates vanish when the separation between them is zero and the two plates fuse into one. But this separation need not be identically zero before they fuse optically and reflections become negligible. Obtain a criterion for when the plates fuse optically.

HINT: Consider a planar air gap between two infinite, identical media and use Eq. (5.24).

6 Multiple Scattering: Advanced

The theory of reflection, transmission, and absorption by a pile of discrete plates, and the two-stream theory for a continuous medium, discussed in the previous chapter, are so mathematically simple and physically transparent that from them you can acquire most of the physical understanding and insight you'll ever need. But sometimes you do need more accurate numbers than these theories can provide. Moreover, the two-stream theory is incapable of accounting for the full directionality of the radiation field. The next step is to generalize from 2 to N streams, which may lead to greater accuracy but at the expense of greater complexity. In the limit as N approaches infinity, we obtain the integro-differential equation of radiative transfer, which is exact (subject to underlying approximations and assumptions). A staggering number of mathematical methods for solving this equation, mostly for plane-parallel media (which don't exist), have been confected, all of them complex, some of them obsolete. We could fill a thick volume with mathematical methods and computations comparing (to nine digits) the results of one method with those of another. But to what end? Instead of drowning in a sea of methods, we consider only two: diffusion theory and the Monte Carlo method. Diffusion theory is the simplest analytical means for escaping from the one-dimensional prison of plane-parallel media. The Monte Carlo method can treat complex, finite, inhomogeneous media with irregular boundaries – the kinds of media that are the rule in Nature.

6.1 N-Stream Theory and Beyond

Equations (5.40) and (5.41) can be written more compactly as a single equation:

$$\mu_k \frac{dF_k}{dz} = -(\kappa + \beta)F_k + \beta \sum_{j=1}^{2} p_{jk} F_j, \quad k = 1, 2, \tag{6.1}$$

where p_{jk} is the probability that a photon in the direction j is scattered in the direction k. Here j and k take on only two values: 1 corresponds to downward, 2 corresponds to upward. The quantity μ_k, the cosine of the angle between the positive z-axis and either of the two allowed directions, takes on only two values, 1 and -1.

This immediately suggests generalizing the two-stream theory to N streams:

$$\mu_k \frac{dF_k}{dz} = -(\kappa + \beta)F_k + \beta \sum_{j=1}^{N} q_{jk} F_j, \quad k = 1, 2, \dots, N, \tag{6.2}$$

Fundamentals of Atmospheric Radiation: An Introduction with 400 Problems. Craig F. Bohren and Eugene E. Clothiaux
Copyright © 2006 Wiley-VCH Verlag GmbH & Co. KGaA, Weinheim
ISBN: 3-527-40503-8

where

$$q_{jk} = \pm p_{jk} \left(\frac{\mu_k}{\mu_j} \right). \tag{6.3}$$

Here p_{jk} is the probability that, given that a photon in direction j is scattered, it is scattered in direction k; the μ_k are direction cosines. The symbol $\pm$ indicates that the sign is chosen to make the entire quantity positive (as it must be). The radiation field is approximated as a set of monodirectional beams, each with zero angular divergence, which, of course, is not possible. Even a laser beam has a small angular divergence. In the N-stream theory we make some assumptions we know are not true. For example, we assume that p_{jk} is the probability that a photon in direction j is scattered into direction k. Such a probability does not strictly exist. Or perhaps we should say that the probability of scattering exactly in one direction is zero because scattering directions (as far as we know) vary continuously. The correct probability is a distribution function giving the probability of scattering into a finite, although possibly small, set of directions (i.e., small solid angle). Note also that if we were to choose one of our directions to be 90°, we'd end up with an infinite quantity in the equation of transfer. In the derivation we assume (strictly) that $\beta \Delta z / \mu \ll 1$. Here $\Delta z / \mu$ is the distance along a particular direction. When $\mu = 0$, this condition cannot be satisfied. The idealizations and approximations that go into the N-stream theory underscore why we have to go beyond it. The right side of Eq. (6.2) is a sum. An integral is the limit of a sum as the number of terms in it goes to infinity. This suggests that an integral will appear in the equation of transfer, which we turn to after making a brief detour to discuss the directional derivative.

6.1.1   Directional Derivative

First consider a function f of a single variable, call it x. The values of x can be represented by points on a line. The derivative of f is defined as

$$\frac{df}{dx} = \lim_{\Delta x \to 0} \frac{f(x + \Delta x) - f(x)}{\Delta x} \tag{6.4}$$

provided this limit exists. The numerator and denominator separately have the limit zero but the limit of their quotient is not necessarily zero. Note that the derivative is *not* the quotient df over dx; this is just a symbol that reminds us how the derivative is defined. The number of possible symbols is indefinite. For example, we could write the derivative as f', a different symbol with the same meaning.

The derivative Eq. (6.4) is, in fact, a *directional derivative*, although rarely called such. It is the rate of change of f along a particular direction (the x-axis). Here there is only one direction, so it is not necessary to call this derivative a directional derivative. What happens, however, when f is defined for points $\mathbf{x}$ in three-dimensional space? Here we have an infinite number of possibilities for the rate of change of f depending on direction. A direction can be specified by a unit vector $\boldsymbol{\Omega}$. Thus the value of the function f at a distance s from x along the direction $\boldsymbol{\Omega}$ is $f(x + s\boldsymbol{\Omega})$, and hence the rate of change of f with distance along this direction is

$$\frac{f(\mathbf{x} + s\boldsymbol{\Omega}) - f(\mathbf{x})}{s}. \tag{6.5}$$

The limit of this quotient, if it exists, is the directional derivative

$$\frac{df}{ds} = \lim_{s \to 0} \frac{f(\mathbf{x} + s\mathbf{\Omega}) - f(\mathbf{x})}{s}. \tag{6.6}$$

Because there are infinitely many directions, there are infinitely many possible directional derivatives: the rate of change of f with distance depends on direction. We might, however, suspect that directional derivatives in three orthogonal directions are sufficient to determine the directional derivative in *any* direction. To show that our suspicion is well founded, assume that f can be expanded in a Taylor series about $\mathbf{x} = (x, y, z)$:

$$f(x + s\Omega_x, y + s\Omega_y, z + s\Omega_z) = f(x, y, z) + s\Omega_x \frac{\partial f}{\partial x} + s\Omega_y \frac{\partial f}{\partial y} + s\Omega_z \frac{\partial f}{\partial z} + O(s^2), \tag{6.7}$$

where the symbol $O(s^2)$ indicates all those terms in the series in powers of 2 or greater in s. Now divide this equation by s and take the limit as s approaches zero to obtain

$$\frac{df}{ds} = \Omega_x \frac{\partial f}{\partial x} + \Omega_y \frac{\partial f}{\partial y} + \Omega_z \frac{\partial f}{\partial z}. \tag{6.8}$$

This is the scalar (or dot) product of $\mathbf{\Omega}$ with the gradient of f:

$$\frac{df}{ds} = \mathbf{\Omega} \cdot \nabla f. \tag{6.9}$$

As we suspected, to find the directional derivative in any direction, we need only the partial derivatives of f along three orthogonal axes.

6.1.2   Equation of Transfer

Once we understand the directional derivative, the equation of transfer is not difficult to derive, but we now must change our allegiance from irradiance to radiance L (Sec. 4.1.2), a function of direction specified by a unit vector $\mathbf{\Omega}$. The rate of change of L along this direction is the corresponding directional derivative $\mathbf{\Omega} \cdot \nabla L(\mathbf{\Omega})$. This rate of change is a consequence of attenuation by absorption and by scattering in all directions other than $\mathbf{\Omega}$, augmentation because of scattering into this direction from all other directions, and emission.

Attenuation of L over a distance Δs

$$\Delta L(\mathbf{\Omega}) = -L(\mathbf{\Omega})(\kappa + \beta)\Delta s \tag{6.10}$$

has the same form as attenuation of monodirectional irradiance in the two-stream theory (Sec. 5.2). And emission

$$\kappa \Delta s P_e / \pi \tag{6.11}$$

has the same form as that in the two-stream theory (Sec. 5.4) with the Planck irradiance P_e replaced by the Planck radiance P_e/π. To obtain the increase in radiance along the direction Δs as a result of scattering consider Fig. 6.1, which depicts all the scatterers in a rectangular

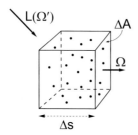

Figure 6.1: Radiation from an arbitrary direction $\mathbf{\Omega}'$ illuminating the scatterers in volume $\Delta s \Delta A$ results in scattering into a unit solid angle about the direction $\mathbf{\Omega}$ and hence contributes to the radiance in this direction.

volume $\Delta s \Delta A$ illuminated by radiation in a small solid angle $\Delta \mathbf{\Omega}'$ about a direction $\mathbf{\Omega}'$. The associated irradiance is $L(\mathbf{\Omega}')\Delta \mathbf{\Omega}'$, and hence the power scattered in all directions is

$$\Delta L(\mathbf{\Omega}')\Delta \mathbf{\Omega}' N \Delta s \Delta A \sigma_\mathrm{s} = L(\mathbf{\Omega}')\Delta \mathbf{\Omega}' \beta \Delta s \Delta A, \tag{6.12}$$

where the scattering coefficient β is the product of the number N of scatterers per unit volume and the scattering cross section σ_s per scatterer. To obtain the scattering per unit solid angle in the direction $\mathbf{\Omega}$, Eq. (6.12) must be multiplied by the probability $p(\mathbf{\Omega}', \mathbf{\Omega})$ that a photon in direction $\mathbf{\Omega}'$ is scattered into unit solid angle about the direction $\mathbf{\Omega}$. But we want the contribution to the radiance in the direction $\mathbf{\Omega}$ so must divide by the area ΔA perpendicular to this direction:

$$L(\mathbf{\Omega}')\Delta \mathbf{\Omega}' \beta p(\mathbf{\Omega}', \mathbf{\Omega})\Delta s. \tag{6.13}$$

The contribution to the radiance in direction $\mathbf{\Omega}$ is then an integral over all directions $\mathbf{\Omega}'$:

$$\beta \Delta s \int_{4\pi} L(\mathbf{\Omega}')p(\mathbf{\Omega}', \mathbf{\Omega}) \, d\mathbf{\Omega}', \tag{6.14}$$

where 4π indicates integration over all directions (4π steradians). The total change of $L(\mathbf{\Omega})$ along Δs is $\mathbf{\Omega} \cdot \nabla L(\mathbf{\Omega})\Delta s$, which when set equal to the sum of attenuation [Eq. (6.10)], emission [Eq. (6.11)], and scattering [Eq. (6.14)] yields

$$\mathbf{\Omega} \cdot \nabla L(\mathbf{\Omega}) = -(\kappa + \beta)L(\mathbf{\Omega}) + \beta \int_{4\pi} L(\mathbf{\Omega}')p(\mathbf{\Omega}', \mathbf{\Omega}) \, d\mathbf{\Omega}' + \kappa P_\mathrm{e}/\pi. \tag{6.15}$$

L in a particular direction decreases along this direction because of absorption and scattering in all other directions, and increases because of emission and scattering into this direction. The Planck function at the temperature of the medium could vary from point to point. Emission may be negligible even if κ is not if at the wavelength of interest the Planck function is negligible. As expected, the generalization of Eq. (6.2) to an infinite number of streams yields an equation with a derivative on the left side, an integral on the right, and hence the name integro-differential equation.

From Eq. (6.15) it follows that in the absence of scattering and absorption, the radiance L along any direction is invariant (i.e., the directional derivative is zero), which we proved in

Section 4.1.3 by geometrical arguments. Pick a point in space and a direction at that point, and the radiance does not change along a line in that direction.

The probability distribution p often is called by astronomers and atmospheric scientists the *phase function*, a singularly inappropriate term given that it has nothing to do with the phases of waves but rather of astronomical bodies. We once were at a scientific meeting at which a fight almost broke out because the speaker was happily pontificating about phase functions (in the sense of angular probability distributions) while a red-faced optical scientist in the audience was excitedly hopping up and down exclaiming "but it doesn't behave that way." Order was finally restored, but it is understandable why phase function, a quantity from which phase has been removed, causes such consternation in the context of optics, a field in which the phases of waves play such a key role.

Because an incident photon becomes a scattered photon, and vice versa, if time is reversed, we must have

$$p(\mathbf{\Omega}', \mathbf{\Omega}) = p(\mathbf{\Omega}, \mathbf{\Omega}'). \tag{6.16}$$

The phase function in Eq. (6.15) is *normalized*:

$$\int_{4\pi} p(\mathbf{\Omega}, \mathbf{\Omega}') \, d\mathbf{\Omega}' = 1. \tag{6.17}$$

Other normalizations are possible so be alert for them. Equation (6.17) is a mathematical statement that if a photon is scattered it has to be scattered in *some* direction, and the set of directions in 4π steradians exhausts all of them. For an isotropic medium p cannot depend on the absolute directions $\mathbf{\Omega}'$ and $\mathbf{\Omega}$, only on the direction $\mathbf{\Omega}$ *relative* to $\mathbf{\Omega}'$, which can be expressed symbolically by writing $p(\mathbf{\Omega}' \cdot \mathbf{\Omega})$, where $\mathbf{\Omega}' \cdot \mathbf{\Omega}$ is the cosine of the angle between the two directions.

For no absorption ($\kappa = 0$) and an isotropic radiation field (L independent of direction) Eqs. (6.15) and (6.17) yield $\mathbf{\Omega} \cdot \nabla L = 0$: the radiance field is the same everywhere. This is consistent with what we obtained in Section 5.2 with the two-stream theory for a nonabsorbing medium: when the radiation field was uniform it also was isotropic.

As an aside we note that Eq. (6.15) is formally identical to the one-velocity neutron transport equation. Neutrons produced in a nuclear reactor are scattered in various directions, absorbed within it, and leak out its boundaries. The primary difference between photons and neutrons is that all photons have the same speed whereas neutrons do not. Also, neutrons, when absorbed by some nuclei, can give rise to more neutrons by way of fission reactions. But the similarities are greater than the differences, which means that mathematical techniques for solving neutron transport problems can be carried over to radiative transfer problems and vice versa.

We defined the *vector irradiance* in Section 4.2 as

$$\mathbf{F} = \int_{4\pi} \mathbf{\Omega} L(\mathbf{\Omega}) \, d\mathbf{\Omega}, \tag{6.18}$$

and the *mean radiance J* is defined as

$$J = \frac{1}{4\pi} \int_{4\pi} L(\mathbf{\Omega}) \, d\mathbf{\Omega}. \tag{6.19}$$

The directional derivative on the left side of Eq. (6.15) can be written

$$\mathbf{\Omega} \cdot \nabla L = \nabla \cdot (\mathbf{\Omega} L). \tag{6.20}$$

Make this substitution and integrate over all directions:

$$\int_{4\pi} \nabla \cdot (\mathbf{\Omega} L) \, d\Omega = \nabla \cdot \int_{4\pi} \mathbf{\Omega} L \, d\Omega = \nabla \cdot \mathbf{F}. \tag{6.21}$$

The integrated first term on the right side of the radiative transfer equation is

$$-(\kappa + \beta) \int_{4\pi} L(\mathbf{\Omega}) \, d\Omega = -4\pi(\kappa + \beta) J. \tag{6.22}$$

The integrated second term is

$$\beta \int_{4\pi} \int_{4\pi} p(\mathbf{\Omega}', \mathbf{\Omega}) L(\mathbf{\Omega}') \, d\Omega' \, d\Omega, \tag{6.23}$$

which, because of the normalization condition Eq. (6.17), is

$$\beta \int_{4\pi} L(\mathbf{\Omega}') \, d\Omega' = 4\pi \beta J. \tag{6.24}$$

The integrated third term is $4\kappa P_\mathrm{e}$, which when combined with Eqs. (6.21), (6.22) and (6.24) yields

$$\nabla \cdot \mathbf{F} = -4\pi \kappa J + 4\kappa P_\mathrm{e}. \tag{6.25}$$

This is a general result, as valid as the equation of transfer. And it makes physical sense. For a nonabsorbing medium ($\kappa = 0$) the divergence of the vector irradiance is zero: all the photons that enter a region come out.

To explain briefly why Eq. (6.15) can lead to such a bewildering array of N-stream approximations we back up just a bit. Consider first approximating an arbitrary one-dimensional integral as a finite sum:

$$\int_a^b f(x) \, dx \approx \sum_{j=1}^{N} w_j f(x_j). \tag{6.26}$$

This is called reducing an integral to *quadratures*; the w_j are quadrature weights and the discrete set of ordinates x_j are quadrature points. Simpson's rule, which you may remember from elementary calculus, is a quadrature formula, one of the simplest imaginable. Note the tremendous latitude of choices with which we are faced: N, w_j, and x_j, a triply-infinite set of possibilities. Thus there are many ways of replacing the integro-differential equation Eq. (6.15) with a set of N coupled ordinary differential equations for L in N discrete directions depending on the many ways of approximating the integral on the right side by a finite sum. This approach goes under the general name of *discrete ordinates*.

At this point, however, we are not going to launch into discussing the many ways of obtaining approximate solutions to Eq. (6.15). There are many methods and countless variations on them, most of them restricted to plane-parallel media (which don't exist). To learn more about all these methods consult the references at the end of this chapter. Instead of discussing even a restricted set of such methods, we devote Section 6.3 to only one, the Monte Carlo method, which has the peculiar characteristic that it enables us to solve equations even if we don't know what they are. By Monte Carlo methods, which are not restricted to plane-parallel media, one can solve Eq. (6.15) without knowing that it exists. Sound like magic? Read on.

6.2   Diffusion Theory: The Elements

Equation (6.25) doesn't do us much good because it contains two unknown quantities, and so we have to go a step further and make some approximations. The quantity J at a point, which we take to be a spectral quantity (i.e., we consider only photons with a narrow range of energies), is, except for a multiplicative factor, the number density of photons. That is, L is the number density of photons moving in a given direction per unit solid angle times the speed of a photon times the photon energy. When we integrate this quantity over all directions we get the total number of photons per unit volume. According to molecular diffusion theory, transport is driven by concentration gradients (Fick's law). The same must be true for photons, so we postulate a diffusion law for photons:

$$\mathbf{F} = -4\pi D \nabla J, \tag{6.27}$$

where D is the photon *diffusion coefficient*, here assumed independent of position, and the factor 4π is introduced for convenience. Now combine Eqs. (6.25) and (6.27) to obtain the diffusion equation

$$\nabla^2 J - \frac{1}{\chi^2} J = 0, \tag{6.28}$$

where the *diffusion length* χ is

$$\chi^2 = \frac{D}{\kappa}. \tag{6.29}$$

We have an equation for the mean radiance J but what about the radiance L? The radiance distribution

$$L = J - 3D\boldsymbol{\Omega} \cdot \nabla J \tag{6.30}$$

is consistent with Eq. (6.27), Fick's law for photons. To show this, multiply both sides of Eq. (6.30) by $\boldsymbol{\Omega}$ and integrate over all directions:

$$\int_{4\pi} L\boldsymbol{\Omega} \, d\Omega = \mathbf{F} = \int_{4\pi} \boldsymbol{\Omega} J \, d\Omega - 3D \int_{4\pi} \boldsymbol{\Omega}(\boldsymbol{\Omega} \cdot \nabla J) \, d\Omega. \tag{6.31}$$

The first integral on the right side vanishes; the second, after a bit of effort, can be shown to be $4\pi\nabla J/3$, which when substituted in Eq. (6.31) yields Eq. (6.27). Thus diffusion theory is based on the assumption that the radiance is a linear function of direction relative to the gradient of the mean radiance.

All that is left is to determine the diffusion coefficient. Multiply both sides of Eq. (6.15) by Ω, assume there is negligible emission and integrate over all directions:

$$\int_{4\pi} \Omega(\Omega \cdot \nabla L) \, d\Omega = -(\kappa+\beta) \int_{4\pi} \Omega L \, d\Omega + \beta \int_{4\pi} \int_{4\pi} p(\Omega', \Omega) \Omega L(\Omega') \, d\Omega' \, d\Omega. \quad (6.32)$$

The first integral on the right side of Eq. (6.32) is the vector irradiance [Eq. (6.18)]. The second integral is a double integral, which we can integrate sequentially. Consider first the integral

$$\int_{4\pi} p(\Omega', \Omega) \Omega \, d\Omega. \quad (6.33)$$

This integral is a vector, call it $\mathbf{G}$, and take its dot product with Ω':

$$\int_{4\pi} p(\Omega', \Omega) \Omega' \cdot \Omega \, d\Omega = \Omega' \cdot \mathbf{G}. \quad (6.34)$$

The integral on the left side of Eq. (6.34) is the asymmetry parameter g, the mean cosine of the scattering angle:

$$g = \Omega' \cdot \mathbf{G}. \quad (6.35)$$

If the medium is isotropic, g does not depend on Ω', which implies that Eq. (6.33) is

$$G = g\Omega'. \quad (6.36)$$

Substitute this result in Eq. (6.32) to obtain

$$\int_{4\pi} \Omega(\Omega \cdot \nabla L) \, d\Omega = -\{\beta(1-g) + \kappa\}\mathbf{F}. \quad (6.37)$$

This result is quite general. To go further we have to make some approximations. For L in Eq. (6.37) take Eq. (6.30); the result for the left side of Eq. (6.37) is

$$\int_{4\pi} \Omega(\Omega \cdot \nabla J) \, d\Omega - 3D \int_{4\pi} \Omega\{\Omega \cdot \nabla(\Omega \cdot \nabla J)\} \, d\Omega. \quad (6.38)$$

The first integral was obtained previously in the derivation of Eq. (6.31). The second integral, with much labor, can be shown to vanish. Thus we have

$$\mathbf{F} = -\frac{4\pi}{3\{\beta(1-g) + \kappa\}} \nabla J, \quad (6.39)$$

which from Eq. (6.27) gives the diffusion coefficient

$$D = \frac{1}{3\{\beta(1-g) + \kappa\}}. \quad (6.40)$$

This is consistent with the two-stream theory for an absorbing medium. We obtained a diffusion equation (one-dimensional) [Eq. (5.69)] identical to Eq. (6.28), satisfied by both upward and downward irradiances. From that result we can infer a diffusion coefficient given by Eq. (6.40) but without the factor 3.

The diffusion equation for the mean radiance [Eq. (6.28)] and the diffusion coefficient [Eq. (6.40)] are not sufficient by themselves. We also need boundary conditions. These are based on irradiances. Consider a surface the normal to which is the unit vector $\mathbf{n}$. The irradiance associated with this surface is

$$\int_{2\pi} L\,\mathbf{n} \cdot \mathbf{\Omega}\, d\Omega = F, \tag{6.41}$$

where 2π indicates integration over a hemisphere of directions centered on $\mathbf{n}$. Substitute Eq. (6.30) in Eq. (6.41) to obtain

$$F = J\mathbf{n} \cdot \int_{2\pi} \mathbf{\Omega}\, d\Omega - 3D\mathbf{n} \cdot \int_{2\pi} \mathbf{\Omega}(\mathbf{\Omega} \cdot \nabla J)\, d\Omega. \tag{6.42}$$

The first integral is π; the second integral is similar to the second integral on the right side of Eq. (6.33). We therefore obtain

$$F = \pi J - 2\pi D\mathbf{n} \cdot \nabla J. \tag{6.43}$$

With this equation we can impose various conditions on the irradiance at boundaries. With Eqs. (6.28) and (6.43) in hand, together with the diffusion coefficient Eq. (6.40), we now can take advantage of the huge literature on solutions to diffusion equations for various geometries and boundary conditions. Given the similarity between diffusion theory and the two-stream theory for an absorbing medium, we expect the two approximations to have the same degree of applicability. The difference is that the diffusion equation [Eq. (6.28)] allows us to solve problems for other than plane-parallel media.

Without solving the diffusion equation we can use it to gain some insight into when vertical attenuation within a laterally finite absorbing medium is dominated by its intrinsic properties (i.e., its diffusion length) rather than by leakage out its sides. Because of the form of the diffusion equation [Eq. (6.28)] we expect attenuation along a vertical line to be determined mostly by the diffusion length if the line is a few diffusion lengths from the lateral boundaries. Within a diffusion length or less, attenuation is more and more determined by leakage out the boundaries.

If the single-scattering albedo is close to 1, as it is for clouds at visible wavelengths, the diffusion length [Eq. (6.29)] is

$$\chi \approx \frac{1}{\beta\sqrt{3(1-\omega)(1-g)}}. \tag{6.44}$$

If we approximate β as $3f/d$, where f is the volume fraction of cloud droplets and d is the droplet diameter, we obtain

$$\chi \approx \frac{d}{f\sqrt{3(1-g)(1-\varpi)}}. \tag{6.45}$$

For cloud droplets g is around 0.85, and so we can take the square root of $3(1 - g)$ to be 1. The volume fraction f is around 10^{-6} or less and d is of order 10 μm, which yields

$$\chi \approx \frac{10}{\sqrt{1 - \varpi}} \tag{6.46}$$

for the diffusion length in meters. If $\varpi = 0.9999$, the diffusion length is 1 km; if $\varpi = 0.999999$, the diffusion length is 10 km. Thus if we wished to determine the diffusion length of a cloud by measuring attenuation within it, we would have to do so along a line a few kilometers from its lateral boundaries.

6.3   The Monte Carlo Method

In the Monte Carlo method applied to radiative transfer, photons are treated statistically. The distance a photon travels before something happens to it (scattering or absorption) is given by a probability distribution. Having traveled this distance it has a specified probability of being either absorbed or scattered. If scattered, the direction of scattering is determined by another probability distribution. One after another photons are imagined to be launched into a medium. As with humans, "time and chance happeneth to them all." The ultimate fate of a photon is either capture by the medium (absorption) or escape from it. Enough photon histories are accumulated to give a statistically valid picture of reflection, transmission, and absorption by a medium illuminated by greatly many photons. The heart of the Monte Carlo method is transforming from probability distributions for physical variables to a uniform probability distribution. Everything else is bookkeeping.

Suppose that we have a probability distribution $p(x)$ for a variable x. Recall that probability distributions are defined by their integral values in that

$$\int_{x_1}^{x_2} p(x) \, dx \tag{6.47}$$

is the probability that x lies in the interval between x_1 and x_2. The integral of p over all x is 1. We want to transform to a variable ξ the values of which are *uniformly* distributed between 0 and 1. That is, ξ is a random number with probability distribution $P(\xi) = 1$. We want to find x as a function of ξ such that we obtain p. From the theorem for transforming variables of integration we have

$$\int_{\xi_1}^{\xi_2} p(x) \frac{dx}{d\xi} \, d\xi = \int_{\xi_1}^{\xi_2} P(\xi) \, d\xi. \tag{6.48}$$

For this always to be true requires that the integrands be equal:

$$p(x) \frac{dx}{d\xi} = P(\xi) = 1. \tag{6.49}$$

This is a first-order differential equation for x as a function of ξ. If we can solve it, we should be able to recover the probability distribution $p(x)$ as follows.

Divide the x-axis into a large set of intervals (or bins). The jth bin is defined by those values lying between x_j and x_{j+1}. Choose a value of ξ randomly and find the corresponding value of x. Assign this value to a bin. After doing this a great many times, divide the number in each bin by the total number of values. This gives the probabilities of x lying in each bin. Divide by the bin width to obtain the probability density. Now plot these probability densities versus x. In the limit of an indefinitely large number of bins and random numbers ξ this result should converge to the probability distribution $p(x)$.

To show that it does consider a simple linear probability distribution $p = 2x$ on the interval $0 \leq x \leq 1$. As required, the integral of p over this interval is 1. The differential equation for the transformation is

$$2x\frac{dx}{d\xi} = \frac{d}{d\xi}\left(x^2\right) = 1, \tag{6.50}$$

the solution to which, subject to the boundary condition $x = 0$ when $\xi = 0$, is

$$x = \sqrt{\xi}. \tag{6.51}$$

Figure 6.2 shows the results of using a random number generator to compute N values of ξ and the corresponding values of x in 100 bins. The computed probability density is shown as dots. As N increases, the dots lie more closely to the exact probability distribution. If for fixed N the same set of calculations is done again and again the distribution of dots will not be exactly the same. This is shown in Fig. 6.3 for 100 bins and two sets of 10^6 random numbers.

The previous paragraphs are general. Now we have to get specific in order to develop a computational method in detail. For this we need probability distributions for path lengths and scattering directions, which we turn to next.

6.3.1   Path Length Distribution

The probability distribution for the path length x a photon travels before it is scattered or absorbed is given in Section 5.1, rewritten here as

$$p(x) = \frac{1}{\ell}\exp(-x/\ell), \tag{6.52}$$

where x is the *physical* path length and ℓ the total mean free path. It is more convenient here to express this distribution in terms of the *optical* path length $\tau = x/\ell$:

$$p(\tau) = \exp(-\tau), \tag{6.53}$$

the integral of which from 0 to ∞ is 1, as it must be if this is a proper probability distribution. As we did previously for the probability distribution $2x$, we find τ as a function of ξ, a random number between 0 and 1 with a uniform probability distribution, such that

$$p(\tau)\frac{d\tau}{d\xi} = -\frac{d}{d\xi}\exp(-\tau) = P(\xi) = 1. \tag{6.54}$$

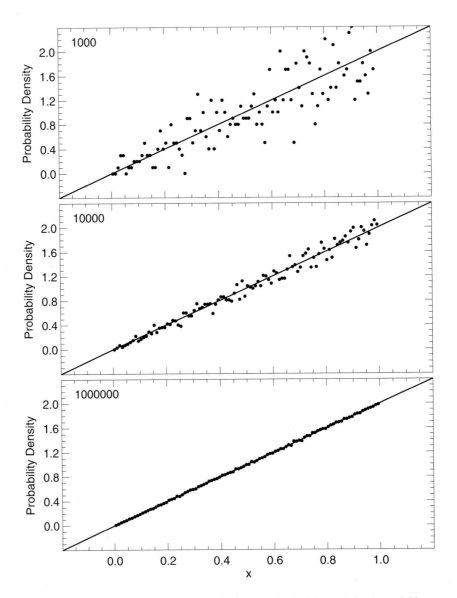

Figure 6.2: The continuous probability distribution $p = 2x$ (solid curve) for the variable x can be approximated by dividing the x-axis into equal intervals (bins); 100 bins were used here. Values are assigned to each bin with a random number generator. The number of values in each bin divided by the total number is the probability of a value lying within a bin. Divide by the bin width to obtain the probability density. The discrete probability densities more closely approximate the continuous distribution the greater the number of values (indicated in the upper left corner of each plot).

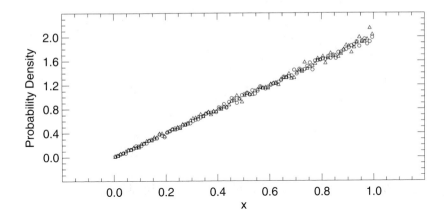

Figure 6.3: The same calculations shown in Fig. 6.2 for 1000000 random numbers but done two times. The probability densities for each run are shown with open circles and open triangles.

The solution to this differential equation, subject to the boundary condition $\tau = 0$ when $\xi = 1$, is

$$\tau = -\ln(1 - \xi). \tag{6.55}$$

With this transformation, the random number ξ is uniformly distributed between 0 and 1 and the optical path length τ is distributed between 0 and ∞ according to the probability distribution Eq. (6.53).

To show this the τ-axis was divided into bins of width 0.01, 10^6 random numbers ξ chosen, the corresponding values of τ calculated from Eq. (6.55) and assigned to appropriate bins, and the fraction of values in each bin divided by the bin width 0.01 and plotted. As expected, this set of discrete probability densities (Fig. 6.4) lies close to the continuous probability distribution $\exp(-\tau)$.

Given that a photon travels an optical path τ before something happens to it, the probability it is scattered is the single-scattering albedo ϖ, and hence the probability it is absorbed is $1 - \varpi$. Thus to determine if a photon is scattered or absorbed, compute a random number. If it is less than or equal to ϖ, the photon is scattered; if not, the photon is absorbed, its death duly recorded, another photon launched, and so on into the long hours of night. If a photon is scattered, the direction of its next path is determined by a probability distribution for scattering directions.

6.3.2 Scattering Direction Distribution

The probability distribution for scattering in a particular direction per unit solid angle is specified by the inaptly-named phase function $p(\vartheta, \varphi)$ (see Sec. 6.1.2), normalized in that

$$\int_0^{2\pi} \int_0^{\pi} p(\vartheta, \varphi) \sin \vartheta \, d\vartheta \, d\varphi = 1, \tag{6.56}$$

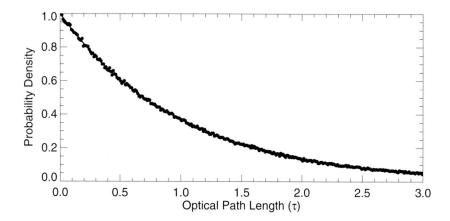

Figure 6.4: The continuous probability distribution $p = \exp(-\tau)$ for the variable τ can be approximated by dividing the τ-axis into equal intervals (bins); 1000 bins were used here. Values are assigned to each bin with a random number generator. The number of values in each bin divided by the total number times the bin width is the probability density for the bin. The discrete probability densities (solid circles) more closely approximate the continuous distribution (solid line) the greater the number of random numbers.

where the z-axis for the spherical polar coordinate system (ϑ, φ) is the direction a photon had *before* it was scattered, ϑ is the scattering angle and φ the azimuthal angle. More often than not, scattering is taken to be azimuthally symmetric (i.e., p is independent of φ), which implies that all azimuthal angles are equally probable, and hence the phase function can be written

$$p(\vartheta, \varphi) = \frac{p(\vartheta)}{2\pi}, \tag{6.57}$$

where $1/2\pi$ is the (uniform) probability distribution for azimuthal angles. By transforming to the variable $\mu = \cos \vartheta$, the normalization condition Eq. (6.56) becomes

$$\int_{-1}^{1} p(\mu) \, d\mu = 1. \tag{6.58}$$

Azimuthal symmetry is strictly valid only for a spherically symmetric medium (i.e., one that doesn't change if it is rotated) and for radiation that is unpolarized.

Determining the azimuthal angle for a scattered photon is straightforward: generate a random number ξ between 0 and 1, and the azimuthal angle is $2\pi\xi$. What about the scattering angle? If the scatterers are spheres, we can calculate $p(\mu)$ from Mie theory. If they are not spheres, recourse may be had to other theories or even measurements. What is often done, however, is to use the simple one-parameter *Henyey–Greenstein* phase function

$$p(\mu) = \frac{1}{2} \frac{1 - g^2}{(1 + g^2 - 2g\mu)^{3/2}}. \tag{6.59}$$

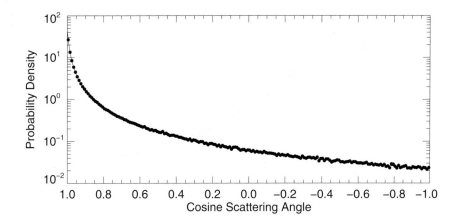

Figure 6.5: Henyey–Greenstein angular scattering probability distribution (solid curve) approximated by discrete probability densities (solid circles) obtained by dividing the range of cosines
into equal intervals (bins), assigning values to each bin randomly, and dividing the fraction of
values in a bin by its width. The asymmetry parameter is 0.85.

We emphasize that this phase function has no physical significance whatsoever. It is a mathematical phase function that bears some resemblance to physical phase functions and has a few
desirable mathematical properties. For example, it is normalized, which is no trick given that
any function can be normalized. More important, g (asymmetry parameter) in Eq. (6.59) is
the mean cosine of the scattering angle:

$$\langle \mu \rangle = \int_{-1}^{1} \mu p(\mu) \, d\mu = g. \tag{6.60}$$

In the limit $g \to 1$, p is an infinite spike in the forward direction ($\mu = 1$), whereas in the limit
$g \to -1$, p is an infinite spike in the backward direction ($\mu = -1$). That is,

$$\lim_{g \to 1} p(1) = \lim_{g \to 1} \frac{1 + g}{(1 - g^2)} = \infty, \quad p(\mu < 1, \; g = 1) = 0, \tag{6.61}$$

$$\lim_{g \to -1} p(-1) = \lim_{g \to -1} \frac{1 - g}{(1 + g^2)} = \infty, \quad p(-1 < \mu, \; g = -1) = 0. \tag{6.62}$$

For $g = 0$, the Henyey–Greenstein function corresponds to isotropic scattering, which, as we
note in Section 5.2 (and explain in Sec. 7.3), does not exist for electromagnetic radiation.

As we did with the probability distribution for photon path lengths, we need to find the
transformation to the random variable ξ by way of a first-order differential equation:

$$p(\mu) \frac{d\mu}{d\xi} = \frac{1 - g^2}{2g} \frac{d}{d\xi} \frac{1}{(1 + g^2 - 2g\mu)^{1/2}} = 1, \tag{6.63}$$

the solution to which, subject to the boundary condition $\mu = -1$ when $\xi = 0$, is

$$\mu = \frac{1}{2g}\left\{1 + g^2 - \left(\frac{1 - g^2}{1 - g + 2g\xi}\right)^2\right\}.$$
(6.64)

The μ-axis was divided into bins of width 0.01, 10^6 random numbers ξ chosen, the corresponding values of μ calculated from Eq. (6.64) for $g = 0.85$ and assigned to appropriate bins, and the fraction of values in each bin divided by the bin width and plotted. As expected, this set of discrete probability densities (Fig. 6.5) lies close to the continuous probability distribution [Eq. (6.59)].

Given the shaky theoretical foundation for the Henyey–Greenstein phase function, how does it compare with one more firmly supported? Figure 6.6 compares the Henyey–Greenstein phase function with calculations using Mie theory for a lognormal distribution of cloud droplets. Agreement is good except close to the forward direction and for scattering angles greater than about 120°. Scattering is much less in these backward directions, however, so the disagreement, although relatively large, more than a factor of 10, is absolutely small. If one were interested in backscattering by an optically thin medium ($\tau \ll 1$), the Henyey–Greenstein function would be inappropriate. For an optically thick medium ($\tau \gg 1$), multiple scattering effaces almost all details of the phase function except the mean cosine of the scattering angle. Less important than the correctness of *each* scattering event is the correctness of the average over *many* events. A corollary of this is that the Henyey–Greenstein phase function is likely to be at its worst for calculations of reflection by media with optical thicknesses around 1.

The sharp rise of the theoretical phase function near the forward direction is almost 100 times greater than the Henyey–Greenstein phase function. But what appears to be a weakness is actually a strength. A narrow peak near the forward direction corresponds to almost no scattering in the sense that photons are hardly diverted from their original direction. This suggests that pesky near-forward scattered photons can be whisked out of sight as follows. Approximate the product of the scattering coefficient β and theoretical phase function $\tilde{p}$ (e.g., from Mie theory) as the weighted sum of a Henyey–Greenstein phase function $\overline{p}$ with asymmetry parameter $\overline{g}$ and an infinitely sharp spike in the forward direction:

$$\beta\tilde{p} \approx \overline{\beta}\overline{p} + \beta_0\delta,$$
(6.65)

where

$$\delta(\mu) = \lim_{g \to 1} p(\mu),$$
(6.66)

and both p and $\overline{p}$ are given by Eq. (6.59). Because $\tilde{p}$, $\overline{p}$, and δ are normalized

$$\beta \approx \overline{\beta} + \beta_0.$$
(6.67)

The asymmetry parameter is

$$g = \int_{-1}^{1} \mu\tilde{p} \, d\mu = \frac{\overline{\beta}}{\beta}\overline{g} + \frac{\beta_0}{\beta}.$$
(6.68)

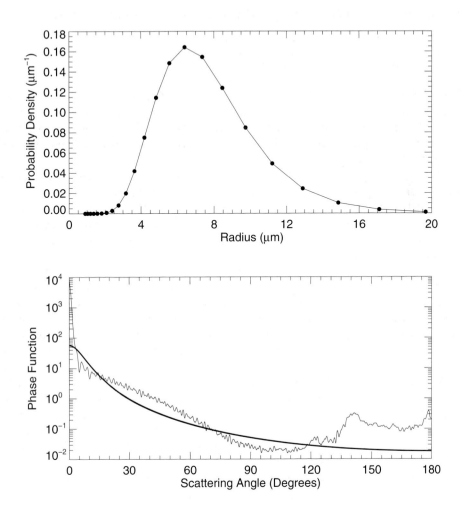

Figure 6.6: Lognormal size distribution of cloud droplets (top panel) for the phase function calculated with Mie theory (thin line) in the bottom panel compared with the Henyey–Greenstein phase function (thick line) with $g = 0.87$, also calculated with Mie theory. The wavelength of the illumination is 550 nm.

From Eqs. (6.67) and (6.68) it therefore follows that a real medium with scattering coefficient β and asymmetry parameter g is approximately equivalent to a fictitious medium with scattering coefficient

$$\bar{\beta} = \beta - \beta_0 \tag{6.69}$$

and a Henyey–Greenstein phase function with asymmetry parameter given by

$$\overline{g} = 1 - \frac{\beta}{\overline{\beta}}(1 - g) \tag{6.70}$$

provided that scattering by the medium is highly peaked near the forward direction (i.e., the scatterers are large compared with the wavelength). Now we have to choose the ratio $\beta/\overline{\beta}$. For example, we argued in Section 3.5.2 that in the limit of infinite radius, half of the scattering cross section of a sphere is associated with scattering in the forward direction, and hence for a sufficiently large sphere $\beta/\overline{\beta} \approx 2$. With this approximation the scattering coefficient of the fictitious medium is $\beta/2$ and its asymmetry parameter is $1 - 2(1 - g)$.

6.3.3 Transforming Coordinate Axes

The trajectory of a photon is a sum of vectors $\tau_1 \mathbf{\Omega}_1$, $\tau_2 \mathbf{\Omega}_2$, ..., where τ_j are optical path lengths and $\mathbf{\Omega}_j$ are unit vectors. In previous sections we showed how to generate τ_j and $\mathbf{\Omega}_j$. The catch is that each direction vector $\mathbf{\Omega}_j$ is specified by a scattering angle ϑ_j and azimuthal angle φ_j relative to a *different* coordinate system. Thus to add these vectors we have to transform them all to a common (reference) coordinate system. How this is done for two successive directions sets the pattern for an arbitrary number of directions.

The first direction vector of a photon launched into a medium is

$$\mathbf{\Omega}_1 = \sin \vartheta_1 \cos \varphi_1 \mathbf{i}_1 + \sin \vartheta_1 \sin \varphi_1 \mathbf{j}_1 + \cos \vartheta_1 \mathbf{k}_1, \tag{6.71}$$

where $\mathbf{i}_1, \mathbf{j}_1, \mathbf{k}_1$ are unit vectors along the coordinate axes of the reference coordinate system. For the second direction we define a new coordinate system with

$$\mathbf{k}_2 = \mathbf{\Omega}_1. \tag{6.72}$$

The three vectors $\mathbf{i}_2, \mathbf{j}_2, \mathbf{k}_2$, where

$$\mathbf{i}_2 = \cos \vartheta_1 \cos \varphi_1 \mathbf{i}_1 + \sin \vartheta_1 \sin \varphi_1 \mathbf{j}_1 + \cos \vartheta_1 \mathbf{k}_1, \tag{6.73}$$

$$\mathbf{j}_2 = - \sin \varphi_1 \mathbf{i}_1 + \cos \varphi_1 \mathbf{j}_1, \tag{6.74}$$

form an orthonormal, right-handed system. That is, they are of unit length, mutually perpendicular and

$$\mathbf{i}_2 \times \mathbf{j}_2 = \mathbf{k}_2, \quad \mathbf{k}_2 \times \mathbf{i}_2 = \mathbf{j}_2, \quad \mathbf{j}_2 \times \mathbf{k}_2 = \mathbf{i}_2. \tag{6.75}$$

Equations (6.72)–(6.74) can be written in matrix form

$$\begin{pmatrix} \mathbf{i}_2 \\ \mathbf{j}_2 \\ \mathbf{k}_2 \end{pmatrix} = \begin{pmatrix} \cos \vartheta_1 \cos \varphi_1 & \cos \vartheta_1 \sin \varphi_1 & - \sin \vartheta_1 \\ - \sin \varphi_1 & \cos \varphi_1 & 0 \\ \sin \vartheta_1 \cos \varphi_1 & \sin \vartheta_1 \sin \varphi_1 & \cos \vartheta_1 \end{pmatrix} \begin{pmatrix} \mathbf{i}_1 \\ \mathbf{j}_1 \\ \mathbf{k}_1 \end{pmatrix}. \tag{6.76}$$

The second direction is specified by the unit vector

$$\mathbf{\Omega}_2 = \sin \vartheta_2 \cos \varphi_2 \mathbf{i}_2 + \sin \vartheta_2 \sin \varphi_2 \mathbf{j}_2 + \cos \vartheta_2 \mathbf{k}_2, \tag{6.77}$$

where ϑ_2 is the scattering angle and φ_2 the azimuthal angle. Given Eqs. (6.76) and (6.77), we can express Ω_2 relative to the reference coordinate system. For the next scattering direction Ω_3, the unit vectors in the local coordinate system are given by Eq. (6.76) with 2 replaced by 3, 1 replaced by 2. The components of this direction in the reference coordinate system are obtained by multiplying two matrices. For the next direction, multiply three matrices, and so on. We can write this compactly in matrix notation:

$$\xi_n = \mathbf{M}_{n-1}\mathbf{M}_{n-2}\ldots\mathbf{M}_1\xi_1, \tag{6.78}$$

where

$$\xi_p = \begin{pmatrix} \mathbf{i}_p \\ \mathbf{j}_p \\ \mathbf{k}_p \end{pmatrix}, \quad \mathbf{M}_p = \begin{pmatrix} \cos\vartheta_p\cos\varphi_p & \cos\vartheta_p\sin\varphi_p & -\sin\vartheta_p \\ -\sin\varphi_p & \cos\varphi_p & 0 \\ \sin\vartheta_p\cos\varphi_p & \sin\vartheta_p\sin\varphi_p & \cos\vartheta_p \end{pmatrix}. \tag{6.79}$$

6.3.4   Surface Reflection

The results so far enable us to determine the statistical behavior of photons *within* a medium: probabilities of distances traveled before an event (scattering or absorption), if the event is scattering or absorption, and, if scattering, the probability of a particular scattering direction. But what happens when a photon encounters a *boundary*? This depends on its reflecting properties. If the reflectivity is zero, the photon is counted as having been removed from the medium. But for a nonzero reflectivity, the photon has a nonzero probability of being reflected back into the medium. Moreover, this probability could depend on the direction of incidence of the photon, and the direction of reflection specified by a different probability distribution for each incident direction.

The simplest idealized reflecting boundary is *diffuse*: the probability distribution for reflected photons is independent of the direction of incidence and the reflected radiation is isotropic. With these assumptions the probability distribution for reflection directions is

$$p(\vartheta, \varphi) = \frac{\cos\vartheta\sin\vartheta}{\pi}, \tag{6.80}$$

where ϑ is the angle between the normal to the reflecting boundary and the reflection direction and φ is its azimuthal angle. The factor π ensures that p is normalized:

$$\int_0^{2\pi}\int_0^{\pi/2} p(\vartheta, \varphi)\, d\vartheta\, d\varphi = 1. \tag{6.81}$$

If a (albedo or reflectivity) is the probability that an incident photon is reflected, to determine if a photon is absorbed at a boundary choose a random number ξ between 0 and 1. If $\xi > a$, the photon is absorbed. If not absorbed, it is reflected, and it follows from Eqs. (6.49) and (6.80) that the distribution of reflected azimuthal angles is specified by

$$\varphi = 2\pi\xi \tag{6.82}$$

and of co-latitudes by

$$\vartheta = \sin^{-1}\left(\sqrt{\xi}\right). \tag{6.83}$$

6.3.5   Emission

As noted in Section 6.1.2, emission may be negligible at a particular wavelength even though absorption is not because of the Planck function multiplying the absorption coefficient in the equation of transfer [Eq. (6.15)]. This function depends on temperature. For terrestrial temperatures emission at visible and near-visible wavelengths is exceedingly small, and so the Planck function can be neglected. But if scattering is negligible whereas emission is not, the equation of transfer becomes a first-order differential equation

$$\mathbf{\Omega} \cdot \nabla L(\mathbf{\Omega}) = -\kappa L(\mathbf{\Omega}) + \kappa P_{\mathrm{e}}/\pi. \tag{6.84}$$

The term on the left is a directional derivative, the rate of change of L with respect to distance s along each direction, and so we can write this equation as

$$\frac{dL}{ds} = -\kappa L + \kappa P_{\mathrm{e}}/\pi. \tag{6.85}$$

If we transform the independent variable from physical distance s to absorption optical thickness along a given path

$$\tau = \int_0^s \kappa(s') \, ds', \tag{6.86}$$

Eq. (6.85) becomes

$$\frac{dL}{d\tau} = -L + P_{\mathrm{e}}/\pi, \tag{6.87}$$

the solution to which is

$$L(\tau) = L_0 \exp(-\tau) + \frac{1}{\pi} \int_0^\tau \exp\{-(\tau - \tau')\} P_{\mathrm{e}}(\tau') \, d\tau'. \tag{6.88}$$

The interpretation of this equation is straightforward: the first term is the contribution to the total radiance by attenuation of a known radiance at some reference point (which could be at a boundary); the second term is the contribution from radiation emitted at every point of a path and attenuated from the emission point to the point of interest. Note the similarity between this equation and the two-stream equations for emission in Section 5.4. Keep in mind that L and P_{e} depend on absorption optical thickness along a path in a given direction between two points in the medium. This dependence is, in general, different for every path and may be quite complicated for heterogeneous media. But in principle we can find the radiance in any direction at any point in the medium by integration along a path in that direction beginning at a point where the radiance is known.

But suppose that we want the *irradiance F*. From the point of interest extrapolate backwards on a straight line to a point where the radiance is known and calculate the radiance with Eq. (6.88). This is for only one direction. To estimate the irradiance requires sampling a sufficient number of directions that the quadrature

$$F = \int_{2\pi} L(\mathbf{\Omega}) \, \mathbf{n} \cdot \mathbf{\Omega} \, d\Omega \approx \sum_i \sum_j w_{ij} L(\vartheta_j, \varphi_i), \tag{6.89}$$

is accurate, where (ϑ_j, φ_i) specifies a direction relative to the normal $\mathbf{n}$ to a plane at the point of interest and the w_{ij} are quadrature weights.

The Planck function in Eq. (6.88) depends on the absolute temperature along the integration path. In Earth's atmosphere, temperature usually decreases with height in the troposphere, although the fractional change in absolute temperature is only about 20%. But because of the strong dependence of emission on absolute temperature (see Sec. 1.2.2), the vertical variation of temperature cannot be neglected.

Scattering of terrestrial radiation by clear air usually is negligible. For example, scattering at $10\,\mu m$ is about 10^4 times smaller than at $1\,\mu m$, and hence Eq. (6.88) is applicable to clear air. Scattering of terrestrial radiation by clouds, although often neglected, is not obviously negligible, as evidenced by the single-scattering albedo of a cloud droplet for such radiation (Fig. 5.15). Equation (6.88) is also applicable to a cloudy atmosphere for which scattering is negligible. Any of the methods for solving the plane-parallel radiative transfer equation can be used for a cloudy atmosphere in which neither scattering nor emission are negligible. So this leaves us with only one kind of emitting and scattering medium – vertically and horizontally inhomogeneous – to address. How do we tackle such a medium with the Monte Carlo method?

To answer this question, first consider a negligibly scattering medium (which can be treated by other methods). Divide this medium into small volumes (boxes). The key step is determining emission in many directions for each box. Pick points on the faces of a box at random, and for each point A choose an outward direction at random. Trace a ray in this direction backwards from A until it exits the box at point B. Calculate the radiance emerging at A by evaluating the integral in Eq. (6.88) along the path BA. Do this for enough points and directions to obtain a good estimate of the radiation emitted by each box. The result is a finite set of sources distributed throughout the medium, the radiation from which can be treated as in previous Monte Carlo calculations. This approach was used to calculate the upward and downward clear-sky terrestrial irradiances shown in Fig. 6.20, which are indistinguishable from irradiances obtained from a solution to the plane-parallel equation.

What if a box contains scatterers? The backwards approach still is used, but because of scattering within a box, a bundle follows a zig-zag path from A to B. As with a negligibly scattering box, the contribution to the source is obtained by integration of Eq. (6.88) along this path. Again, the end result is a set of boxes, each a source of radiation to which the usual Monte Carlo method is applied.

Although this scheme for radiative transfer in emitting and scattering media can be described simply and briefly, writing computer programs to obtain numbers requires pain and sweat followed by calculations that may make take weeks using hundreds of billions of photons, as was done for the terrestrial radiation calculations for the horizontally heterogeneous cloud field discussed in Section 6.4.4.

6.3.6 Irradiance, Flux Divergence, Radiance, and Path Lengths

Previous subsections have dealt mostly with how to trace the life histories of photons imagined to be injected into a medium, then scattered or absorbed according to probabilistic equations. These histories then lead to the measurable physical quantities to which we now turn.

Irradiance

For problems in which a scattering and absorbing medium is illuminated by a source of radiation external to it, the general method for estimating irradiances is as follows. The medium of interest could lie within a domain one bounding surface of which is illuminated by irradiance F_i within some spectral interval. Illumination of more than one surface poses no special difficulties, just more bookkeeping. For example, for solar radiation in the atmosphere, F_i could be the solar irradiance in a spectral interval times the cosine of the solar zenith angle, the illuminated surface a horizontal plane at an altitude taken to be the top of the atmosphere. A finite cloud would in general be illuminated on more than one surface. The total spectral power incident on an illuminated boundary with area A_i is $F_i A_i$. Although the incident illumination often is taken to be uniform across A_i, this restriction could be removed by dividing A_i into smaller areas over each of which the illumination is uniform. For example, the bounding surface of a spherical cloud (why not?) could be divided into small areas over each of which the irradiance is different but approximately uniform. If N_b photons are considered to be launched at random across A_i, the incident power per photon is

$$P_b = \frac{F_i A_i}{N_b}. \tag{6.90}$$

This is not the power *of* a photon, a meaningless concept. Although we say that we inject photons into a medium, an intuitively helpful way of speaking, in a strict sense they are bundles of energy or power, and so in what follows we sometimes use the term bundle when we want to make the distinction. Each time a bundle crosses a specified area A within the domain, P_b is added to the bank account for that area. We keep track of the sense in which bundles cross A because in general the irradiance there has two senses defined by two hemispheres of directions. If at the end of a calculation N_A bundles crossed A in a given sense, the total irradiance there is

$$F_A = \frac{N_A P_b}{A} = \frac{N_A F_i A_i}{N_b A}. \tag{6.91}$$

If more than one bounding surface of the domain is illuminated, $F_i A_i$ becomes a sum over each surface. Moreover, one could keep track of the separate contributions of bundles that cross each bounding surface to the physical quantity of interest. If the incident radiation is monodirectional, a single bounding surface is sufficient, and is mathematically equivalent to many surfaces, provided the surface is sufficiently large that the bundles that cross it are equally likely to enter the medium anywhere.

Flux Divergence

Because the rate at which radiant energy is absorbed per unit volume is the negative divergence of the vector irradiance [Eq. (6.18)], we can calculate this quantity directly with the Monte Carlo method. If N_V power bundles out of a total of N_b are absorbed within a volume V

$$-(\nabla \cdot F_V)_V = \frac{N_V P_b}{V} = \frac{N_V F_i A_i}{N_b V}. \tag{6.92}$$

Again, if illumination is over more than one bounding surface, $F_i A_i$ becomes a sum. Equations (6.91) and (6.92) are similar in form. The quantity $F_i A_i$ is a boundary condition, A and V are geometrical quantities specified to taste, and hence the only quantity to be calculated is a ratio of numbers, N_A/N_b or N_V/N_b.

Radiance

Radiances and path lengths are more difficult to calculate by Monte Carlo methods than are irradiances and flux divergences. If we were to naively attempt to estimate the radiance in a particular direction at a point by counting the bundles that cross a small area around that point in a small set of directions around that direction, the computations would take a very long time to yield statistically meaningful results. This is because only a small fraction of bundles, in general, would lie within this set of directions, and so a great number of incident bundles would be needed to reduce the statistical spread (variance). To avoid this problem we use a variance reduction technique: a method for extracting more information from each bundle than by simply tracing each bundle through the domain, thereby reducing noise in the results.

The easiest way to explain the essence of this technique is to consider a simple problem: estimating the vertical radiance distribution at the top of a rectangular, negligibly absorbing cloud illuminated at normal incidence. Power bundles are incident at random points on the top surface A_i of this cloud. An incident bundle travels a certain distance specified by choosing a random number, then scatters in a direction determined by choosing other random numbers. Denote by ϑ_{s1} the first scattering angle and by φ_{s1} the first azimuthal angle. We also can determine a scattering angle ϑ_1 and azimuthal angle φ_1 such that the bundle would be scattered vertically upward. The power scattered per unit solid angle in this hypothetical scattering event and transmitted to the top surface is

$$P_b\, p(\vartheta_1, \varphi_1)\exp(-\tau_1), \tag{6.93}$$

where τ_1 is the optical thickness of the path between the point of scattering and the top surface. The bundle trajectory intersects this surface at a point, and we record its coordinates and the associated power per unit solid angle [Eq. (6.93)], which is proportional to its contribution to the vertical radiance. Now return to the point of first scattering, pick random numbers to determine the distance to the next scattering event and the next scattering direction, and again determine the amount of radiation per unit solid angle scattered vertically and transmitted to the surface. Continue in this vein until the bundle exits the medium. If many bundles are incident randomly across the top surface, the mathematical result is a surface non-uniformly peppered with points (Fig. 6.7). Any planar area A_k on this surface encloses a subset of these points. To obtain an estimate of the average vertical radiance over this area, divide the sum of all terms of the form Eq. (6.93) by A_k:

$$L_k = \frac{P_b}{A_k}\sum_{j=1}^{N_k} p(\vartheta_j, \varphi_j)\exp(-\tau_j) = \frac{F_i A_i}{A_k N_b}\sum_{j=1}^{N_k} p(\vartheta_j, \varphi_j)\exp(-\tau_j), \tag{6.94}$$

where the sum is over all N_k points lying within A_k. This sum embodies all the different ways bundles can contribute to the vertical radiance within A_k.

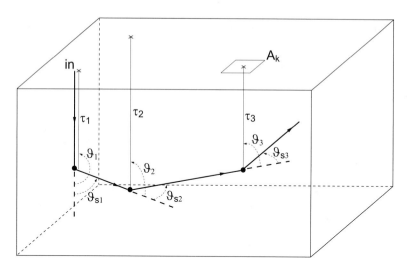

Figure 6.7: Incident power bundles are scattered at various points and in various directions ϑ_{sj} determined by choosing random numbers. At each of these points we can determine the contribution to the radiance at the top surface, an optical distance τ_j, if a bundle had been scattered in the direction ϑ_j such that it is perpendicular to this surface. Many incident bundles result in a surface peppered with points, only three of which are shown. The number of points enclosed within any area A_k is proportional to the average radiance (here vertical) over this area.

To make this clearer, we reconsider the first scattering event. Although the scattering angle was ϑ_{s1}, it *could* have been ϑ_1, because both angles are chosen in the same way, with a random number. Similarly, we could have chosen a random number that gave at least a value τ_1, an optical thickness such that a vertical bundle would be transmitted to the surface. The same arguments hold at the point of the second scattering event, the third, and so on. Thus the sum in Eq. (6.94) is taken over many physically possible ways in which a bundle could emerge vertically from the top surface. Choose enough random numbers specifying directions and distances and most of these would be realized.

For N_b sufficiently large, N_k is approximately proportional to N_b, and to A_k if it is sufficiently large. We must choose the ratio A_k/A_i to be sufficiently large so that we obtain a good estimate of the radiance, but not so large that this is not a local radiance. For example, if $A_k/A_i = 1$, Eq. (6.94) would yield an average vertical radiance over the entire top surface. The downward radiance below cloud is calculated in the same way. For oblique incidence and radiances other than in the vertical, the procedure is the same, the bookkeeping more cumbersome. Choose a desired direction for the radiance at the upper or lower surface, determine the intersection points of bundles in that direction with these surfaces, calculate quantities of the form Eq. (6.93), and sum over all of them associated with points lying within some area. To obtain the radiance divide this sum by this area projected onto the radiance direction. More complicated geometries can be treated at the expense of more complicated bookkeeping. Clouds of strange and wondrous shapes could be treated by imagining them to be enclosed in rectangular boxes illuminated by a specified radiation field. Calculate the radiance distribution over the box and use the radiance invariance principle to extrapolate to points on the cloud surface.

Photon Path Length Distributions

Many photon path length distributions are possible depending on whatever conditions we impose. The simplest example is a plane-parallel medium illuminated at an arbitrary angle. With the Monte Carlo method we can keep track of the total path lengths, from entry point to exit point, of bundles that emerge from the medium at its upper or lower boundaries. Each bundle yields a different path length. We pay no heed to the directions at which the bundles emerge. We keep track of these path lengths and assign them to bins of specified width. Divide the number of times a path length falls within a particular bin by the bin width to obtain an estimate of the continuous probability distribution for photon paths. All these path lengths are equally weighted.

But suppose we wanted to restrict ourselves to path lengths of only those bundles that emerge in a particular direction, the vertical, say, for a given illumination. We could proceed as we did in the previous subsection about radiances. At each scattering point, we know the total path length up to that point and the vertical path length to the surface. Weight the sum of the total path length to the scattering point and the vertical path length by the associated radiance. Do this for an incident bundle until it emerges from the medium. Repeat again and again, and then divide the set of path lengths by the sum of all radiances. This gives a radiance-weighted set of path lengths for those bundles that contribute to the vertical radiance at the top (or bottom) surface. Bin all these path lengths and divide the number in each bin by the bin width to obtain the probability distribution.

We could be even more specific and determine the path length distribution corresponding to all bundles that enter a medium, which need not be plane-parallel, only within a specified area, and emerge in a particular direction through yet another specified area. Thus there is no such thing as *the* photon path length distribution but rather many such distributions, one for each set of constraints imposed on incident and exiting photons. All these distributions may bear a family resemblance for a given medium but they will not be identical.

6.4   Atmospheric Applications of the Monte Carlo Method

The Monte Carlo method can be applied to many problems of radiative transfer in planetary atmospheres. In the following sections we give a sample of such applications. We begin with irradiances in plane-parallel media, which are amenable to treatment by other methods, then show photon path length distributions, reflection and transmission by finite clouds, and finally irradiance, flux divergence, and heating rate profiles in a clear and in a horizontally heterogeneous cloudy atmosphere. For all these Monte Carlo calculations, the Henyey–Greenstein phase function (Sec. 6.3.2) was used.

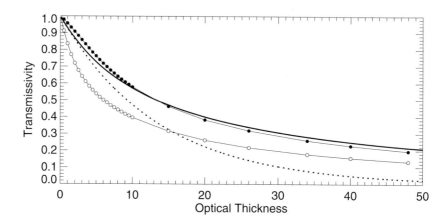

Figure 6.8: Transmissivity of a plane-parallel medium above a nonreflecting surface. The dotted curve is exponential attenuation, the thick solid curve with no circles is from the two-stream theory, and the solid circles are Monte Carlo calculations, all for an overhead sun. Open circles are Monte Carlo calculations for a 60° solar zenith angle. The asymmetry parameter is 0.85.

6.4.1   Irradiances in Plane-Parallel Media

Perhaps the simplest Monte Carlo calculations are of irradiances reflected by , transmitted by, and within a plane-parallel medium. An incident photon that emerges at any direction from the illuminated (top) boundary of such a medium is counted as contributing to the reflectivity, and we are spared the bother of keeping track of this direction. And similarly for photons emerging from the bottom boundary.

Figure 6.8 shows the transmissivity of a negligibly absorbing medium with properties similar to those of clouds at visible wavelengths calculated by the Monte Carlo method and compared with that calculated by the two-stream theory of Section 5.2. The two methods yield almost identical results, which gives us some confidence in the simple two-stream theory. Of course, this theory is limited to normal incidence, whereas the Monte Carlo method is indifferent to the direction of the incident photons. Figure 6.8 also shows transmissivity for a 60° solar zenith angle, which is never greater than that for overhead illumination. Can you give a simple physical explanation for this?

Monte Carlo calculations of the diffuse irradiance transmitted by a cloud are no more difficult than calculations of the total transmitted irradiance: incident photons that make a direct transit through the cloud are not counted as contributing to the diffuse irradiance. Figure 6.9 shows the diffuse downward irradiance below a plane-parallel medium (cloud) calculated by the Monte Carlo method and compared with that calculated by the two-stream theory. The agreement is quite good. Again, the two-stream theory is limited to normal incidence whereas the Monte Carlo method is not. This figure also shows the diffuse downward irradiance for a 60° solar zenith angle, which is greater than that for 0° illumination except at the smallest optical thicknesses.

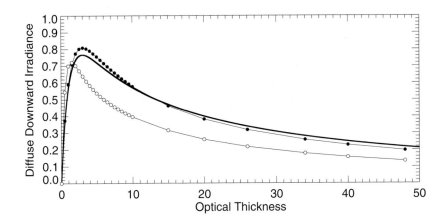

Figure 6.9: Diffuse downward irradiance normalized by the incident irradiance from the two-stream theory (thick solid line with no circles) and Monte Carlo calculations (solid circles) for an overhead sun. Open circles are Monte Carlo calculations for a $60°$ solar zenith angle. The asymmetry parameter is 0.85 and the medium is above a non-reflecting surface.

Figure 6.8 shows total transmissivity and Fig. 6.9 diffuse transmissivity. For small optical thickness (≈ 0) the two are quite different: ≈ 1 for the former, ≈ 0 for the latter. With increasing optical thickness the two approach each other, as expected. For sufficiently large optical thickness essentially all the transmitted radiation is diffuse because of exponential attenuation of the direct radiation.

According to the two-stream theory, both upward and downward irradiances within a negligibly absorbing medium decrease linearly with increasing optical depth. The Monte Carlo calculations in Fig. 6.10, however, show that both irradiance profiles display some curvature, especially near the upper boundary. Moreover, for an optically thick medium, the curvature can be so extreme that the downward irradiance within the cloud can exceed the incident irradiance. This does not violate conservation of energy, which requires only that the difference between the two irradiances be constant with optical depth and the downward irradiance never be less than the upward, conditions the irradiances in Fig. 6.10 do indeed satisfy. But as Fig. 6.11 shows, this increase of the downward irradiance above the incident irradiance near the upper boundary occurs only for illumination near normal incidence.

Real clouds – horizontally and vertically inhomogeneous and without sharp boundaries – are vastly more complicated than mathematical clouds, and calculating irradiances within and outside such clouds is much easier than measuring these quantities for real clouds. How well do calculated irradiances compare with measured irradiances? Figure 6.12 shows irradiances in the middle of the visible spectrum for a stratocumulus cloud about 700 m thick and with an estimated optical thickness of 10–20. Unlike the mathematical clouds for which previous calculations were done, this real cloud is not vertically homogeneous. Nevertheless, the upward and downward measured irradiances are not jarringly different from the calculated ones. Both upward and downward irradiances decrease approximately linearly with increasing op-

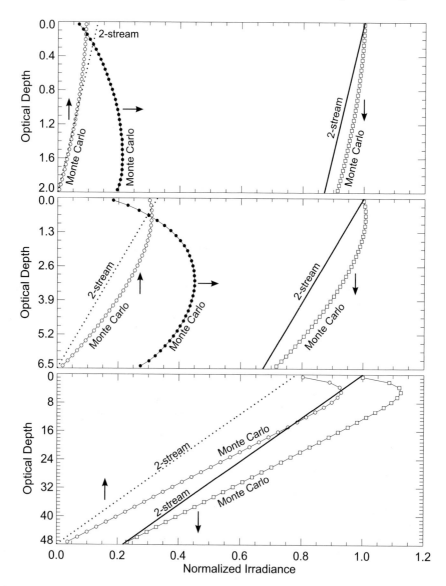

Figure 6.10: Downward (↓) and upward (↑) normalized irradiances within negligibly absorbing clouds of optical thickness 2 (top), 6.5 (center), and 48 (bottom) from two-stream theory and the Monte Carlo method. The asymmetry parameter is 0.85, the sun is overhead, and the surface below the clouds is nonreflecting. Horizontal irradiances are denoted by →.

tical depth, the downward irradiance is greater than the upward, and the difference between them is approximately constant. And the relative difference between them is in rough accord with what we would expect from Fig. 6.10, which shows a steadily decreasing relative differ-

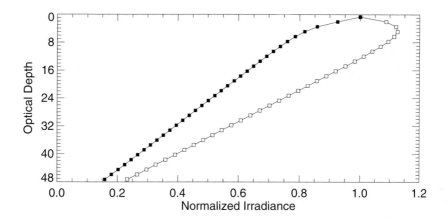

Figure 6.11: Downward normalized irradiance within a negligibly absorbing cloud with total optical thickness 48 and asymmetry parameter 0.85 calculated by the Monte Carlo method. Open squares are for the sun overhead, solid squares for $60°$ solar zenith angle. The surface below cloud is nonreflecting.

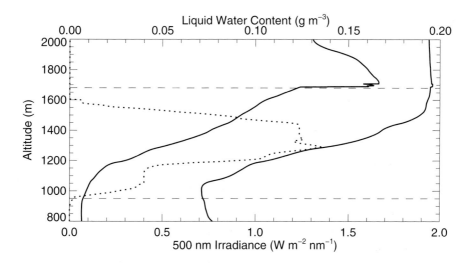

Figure 6.12: Measured downward (right solid curve) and upward (left solid curve) irradiances at 500 nm within a stratocumulus cloud between about 950 m and 1700 m (dashed lines). The solar zenith angle was approximately $10°$. The dotted curve is liquid water content. These curves were obtained from as-yet unpublished measurements provided by Peter Pilewskie.

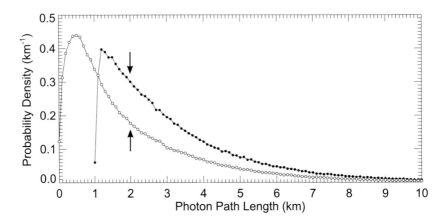

Figure 6.13: Probability density for path lengths of photons that contribute to the downward radiance below cloud (↓) and to the upward radiance above cloud (↑). The incident radiation is normal to a plane-parallel cloud of physical thickness 1 km, scattering mean free path 1/16 km, and asymmetry parameter 0.75. The cloud overlies nonreflecting ground.

ence with increasing optical thickness, about 80% for an optical thickness of 2, about 60% for an optical thickness 6.5, and about 20% for an optical thickness of 48, whereas the relative difference is about 40% for the real cloud with an estimated optical thickness of 10–20.

6.4.2   Photon Path Lengths

Everything takes time. Nothing occurs in an instant. And yet up to this point time has been absent from our analyses. We have implicitly assumed, for example, that reflected photons appear instantaneously at a detector when a cloud is illuminated. This is usually an acceptable fiction because relevant distances, say of order kilometers, divided by the speed of light correspond to times of order millionths of a second. But suppose that a cloud is illuminated by a short duration pulse, as from a laser beam. A pulse width in length units may be of order 100 m, which corresponds to a pulse width in time units of order 0.1 μs. We may think of an idealized pulse as having a square shape: the beam is turned on instantaneously, is constant for a fixed time (pulse width), then is turned off instantaneously. Suppose that such a pulse illuminates a cloud illuminated at normal incidence. What is the shape of the reflected pulse in the backward direction and the transmitted pulse in the forward direction? All photons that contribute to the reflected and transmitted radiances are born equal but do not suffer the same fates in the cloud. Because "time and chance happeneth to them all" pulses are broadened or stretched by an amount depending on the scattering properties of the cloud. Calculating this pulse broadening is a task to which the Monte Carlo method is well suited.

Figure 6.13 shows calculated probability distributions of path lengths for those normally incident photons that contribute to the upward radiance above a plane-parallel cloud and to the downward radiance below the cloud; the negligibly absorbing cloud is 1 km thick with a scattering mean free path 1/16 km and asymmetry parameter 0.75. The most probable path length of photons contributing to the downward radiance is close to 1 km. Most of this ra-

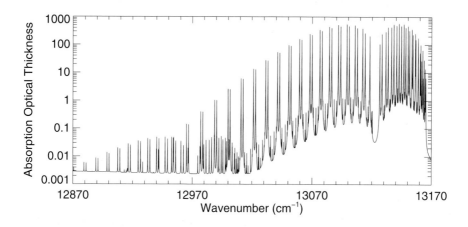

Figure 6.14: Absorption optical thickness of oxygen for 100 km of Earth's atmosphere. From Qilong Min.

diance comes from photons that make a direct path through the cloud. The most probable path length of photons contributing to the upward radiance is about half the thickness of the cloud. For both radiances the pulse is stretched considerably, about a factor of 20 or so for a pulse width of 100 m. Pulse broadening sets an upper limit on the pulse repetition rate (time between pulses) but also contains information about the cloud.

Photon path length distributions also can help us understand atmospheric radiative transfer. As an example, we consider upward and downward radiances at two adjacent wavenumbers in the near-infrared (wavelengths near 770 nm) where absorption in Earth's clear atmosphere is mostly by molecular oxygen. The absorption optical thickness of oxygen up to an altitude of 100 km is shown in Fig. 6.14. At wavenumber 12970 cm^{-1} the absorption optical thickness is negligible whereas at the nearby wavenumber 12965 cm^{-1} the absorption optical thickness is about 0.11. Monte Carlo calculations of the relative radiance difference $(L_a - L_n)/L_n$, where L_a is the radiance at the absorbing wavenumber and L_n is the radiance at the negligibly absorbing wavenumber, for the upward radiance at 100 km and the downward radiance at 0 km are shown in Fig. 6.15 for clear sky, a cloudy sky with a single cloud of scattering optical thickness 16 between 1.0 km and 1.6 km, and a cloudy sky with same total optical thickness but distributed differently: a low cloud of scattering optical thickness 8 between 1.0 km and 1.6 km and a high cloud of optical thickness 8 between 8.6 km and 10 km. The cloud is negligibly absorbing and has an asymmetry parameter 0.75.

For the clear sky the relative radiance difference at the two adjacent wavenumbers is almost the same, about 10%, for the upward (100 km) and downward (0 km) radiances and is negative because radiances are less at the absorbing wavenumber. A single low cloud changes these results: both relative radiance differences increase in magnitude and by approximately the same amount. But the same total cloud, when distributed equally between high and low altitudes, decreases the magnitude of the relative upward radiance difference and markedly increases the magnitude of the relative downward radiance difference. Why?

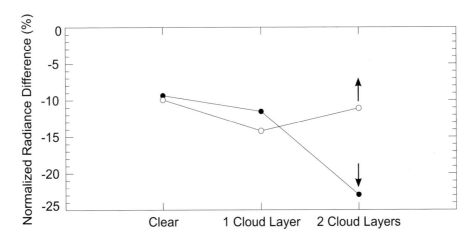

Figure 6.15: Relative difference of radiances at two adjacent wavenumbers in the near-infrared, one for which absorption is negligible and one for which the absorption optical thickness, a consequence of absorption by molecular oxygen, is 0.11. This difference for the upward radiance at 100 km is denoted by ↑, that for the downward radiance at 0 km is denoted by ↓. The cloud layers are negligibly absorbing and have the same total optical thickness (16) but are distributed differently: a single layer between 1.0 km and 1.6 km and two layers of equal optical thickness, one between 1.0 km and 1.6 km, one between 8.6 km and 10 km.

To answer this we appeal to photon path length distributions. Figure 6.16 shows distributions for photons that contribute to the downward radiance at the negligibly absorbing wavenumber for the clear and two cloudy skies and at the absorbing waveneumber for the cloudy sky with equal optical thickness low and high clouds. The source of illumination is at 100 km and is normal to the plane-parallel atmosphere.

For the clear sky almost all path lengths of photons that contribute to the downward radiance are 100 km and hence so is the average path length. A single cloud layer changes the distribution but not by much. Two widely separated cloud layers, however, markedly change the photon path length distribution because of multiple scattering within each cloud and, more important, multiple scattering between clouds. This is why the downward radiance at the absorbing wavelength decreases. Longer path lengths expose photons to more chances of being absorbed by molecular oxygen.

The magnitude of the upward relative radiance difference with two cloud layers is less than that with one cloud because scattering by the high altitude cloud shields some incident photons from the most absorbing part of the atmosphere. Also shown in Fig. 6.16 is the photon path length distribution for the downward radiance at the absorbing wavenumber, which is not appreciably different from that at the negligibly absorbing wavenumber. Thus although we show mostly photon path length distributions at the negligibly absorbing wavenumber, our conclusions are valid for both wavenumbers.

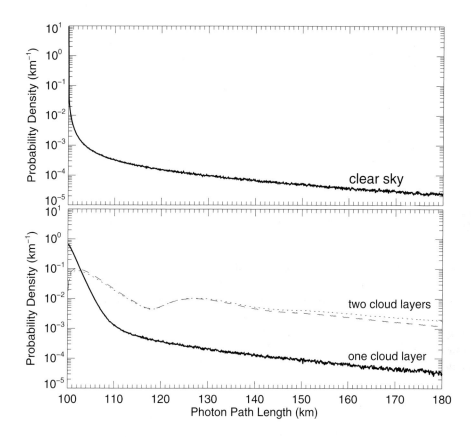

Figure 6.16: Path length distributions for photons contributing to the downward radiance at the ground for clear sky and two cloudy skies, one with a negligibly absorbing cloud between 1.0 km and 1.6 km, and one with two clouds of the same total optical thickness (16) equally shared by a low cloud (1.0–1.6 km) and a high cloud (8.6–10 km). Radiation is normally incident on a plane-parallel atmosphere at 100 km. Except for the dashed curve in the bottom panel, the wavenumber is that for which absorption is negligible. For the dashed curve, the molecular oxygen optical thickness is 0.11.

6.4.3 Three-dimensional Clouds

The qualifier "three-dimensional" is in a sense redundant give that *all* clouds are three-dimensional. Their properties vary in all three spatial directions. A cloud with properties varying in only one direction is a figment of the imagination of modelers. A one-dimensional cloud is an idealization never realized in nature. This should always be kept in mind when assessing the pronouncements of modelers, who sometimes confuse model clouds with real ones. Plane-parallel homogeneous clouds are so much easier to deal with. Even though they don't exist, they ought to. Nor do rectangular clouds exist, but they are a closer match to those that do.

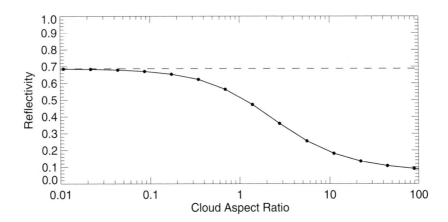

Figure 6.17: Reflectivity (solid curve) for normal incidence of a negligibly absorbing columnar cloud with square cross section, vertical optical thickness 16, overlying a nonreflecting surface. The aspect ratio is the vertical geometrical thickness of the cloud relative to the length of its side. Reflectivity (dashed curve) of a cloud with the same vertical optical thickness but infinite in lateral extent. The asymmetry parameter is 0.75.

Figure 6.17 shows the reflectivity of a square columnar cloud of fixed vertical optical thickness but varying aspect ratio, the ratio of its geometrical thickness to the length of one of its sides. Absorption is negligible, the ground below cloud is non-reflecting, and incident radiation is vertically downward. For a sufficiently small aspect ratio, the reflectivity is close to that for a cloud infinite in lateral extent. For aspect ratios greater than about 1, the difference between a finite and an infinite cloud, otherwise identical, becomes appreciable. For large aspect ratios the reflectivity plunges because of leakage of photons out the sides of the cloud. You can observe the consequences of this while flying over a field of broken clouds, all with about the same vertical thickness but different lateral dimensions. The larger clouds are brighter than the smaller clouds, not because they mysteriously have become corrupted by some highly absorbing pollutant (an explanation we have encountered) but because leakage out the sides of a cloud is equivalent to absorption within it.

Figure 6.18 shows how leaked photons contribute to the irradiance distribution below cloud. For the highest aspect ratio (11), little radiation is transmitted into the geometrical shadow of the cloud, most incident radiation either reflected or transmitted through the sides. For the smallest aspect ratio (0.09), the irradiance in the shadow of the cloud is almost what would be transmitted if the cloud were infinite in lateral extent. For an appreciable distance beyond the geometrical shadow of this cloud the downward irradiance is 5–10% greater than the clear sky value. See Problem 4.15 for more about irradiances greater than clear sky values on days with broken clouds.

Now we turn to vertical radiances above and below cloud. Again, the physical thickness is 1 km, the optical thickness 16. Figure 6.19 shows radiances for clouds with four aspect ratios. For small aspect ratios ($\ll 1$), the radiance hardly varies from cloud center to edge. And this is also true for large aspect ratios (> 1), although the magnitude of the radiance is

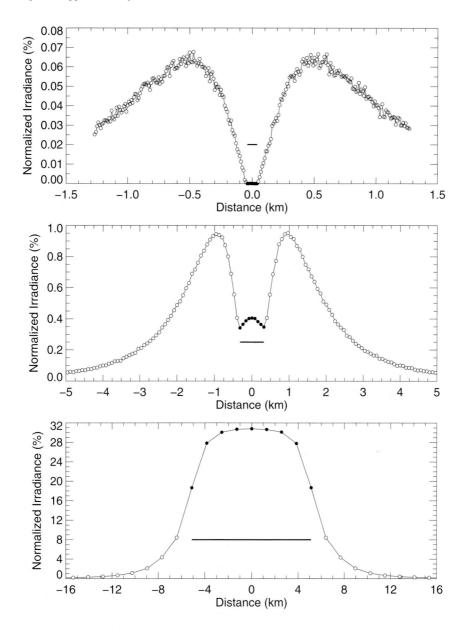

Figure 6.18: Downward irradiance distribution along a line formed by the intersection with the ground of a vertical plane through the center of columnar clouds 1 km thick with vertical optical thickness 16 and asymmetry parameter 0.75 but different aspect ratios: 11 (top), 1.4 (middle), 0.09 (bottom). The incident radiation is directly overhead and the irradiances are normalized by the incident irradiance. Outside the geometrical shadow region, shown by horizontal lines, the incident irradiance is subtracted from the total (open circles).

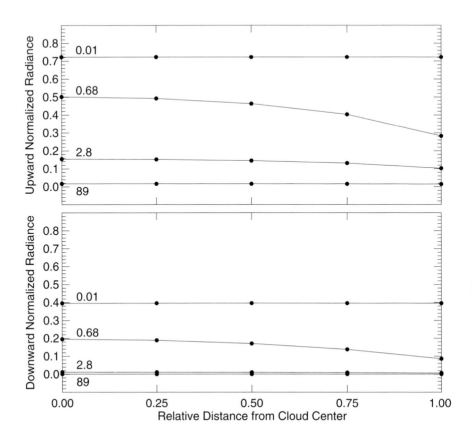

Figure 6.19: Upward radiance above and downward radiance below square clouds 1 km thick with vertical optical thickness 16 and asymmetry parameter 0.75 and with the aspect ratios indicated. The horizontal axis is the position along a line formed by the intersection of a vertical plane through cloud center with the top and bottom boundaries. Cloud center is indicated by 0, cloud edge by 1. The downward radiances for the two largest aspect ratios are nearly identical. The normalization factor is the radiance of a diffuse reflector with reflectivity 1. Radiances calculated by Jonathan Petters for a plane-parallel cloud and a one-dimensional solution to the radiative transfer equation agree to at least three digits with Monte Carlo calculations for the 0.01 aspect ratio cloud.

much smaller. For an aspect ratio near 1, however, the gradient of radiance is appreciable, with higher radiances near the center than near the edge. This can be observed flying over broken clouds with comparable vertical and horizontal dimensions, and has nothing fundamental to do with a possibly different drop size distribution and liquid water content near cloud edge. For the almost plane-parallel cloud (smallest aspect ratio), we can estimate the vertical radiance by other than the Monte Carlo method. Radiances obtained both ways are in agreement, which gives us confidence that the scheme outlined in Section 6.3.6 for estimating radiances is sound.

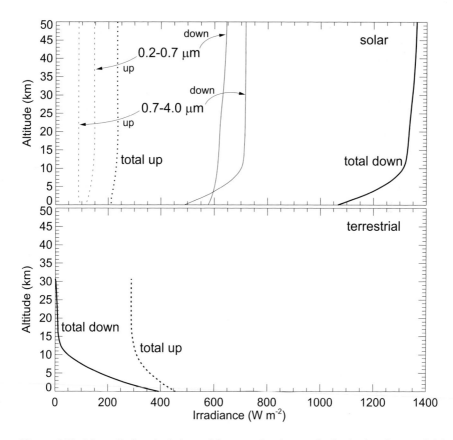

Figure 6.20: Monte Carlo calculations of downward and upward solar (top) and terrestrial (bottom) irradiance profiles for a typical tropical clear-sky atmosphere. Temperature and moisture profiles from Barker *et al.* (2003). The sun is overhead and the reflectivity of the diffusely reflecting ground is 0.2 over the solar spectrum. Terrestrial irradiances calculated using a standard method for plane-parallel media are indistinguishable from these Monte Carlo calculations. Terrestrial radiation results from Cole (2005).

6.4.4   Solar and Terrestrial Irradiances, Flux Divergences, and Heating Rate Profiles

For Earth's thermodynamic internal energy to be constant over a sufficiently long time (e.g., a year) solar radiation absorbed by Earth must be equal to terrestrial radiation emitted by it. This was expressed by the radiative equilibrium equations in Section 1.6. These are what might be called zero-dimensional or global equations, specifying how *much* is absorbed and emitted but not *where*. Now we turn to "where" in the atmosphere but only in one dimension, the vertical.

Figure 6.20 shows Monte Carlo calculations of downward and upward solar and terrestrial irradiance profiles in a clear sky for a temperature and moisture profile typical of the tropics.

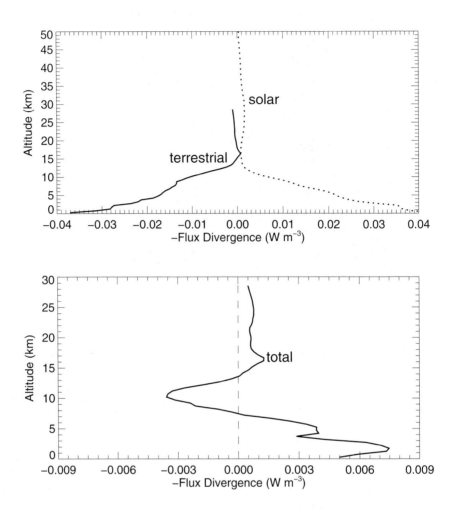

Figure 6.21: Solar, terrestrial, and total *negative* flux divergences corresponding to the irradiances in Fig. 6.20. Positive values correspond to net heating, negative to net cooling. Note the change in both horizontal and vertical scales between the two panels. Terrestrial radiation results provided by Jason Cole.

Separate profiles are shown for the solar irradiance between 0.2 μm and 0.7 μm, denoted for brevity as visible, and between 0.7 μm and 4.0 μm, denoted as infrared. The sharp decrease in the downward solar infrared irradiance below 10 km must be a consequence of absorption by something because the scattering optical thickness of the atmosphere is so small at these wavelengths. Figure 2.12 indicates that the likely culprit is water vapor. Between about 50 km and 15 km the visible downward irradiance decreases faster than the infrared irradiance because of absorption of ultraviolet (0.2–0.3 μm) radiation by ozone. The visible upward irradiance

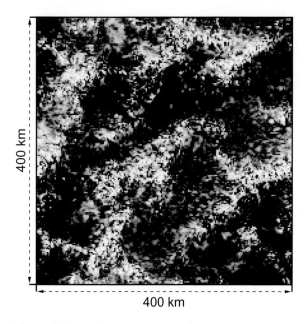

Figure 6.22: Depiction by shading of spatially varying integrated cloud liquid water content, from the surface to about 100 km, obtained from model calculations by Grabowski et al. (1998). The higher the brightness, the higher the liquid water content. The horizontal resolution is 2 km.

from 15 km to 50 km does not decrease with increasing altitude because most of the photons that ozone could absorb are removed on their downward trip through the atmosphere.

The irradiances in Fig. 6.20 as well as the flux divergences and heating rates in following paragraphs are spectrally *integrated* quantities. Easier said than done. Any of the absorption spectra in Chapter 2 or, closer to hand, Fig. 6.14, convey the unavoidable fact that Monte Carlo calculations – indeed, any calculations – of absorption at wavenumber intervals sufficiently small to resolve all absorption peaks and troughs over solar and terrestrial spectra would require computing time not measured in seconds, minutes, or even hours but days, weeks, months, years, possibly lifetimes. So we have to compromise and use absorption coefficients suitably averaged over a small number of bands: 32 solar bands, 12 terrestrial bands for the results in this section.

It is not irradiances *per se* that result in local atmospheric heating or cooling but rather their spatial rate of change, that is, solar and terrestrial flux divergences, which are shown in Fig. 6.21 for the irradiance profiles in Fig. 6.20. A negative flux divergence corresponds to heating (net transfer of radiant energy into a volume), a positive flux divergence corresponds to cooling (net transfer of radiant energy out of a volume). Almost all the action occurs in the lower 30 km of the atmosphere. Below about 15 km heating by absorption of solar radiation rapidly increases as does cooling by emission of terrestrial radiation. The two divergences are almost equal and opposite. The total flux divergence is shown on an expanded horizontal scale. Between about 8 km and 14 km the total flux divergence is positive, negative between about 8 km and the ground. The vertically integrated total flux divergence is negative because this profile is for a clear sky at solar noon.

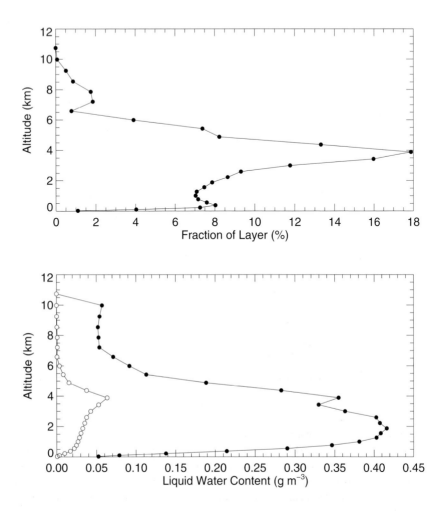

Figure 6.23: Liquid water profile for the cloud field depicted in Fig. 6.22. The top panel shows, for each layer (altitude) of boxes, the fraction of the total number of boxes that contains liquid water. The bottom panel shows the average liquid water content calculated two ways: an average over all boxes at each altitude (open circles); an average over only those boxes at each altitude that contain liquid water (closed circles).

Although the profiles in Figs. 6.20 and 6.21 were obtained by Monte Carlo calculations, they could have been obtained, and with much less time and effort, using any of the many methods for solving plane-parallel problems. Such methods, however, cannot be used to calculate profiles for a horizontally heterogeneous cloud field such as that shown in Fig. 6.22. This figure depicts by brightness differences (shading) horizontal variations in integrated liquid water content, obtained from model calculations, in a region 400 km on a side and 100 km deep

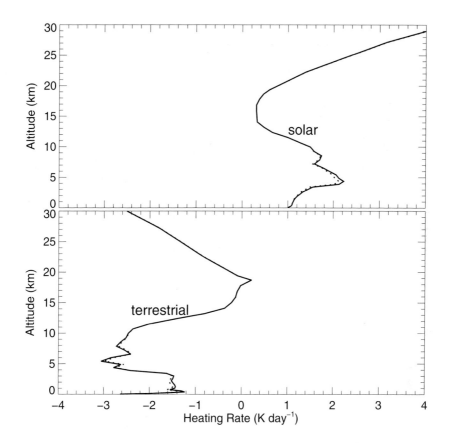

Figure 6.24: Horizontally averaged heating rates for solar (top) and terrestrial radiation (bottom) calculated by the Monte Carlo method for the cloud field depicted in Fig. 6.22. The solar zenith angle is $60°$ and the reflectivity of the diffusely reflecting ground is 0.2 at all solar wavelengths. The dots show results obtained from one-dimensional radiative transfer calculations in which each 2 km by 2 km vertical column with 44 levels is treated as infinite in lateral extent and an average calculated for the 40,000 vertical columns. Terrestrial radiation results from Cole (2005).

composed of 40,000 columns with 44 levels. Thus the region is subdivided into 1,760,000 rectangular boxes each 2 km on a side but with variable thicknesses because equal vertical distances do not correspond to equal masses. Fewer than about 4% of the boxes contain liquid water, and as evidenced by Fig. 6.22, the horizontal distribution of liquid water is not uniform. Figure 6.23 shows the vertical distribution of liquid water, which peaks at a few kilometers and vanishes above about 10 km.

Vertical heating rates, expressed as a rate of temperature change (assuming no net evaporation or condensation), arithmetically averaged at each altitude over the entire region, are shown in Fig. 6.24 for both solar and terrestrial radiation. For each of the 40,000 vertical columns with 44 levels profiles also were calculated by the Monte Carlo method treating each

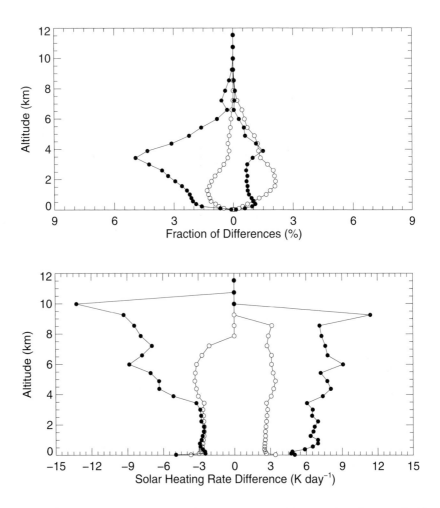

Figure 6.25: Fraction at each altitude of the total number of boxes with statistically significant differences between solar radiation heating rates computed by the Monte Carlo method applied to the full domain and applied to each vertical column as if it were infinite in lateral extent for the cloud field depicted in Fig. 6.22 (top). The arithmetic average of the significant heating rate differences for the boxes in each layer (bottom); the bin size is $0.5\,\mathrm{K\,day^{-1}}$. Solid circles are for boxes containing cloud liquid water, and open circles are for boxes with no liquid water.

column as if it were infinite in lateral extent, and hence independent of all other columns, then all these profiles arithmetically averaged. Agreement between the two profiles is surprisingly good – on average.

The Monte Carlo method is a statistical sampling technique that gives only *estimates* of quantities (e.g., heating rates) that in principle could be obtained by solving deterministic

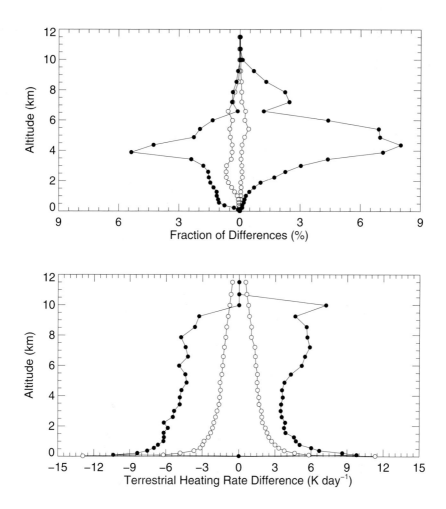

Figure 6.26: Fraction at each altitude of the total number of boxes with statistically significant differences between terrestrial radiation heating rates computed by the Monte Carlo method applied to the full domain and applied to each vertical column as if it were infinite in lateral extent for the cloud field depicted in Fig. 6.22 (top). The arithmetic average of the significant heating rate differences for the boxes in each layer (bottom); the bin size is $0.5\,\mathrm{K\,day^{-1}}$. Solid circles are for boxes containing cloud liquid water, and open circles are for boxes with no liquid water. Terrestrial radiation results provided by Jason Cole.

equations such as Eq. (6.15). Wherever there are estimates, uncertainties in their values are lurking in the background. A single Monte Carlo calculation gives an estimate, similar calculations give different estimates (see Fig. 6.3), and hence curves obtained from such calculations are plots of estimated *mean* values even if not explicitly noted. If one does a Monte

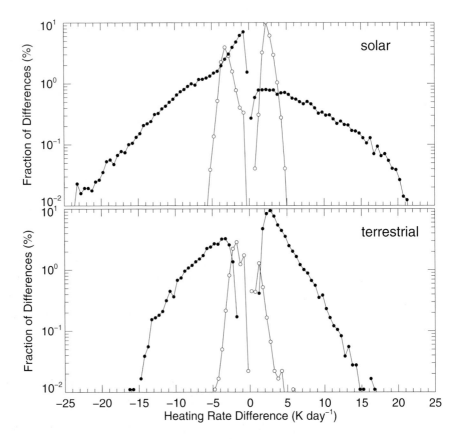

Figure 6.27: Fraction of the total number of boxes with statistically significant differences between radiative heating rates computed by the Monte Carlo method applied to the full domain and applied to each vertical column as if it were infinite in lateral extent for the cloud field depicted in Fig. 6.22. Solid circles are for boxes containing cloud liquid water, open circles are for boxes with no liquid water. The bin size is $0.5\,\mathrm{K\,day^{-1}}$. Terrestrial radiation results provided by Jason Cole.

Carlo calculation of, say, reflected irradiance by following the history of 10^8 photons, then does this again and again and again, each result will be slightly different. By doing several sets of such calculations, say 10 or more, one can calculate a mean and its standard deviation.

This was done to obtain the solar and terrestrial heating rate profiles (Fig. 6.24) for the heterogeneous cloud field depicted in Fig. 6.22. And the horizontally averaged heating rate profiles were surprisingly close – on average. But, how do the heating rates computed the two different ways compare from one box to the next? In each solar wavelength band 10^8 photons were injected at random points into the top (at 100 km) of the 400 km square cloud field; the solar zenith angle was $60°$. A set of calculations was repeated 10 times and a heating rate mean and its standard deviation determined for each of the 1,760,000 boxes. Then we did 10 sets of Monte Carlo calculations for each of the 40,000 vertical columns taken to be infinite in lateral extent and with $10^8/4 \times 10^4 = 2.5 \times 10^3$ incident photons in each solar band and again computed a heating rate mean and standard deviation. Because Monte Carlo calculations give only estimated means with their standard deviations we have to decide whether a calculated difference for a box between the two calculations is statistically significant.

Let $\overline{Q}$ be the mean heating rate and σ its standard deviation for any of the 1,760,000 boxes calculated by assuming that each column is infinite in lateral extent. Let Q_c be the mean heating rate and σ_c its standard deviation for that same box calculated by the Monte Carlo method for the three-dimensional heterogeneous cloud field. If $\overline{Q} + 3\sigma < Q_c + 3\sigma_c$, we take the statistically significant difference to be $\overline{Q} - Q_c$ (negative). If $\overline{Q} - 3\sigma > Q_c + 3\sigma_c$, we take the statistically significant difference to be $\overline{Q} - Q_c$ (positive). Solar heating rate differences are statistically significant for 56,690 boxes, 35,761 in cloudy sky, 20,292 in clear sky, all below the highest cloudy boxes. Terrestrial heating rate differences are statistically significant for 18,098 boxes, 15,885 in cloudy sky, 2,213 in clear sky.

Figure 6.25 shows the vertical distribution of the solar heating rate differences for both cloudy and clear sky and Fig. 6.26 shows the vertical distribution of the terrestrial heating rate differences. Figure 6.27 shows how the statistically significant solar and terrestrial heating rate differences are apportioned over all boxes, those with and without cloud water. The conclusions to be drawn from all these figures are that although treating each column in a heterogeneous cloud field as if it were infinite in lateral extent and then averaging over all columns agrees well with the average for the three-dimensional cloud field, box by box differences can be large, are mostly below the highest clouds, are greater for cloudy than for clear regions, and greater for solar radiation than for terrestrial radiation. All of these results are expected on physical grounds: clouds throw a monkey wrench into one-dimensional calculations, the more so at solar wavelengths because scattering by clouds at these wavelengths is not negligible. The average over columns does not, in general, represent any particular box well. The extent to which this affects climate and weather prediction by models is difficult to say, a variation on the perennial question, Do many wrongs make a right?

References and Suggestions for Further Reading

The classic – a much overused word – treatise on radiative transfer is by Subrahmanyan Chandrasekhar, 1960: *Radiative Transfer*. Dover. This is mathematically formidable, not the first place for a neophyte to go to learn the rudiments. Less formidable are the two volumes by another of the giants of the field, Hendrick C. van de Hulst, 1980: *Multiple Light Scattering: Tables, Formulas, and Applications*. 2 Vols., Academic.

Two textbooks at about the same level as this one are Jacqueline Lenoble, 1993: *Atmospheric Radiative Transfer*. A. Deepak; Grant William Petty, 2004: *A First Course in Atmospheric Radiation*. Sundog Pub. More advanced textbooks are Gary E. Thomas and Knut Stamnes, 1999: *Radiative Transfer in the Atmosphere and Ocean*. Cambridge University Press, and K. N. Liou, 2002: *An Introduction to Atmospheric Radiation*, 2nd ed., Academic Press.

For atmospheric radiative transfer applied to remote sensing see Graeme L. Stephens, 1994: *Remote Sensing of the Lower Atmosphere: An Introduction*. Oxford University Press.

Early papers by such household (if inhabited by astronomers) names as Schuster, Eddington, Schwarzschild, Rosseland and Milne are reprinted in Donald H. Menzel, 1966: *Selected Papers on Transfer of Radiation*. Dover.

One of the methods most widely used in atmospheric science for solving the one-dimensional radiative transfer equation for scattering and emitting media is described by Knut Stamnes, Si-Chee Tsay, Warren Wiscombe, and Kolf Jayaweera, 1988: Numerically stable algorithm for discrete-ordinate-method radiative transfer in multiple scattering and emitting layered media. *Applied Optics*, Vol. 27, pp. 2502–9.

Radiative transfer for which the diffusion approximation is adequate can take advantage of the great number of solutions to problems of diffusion of matter and of energy, for example, see Horatio S. Carslaw and John C. Jaeger, 1959: *Conduction of Heat in Solids*, 2nd ed., Oxford , John Crank, 1975: *The Mathematics of Diffusion*, 2nd ed., Oxford University Press.

Transport of neutrons is formally similar to transport of photons, and the diffusion approximation for neutrons has been used extensively for more than half a century in the design of nuclear reactors. For two good treatments of neutron diffusion theory see Samuel Glasstone and Milton C. Edlund, 1952: *The Elements of Nuclear Reactor Theory*. D. Van Nostrand, Ch. V; Robert V. Meghreblian and David K. Holmes, 1960: *Reactor Analysis*, McGraw-Hill, Ch. 5.

For a brief, elementary treatment of the Monte Carlo method (although not applied to radiative transfer) see I. M. Sobol', 1974: *The Monte Carlo Method*. University of Chicago Press. For an advanced treatise on this method applied to atmospheric radiation see Guri I. Marchuk., Gennadi A. Mikhailov, Magamedshafi A. Nazaraliev, Radzmik A. Darbinjan, Boris A. Kargin, and Boris S. Elepov, 1980: *The Monte Carlo Methods in Atmospheric Optics*. Springer.

For a clear discussion of the Monte Carlo method applied to engineering radiative transfer see Michael F. Modest, 1993: *Radiative Heat Transfer*. McGraw-Hill, Ch. 19. This book also has chapters on other methods for solving the radiative transfer equation in emitting and scattering media.

For more about subtracting the forward peak from angular scattering and reducing the scattering coefficient correspondingly (Sec. 5.3.2) see John F. Potter, 1970: The delta function approximation in radiative transfer theory. *Journal of the Atmospheric Sciences*, Vol. 27, pp. 943–9, and Warren J. Wiscombe, 1977: The delta-M method: Rapid yet accurate radiative flux calculations for strongly asymmetric phase functions. *Journal of the Atmospheric Sciences*, Vol. 34, pp. 1408–22.

For an application of the Monte Carlo method to the transport of radiation from lightning within a cloud see Larry W. Thomason and E. Philip Krider, 1982: The effects of clouds on the light produced by lightning. *Journal of the Atmospheric Sciences*, Vol. 39, pp. 2051–65.

The instrument used to make the irradiance measurements in Fig. 6.12 is described in Peter Pilewskie, John Pommier, Robert Bergstrom, Warren Gore, Steve Howard, Maura Rabbette, Beat Schmid, Peter V. Hobbs, and Si-Chee Tsay, 2003: Solar spectral radiative forcing during the Southern African Regional Science Initiative. *Journal of Geophysical Research*, Vol. 108, pp. 8486–92. The instrument used to measure droplet concentrations for deriving the liquid water content profile in Fig. 6.12 is described in Darrel Baumgardner, Haflidi Jonsson, William Dawson, Darren O'Connor, and R. Newton, 2001: The cloud, aerosol and precipitation spectrometer: A new instrument for cloud investigations, *Atmospheric Research*, Vol. 59, pp. 251–64. The experiment in which the measurements in Fig. 6.12 were obtained is described by Graham Feingold, Reinhold Furrer, Peter Pilewskie, Lorraine A. Remer, Qilong Min, and Haflidi Jonsson, 2005: Aerosol indirect effect studies at Southern Great Plains during the May 2003 intensive operations period: Optimal estimation of drop-size from multiple instruments. *Journal of Geophysical Research* (in press).

The Monte Carlo method for horizontally and vertically inhomogeneous media that emit and scatter, outlined in Section 6.3.5, is described in more detail by Jason N. S. Cole, 2005: Assessing the importance of unresolved cloud-radiation interactions in atmospheric global climate models using the multiscale modelling framework. Doctoral Thesis, Department of Meteorology, The Pennsylvania State University. Computer codes described in this thesis were used for the terrestrial radiation calculations shown in Figs. 6.20, 6.21, 6.24, 6.26 and 6.27.

The method within the atmospheric sciences in widest use for treating three-dimensional emitting and scattering media is by K. Franklin Evans, 1998: The spherical harmonics discrete ordinate method for three-dimensional atmospheric radiative transfer. *Journal of the Atmospheric Sciences*, Vol. 55, pp. 429–46.

The scheme for efficiently calculating radiances with the Monte Carlo method is discussed by K. Franklin Evans and Alexander Marshak, 2005: Numerical methods, in Alexander Marshak and Anthony B. Davis Eds., *3D Radiative Transfer in Cloudy Atmospheres*, Ch. 4, Springer.

Time-resolved radiative transfer is not a pipe dream, not only possible but with practical applications. For example, the shadow of an illuminated object embedded in a turbid medium is blurred because of multiply scattered radiation. But this radiation takes longer to reach a detector, and hence can be partly eliminated by ultrafast illumination and detection. See, for example, K. M. Yoo, B. B. Das, and R. R. Alfano, 1992: Imaging of a translucent object hidden in a highly scattering medium from the early portion of the diffuse component of a transmitted ultrafast laser pulse. *Optics Letters*, Vol. 17, pp. 958–60 and references cited therein.

For an overview of photon path lengths and their relevance to atmospheric radiative transfer see Graeme L. Stephens, Andrew K. Heidinger, and Philip M. Gabriel, 2005: Photon paths and cloud heterogeneity: An observational strategy to assess effects of 3D geometry on radiative transfer, in Alexander Marshak and Anthony B. Davis, Eds., *3D Radiative Transfer in Cloudy Atmospheres*, Ch. 13, Springer.

Some of the first surface-based instruments developed specifically for photon path length studies in clear and cloudy atmospheres are discussed by Klaus Pfeilsticker, Frank Erle, Oliver Funk, Hansjörg Veitel, and Ulrich Platt, 1998: First geometrical pathlengths probability density function derivation of the skylight from spectroscopically highly resolving oxygen A-band observations – 1. Measurement technique, atmospheric observations and model calculations. *Journal of Geophysical Research-Atmospheres*, Vol. 103, pp. 11483–504, and Qilong Min and Lee C. Harrison, 1999: Joint statistics of photon pathlength and cloud optical depth. *Geophysical Research Letters*, Vol. 26, pp. 1425–8.

The horizontally inhomogeneous cloud field depicted in Fig. 6.22 was obtained from model calculations by Wojciech W. Grabowski, Xiaoqing Wu, Mitchell W. Moncrieff, and William D. Hall, 1998: Cloud-Resolving modeling of cloud systems during Phase II of GATE. Part II. Effects of resolution and the third spatial dimension. *Journal of the Atmospheric Sciences*, Vol. 55, pp. 3264–82.

The moisture and temperature profiles used for the calculations in Fig. 6.20 are from Howard W. Barker and 31 co-authors, 2003: Assessing 1D atmospheric solar radiative transfer models: interpretation and handling of unresolved clouds. *Journal of Climate*, Vol. 16, pp. 2676–99.

What are called correlated-k methods often are used in atmospheric science for averaging over fine-structured absorption spectra to obtain broad-band absorption coefficients. The variation on this method used for the solar calculations in this chapter is from Seiji Kato, Thomas P. Ackerman, James H. Mather, and Eugene E. Clothiaux, 1999: The k-distribution method and correlated-k approximation for a shortwave radiative transfer model. *Journal of Quantitative Spectroscopy and Radiative Transfer*. Vol. 62, pp. 109–21. The method used at terrestrial wavelengths is that by Qiang Fu and Kuo-Nan Liou, 1992: On the correlated k-distribution method for radiative transfer in non-homogeneous atmospheres. *Journal of the Atmospheric Sciences*, Vol. 49, pp. 2139–56. Another variation on this method is by Eli Mlawer, Steven J. Taubman, Patrick D. Brown, Michael J. Iacono, and Shepard A. Clough, 1997: Radiative transfer for inhomogeneous atmospheres: RRTM, a validated correlated-k model for the longwave. *Journal of Geophysical Research-Atmospheres*, Vol. 102, pp. 16,663–82.

Correlated-k methods are not the only game in town for reducing the complexity of spectral calculations. For other methods see Andrew A. Lacis and James E. Hansen, 1974: A parameterization for the absorption of solar radiation in the Earth's atmosphere. *Journal of the Atmospheric Sciences*, Vol. 31, pp. 118–33 and Bruce P. Briegleb, 1992: Delta-Eddington approximation for solar radiation in the NCAR community climate model. *Journal of Geophysical Research-Atmospheres*, Vol. 97, pp. 7603–12.

Problems

6.1. To gain confidence that the N-stream equation [Eq. (6.2)] is correct (within the limits of the underlying approximations), show that it is correct for four streams. For simplicity take two directions in the downward hemisphere, two in the upward hemisphere, and the cosines in the downward direction equal in magnitude but opposite in sign to those in the upward hemisphere. Don't forget that attenuation is along a direction of propagation, which corresponds to the z direction only for light directed upward or downward.

6.2. Show that the Henyey–Greenstein phase function [Eq. (6.59)] is normalized.

6.3. Verify Eq. (6.60).

HINT: This is perhaps done most easily by using the theorem for differentiation under the integral sign:

$$\frac{d}{da} \int f(a, x)\, dx = \int \frac{\partial f}{\partial a}\, dx,$$

where f and its partial derivative are continuous and the limits of integration do not depend on a.

6.4. For the phase function for scattering by a spherical dipole (Prob. 7.18), sometimes called the Rayleigh phase function, derive $\mu(\xi)$ in the same way that Eq. (6.64) is derived.

HINT: Cubic equations are exactly soluble.

6.5. Show that the vectors Eqs. (6.72)–(6.74) form an orthonormal, right-handed system.

6.6. Within the framework of the diffusion approximation, find the rate (volumetric) at which radiant energy is absorbed in an infinite, uniform, isotropic medium at any distance r from a point source (isotropic). A check on your solution is conservation of radiant energy: the rate of radiant energy from the source must be equal to the rate at which energy is absorbed in the entire medium.

HINT: The most difficult part of this problem is determining how to incorporate the point source as a boundary condition. Imagine a small spherical cavity to be carved out of the medium and determine the net rate at which radiant energy leaves the cavity in the limit as its radius approaches zero.

6.7. Use the result of the previous problem to determine the mean distance from the origin a photon travels (*not* the total path length of the photon) before it is absorbed. You can probably make a good guess at the answer without doing any calculations.

6.8. Derive the reflectivity of an infinite, plane-parallel, absorbing medium using diffusion theory and compare this reflectivity with Eq. (5.72) for the two-stream theory. By inspection of the diffusion theory reflectivity you should be able to give rough criteria (i.e., the range of medium properties) for when diffusion theory is definitely *not* a good approximation. After you have done so, try to obtain these criteria solely by physical arguments.

HINT: Reflectivities must be less than 1.

6.9. For what function, and only what function (of three space variables), is the directional derivative the same in all directions?

6.10. Equation (6.25) is a continuity equation for radiant energy. But it cannot be correct in general if the radiation field is explicitly time-dependent. Derive a more general form of this continuity equation. Why is it not, in general, identical in form to the continuity equation in fluid mechanics? When is it identical and why?

HINTS: Use the definition of the vector irradiance, the divergence theorem, and the same kinds of arguments used to derive the continuity equation of fluid mechanics.

6.11. The continuity equation derived in the previous problem is completely general. It does not depend on the equation of radiation transfer Eq. (6.15). To the contrary, this equation must be consistent with the continuity equation, and as it stands it is not. What time-dependent term must be added to the left side of Eq. (6.15) so that it is consistent with the general continuity equation?

6.12. Beginning with the general continuity equation obtained in Problem 6.10, derive a time-dependent diffusion equation for photons. Consider the special case of no absorption. Have you seen this equation before?

6.13. It is not necessary to solve a differential equation in order to obtain some insight from it. For example, beginning with the diffusion equation derived in Problem 6.12 for a nonabsorbing medium you should be able to answer the following question. Suppose that a plane-parallel medium is suddenly illuminated at its upper boundary. Everything takes time. Approximately how long after the illumination is turned on will the medium at a distance h from the boundary be illuminated? If your answer is not what might be expected at first glance, explain.

HINT: This problem entails what the fluid mechanics folks call scale analysis.

6.14. Consider a cylindrical medium of radius a extending from $z = 0$ to ∞. The scattering and absorption properties of the medium are uniform and it is illuminated by a uniform and isotropic source at $z = 0$. The medium is surrounded by empty space, which means that no photons that leak out the sides can return. Using the diffusion approximation, find the rate at which irradiance is attenuated deep within the medium. By deep is meant that z is sufficiently large that attenuation is dominated by a *single* exponential term. Compare this attenuation rate with that for the same medium but infinite in lateral extent ($a \to \infty$). Interpret your result physically. Also, consider the limiting case in which the medium is nonabsorbing.

HINT: This is an advanced problem. To solve it requires knowing how to solve partial differential equations in more than one variable using the method of separation of variables. An outline of the solution is given by Craig F. Bohren and Bruce R. Barkstrom, 1974: Theory of the optical properties of snow. *Journal of Geophysical Research*, Vol. 79, pp. 4527–35.

6.15. If you assume that the area of a circle is proportional to the area of the square that circumscribes it, you can estimate the value of π by generating many points randomly in a square with side one unit long and counting the fraction of points that lie within the circle. Try it.

6.16. By computations similar to those used to obtain π in Problem 6.15 you can estimate the circumference of a circle. Does an additional source of error enter into the estimate for this problem that is absent from Problem 6.15?

6.17. To do Problems 6.20, 6.21, and 6.22 you need a correct algorithm for choosing source points at random on a circular disc. You might think that the way to do this is to choose cylindrical polar coordinates (r, φ) at random. That is, choose r by picking a random number between 0 and 1 (the disc has radius 1) and φ by choosing a random number between 0 and 1 and multiplying it by 2π. Try this. Plot enough points to see a pattern. It is not likely to look random. Why? Can you come up with an algorithm (or even two) that does result in a distribution of points on the disc that at least looks random?

6.18. Use the expression for a small solid angle in spherical coordinates to derive Eq. (6.80). It may be easier to derive this result by considering isotropic emission by a surface.

6.19. Derive equations Eqs. (6.82) and (6.83). Write a Monte Carlo code that demonstrates that these equations do indeed produce the correct distributions.

6.20. Redo Problem 4.63 using the Monte Carlo method to determine the (average) irradiance at any depth z in the black tube. Compare your computational results with the (approximate) analytical expression obtained in that problem.

6.21. As a variation on Problem 6.20, suppose that the walls of the tube are specularly reflecting with a reflectivity less than 100%. For simplicity take this reflectivity to be independent of direction of incidence. Again, determine the average irradiance at a depth z in the tube as a function of reflectivity.

6.22. As a variation on Problem 6.20, suppose that the walls of the tube are diffusely reflecting with a reflectivity less than 100%. For simplicity take this reflectivity to be independent of direction of incidence. Again, determine the average irradiance at a depth z in the tube as a function of reflectivity.

6.23. Within the framework of diffusion theory, find the angular dependence of the radiance reflected by an infinite, absorbing, plane-parallel medium illuminated by irradiance F_0.

6.24. Within the framework of diffusion theory, find the angular dependence of the radiance reflected by a finite, nonabsorbing, plane-parallel medium overlying a medium with an absorptivity of 1 and illuminated by irradiance F_0.

6.25. Suppose that you propose to measure absorption by a medium by measuring the angular dependence of the reflected radiance from it. You argue that the greater the slope of the radiance (relative to the radiance at $\vartheta = \pi/2$) versus $\cos \vartheta$, the more absorbing the medium. Based on inspection of the solution to the two previous diffusion theory problems, what is one of the major drawbacks to your idea?

6.26. The two-stream theory is fundamentally incapable of yielding reflected radiances. What we often do, therefore, is assume that the radiance is isotropic and hence can be obtained from the irradiance. The diffusion approximation is not quite so limited, but can at best yield only an approximate radiance of simple form. Because of these limitations of the two (similar) theories, explain why they are not likely to give good results for optically thin, negligibly absorbing media and for strongly absorbing media.

6.27. Show that within the framework of the diffusion approximation it would be possible to infer ground albedo under a completely overcast sky by measuring the (relative) slope of the curve of radiance versus cosine of the radiance zenith angle (equivalent to the angle between the downward directed normal at cloud base and the direction of the radiance as it leaves the bottom of the cloud). Assume that the clouds are nonabsorbing. This problem was inspired by a discussion on pages 653–4 of Hendrik C. van de Hulst, 1980: *Multiple Light Scattering: Tables, Formulas, and Applications.* Vol. 2, Academic, who in turn was inspired by work (in Russian) by K. S. Shifrin and D. A. Kozhaev.

6.28. The equation of transfer in Section 6.1.2 is for incoherent scattering. We pointed out in Section 3.4.8 that scattering in the exact forward direction by two (or more) particles is in phase regardless of their separation. Thus the theory in this chapter is not applicable to this direction. And the same is true, although perhaps not as obvious, for the exact backward direction. Even multiple scattering by a suspension of randomly distributed particles can give rise to *coherent backscattering*. This has essentially nothing to do with the glory (Sec. 8.4.3) because coherent backscattering can be obtained with particles too small to yield the glory. What is special about the backward direction such that multiply scattered waves can interfere constructively for this direction regardless of the number of scatterings and the separation of the scatterers?

HINT: A sketch of various rays scattered by two or more scatterers is essential as is invoking time reversal (Sec. 1.3). A simple figure is all that is needed. For a good expository article see R. Corey, M. Kissner, and P. Saulnier, 1995: Coherent backscattering of light. *American Journal of Physics*, Vol. 63, pp. 560–4.

6.29. In Problem 6.28, what is the ratio of the radiance in the exact backscattering direction taking into account coherent backscattering relative to the radiance in this direction assuming complete incoherence? What is the error in the reflected irradiance as a result of ignoring coherent backscattering? Is this error of any consequence for irradiances given that the angular region of coherent backscattering is a few mrad.

HINT: For the second part of this problem assume that the incoherent reflected radiance is constant and that the coherent reflected radiance is constant everywhere and equal to the incoherent value except within an angle δ of the backward direction, where it is equal to the value determined in the first part of this problem.

6.30. According to the two-stream equations in Section 5.2 for a nonabsorbing medium, the net irradiance (difference between upward and downward irradiances) is the same for every altitude z. Show that this is a general result for any plane-parallel, nonabsorbing medium.

HINT: Use the equation satisfied by the vector irradiance and the divergence theorem.

6.31. Figure 6.15 shows Monte Carlo calculations of the relative difference in radiances at two closely-spaced frequencies in the near infrared (around 770 nm). The absorption optical thickness is 0.11 at one frequency (moderate absorption), but much smaller at the other frequency (weak absorption). Show that this relative difference for both the upward radiance at the top of the clear atmosphere and the downward radiance at the bottom is approximately equal to the negative of the absorption optical thickness. Does this result square with the detailed Monte Carlo calculations?

HINTS: No elaborate derivation is necessary. Assume negligible multiple scattering. You can derive the radiances by evaluating a path integral of radiance similar to Eq. (8.3) with the addition of absorption. Assume the uniform atmosphere approximation of Section 8.1.1. Both the scattering optical thickness and absorption optical thickness are $\ll 1$.

6.32. According to Fig. 6.20 the upward irradiance for a clear sky is approximately constant with altitude for the infrared part of the solar spectrum but increases slightly with altitude for the combined visible and ultraviolet. Try to explain this by physical arguments. You can check your intuition with the two-stream theory of Section 5.2.

6.33. Show, using the two-stream theory, that the relative values of the upward and downward radiances in Fig. 6.19 for the cloud with the smallest aspect ratio are plausible.

6.34. Explain the sudden rise in the heating rate by solar radiation at altitudes greater than about 20 km in Fig. 6.24.

6.35. We state in Section 6.3.5 that at terrestrial temperatures emission at visible and near-visible wavelengths is exceedingly small. To support this statement determine the ratio of spectral emission at the long-wavelength end of the solar spectrum (often taken to be $2.5\,\mu m$) to emission at the middle of the terrestrial spectrum (often taken to be $10\,\mu m$) at terrestrial temperatures.

6.36. Write a Monte Carlo program for investigating absorption by a cloudy sphere, large compared with the wavelength. See Problem 5.57.

6.37. Why are the Monte Carlo method upward and downward irradiances near cloud top [see Fig. 6.10 (bottom) and Fig. 6.11] greater than the incident irradiance and why does the two-stream theory not predict this?

7 Polarization: The Hidden Variable

We call polarization the hidden variable because it is a property of light not readily observed with the unaided eye. Everyone is aware of variations in color and brightness, so in teaching about these properties of light we can appeal to observations that everyone has made or can make with little effort. Alas, this is not so with the polarization of light, which to observe requires a bit of effort and, more important, a few simple tools. Once you have fully grasped polarized light, however, you sometimes can observe its manifestations with nothing but your eyes. We recommend that you have polarizing sunglasses or a polarizing filter near to hand as you read this chapter so that you can observe for yourselves some of the consequences of polarization we discuss.

Treatments of polarization in elementary physics textbooks sometimes are misleading, usually incomplete, and sometimes wrong or incomprehensible. And as one moves further from physics, into chemistry, biology, geology, and meteorology, what is bad becomes progressively worse. We once surveyed geology textbooks and reference works for discussions of the colored patterns seen when transparent crystals are interposed between crossed polarizing filters (see Sec. 7.1.6). What we found was mostly confusing and often erroneous.

In previous chapters we noted in passing that electromagnetic waves are vector waves but were able to sidestep this and make physical arguments based on scalar waves. For example, the simple phase difference arguments in Chapter 3, so helpful for understanding scattering by particles, are essentially independent of the vector nature of electromagnetic waves. To understand polarization, however, requires us to face this head on.

7.1 The Nature of Polarized Light

The more times you see an explanation of a physical phenomenon or a statement about physical reality, especially in the form of an invariable mantra, especially in a textbook unaccompanied by any qualifications, the more certain you can be that it is wrong. Stated more succinctly, repetition increases the probability of incorrectness. This is a law of almost universal validity. One example is the assertion that the electric and magnetic fields of light waves are *always* perpendicular to each other and to the direction of propagation. Because this assertion has been made so often, without qualification, you can be certain it is wrong. And indeed it is. Ask any electrical engineer who knows something about near fields.

In Section 4.1 we noted that the Poynting vector

$$\mathbf{S} = \mathbf{E} \times \mathbf{H} \qquad (7.1)$$

Fundamentals of Atmospheric Radiation: An Introduction with 400 Problems. Craig F. Bohren and Eugene E. Clothiaux
Copyright © 2006 Wiley-VCH Verlag GmbH & Co. KGaA, Weinheim
ISBN: 3-527-40503-8

specifies the magnitude and direction of energy transport by any electromagnetic field at any point. Both the electric field $\mathbf{E}$ and the magnetic field $\mathbf{H}$ are necessarily perpendicular to $\mathbf{S}$, although they are not necessarily perpendicular to each other or to the direction of propagation (if by which is meant the normal to a surface of constant phase). For example, the fields within an illuminated sphere (indeed, any particle) are not perpendicular to each other, and the concept of a surface of constant phase is meaningless. The fields scattered by a sphere also are not perpendicular to each other except approximately at sufficiently large (compared with the wavelength) distances; and the concept of a surface of constant phase has its limitations. All we can be certain of is that the electric and magnetic fields lie in a plane to which the Poynting vector is perpendicular. It has become the custom to specify the polarization properties of electromagnetic waves by the *electric* field, although the magnetic field would serve just as well, and you occasionally come across works (especially by British authors) in which polarization is based on the magnetic field.

7.1.1   Vibration Ellipse and Ellipsometric Parameters

The only assumption we make at this point is that the electric field is time-harmonic:

$$\mathbf{E}(\mathbf{x}, t) = \mathbf{E}(\mathbf{x}) \exp(-i\omega t), \tag{7.2}$$

where $\mathbf{E}$ is the complex representation of the electric field (see Sec. 2.5). To specify the polarization state of $\mathbf{E}$, however, we need the real field. Because $\mathbf{E}$ lies in a plane perpendicular to $\mathbf{S}$, only two components are needed. We denote two orthogonal unit vectors as $\mathbf{e}_\perp$ and $\mathbf{e}_\parallel$ chosen such that $\mathbf{e}_\perp \times \mathbf{e}_\parallel$ is in the direction of the Poynting vector. It will become apparent when we discuss applications why the coordinate axes are denoted as perpendicular ($\perp$) and parallel ($\parallel$). The field components are the real parts of

$$E_\perp = a_\perp \exp\{-i(\vartheta_\perp + \omega t)\}, \ \ E_\parallel = a_\parallel \exp\{-i(\vartheta_\parallel + \omega t)\}, \tag{7.3}$$

where the amplitudes a and phases ϑ are real functions that may depend on position but not time. Without loss of generality we may take the amplitudes as positive because the field components can be negative by virtue of the phases (i.e., $\cos \pi = -1$).

At a fixed point in space the tip of the electric vector (the point with coordinates given by the real parts of $E_\perp$ and $E_\parallel$) endlessly traces out a closed, bounded curve. When the two phases are equal, this curve is a straight line with slope equal to the ratio of amplitudes. When the phases differ by $\pi/2$, the curve is an ellipse with principal axes aligned along the coordinate axes, where the lengths of the two semi-axes are $a_\perp$ and $a_\parallel$. A circle results when these two amplitudes are equal.

In general, Eq. (7.3) describes an arbitrarily oriented ellipse of arbitrary *ellipticity* (not to be confused with eccentricity), defined as the ratio of the minor to major axis lengths (Fig. 7.1). The *azimuth* of this *vibration ellipse* is the angle between the major axis and a reference axis (e.g., one of the coordinate axes). One more *ellipsometric parameter* of the vibration ellipse is its *handedness*, the rotational sense in which it is traced out in time. There is no universal convention for what is meant by right- and left-handed rotation. Moreover, investigators in a particular field often assume that everyone knows what their convention is so feel no need to state it. We adopt the convention of calling a field right-handed if the vibration ellipse is traced

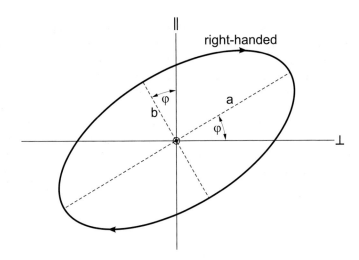

Figure 7.1: A time-harmonic electric field traces out an ellipse specified by its handedness, azimuth φ, and ellipticity b/a.

out clockwise as imagined to be seen looking into the Poynting vector. With this convention the helix traced out in space by the tip of the electric field vector is what all the world calls right-handed.

The electric field described by Eq. (7.3) is 100% or completely polarized in that it has a definite and fixed vibration ellipse. The general state of complete polarization is elliptical, special cases being linear and circular. But some textbooks, and, even more so, books on popular science, convey the notion that by polarization is meant linear polarization, no other kind being conceivable. To make matters worse, linearly polarized light is sometimes called plane polarized, especially in older works. This is a poor choice of terminology on several grounds. If a plane electromagnetic wave [Eq. (7.4)] is linearly polarized we would have the awkward designation plane-polarized plane wave (and polarized parallel or perpendicular to yet another plane). The first plane is defined by the electric field vector and the direction of propagation (equivalently, the plane surface traced out by the field as it propagates), the second plane is a surface of constant phase. To be consistent we would have to describe elliptically polarized light as elliptical-helicoidally polarized light because its electric field traces out an elliptical helicoid in space. Yet this is unnecessary because the polarization state of a plane wave (indeed, any wave) is specified by the ellipsometric parameters, which have nothing essential to do with surfaces.

Our experience has been that people who were taught at an impressionable age that light is plane polarized then find it difficult to understand elliptically polarized light and even more difficult to understand *partially polarized* light. Indeed, they sometimes confuse unpolarized light with circularly polarized light. And yet partially polarized light is readily understood beginning with a firm grasp of completely polarized light. The essential property of such light is *complete* correlation between two orthogonal components of the electric field. They may fluctuate in time, but if they do so synchronously (i.e., the ratio of amplitudes is constant as is the phase difference), the vibration ellipse has a definite and fixed form. Partially polarized light results when there is *partial* correlation between the two orthogonal components; unpolarized light results when there is *no* correlation.

We are aware of the existence of polarization only because two beams, identical in all respects except in one or more ellipsometric parameters (ellipticity, azimuth, handedness) can interact with matter in observably different ways. Were it not for this, the polarization state of time-harmonic fields would be a kind of non-functional adornment, like the fins on 1959 Cadillacs. Only two things can happen to a field when it interacts with matter: its amplitude or phase (or both) are changed. If two orthogonal components of the field are changed *differently*, the polarization state is changed. By "changed" here is meant that the polarization state of an incident (or exciting) wave is different from the polarization state of waves it gives rise to.

7.1.2   Orthogonally Polarized Waves do not Interfere

The general plane harmonic (complex) electric wave has the form

$$\mathbf{E} = \mathbf{E}_0 \exp(i\mathbf{k} \cdot \mathbf{x} - i\omega t), \tag{7.4}$$

where $\mathbf{E}_0$ is constant in space and time and the wave vector $\mathbf{k}$ may be complex; the magnetic field $\mathbf{H}$ is given by a similar expression. These fields must satisfy

$$\mathbf{k} \cdot \mathbf{E} = 0, \ \mathbf{H} = C\mathbf{k} \times \mathbf{E}, \ \mathbf{k} \cdot \mathbf{H} = 0, \tag{7.5}$$

where C is a frequency-dependent parameter (possibly complex if the propagation medium is absorbing) characteristic of the medium in which the wave propagates; its value is of no consequence here. Keep in mind that the real and imaginary parts of $\mathbf{k}$ need not be parallel to each other; that is, the surfaces of constant phase and the surfaces of constant amplitude need not coincide (see Prob. 7.52). If they do, the wave is said to be *homogeneous*; if not, it is *inhomogeneous*. Inhomogeneous waves are not the product of an unbridled imagination. They can be produced readily by illuminating an absorbing medium at oblique incidence. Only if a wave is homogeneous, that is, its wave vector has the form $\mathbf{k} = k\,\mathbf{e}$, where k may be complex but $\mathbf{e}$ is a real unit vector, are the (real) electric and magnetic fields perpendicular to $\mathbf{e}$, the direction of propagation. And only if k is real are the (real) electric and magnetic fields perpendicular to each other. From now on we assume, unless stated otherwise, that the waves of interest are homogeneous ($\mathbf{k} = k\,\mathbf{e}$) and the medium in which they propagate is nonabsorbing (k is real). Absolutely plane waves and nonabsorbing media do not exist. We can get away with assuming they do because measurements of polarization are almost always made in negligibly absorbing media (e.g., air) and at sufficiently large distances (compared with the wavelength) from finite sources (e.g., bounded scatterers) that the fields from them are approximately planar over the detector. But if we were to inquire about polarization of waves in absorbing media or close to sources, much of the following analysis would not be strictly applicable.

Given the assumptions in the previous paragraph, the Poynting vector is

$$\mathbf{S} = C\mathbf{k}(\mathbf{E} \cdot \mathbf{E}), \tag{7.6}$$

which follows from Eqs. (7.1) and (7.5) and the identity

$$\mathbf{A} \times (\mathbf{B} \times \mathbf{C}) = \mathbf{B}(\mathbf{A} \cdot \mathbf{C}) - \mathbf{C}(\mathbf{A} \cdot \mathbf{B}). \tag{7.7}$$

Equation (7.6) is a generalization to vector waves of a result in Section 3.3.2, namely, energy propagation by a scalar plane harmonic wave (on a string) is proportional to the square of a wave function. To avoid cluttered notation we do not use different symbols for fields and their complex representations, trusting that context indicates which is meant.

Two plane harmonic waves are said to be *orthogonally polarized* if they are opposite in handedness and the azimuths of their vibration ellipses are perpendicular. Orthogonally polarized waves do not interfere in that the Poynting vector of their sum is the sum of their Poynting vectors. To prove this, consider two such waves:

$$\mathbf{E}_1 = a_1 \cos \omega t \, \mathbf{e}_\parallel + b_1 \sin \omega t \, \mathbf{e}_\perp, \quad \mathbf{E}_2 = b_2 \sin \omega t \, \mathbf{e}_\parallel + a_2 \cos \omega t \, \mathbf{e}_\perp, \tag{7.8}$$

where a_j and b_j are positive but otherwise arbitrary. The Poynting vector corresponding to the sum of these two waves is

$$\mathbf{S} = C\mathbf{k}[(a_1^2 + a_2^2) \cos^2 \omega t + (b_1^2 + b_2^2) \sin^2 \omega t + 2(a_1 b_2 + a_2 b_1) \sin \omega t \cos \omega t]. \tag{7.9}$$

If the two fields are orthogonal in the restricted sense that $\mathbf{E}_1 \cdot \mathbf{E}_2 = 0$ then $a_1 b_2 + a_2 b_1 = 0$ and the waves do not interfere at *any* instant. Regardless of their state of orthogonal polarization the *time-averaged* Poynting vectors are additive because $\langle \sin \omega t \cos \omega t \rangle = 0$:

$$\langle \mathbf{S} \rangle = \langle \mathbf{S}_1 \rangle + \langle \mathbf{S}_2 \rangle. \tag{7.10}$$

We are usually interested in Poynting vectors averaged over times large compared with the period (inverse frequency) of waves. Although the fields in Eq. (7.1) must be real, we can determine time-averaged Poynting vectors directly from the complex representations of fields:

$$\langle \mathbf{S} \rangle = \frac{1}{2} \Re\{\mathbf{E} \times \mathbf{H}^*\}. \tag{7.11}$$

This equation is valid for any time-harmonic electromagnetic field. A more restricted version, applicable only to plane homogeneous waves in nonabsorbing media, is

$$\langle \mathbf{S} \rangle = \frac{1}{2} C\mathbf{k}(\mathbf{E} \cdot \mathbf{E}^*). \tag{7.12}$$

This equation is at the heart of what follows.

7.1.3   Stokes Parameters and the Ellipsometric Parameters

Although we can imagine watching the tip of an electric vector rotating at, say, 10^{15} Hz, to think that we could actually do so is pure fantasy. All that we can measure, usually, is time-averaged irradiances. Such measurements of the magnitude of the Poynting vector must therefore be the route to ellipsometric parameters, and given that Eq. (7.3) is the equation of an ellipse, they must depend only on the amplitudes $a_\parallel$ and $a_\perp$ and the phases $\vartheta_\parallel$ and $\vartheta_\perp$.

From Eqs. (7.3) and (7.12) the time-averaged irradiance of a beam, denoted here by I, is the sum of squares of amplitudes

$$I = |\langle \mathbf{S} \rangle| = E_\| E_\|^* + E_\perp E_\perp^* = a_\|^2 + a_\perp^2.\tag{7.13}$$

Missing from this equation is a constant factor, which we ignore here because absolute measurements are not needed to determine ellipsometric parameters. To obtain the separate amplitudes we need the help of an *ideal linear polarizer* (or *linear polarizing filter*). Such a filter completely transmits light linearly polarized in a particular direction but does not transmit light linearly polarized in the orthogonal direction. As its name implies, an ideal linear polarizer does not exist but we can come close, at least over a restricted range of wavelengths. An example is the sheet polarizers used in polarizing sunglasses or in polarizing filters for cameras (the function of which is explained in Sec. 7.4). Absorption by such a sheet polarizer is asymmetric in that $\kappa d \ll 1$, where κ is the absorption coefficient and d the thickness of the sheet, for light linearly polarized along the *transmission axis*, whereas $\kappa d \gg 1$ for light linearly polarized perpendicular to this axis. At visible and near-visible wavelengths this difference in absorption coefficients is a consequence of anisotropy of the sheet material on a molecular scale. We cannot see the transmission axis, although we might be able to see that of a polarizing filter for microwave radiation. A medium with different absorption coefficients for different orthogonal linear states of polarization is said to be *linearly dichroic*.

Now we imagine inserting an ideal linear polarizing filter in the beam and measuring transmitted irradiances, first for the transmission axis along $\mathbf{e}_\|$, then along $\mathbf{e}_\perp$, and then subtracting these two irradiances:

$$Q = E_\| E_\|^* - E_\perp E_\perp^* = a_\|^2 - a_\perp^2.\tag{7.14}$$

We now have done enough to obtain the amplitudes:

$$a_\|^2 = \frac{1}{2}(I + Q), \; a_\perp^2 = \frac{1}{2}(I - Q).\tag{7.15}$$

What about the phases? From Eqs. (7.3) and (7.12) it would seem that to obtain phases we must transmit a bit of both orthogonal components of an electric field. For example, if we align a linear polarizing filter with its transmission axis at $45°$ to $\mathbf{e}_\|$, the transmitted amplitude is

$$\frac{1}{\sqrt{2}}(E_\| + E_\perp).\tag{7.16}$$

Rotate the filter by $90°$ and the transmitted amplitude is

$$\frac{1}{\sqrt{2}}(E_\perp - E_\|).\tag{7.17}$$

The difference in the irradiances corresponding to Eqs. (7.16) and (7.17) is

$$U = E_\| E_\perp^* + E_\perp E_\|^* = 2a_\| a_\perp \cos \delta,\tag{7.18}$$

where $\delta = \vartheta_\| - \vartheta_\perp$.

Measurement of I, Q, and U is sufficient to obtain $\cos \delta$, but because $\cos \delta = \cos(-\delta)$ is not sufficient to determine the handedness of the wave. Given $\cos \delta$ we cannot say if δ is positive or negative, which determines handedness. To find this quantity requires the help of *ideal circular polarizers* (or *circular polarizing filters*), devices that completely transmit circularly polarized light of one handedness but do not transmit circularly polarized light of the opposite handedness. Such circular polarizers are much more difficult to find than linear polarizers. Media with different absorption coefficients for different states of circularly polarized light, said to be *circularly dichroic*, exist. For example, our bodies and all organic matter are chock full of helical molecules (e.g., the double helix of DNA), and helices are not superposable on their mirror images: the reflection of a right-handed helix is a left-handed helix. Because of this mirror asymmetry we expect absorption by such molecules to be different for different states of circular polarization. And indeed this is so, but the difference is usually greatest at ultraviolet frequencies and, moreover, media with greatly different absorption coefficients are difficult to find. Nevertheless, we can imagine thought experiments with ideal circular polarizers.

To discuss circularly polarized light it is convenient to introduce a set of complex basis vectors:

$$\mathbf{e}_R = \frac{1}{\sqrt{2}}(\mathbf{e}_\parallel + i\mathbf{e}_\perp), \ \mathbf{e}_L = \frac{1}{\sqrt{2}}(\mathbf{e}_\parallel - i\mathbf{e}_\perp), \tag{7.19}$$

which are orthonormal in that

$$\mathbf{e}_R \cdot \mathbf{e}_R^* = 1, \ \mathbf{e}_L \cdot \mathbf{e}_L^* = 1, \ \mathbf{e}_R \cdot \mathbf{e}_L^* = 0. \tag{7.20}$$

$\mathbf{e}_R$ corresponds to a right-circularly polarized wave of unit amplitude, $\mathbf{e}_L$ to a left-circularly polarized wave. An arbitrary electric field thus can be written

$$\mathbf{E} = E_R \mathbf{e}_R + E_L \mathbf{e}_L, \tag{7.21}$$

where the circularly polarized (complex) amplitudes are related to the linearly polarized amplitudes by

$$E_R = \frac{1}{\sqrt{2}}(E_\parallel - iE_\perp), \ E_L = \frac{1}{\sqrt{2}}(E_\parallel + iE_\perp). \tag{7.22}$$

Now imagine that an ideal right-circular polarizer is inserted in the beam and the transmitted irradiance $E_R E_R^*$ is measured, then the irradiance $E_L E_L^*$ transmitted by an ideal left-circular polarizer is measured, and the second irradiance subtracted from the first:

$$V = E_R E_R^* - E_L E_L^* = i(E_\parallel E_\perp^* - E_\perp E_\parallel^*) = 2\Im\{E_\perp E_\parallel^*\} = 2a_\parallel a_\perp \sin \delta. \tag{7.23}$$

Knowing both $\sin \delta$ and $\cos \delta$ we can determine the sign of δ and hence the handedness of the wave.

The four quantities $\{I, Q, U, V\}$, which are no more than sums and differences of irradiances, are called the *Stokes parameters*, first set down by Sir George Gabriel Stokes in 1852. Even more than 150 years later his paper "On the composition and resolution of streams of

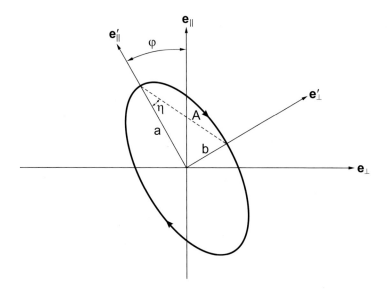

Figure 7.2: The unprimed coordinate system is rotated relative to the primed coordinate system, the axes of which are along the minor axis b and major axis a of the vibration ellipse.

polarized light from different sources" is still worth reading. You are likely to encounter different symbols for the Stokes parameters (he used A, B, C, and D), and linear combinations of Stokes parameters are also valid Stokes parameters. They sometimes are written compactly as a column matrix

$$
\begin{pmatrix} I \\ Q \\ U \\ V \end{pmatrix} = \begin{pmatrix} E_\| E_\|^* + E_\perp E_\perp^* \\ E_\| E_\|^* - E_\perp E_\perp^* \\ E_\| E_\perp^* + E_\perp E_\|^* \\ i(E_\| E_\perp^* - E_\perp E_\|^*) \end{pmatrix} = \begin{pmatrix} a_\|^2 + a_\perp^2 \\ a_\|^2 - a_\perp^2 \\ 2a_\| a_\perp \cos\delta \\ 2a_\| a_\perp \sin\delta \end{pmatrix}, \tag{7.24}
$$

called the *Stokes vector*, although it does not have the same rights and privileges as proper vectors. For example, the Stokes parameters are not independent in that

$$
I^2 = Q^2 + U^2 + V^2. \tag{7.25}
$$

Stokes's paper appeared a dozen years before the publication of Maxwell's famous electromagnetic theory of light and 32 years before the publication of Poynting's work. Much was known about the properties of light waves even before they were fully grounded in an adequate theory.

Now we have to show that the Stokes parameters determine the ellipsometric parameters. In what follows fields are real. The real parts of the fields in Eq. (7.3) can be expanded using the identity for the cosine of the sum of angles and written in matrix form as

$$
\begin{pmatrix} E_\| \\ E_\perp \end{pmatrix} = \begin{pmatrix} a_\| \cos\vartheta_\| & -a_\| \sin\vartheta_\| \\ a_\perp \cos\vartheta_\perp & -a_\perp \sin\vartheta_\perp \end{pmatrix} \begin{pmatrix} \cos\omega t \\ \sin\omega t \end{pmatrix}. \tag{7.26}
$$

In a coordinate system for which the field components are

$$E'_\| = a \cos \omega t, \quad E'_\perp = b \sin \omega t, \tag{7.27}$$

the field traces out a right-handed ellipse with minor axis b, major axis a (if $a > b$), which follows from solving Eq. (7.27) for $\cos \omega t$ and $\sin \omega t$, then squaring and adding them. The transformation from the original coordinate system to the primed coordinate system is (Fig. 7.2)

$$\begin{pmatrix} E_\| \\ E_\perp \end{pmatrix} = \begin{pmatrix} \cos \varphi & \sin \varphi \\ -\sin \varphi & \cos \varphi \end{pmatrix} \begin{pmatrix} E'_\| \\ E'_\perp \end{pmatrix}$$

$$= \begin{pmatrix} a \cos \varphi & b \sin \varphi \\ -a \sin \varphi & b \cos \varphi \end{pmatrix} \begin{pmatrix} \cos \omega t \\ \sin \omega t \end{pmatrix}. \tag{7.28}$$

Equality of Eqs. (7.26) and (7.28) requires that

$$a_\| \cos \vartheta_\| = a \cos \varphi, \tag{7.29}$$

$$-a_\| \sin \vartheta_\| = b \sin \varphi, \tag{7.30}$$

$$a_\perp \cos \vartheta_\perp = -a \sin \varphi, \tag{7.31}$$

$$-a_\perp \sin \vartheta_\perp = b \cos \varphi. \tag{7.32}$$

Square and add the left sides of these equations and set the result equal to the sum of the squares of the right side:

$$a_\|^2 + a_\perp^2 = a^2 + b^2 = I. \tag{7.33}$$

We may write

$$a = A \cos \eta, \quad b = A \sin \eta, \tag{7.34}$$

where $0 \le \eta \le \pi/4$ and $\tan \eta = b/a$. Now multiply Eq. (7.29) by Eq. (7.30), Eq. (7.31) by Eq. (7.32), add the result, and use the identities for the sine and cosine of the sum of angles:

$$U = 2a_\| a_\perp \cos \delta = (b^2 - a^2) \sin 2\varphi = -A^2 \cos 2\eta \sin 2\varphi. \tag{7.35}$$

Now square Eqs. (7.32) and (7.30), take their difference, square Eqs. (7.29) and (7.31), take their difference, and finally take the difference of the result:

$$a_\perp^2 - a_\|^2 = (b^2 - a^2) \cos 2\varphi, \tag{7.36}$$

which yields

$$Q = a_\|^2 - a_\perp^2 = A^2 \cos 2\eta \cos 2\varphi. \tag{7.37}$$

Finally, to obtain V, multiply Eqs. (7.29) and (7.32), Eqs. (7.30) and (7.31), and add to obtain

$$a_\| a_\perp \sin \delta = ab, \tag{7.38}$$

which from Eq. (7.34) yields

$$V = 2a_\parallel a_\perp \sin\delta = A^2 \sin 2\eta. \tag{7.39}$$

If we go through the same steps for a left-circularly polarized wave, I, Q, and U are unchanged whereas V becomes

$$V = -A^2 \sin 2\eta. \tag{7.40}$$

Instead of having separate sets of equations for left- and right-circularly polarized light we can combine them by allowing η to lie in the range $(-\pi/4, \pi/4)$, where negative angles correspond to left-circularly polarized light, positive angles to right-circularly polarized light. To recapitulate:

$$I = A^2, \; Q = A^2 \cos 2\eta \cos 2\varphi, \; U = -A^2 \cos 2\eta \sin 2\varphi, \; V = A^2 \sin 2\eta, \tag{7.41}$$

$$\frac{U}{Q} = -\tan 2\varphi, \quad \frac{V}{\sqrt{Q^2 + U^2}} = \tan 2\eta, \quad \frac{a}{b} = |\tan\eta|, \tag{7.42}$$

where $0 \le \varphi \le \pi$ and $-\pi/4 \le \eta \le \pi/4$. Because $\tan 2\varphi$ does not uniquely determine φ we need additional information: if $U < 0, 0 < \varphi < \pi/2$, whereas if $U > 0, \pi/2 < \varphi < \pi$. From Eq. (7.41) it follows that I and V do not depend on the coordinate system (i.e., on φ), Q and U do, but the sum of their squares does not.

The surfaces of constant phase and amplitude for the plane wave Eq. (7.4) are infinite in extent. Thus the electric field of this wave occupies all space, which, of course, is physically unrealistic. To apply the previous analysis to real beams finite in lateral extent their properties have to be more or less laterally uniform. The Stokes parameters [Eq. (7.24)] were obtained by way of thought experiments that are easy to state but not all of them readily done in practice. Nevertheless, once we know the form of these parameters we can devise feasible ways of measuring them with readily available linear retarders and polarizing filters (see Prob. 7.36 at the end of this chapter).

7.1.4   Unpolarized and Partially Polarized Light

An electric wave described by Eq. (7.2) or Eq. (7.3) is necessarily completely polarized in that its vibration ellipse is traced out with monotonous regularity from the beginning until the end of time (actually, this time interval need not span eternity, just much longer than the period of the wave). Radiation from a microwave or radio antenna might closely fit this description because an antenna is a coherent object, its parts fixed relative to each other (on the scale of the wavelength), driven by electric currents that are more or less time-harmonic. It would take some ingenuity to make a microwave or radio antenna that *did not* radiate completely polarized waves. Radiation at much shorter wavelengths, however, often originates from vast arrays of tiny antennas (molecules) emitting more or less independently of each other, and hence we would not expect the same degree of regularity of the radiation from such sources. The extreme example of irregularity is unpolarized light whereas the extreme example of regularity is completely polarized light, both idealizations never strictly realized in nature. But what is unpolarized light?

Perhaps the simplest way to define such light is operationally, subject to previous caveats about ideal linear and circular polarizers. What kind of experimental tests can we devise to determine if a beam is unpolarized? Suppose that we transmit it through an ideal linear polarizer and discover that regardless of the orientation of its transmission axis, the transmitted irradiance is the same. This implies that there is no preferred direction of the electric field, for if there were the irradiance would vary. According to our operational definition of the Stokes parameters, $Q = U = 0$ for this beam. But wait! A circularly-polarized beam would yield the same result. So we now have to determine if the beam exhibits a preferential handedness. First transmit the beam through an ideal left-circular polarizer, then through a right-circular polarizer. If the two transmitted irradiances are equal, $V = 0$, and the electric field of the beam exhibits no preference for left-handed over right-handed rotation. Thus our operational definition of unpolarized light is that for which $Q = U = V = 0$. The Stokes parameters of *partially polarized* light also do not satisfy Eq. (7.25) but Q, U, V are not all zero.

We can put more theoretical flesh onto these bare bones by extending Eq. (7.3) to *quasi-monochromatic* radiation with (real) electric field components

$$E_\parallel(t) = a_\parallel(t)\cos\{\vartheta_\parallel(t) + \omega t\}, \ \ E_\perp(t) = a_\perp(t)\cos\{\vartheta_\perp(t) + \omega t\}, \tag{7.43}$$

where the amplitudes and phases now vary with time but much more slowly than $\cos\omega t$. With this restriction the electric field Eq. (7.43) and its associated magnetic field *approximately* satisfy Eq. (7.5). The instantaneous Poynting vector corresponding to Eq. (7.43) is

$$\mathbf{S} = C\mathbf{k}(E_\parallel^2 + E_\perp^2), \tag{7.44}$$

the magnitude of which (within a constant factor) is

$$\begin{aligned} |\mathbf{S}| = &\ (a_\parallel^2 \cos^2\vartheta_\parallel + a_\perp^2 \cos^2\vartheta_\perp)\cos^2\omega t \\ &+ (a_\parallel^2 \sin^2\vartheta_\parallel + a_\perp^2 \sin^2\vartheta_\perp)\sin^2\omega t \\ &- 2(a_\parallel^2 \cos\vartheta_\parallel \sin\vartheta_\parallel + a_\perp^2 \cos\vartheta_\perp \sin\vartheta_\perp)\sin\omega t \cos\omega t, \end{aligned} \tag{7.45}$$

where the amplitudes and phases may depend on time (not explicit to keep the notation uncluttered). To determine the time average of Eq. (7.45) requires evaluating integrals of the form

$$\frac{1}{\tau}\int_0^\tau f(t)\cos^2\omega t \, dt, \ \ \frac{1}{\tau}\int_0^\tau g(t)\sin^2\omega t \, dt, \ \ \frac{1}{\tau}\int_0^\tau h(t)\sin\omega t\cos\omega t \, dt. \tag{7.46}$$

We need consider only the first of these integrals because it sets the pattern for the other two.

Divide the range of integration into N equal intervals Δt:

$$\langle f\cos^2\omega t\rangle = \frac{1}{\tau}\int_0^\tau f(t)\cos^2\omega t \, dt = \frac{1}{N\Delta t}\sum_{i=1}^N \int_{t_i}^{t_i+\Delta t} f(t)\cos^2\omega t \, dt. \tag{7.47}$$

From the mean-value theorem of integral calculus

$$\int_{t_i}^{t_i+\Delta t} f(t)\cos^2\omega t \, dt = f(\overline{t_i})\int_{t_i}^{t_i+\Delta t}\cos^2\omega t \, dt, \tag{7.48}$$

where $t_i \leq \bar{t}_i \leq t_i + \Delta t$. By the definition of quasi-monochromatic light we can choose $\Delta t \gg 1/\omega$ such that $f(t)$ is approximately constant [call the value $f(\bar{t}_i)$] over this time interval, and hence the integral of the cosine squared is approximately $\Delta t/2$ and the time average is approximately

$$\langle f \cos^2 \omega t \rangle \approx \frac{1}{2N} \sum_{i=1}^{N} f(\bar{t}_i). \tag{7.49}$$

Similarly,

$$\langle g \sin^2 \omega t \rangle \approx \frac{1}{2N} \sum_{i=1}^{N} g(\bar{t}_i), \quad \langle h \sin \omega t \cos \omega t \rangle \approx 0. \tag{7.50}$$

From Eqs. (7.45), (7.49), and (7.50) it therefore follows that the time-averaged irradiance for quasi-monochromatic light is

$$I = \langle |\mathbf{S}| \rangle = \langle a_\parallel^2 + a_\perp^2 \rangle = \langle E_\parallel E_\parallel^* + E_\perp E_\perp^* \rangle. \tag{7.51}$$

As previously, all common factors are omitted. Because the Stokes parameters are sums and differences of irradiances, the other three parameters for such light are given by similar expressions:

$$Q = \langle E_\parallel E_\parallel^* - E_\perp E_\perp^* \rangle = \langle a_\parallel^2 - a_\perp^2 \rangle, \tag{7.52}$$

$$U = \langle E_\parallel E_\perp^* + E_\perp E_\parallel^* \rangle = \langle 2a_\parallel a_\perp \cos \delta \rangle, \tag{7.53}$$

$$V = \langle i(E_\parallel E_\perp^* - E_\perp E_\parallel^*) \rangle = \langle 2a_\parallel a_\perp \sin \delta \rangle. \tag{7.54}$$

An example of a quasi-monochromatic beam of light is collimated sunlight passed through an ordinary (as opposed to a polarizing) filter, a device that transmits light only over a band of frequencies. The spectral width of this transmitted light may be quite narrow but the amplitudes and phases of its orthogonal field components fluctuate over times large compared with the period and small compared with the response time of the detector.

According to Eq. (7.42) the ellipsometric parameters depend only on ratios of Stokes parameters, which in turn implies that they depend only on the *ratio* of the amplitudes $a_\parallel$ and $a_\perp$ and the *difference* in phases $\delta = \vartheta_\parallel - \vartheta_\perp$. Suppose that these amplitudes and phases fluctuate in time but do so synchronously, that is, they are *correlated*, the ratio of amplitudes and the difference in phases constant in time. It then follows from Eqs. (7.51)–(7.54) that $I^2 = Q^2 + U^2 + V^2$, and hence the light is completely polarized despite the fluctuations. Correlation is essential to understanding polarized light. Complete correlation corresponds to completely polarized light, no correlation to unpolarized light, and partial correlation to partially polarized light.

We may visualize Eq. (7.43) as follows. Over a time interval of several periods the electric field vector traces out a more or less definite vibration ellipse, but with the passage of time the vibration ellipse changes. If all vibration ellipses are traced out over the response time of the detector the light is unpolarized.

7.1.5   Degree of Polarization

Any beam with Stokes parameters I, Q, U, V may be considered the incoherent superposition of two beams, one unpolarized , one completely polarized:

$$
\begin{pmatrix} I \\ Q \\ U \\ V \end{pmatrix} = \begin{pmatrix} I_{\mathrm{u}} \\ 0 \\ 0 \\ 0 \end{pmatrix} + \begin{pmatrix} I_{\mathrm{p}} \\ Q \\ U \\ V \end{pmatrix},
\tag{7.55}
$$

where

$$
I_{\mathrm{p}}^2 = Q^2 + U^2 + V^2.
\tag{7.56}
$$

Because $I_{\mathrm{p}} \leq I$ it follows (but see Prob. 7.27) that

$$
Q^2 + U^2 + V^2 \leq I^2,
\tag{7.57}
$$

equality holding for completely polarized light. We define the *degree of (elliptical) polarization* of this beam as the ratio of the irradiance of the polarized component to the total irradiance:

$$
\frac{I_{\mathrm{p}}}{I_{\mathrm{p}} + I_{\mathrm{u}}} = \frac{\sqrt{Q^2 + U^2 + V^2}}{I},
\tag{7.58}
$$

often multiplied by 100 and expressed as a percentage ($\leq 100\%$).

We can go further and imagine the beam to be a superposition of three beams, one unpolarized, one linearly polarized, and one circularly polarized:

$$
\begin{pmatrix} I \\ Q \\ U \\ V \end{pmatrix} = \begin{pmatrix} I_{\mathrm{u}} \\ 0 \\ 0 \\ 0 \end{pmatrix} + \begin{pmatrix} I_{\mathrm{lp}} \\ Q \\ U \\ 0 \end{pmatrix} + \begin{pmatrix} I_{\mathrm{cp}} \\ 0 \\ 0 \\ V \end{pmatrix},
\tag{7.59}
$$

where

$$
I_{\mathrm{lp}} = \sqrt{Q^2 + U^2}, \ \ I_{\mathrm{cp}} = |V|.
\tag{7.60}
$$

This naturally leads to definitions of the *degree of linear polarization*

$$
\frac{I_{\mathrm{lp}}}{I} = \frac{\sqrt{Q^2 + U^2}}{I},
\tag{7.61}
$$

and the *degree of circular polarization*

$$
\frac{I_{\mathrm{cp}}}{I} = \frac{V}{I},
\tag{7.62}
$$

a signed quantity: positive values correspond to right-circular polarization, negative to left-circular polarization.

To determine the degree of linear polarization of a beam, insert an ideal linear polarizer in it and measure the irradiance of the transmitted light. Suppose that the transmission axis of the polarizer makes an angle ξ (between 0 and π) with the $e_\perp$ axis. The transmitted amplitude along this axis is

$$E_{\|i} \sin \xi + E_{\perp i} \cos \xi, \tag{7.63}$$

where the subscript i denotes components of the incident field. The components of this transmitted field are therefore

$$E_{\perp t} = E_{\|i} \sin \xi \cos \xi + E_{\perp i} \cos^2 \xi, \tag{7.64}$$

$$E_{\|t} = E_{\|i} \sin^2 \xi + E_{\perp i} \sin \xi \cos \xi. \tag{7.65}$$

From these equations and the definition of the Stokes parameters it follows that the transmitted irradiance is

$$I_t = \frac{1}{2}(I_i - Q_i \cos 2\xi + U_i \sin 2\xi). \tag{7.66}$$

The maximum and minimum of I_t occur for

$$\tan 2\xi = -\frac{U_i}{Q_i}. \tag{7.67}$$

The two solutions to Eq. (7.67) are separated by $\pi/2$. Without loss of generality we may take U_i and Q_i to be positive, in which instance the maximum occurs when the cosine is negative and the sine positive, whereas the minimum occurs when the cosine is positive and the sine negative:

$$I_{max} = \frac{1}{2}(I_i - Q_i \cos 2\xi + U_i \sin 2\xi), \tag{7.68}$$

$$I_{min} = \frac{1}{2}(I_i + Q_i \cos 2\xi - U_i \sin 2\xi), \tag{7.69}$$

where ξ is the solution to Eq. (7.67) for which the cosine is negative. Subtract these two equations, add them, and take their ratio:

$$\frac{I_{max} - I_{min}}{I_{max} + I_{min}} = \frac{-Q_i \cos 2\xi + U_i \sin 2\xi}{I_i}. \tag{7.70}$$

Because of Eq. (7.67) we can write

$$Q_i = -A \cos 2\xi, \quad U_i = A \sin 2\xi, \tag{7.71}$$

where

$$A = \sqrt{Q_i^2 + U_i^2}. \tag{7.72}$$

This then yields

$$\frac{\sqrt{Q_i^2 + U_i^2}}{I_i} = \frac{I_{max} - I_{min}}{I_{max} + I_{min}}, \tag{7.73}$$

the degree of linear polarization Eq. (7.61). Now we have a procedure for measuring it: rotate a linear polarizing filter in the beam and measure the minimum and maximum irradiances. We don't need absolute irradiances because the degree of polarization is a ratio.

7.1.6   Linear Retarders and Birefringence

We noted previously that circular polarizing filters are rare. But this does not mean that we cannot readily transform unpolarized light into circularly polarized light. We can by using a sandwich composed of an ideal linear polarizing filter and an ideal *linear retarder*. A linear polarizing filter has different absorption coefficients for different orthogonal states of linear polarization. Stated another way, the imaginary parts of its refractive index are different (see Sec. 3.5.2). Thus it is hardly a stretch to imagine a medium for which the *real* parts are different, called *linear birefringence*.

We take the retarder to be a transparent (at the wavelength of interest) plate of uniform thickness illuminated at normal incidence. One direction in the plane of this plate is called the *slow axis*, the other the *fast axis* (for reasons that will become apparent). With the assumption that reflection by the plate is negligible, the electric field components in the plate along the fast and slow axes are

$$E_f = a_f \exp(ik_f z - i\omega t), \ E_s = a_s \exp(ik_s z - i\omega t). \tag{7.74}$$

The wavenumbers are

$$k_f = \frac{2\pi}{\lambda} n_f, \ k_s = \frac{2\pi}{\lambda} n_s, \tag{7.75}$$

where n_f and n_s are the corresponding (real) refractive indices. Because of the inverse relation between phase speed and refractive index, the fast axis is so named because $n_f < n_s$. The phase difference of the two components after transmission a distance h is

$$\Delta\vartheta = \vartheta_s - \vartheta_f = \frac{2\pi h}{\lambda}(n_s - n_f), \tag{7.76}$$

whence the name retarder: the plate retards one phase relative to the other.

At visible and near-visible wavelengths the difference in refractive indices is a consequence of anisotropy at the molecular level, either because the medium is a crystal (with other than cubic symmetry), which gives *natural linear birefringence*, or because of non-uniform stresses in the medium, which gives *induced linear birefringence*. And, of course, birefringence could be both natural and induced (e.g., a stressed crystal). Ice is a naturally birefringent material; transparent adhesive tape is an induced birefringent material (stresses along the axis of the tape are different from those perpendicular to this axis). Although the difference in refractive indices may be, and usually is $\ll 1$, the phase difference Eq. (7.76) may be appreciable, $\pi/2$ or greater, if the plate is much thicker than the wavelength.

We now have set the stage for showing how to produce circularly polarized light. First an unpolarized beam is transmitted by an ideal linear polarizer, resulting in light linearly polarized along a direction we take to be the $e_\parallel$ axis. Following this polarizer is an ideal linear retarder with its slow and fast axes oriented at $45°$ to the $e_\parallel$ axis (Fig. 7.3). The real field components along the fast and slow axes transmitted by the retarder are

$$E_{ft} = \frac{a}{\sqrt{2}}\cos(\vartheta_f - \omega t), \ E_{st} = \frac{a}{\sqrt{2}}\cos(\vartheta_f - \omega t + \Delta\vartheta). \tag{7.77}$$

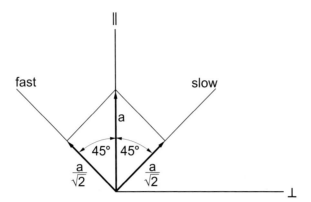

Figure 7.3: An electric field with amplitude a linearly polarized along the $\mathbf{e}_\parallel$ axis is incident on an ideal linear retarder with its slow and fast axes oriented at $45°$ to the $\mathbf{e}_\parallel$ axis. The equal-amplitude components of the incident field along the fast and slow axes undergo different phase shifts upon transmission.

If the thickness h of the retarder is such that $\Delta\vartheta = \pi/2$, these field components are

$$E_{\mathrm{ft}} = \frac{a}{\sqrt{2}} \cos(\vartheta_{\mathrm{f}} - \omega t), \ \ E_{\mathrm{st}} = -\frac{a}{\sqrt{2}} \sin(\vartheta_{\mathrm{f}} - \omega t), \tag{7.78}$$

which corresponds to a right-circularly polarized beam. Rotate the retarder by $90°$ around the axis defined by the beam and the transmitted light is left-circularly polarized. The requirement that the phase difference Eq. (7.76) be $\pi/2$ implies that

$$h(n_{\mathrm{s}} - n_{\mathrm{f}}) = \frac{\lambda}{4}, \tag{7.79}$$

and hence such a retarder is sometimes called a *quarter-wave plate*.

A sandwich composed of a linear polarizing filter and a quarter-wave plate oriented at $45°$ to the transmission axis of the polarizing filter is not a circular polarizing filter. And it is a one-way device: light incident from the polarizer side is transformed into circularly polarized light, whereas light incident from the retarder side is transformed into linearly polarized light.

What is the difference between birefringence and *double refraction*? All doubly refracting media are birefringent but the converse is not necessarily true. A doubly refracting medium is one with a difference in refractive indices so large that perceptible double images can be seen through it. Keep in mind that *all* solid objects made of amorphous materials (e.g., glass, plastic) always have some residual non-uniform stresses, and hence exhibit some induced birefringence. About 30 years ago one of the authors attempted to make a glass container free of birefringence. Even after many hours of annealing, the container still exhibited measurable birefringence, although to the eye it certainly was not doubly refracting.

Now consider another sandwich, a triple-decker composed of a polarizing filter, a linear retarder, and another polarizing filter with its transmission axis perpendicular to that of the first filter. Transmission of unpolarized light by the first filter results in light linearly polarized

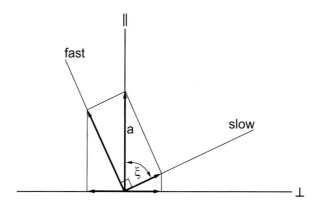

Figure 7.4: The orthogonal slow and fast axes of a linear retarder are rotated by an angle ξ relative to the reference coordinates axes $\perp$ and $\parallel$.

along the $e_\parallel$ axis, say, with amplitude a. The field components along the slow and fast axes of the retarder are

$$a \cos \xi, \ a \sin \xi, \tag{7.80}$$

where ξ is the angle between the slow axis and $e_\parallel$ (Fig. 7.4). After transmission by the retarder these components are

$$a \cos \xi \exp(ik_s h), \ a \sin \xi \exp(ik_f h). \tag{7.81}$$

Only the projections of these components onto the transmission axis of the final polarizing filter are transmitted, and hence the transmitted field is

$$E_t = a \sin \xi \cos \xi \{\exp(ik_s h) - \exp(ik_f h)\}. \tag{7.82}$$

The transmitted irradiance is

$$I_t = E_t E_t^* = 2a^2 \sin^2 \xi \cos^2 \xi \{1 - \cos[(k_s - k_f)h]\}. \tag{7.83}$$

By using the identity

$$1 - \cos x = 2 \sin^2 (x/2), \tag{7.84}$$

we can write Eq. (7.83) as

$$I_t = 4a^2 \sin^2 \xi \cos^2 \xi \sin^2 \delta, \tag{7.85}$$

where the *retardance* is defined as

$$\delta = \frac{\pi h}{\lambda} (n_s - n_f). \tag{7.86}$$

For zero retardance, no light is transmitted by this triple-decker sandwich. Note that the retardance explicitly depends on wavelength by way of the factor $1/\lambda$ and implicitly by way of the wavelength dependence of the two refractive indices. Because of this wavelength dependence colored patterns can be seen when a birefringent medium is between crossed polarizing filters. Charles and Nancy Knight have photographed hailstones between crossed polarizing filters as a way of elucidating how hail is formed. Hailstones are agglomerations of many small crystals of different size and orientation, and hence from Eqs. (7.85) and (7.86) the wavelength dependence of transmission by each crystal is different. As a consequence, a hailstone seen through crossed polarizing filters displays a striking multi-colored mosaic, and the photographs taken by the Knights are as much art as science. Sun dogs (see Sec. 8.5.1) result from scattering of sunlight by ice crystals falling in air. Günther Können has noted that because of the birefringence of ice the (angular) position of a sun dog shifts slightly (a fraction of a degree) when observed through a polarizing filter while it is rotated.

Another observable consequence of birefringence in which the atmosphere plays a role is provided by airplane windows, which are made of tough plastic, several millimeters thick, and highly stressed. As we show in Section 7.3, skylight is partially polarized. Such light transmitted through an airplane window (i.e., retarder), then through a polarizing filter such as polarizing sunglasses or a polarizing filter for a camera, can result in striking colored patterns. You can observe them even without such a filter. As we show in the following section, a specular (mirror-like) reflecting interface is a kind of polarizing filter in that it reflects light of linear orthogonal polarization states differently. Thus a passenger in an airplane may see colored patterns in the reflected image of a window illuminated by skylight. Although we have called polarization the hidden variable, it is not completely hidden from those who know where to look.

7.2   Polarization upon Specular Reflection

As noted at the end of the previous section, specular reflection can change the polarization state of light. Consider two optically homogeneous, isotropic media separated by an optically smooth planar interface. Strictly speaking both media should be infinite for theory to be applicable, but this ideal (and unobtainable) condition is satisfied to good approximation if the dimensions of the media are much larger than the wavelength of the light of interest. A plane wave with wave vector

$$\mathbf{k}_i = -k(\sin\vartheta_i\mathbf{e}_y + \cos\vartheta_i\mathbf{e}_z) \tag{7.87}$$

is incident on the interface from a negligibly absorbing medium (e.g., air) with wavenumber k. This wave gives rise to (i.e., excites) a reflected wave with wave vector $\mathbf{k}_r$ and a transmitted wave with wave vector $\mathbf{k}_t$:

$$\mathbf{k}_r = -k(\sin\vartheta_r\mathbf{e}_y - \cos\vartheta_r\mathbf{e}_z), \;\; \mathbf{k}_t = -k_t(\sin\vartheta_t\mathbf{e}_y + \cos\vartheta_t\mathbf{e}_z), \tag{7.88}$$

where k_t is the wavenumber in the transmitting medium. The *plane of incidence*, which we take to be the yz-plane (Fig. 7.5), is determined by the normal to the interface and the incident

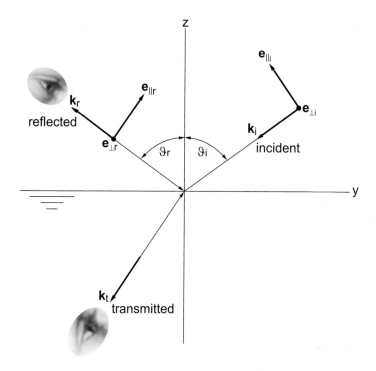

Figure 7.5: A plane wave illuminating the optically smooth interface between optical homogeneous (infinite) media gives rise to (specularly) reflected and transmitted (refracted) waves.

wave vector. All three wave vectors lie in the plane of incidence. Moreover, the angle of reflection equals the angle of incidence (law of specular reflection)

$$\vartheta_r = \vartheta_i \tag{7.89}$$

and ϑ_t is given by the law of refraction (Snel's law)

$$\sin \vartheta_i = \frac{k_t}{k} \sin \vartheta_t = m \sin \vartheta_t, \tag{7.90}$$

where m is the refractive index of the transmitting medium relative to that of the incident medium. Because m may be complex, so may be the angle of refraction, but the real part of this complex angle is *not* the angle of refraction defined as the angle between the real part of the complex wave vector and the normal to the interface. When m is complex, the transmitted wave is inhomogeneous (except for normal incidence, $\vartheta_i = 0$): the real and imaginary parts of the complex wave vector are not parallel.

   Despite what you see elsewhere, the spelling of Snel here is correct. Whatever fame Snel may deserve for discovering (empirically) the law of refraction, which has been disputed, he likely has the dubious honor of having his name misspelled more than that of any other scientist in history. Kirchhoff probably runs a close second. A resurrected Snel, upon seeing his name in hundreds of textbooks and thousands of papers, might exclaim, "What the l!", whereas Kirchhoff's reaction might be, "Where the h?"

The incident and reflected electric field components are specified relative to orthogonal basis vectors parallel and perpendicular to the plane of incidence (Fig. 7.5), defined such that $\mathbf{e}_\perp \times \mathbf{e}_\parallel$ is in the direction of the wave vector. Note that the parallel basis vector for the incident field is *not* the same as that for the reflected field. The complex field components of the reflected field relative to those of the incident field are

$$\tilde{r}_\parallel = \frac{E_{\parallel r}}{E_{\parallel i}} = \frac{m \cos \vartheta_i - \cos \vartheta_t}{m \cos \vartheta_i + \cos \vartheta_t} = \frac{\tan(\vartheta_i - \vartheta_t)}{\tan(\vartheta_i + \vartheta_t)}, \tag{7.91}$$

$$\tilde{r}_\perp = \frac{E_{\perp r}}{E_{\perp i}} = \frac{\cos \vartheta_i - m \cos \vartheta_t}{\cos \vartheta_i + m \cos \vartheta_t} = \frac{\sin(\vartheta_i - \vartheta_t)}{\sin(\vartheta_i + \vartheta_t)}, \tag{7.92}$$

where subscripts i and r denote incident and reflected, respectively. Equations (7.91) and (7.92) are the *Fresnel coefficients*. Derived before the electromagnetic theory of light had been developed (Fresnel died in 1827), they specify the amplitude and phase of the reflected field for any angle of incidence and illuminated medium. The corresponding reflectivities for the two orthogonal polarization states are

$$R_\parallel = \left| \tilde{r}_\parallel \right|^2, \ R_\perp = \left| \tilde{r}_\perp \right|^2 . \tag{7.93}$$

Underlying Eqs. (7.91) and (7.92) is the additional assumption that both media are non-magnetic at the wavelength of interest. At normal incidence ($\vartheta_i = 0°$) both reflectivities are equal,

$$R_\parallel(0°) = R_\perp(0°) = \left| \frac{m - 1}{m + 1} \right|^2, \tag{7.94}$$

and at glancing incidence ($\vartheta_i = 90°$) both are 1 for arbitrary m. But for all intermediate angles of incidence the two are different, as, for example, those for an air–water interface illuminated by visible light (Fig. 7.6). In particular, $R_\parallel = 0$ if

$$\vartheta_i + \vartheta_t = \frac{\pi}{2}, \tag{7.95}$$

whereas $R_\perp$ does not vanish for any angle of incidence. From Eq. (7.90) it follows that Eq. (7.95) is equivalent to

$$\tan \vartheta_i = m. \tag{7.96}$$

Equation (7.96) can be satisfied strictly only for m real.

Because $R_\parallel = 0$ for the angle of incidence satisfied by Eq. (7.96), unpolarized light reflected at this angle is 100% linearly polarized perpendicular to the plane of incidence. This angle is called the *polarizing angle* or the *Brewster angle* to honor Sir David Brewster, who first discovered it empirically. Brewster's paper, published in 1815, has aged well. Written with a clarity almost banned from scientific writing today, it ends with a touching homage to Etienne-Louis Malus, who discovered polarization upon reflection in 1809, coined the term

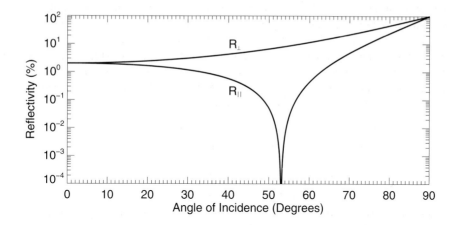

Figure 7.6: Reflectivities of water at visible wavelengths ($n = 1.33$) for incident light polarized parallel ($\parallel$) and perpendicular ($\perp$) to the plane of incidence.

polarization by way of a (faulty) analogy with magnetic poles, but failed to recognize the regular law later enunciated by Brewster: "*The index of refraction is the tangent of the angle of polarisation.*" Brewster measured the polarizing angle for more than a dozen transparent substances and compared its value calculated from the tangent equation, finding agreement, on average, to within 15'. But Brewster knew that he hadn't derived this equation: "In these enquiries I have made use of no hypothetical assumptions... the language of theory has been occasionally employed, but the terms thus introduced are merely expressive of experimental results... When discoveries shall have accumulated a greater number of facts, and connected them together with general laws, we may then safely begin... to speculate respecting the cause of those wonderful phenomena which light exhibits under all its various modifications." These "general laws" had to wait a few more years for Fresnel and half a century for the glorious synthesis by Maxwell.

In a paper with the provocative title "Would Brewster recognize today's Brewster angle?", Akhlesh Lakhtakia critically examined the ways in which the term Brewster angle is used today. In modern textbooks it is defined most often by the condition $R_\parallel = 0$, but Brewster would not have known this and to him the angle now bearing his name is the angle of incidence (polarizing angle) for which reflected light is completely linearly polarized given incident unpolarized light. Moreover, Lakhtakia showed that there are isotropic media for which the polarizing angle and the angle of zero reflection for the parallel component are *not* the same.

You sometimes encounter the assertion that reflection by metals does not polarize incident unpolarized light. Although Eq. (7.96) does not have a real solution when the imaginary part of m is not negligible, metals do exhibit an angle of reflection, sometimes called the *pseudo-Brewster angle*, at which the degree of polarization is a maximum. And it can be surprisingly high, well over 50%, especially for metals (e.g., iron and chromium) with reflectivities lower than those of more conductive metals such as silver and aluminum. One reason for the misconception that light reflected by metals is unpolarized is that the pseudo-Brewster angle is typically within $10°$ or so from glancing incidence.

Yet another misconception about polarization is that light emitted by incandescent bodies is unpolarized, which may have its origins in the fact that radiation emitted by blackbodies (which don't exist) is unpolarized. Consider an optically smooth and homogeneous medium in air, sufficiently thick that transmission by it is negligible. With this assumption, emissivity is 1 minus reflectivity (see Sec. 1.4.1), and because reflectivity depends on polarization so does emissivity:

$$\varepsilon_\parallel = 1 - R_\parallel, \ \varepsilon_\perp = 1 - R_\perp. \tag{7.97}$$

These two emissivities are equal for normal and glancing directions, but for real bodies are unequal in all intermediate directions. Both are 1 in all directions for a blackbody ($R_\parallel = R_\perp = 0$).

Although the emphasis in this section is on polarization upon specular reflection, because it is relatively easy to observe, the state of polarization of incident radiation also can be changed upon transmission. The Fresnel coefficients for reflection, Eqs. (7.91) and (7.92), are accompanied by two for transmission:

$$\tilde{t}_\parallel = \frac{E_{\parallel t}}{E_{\parallel i}} = \frac{2\cos\vartheta_i}{m\cos\vartheta_i + \cos\vartheta_t}, \tag{7.98}$$

$$\tilde{t}_\perp = \frac{E_{\perp t}}{E_{\perp i}} = \frac{2\cos\vartheta_i}{\cos\vartheta_i + m\cos\vartheta_t}, \tag{7.99}$$

where the subscript t denotes transmission.

7.2.1   Scattering Interpretation of Specular Reflection and the Brewster Angle

According to Eq. (7.95) the degree of polarization of specularly reflected light is a maximum (100%) when the angle between the reflected and transmitted waves in a negligibly absorbing medium is 90°. As we show in the following section, the degree of polarization of light scattered by a spherically symmetric dipole, again for unpolarized incident light, is also a maximum (100%) when the angle between the incident and scattered waves (scattering angle) is 90°. This is not a coincidence but rather a consequence of the same underlying physics.

There are important differences between electromagnetic waves and, say, acoustic or water waves. The mathematics may be similar but the physics is not. People who study only the mathematics of wave motion see no fundamental difference between electromagnetic waves and other kinds, but the physical differences are profound. Acoustic and water waves require a material medium whereas electromagnetic waves do not: they propagate even in free space. And matter, appearances to the contrary, is almost entirely free space sparsely populated by charges, which can be acted upon by electromagnetic fields. What this means is that an incident electromagnetic wave exists everywhere in an illuminated medium. Contrast this with, say, acoustic waves incident on a wall in air. These waves do not literally penetrate the wall because they cannot carry their propagating medium (air) with them. But an incident electromagnetic wave penetrates an illuminated medium and excites *all* the molecules in it to radiate (scatter) waves. These scattered waves superpose in such a way that a total scattered wave is produced that interferes with the incident wave. We cannot observe these two waves separately, only their coherent superposition. When the illuminated medium is optically smooth and homogeneous the net result of this superposition is a reflected wave and a refracted wave with wavenumber different from that of the incident wave. This is only approximately true because a small amount of light also is scattered in directions other than the two special directions of reflection and refraction.

The Fresnel coefficients, Eqs. (7.91) and (7.92), usually are derived from continuum electromagnetic theory in which the graininess of matter is hidden from view. But that doesn't mean that this graininess isn't operating behind the scenes. Indeed, Bill Doyle showed that despite their macroscopic derivation the Fresnel coefficients can be made to betray their microscopic origins if properly interrogated. In particular, he showed that the Fresnel coefficients can be written as the product of a scattering function characteristic of an isolated dipole and a function characteristic of the coherently excited array of dipoles. Thus an illuminated medium is a *phased array* of dipolar antennas, which accounts for the highly directional radiation pattern of reflected and refracted waves. Radio antennas often are *designed* to produce highly directional beams; every specular reflector does this naturally.

Appearances to the contrary, specular reflection does not occur *at* an interface but rather *because* of it. An incident electromagnetic wave excites every molecule in a medium, not just those at its surface. As Doyle noted in his superb expository paper on the scattering interpretation of specular reflection, "In most discussions of Brewster's law the location of the dipoles responsible for the creation of the reflected beam, while usually not well defined, is put somewhere in the neighborhood of the interface. Actually, *all* of the dipoles together create the reflected beam." Thus notions about electromagnetic waves bouncing and bending are metaphorical, not to be taken literally. To do otherwise is to hobble one's thinking, not only getting the physics wrong but missing opportunities. Although it is difficult to derive the Fresnel coefficients from a microscopic point of view, knowing that scattering is ultimately the result of excitation of coherent arrays of dipoles can lead to solutions to problems of scattering by particles. Mie theory, discussed briefly in Section 3.5.3, is a continuum theory of scattering by a sphere, the solution to a boundary-value problem. But results similar to those given by Mie theory can be obtained from a discrete theory in which boundaries do not exist.

7.2.2 Transformations of Stokes Parameters: The Mueller Matrix

Up to this point we have determined the state of polarization of light transmitted by polarizers and retarders and their combinations by considering how these optical elements transform field components (amplitudes and phases). This was done purposely to give insight into what these optical elements do. But because we usually measure irradiances, we could have skipped fields and gone directly to irradiances. Thus if we denote by $\mathbf{I}_{in}$ the Stokes parameters of an input beam and by $\mathbf{I}_{out}$ those of an output beam that results from interaction of the input beam with any optical element, the two sets of Stokes parameters, written as column matrices, are related by

$$\mathbf{I}_{out} = \mathbf{MI}_{in}, \tag{7.100}$$

where $\mathbf{M}$ is a 4×4 matrix, sometimes called the *Mueller matrix*, characteristic of the optical element. The advantage of this approach is that for an input beam of *any* state of polarization we can determine that of the output beam by matrix multiplication without getting ensnarled in amplitudes and phases. Another use of this matrix algebraic approach is to determine how a series of optical elements (polarizing filters, retarders, reflectors, suspensions of particles, etc.) transforms the polarization state of a beam. Thus if $\mathbf{M}_1$, $\mathbf{M}_2,\ldots$, $\mathbf{M}_n$ are the Mueller matrices for a series of optical elements encountered by an input beam in the order of the subscripts, by matrix multiplication we have

$$\mathbf{I}_{\text{out}} = \mathbf{M}_n\mathbf{M}_{n-1}\ldots\mathbf{M}_2\mathbf{M}_1\mathbf{I}_{\text{in}}. \tag{7.101}$$

Underlying the strict validity of Eq. (7.101) is the assumption that the optical elements act independently of each other. Even if they are mutually incoherent, they still act in concert to some degree. For example, we show in Section 5.1 that the transmissivity of two identical (incoherent) plates is not exactly equal to the product of their transmissivities, although this may be a good approximation if the square of the reflectivity of a single plate is $\ll 1$. And if the optical elements are coherent, Eq. (7.101) may not even be approximately correct. An example is a multi-layer interference filter, a set of N negligibly absorbing identical double layers, each member with a suitably chosen (different) thickness (less than the wavelength) and refractive index. The transmissivity of the set decreases much more rapidly with N than that of a double layer raised to the power N. Indeed, this is how highly reflecting mirrors are made. Even the most reflective metallic mirror can't come nearly as close to 100% reflectivity as can a multi-layer interference filter the layers of which have optical properties quite different from those of metals. The key word here is "interference". If interference between optical elements is not negligible, Eq. (7.101) may not be valid. Nevertheless, if used with caution Mueller matrix algebra can save considerable effort and reduce the chances of errors in analysis. Its disadvantage is that one loses sight of what is happening physically, getting answers at the expense of understanding.

One of the simplest Mueller matrices is for specular reflection. From the Fresnel coefficients and the definition of the Stokes parameters we can obtain the reflected Stokes parameters as linear functions of the incident Stokes parameters. For example, it follows from Eqs. (7.15), (7.91) and (7.92) that

$$I_{\text{r}} + Q_{\text{r}} = |\tilde{r}_{\parallel}|^2\,(I_{\text{i}} + Q_{\text{i}}), \quad I_{\text{r}} - Q_{\text{r}} = |\tilde{r}_{\perp}|^2\,(I_{\text{i}} - Q_{\text{i}}). \tag{7.102}$$

Similar algebraic manipulation yields the entire Mueller matrix for specular reflection:

$$\begin{pmatrix} I_{\text{r}} \\ Q_{\text{r}} \\ U_{\text{r}} \\ V_{\text{r}} \end{pmatrix} = \begin{pmatrix} R_{11} & R_{12} & 0 & 0 \\ R_{12} & R_{11} & 0 & 0 \\ 0 & 0 & R_{33} & R_{34} \\ 0 & 0 & -R_{34} & R_{33} \end{pmatrix} \begin{pmatrix} I_{\text{i}} \\ Q_{\text{i}} \\ U_{\text{i}} \\ V_{\text{i}} \end{pmatrix}, \tag{7.103}$$

where

$$R_{11} = \frac{1}{2}(|\tilde{r}_{\parallel}|^2 + |\tilde{r}_{\perp}|^2) = \frac{1}{2}(R_{\parallel} + R_{\perp}), \tag{7.104}$$

$$R_{12} = \frac{1}{2}(|\tilde{r}_{\parallel}|^2 - |\tilde{r}_{\perp}|^2) = \frac{1}{2}(R_{\parallel} - R_{\perp}), \tag{7.105}$$

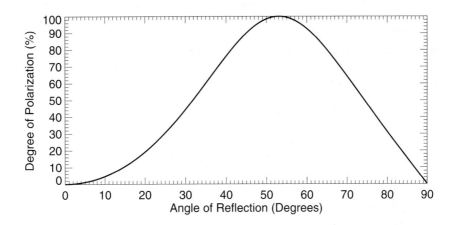

Figure 7.7: Degree of polarization of reflected light for unpolarized visible light incident at an air–water interface.

$$R_{33} = \frac{1}{2}(\tilde{r}_\parallel \tilde{r}_\perp^* + \tilde{r}_\parallel^* \tilde{r}_\perp) = \Re\{\tilde{r}_\parallel \tilde{r}_\perp^*\}, \tag{7.106}$$

$$R_{34} = \frac{i}{2}(\tilde{r}_\parallel^* \tilde{r}_\perp - \tilde{r}_\parallel \tilde{r}_\perp^*) = \Im\{\tilde{r}_\parallel \tilde{r}_\perp^*\}. \tag{7.107}$$

From these equations various results follow readily. For example, R_{11}, the arithmetic average of the two reflectivities for orthogonal polarization states, is the reflectivity for unpolarized incident light, which we could have guessed. The degree of linear polarization of reflected light, given incident unpolarized light, is

$$-\frac{R_{12}}{R_{11}} = \frac{R_\perp - R_\parallel}{R_\perp + R_\parallel}, \tag{7.108}$$

where the minus sign is introduced to make this quantity positive for reflection described by Eqs. (7.91) and (7.92).

Textbooks often are strangely silent about reflected light at angles other than the Brewster angle, the implication being that at such angles the reflected light is still unpolarized. But if reflected light were polarized only at the Brewster angle, polarizing sunglasses would not be very useful. Their function is to reduce reflection (glare) by non-metallic interfaces (e.g., water, painted hoods of automobiles) for directions in which the wearers of such sunglasses are likely to be looking. Brewsterists, members of an obscure religious sect, wear polarizing sunglasses and tilt their heads so that they see specular reflections only at the Brewster angle. But sunglasses are effective even for non-Brewsterists (called infidels) because the degree of polarization of specularly reflected light is high ($> 50\%$) over a wide range of angles around the Brewster angle. An example is shown in Fig. 7.7, the degree of polarization of visible light, for incident unpolarized light, reflected by water. You can verify the qualitative correctness of this figure from the waxing and waning of sun glint from water or glare from all kinds of smooth surfaces (polished floors, furniture, etc.) seen through a rotated polarizing filter.

Regardless of the illuminated medium and direction of incident light, it follows from Eq. (7.103) that reflection of unpolarized light cannot yield light with a degree of circular polarization ($V_r \neq 0$). To obtain such light requires incident light (partially) polarized *obliquely* to the plane of incidence ($U_i \neq 0$):

$$V_r = -R_{34}U_i. \tag{7.109}$$

But $R_{34} = 0$ for a negligibly absorbing medium (m real). Moreover, even for an arbitrary absorbing medium, $R_{34} = 0$ for normal (and glancing) incidence. Thus to obtain elliptically polarized reflected light requires obliquely polarized light incident on a metal, or material with appreciable absorption at the wavelength of interest, at non-normal incidence (but see Prob. 7.33 for an exception).

7.3 Polarization by Dipolar Scattering: Skylight

In the previous section, in which we consider specular reflection by an infinite interface, the plane of incidence is defined by the normal to the interface and the direction (wave vector) of the incident wave. But we could define this plane by the directions of the incident and reflected (or refracted) waves. Indeed, this is what is done in problems of scattering by finite objects: the *scattering plane* is defined by the direction of the incident wave and the Poynting vector of the scattered wave. In Sections 3.4 and 3.5 we discuss the frequency and angular but not polarization dependence of scattering. Here we tie up this loose end.

Consider a spherically symmetric electric dipole illuminated by a plane harmonic wave. We take the surrounding medium to be negligibly absorbing. We may consider the dipole to be a sphere of vanishingly small dimensions relative to the wavelength. The incident wave excites the dipole to radiate (i.e., scatter) in all directions. The direction of the incident wave and that of any Poynting vector of the scattered wave not parallel to the incident wave define a unique scattering plane, and we confine ourselves to all scattering directions lying in that plane. The electric fields of the incident and scattered waves lie in different planes perpendicular to the scattering plane (Fig. 7.8). Denote by $E_{\perp i}$ and $E_{\parallel i}$ the two components of the incident wave, perpendicular and parallel, respectively, to the scattering plane; $E_{\perp s}$ and $E_{\parallel s}$ are the components of the scattered field. Because of the assumed spherical symmetry of the dipole, incident light polarized perpendicular to the scattering plane can give rise to only a scattered field perpendicular to this plane. The same is true for incident light polarized parallel to the scattering plane. And because of the linearity of the equations of the electromagnetic field these scattered field components, like specularly reflected components, are proportional to the incident field components that excite them:

$$E_{\perp s} \propto E_{\perp i}, \ E_{\parallel s} \propto E_{\parallel i}. \tag{7.110}$$

When the incident field is perpendicular to the scattering plane the field scattered by a dipole cannot depend on the scattering angle within that plane. You can grasp this by looking at a needle perpendicular to a sheet of paper. As you change your viewing direction within this sheet, the needle always looks the same. But two or more parallel needles separated by a distance that is not small compared with the wavelength would look different in different

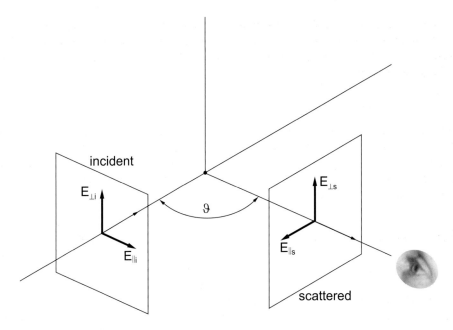

Figure 7.8: Two directions, those of incident and scattered waves, determine a plane, the scattering plane. Incident and scattered fields can be resolved into components perpendicular ($\perp$) and parallel ($\parallel$) to this plane. The basis vectors for the incident electric field are fixed whereas those for the scattered field change with scattering angle ϑ.

directions (see also Sec. 3.4). This independence of the field scattered by a dipole on direction when the incident field is perpendicular to the scattering plane is completely general, but when the incident field is parallel to the scattering plane, we have to impose an additional condition. Imagine a spherical polar coordinate system (r, ϑ, φ) to be centered on the dipole. At any point, the scattered field has a radial component (r-component) as well as tangential components. But the radial component decreases more rapidly with increasing r than do the tangential components. At sufficiently large distances relative to the wavelength, in what is called the *far field*, the radial component becomes negligible. We can make a simple, but not rigorous, argument for the angular dependence of the scattered far field when the incident radiation is parallel to the scattering plane. Figure 7.9 depicts the field $\mathbf{E}_d$ at a dipole, excited by an incident parallel field, which gives rise to scattered field components both parallel (radial) E_{dr} and perpendicular (tangential) E_{dt} to the scattering direction. But in the far field the radial component is negligible, leaving only the tangential component, which is proportional to $E_d \cos \vartheta$, where ϑ is the scattering angle. By this crude argument we conclude that in the far field

$$E_{\parallel s} \propto E_{\parallel i} \cos \vartheta. \tag{7.111}$$

We must mention an additional dependence of the field components even though it plays no essential role in our discussion of polarization. Imagine a sphere of arbitrary radius r

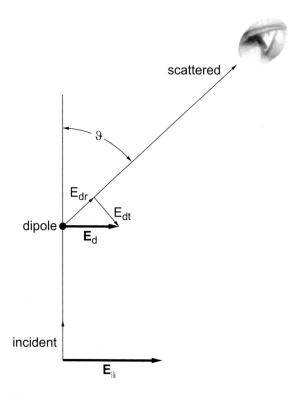

Figure 7.9: An incident electric field (in the plane of the diagram) excites a parallel field $\mathbf{E}_d$ scattered by a spherically symmetric dipole. The radial (r) and tangential (t) components of this scattered field decrease with increasing distance at different rates. At sufficiently large distances (compared with the wavelength), the radial component is negligible compared with the tangential component.

centered on the dipole. The integral

$$\int_0^{2\pi} \int_0^{\pi} S_r r^2 \sin\vartheta \, d\vartheta \, d\varphi, \tag{7.112}$$

over this spherical surface, where S_r is the radial component of the scattered Poynting vector, is the radiant energy scattered in all directions. Because the surrounding medium is negligibly absorbing, this energy is conserved, and hence Eq. (7.112) must be independent of r. A sufficient condition that this requirement be satisfied is that S_r be inversely proportional to r^2, which it is in the far field. This also implies that the electric field components are inversely proportional to r. But for our purposes we can write the relation between incident and scattered field components as

$$E_{\perp s} = C E_{\perp i}, \ E_{\|s} = C \cos\vartheta E_{\|i}, \tag{7.113}$$

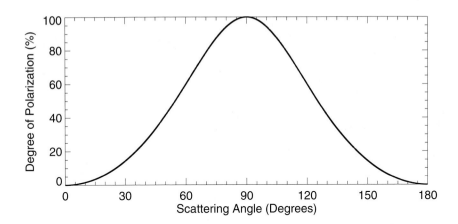

Figure 7.10: Degree of polarization of light scattered by a spherically symmetric dipole illuminated by unpolarized light.

where C incorporates all factors irrelevant to the polarization state (but not magnitude) of the scattered field. C is analogous to $\tilde{r}_\perp$, $C\cos\vartheta$ to $\tilde{r}_\parallel$, and so the Mueller matrix has the same form as Eq. (7.103):

$$\begin{pmatrix} \frac{1}{2}(\cos^2\vartheta + 1) & \frac{1}{2}(\cos^2\vartheta - 1) & 0 & 0 \\ \frac{1}{2}(\cos^2\vartheta - 1) & \frac{1}{2}(\cos^2\vartheta + 1) & 0 & 0 \\ 0 & 0 & \cos\vartheta & 0 \\ 0 & 0 & 0 & \cos\vartheta \end{pmatrix}, \tag{7.114}$$

where we omit any factor common to all matrix elements. The Mueller matrix for scattering by a particle or molecule is often called the *scattering matrix*.

From Eq. (7.114) it follows that the degree of linear polarization of scattered light, for incident unpolarized light, is

$$\frac{1 - \cos^2\vartheta}{1 + \cos^2\vartheta}, \tag{7.115}$$

which is shown in Fig. 7.10. At a scattering angle of 90° the scattered light is 100% polarized perpendicular to the scattering plane and is more than 50% polarized over a swath about 70° wide. For incident unpolarized light, the scattered irradiance varies with scattering angle according to

$$\frac{1}{2}(1 + \cos^2\vartheta). \tag{7.116}$$

Scattering described by Eq. (7.116) is not isotropic, but is equal in the forward and backward hemispheres. If the incident light is polarized perpendicular to a particular scattering plane, scattering in that plane is indeed isotropic – but only in directions in that plane, not in all directions. As noted previously (Sec. 5.2), isotropic light scatterers do not exist, and now we can better understand why: light is a vector wave.

In Section 3.4 we make simple interference arguments that scattering in the forward direction should increase more rapidly with particle size than in any other direction. In Section 3.5 we support these arguments with calculations for spheres. Because Eq. (7.116) for a spherically symmetric electric dipole shows equal forward and backward scattering, we might be tempted to conclude that forward scattering is *never* less than backward scattering. But here again there is an exception, and it is related to the term "electric" qualifying dipole (absent in Ch. 3). Incident light also can excite *magnetic dipole radiation* by a molecule or particle small compared with the wavelength. An electric dipole (Sec. 2.6) is two charges equal in magnitude, opposite in sign, its (electric) dipole moment the magnitude of the charge times the distance between them. A magnetic dipole is a small current loop, its magnetic dipole moment the current times the area enclosed by the loop. Usually, magnetic dipole radiation from molecules and small particles is much smaller than electric dipole radiation. One exception is tiny metallic particles at far infrared frequencies. The conductivity of metals depends on frequency, and as a rule the lower the frequency the higher the conductivity. At low frequencies the high conductivity of, and hence high current in, a small metallic sphere may result in a magnetic moment sufficiently large to give rise to magnetic dipole radiation comparable with electric dipole radiation. Moreover, these two dipole fields interfere in such a way that forward scattering can be considerably *less* than backward scattering, almost a factor of 10. As far as we know, this has no consequences for the atmosphere, but it is well to be aware of at least one exception to the general rule that forward scattering is greater than or at most equal to backward scattering. There is always an exception to every rule, even the rule that there is always an exception.

7.3.1 Polarization of Skylight

You sometimes encounter assertions that skylight is 100% polarized at 90° from the sun, presumably because of Eq. (7.115). Skylight is *never* 100% polarized in any direction, and there is more to the story of skylight polarization than this simple equation tells. All the reasons for the departure from 100% are variations on the same theme, and hence we can explain them with the same analysis.

Suppose that two beams with Stokes parameters $\{I_k, Q_k, U_k, V_k\}$ and degree of polarization P_k, where $k = 1, 2$, are superposed incoherently. The degree of polarization of the resultant beam is

$$P_{12} = \frac{\sqrt{(Q_1 + Q_2)^2 + (U_1 + U_2)^2 + (V_1 + V_2)^2}}{I_1 + I_2}. \qquad (7.117)$$

The Stokes parameters can be looked upon as specifying the coordinates of a point in a 4-dimensional space (Stokes space). If we denote a position vector in this space by $\mathbf{A}$ we can write Eq. (7.117) as

$$P_{12}^2 = \frac{P_1^2 I_1^2 + P_2^2 I_2^2 + 2\mathbf{A}_1 \cdot \mathbf{A}_2 - 2I_1 I_2}{(I_1 + I_2)^2}, \qquad (7.118)$$

where $\mathbf{A}_1 \cdot \mathbf{A}_2 = I_1 I_2 + Q_1 Q_2 + U_1 U_2 + V_1 V_2$ is the scalar (dot) product of the two vectors. From the identity

$$2I_1 I_2 = (I_1 + I_2)^2 - I_1^2 - I_2^2 \tag{7.119}$$

and the inequality

$$\mathbf{A}_1 \cdot \mathbf{A}_2 \le |\mathbf{A}_1||\mathbf{A}_2| = I_1 I_2 \sqrt{P_1^2 + 1} \sqrt{P_2^2 + 1} \tag{7.120}$$

it follows that

$$\sqrt{P_{12}^2 + 1} \le R_1 \sqrt{P_1^2 + 1} + R_2 \sqrt{P_2^2 + 1}, \tag{7.121}$$

where $R_k = I_k/(I_1 + I_2)$. Equality holds only if both beams are polarized in the same way, that is, have identical vibration ellipses and degrees of polarization. If the two beams are not identical in this sense and we take

$$P_2 \ge P_1, \tag{7.122}$$

it follows from Eq. (7.121) that

$$\sqrt{P_{12}^2 + 1} < \sqrt{P_2^2 + 1}, \tag{7.123}$$

and because the degree of polarization is positive,

$$P_{12} < P_2. \tag{7.124}$$

Thus if two beams (other than those with the same polarization state) are combined incoherently, the degree of polarization of the resultant is less than that of the beam with the highest degree of polarization. And what is true of two beams is true for any number of beams. We now have the tools necessary to explain all the reasons why skylight is not 100% polarized at 90° from the sun or in any other direction.

First consider the least important reason. An observer looking in a direction at 90° to light in a particular direction from the sun receives sunlight scattered over a small range of angles (the angular width of the sun) near 90°. Even if this scattered light were 100% polarized at 90°, it is not at neighboring angles. Scattering at all these angles contributes to the light received by the observer, and hence from Eq. (7.124) the degree of polarization must be less than 100%.

Equation (7.113) is valid only for a spherically symmetric dipolar scatterer. But the dominant molecules in Earth's atmosphere are the *linear* molecules O_2 and N_2, which are not spherically symmetric. A linear molecule oriented perpendicular to the scattering plane does indeed result in 100% polarized scattered light at 90°. But if the orientation of the molecule is changed, the degree of polarization of the scattered light in that direction is less than 100%. An observer looking in a particular direction receives the incoherent sum of light scattered by an ensemble of randomly oriented asymmetric molecules, because of which the degree of polarization must be less than 100%. Molecular asymmetry alone reduces the maximum degree of polarization of sunlight scattered by air to around 94%.

sunlight

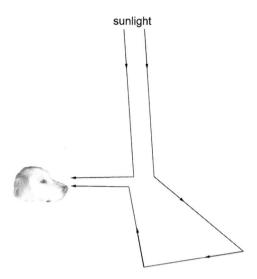

Figure 7.11: Light from the sky is the sum of successively decreasing contributions from single scattering, double scattering, triple scattering, and so on, of sunlight. Depicted here are the contributions to the total light at 90° from the direction of the sun as a result of single and quadruple scattering.

Multiple scattering of sunlight by the atmosphere is never completely absent. An observer looking in a direction 90° to light from the sun receives singly scattered light, doubly scattered light, triply scattered light, and so on (see Fig. 7.11). The degree of polarization of the singly scattered light is a maximum, that of the multiply scattered components being, in general, less. So again, by Eq. (7.124), multiple scattering in the atmosphere can only reduce the degree of polarization below the theoretical maximum value. Reflection by the ground is just another form of multiple scattering.

The atmosphere is rarely, if ever, entirely free of particles. Even scattering of unpolarized light by spheres (Sec. 7.4), if they are comparable with or larger than the wavelength, does not result in 100% polarization at 90° or any scattering angle. If the particles are small compared with the wavelength but nonspherical and randomly oriented, the degree of polarization of the light scattered by them is less than 100% at 90°. Moreover, *all* particles increase the scattering optical thickness of the atmosphere, and hence increase multiple scattering.

The finite angular width of the sun, molecular asymmetry, multiple scattering, reflection by the ground, and scattering by particles are all variations on the same underlying cause of why sunlight scattered at 90° is not 100% polarized: the incoherent superposition of beams with different degrees of polarization can only reduce the degree of polarization.

If we set $m = 1$ in Brewster's law [Eq. (7.96)] $\vartheta_i = 45°$. The angle between the incident and reflected waves can be considered a kind of scattering angle. Thus a scattering angle of 90° at which polarization is a maximum, expressed in the language of specular reflection, corresponds to a polarizing angle of 45°. Indeed, Brewster noted that "if we take the refractive power of air at 1.00031 the polarizing angle will be $45° \, 00' \, 32''$, a result which agrees most strikingly with the observed angle." Well, yes and no. According to the Fresnel coefficients, which underlie Brewster's law, when $m = 1$ the interface disappears and reflection is zero. So although these equations correctly predict the polarizing angle for air, as we might expect given the scattering interpretation of reflection, they are powerless to say anything about the magnitude of scattering by air. Equation (7.115) applies to a *single*, spherically symmetric dipolar scatterer, whereas Eq. (7.108) applies to a coherent array of *many* dipolar scatterers. In the limit $m \rightarrow 1$ Eq. (7.115) approaches Eq. (7.108) if we call the scattering angle for specular reflection that between the reflected and transmitted waves. Brewster got the right answer by (slightly) wrong reasoning because of the similar underlying physical mechanisms for specular reflection (scattering by a coherent dipolar array) and for scattering of sunlight by air (scattering by an incoherent dipolar array).

An apparent conflict between polarization upon scattering by small particles and Brewster's law troubled John Tyndall, who did many of the early experimental investigations of light scattering by suspensions of particles. One of the opening paragraphs of Lord Rayleigh's famous 1871 paper "On the light from the sky, its polarization and colour" begins by addressing what Tyndall "felt as a difficulty": "Tyndall says, '...the polarization of the beam by the incipient cloud has thus far proved to be *absolutely independent of the polarizing-angle*. The law of Brewster does not apply to matter in this condition; and it rests with the undulatory theory to explain why. Whenever the precipitated particles are sufficiently fine, no matter what the substance forming the particles may be, the direction of maximum polarization is at right angles to the illuminating beam, the polarizing angle for matter in this condition being invariably 45°. This I consider to be a point of capital importance...' Rayleigh responds: "As to the importance there will not be two opinions; but I venture to think that the difficulty is imaginary and is caused mainly by the misuse of the word reflection. Of course there is nothing in the etymology of reflection or refraction to forbid their application in this sense; but the words have acquired technical meanings, and become associated with certain well-known laws called after them. Now a moment's consideration of the principles according to which reflection and refraction are explained in the wave theory is sufficient to show that they have no application unless the surface of the disturbing body is larger than many square wave-lengths; whereas the particles to which the sky is supposed to owe its illumination must be *smaller* than the wave-length... The idea of polarization by reflection is therefore out of place;" and that 'the law of Brewster does not apply to matter in this condition' (of extreme fineness) is only what might have been inferred from the principles of the wave theory."

For this reason we are careful not to use reflection as a synonym for scattering by particles. Moreover, we are also careful to qualify reflection as specular when we have this kind in mind. The laws of specular reflection and refraction, the Fresnel coefficients, and Brewster's law, all have a limited range of validity.

Reduction of the maximum degree of polarization is not the only consequence of multiple scattering. According to Eq. (7.115) there should be two *neutral points* in the sky, directions in which skylight is unpolarized: directly toward the sun (*solar point*) and directly away from the sun (*antisolar point*). Because of multiple scattering, however, there are three such points. When the sun is higher than about 20° above the horizon there are neutral points within 20° of the sun, the *Babinet point* above it, the *Brewster point* below. They coincide when the sun is directly overhead and move apart as the sun descends. When the sun is lower than 20°, the *Arago point* is about 20° above the antisolar point.

7.4 Particles as Polarizers and Retarders

All the simple rules about polarization upon scattering are broken when we turn from molecules and small particles to particles comparable with or larger than the wavelength. The degree of polarization of light scattered by small particles is a simple function of scattering angle [Eq. (7.115)] but simplicity gives way to complexity as particles grow. The Mueller matrix for scattering (the scattering matrix) by an arbitrary homogeneous sphere must have the same symmetry as that for specular reflection [Eq. (7.103)] because incident light polarized perpendicular (parallel) to the scattering plane gives rise only to scattered light polarized perpendicular (parallel) to the scattering plane. So we immediately can write down the form of the scattering matrix for a sphere

$$
\begin{pmatrix}
S_{11} & S_{12} & 0 & 0 \\
S_{12} & S_{11} & 0 & 0 \\
0 & 0 & S_{33} & S_{34} \\
0 & 0 & -S_{34} & S_{33}
\end{pmatrix},
\tag{7.125}
$$

where the matrix elements S_{ij} depend on the size of the sphere relative to the wavelength of the illumination and its composition (i.e., complex refractive index). Unlike the Mueller matrix elements for specular reflection, these matrix elements are complicated functions not so readily calculated as sines and cosines. But many of the general statements we made about specular reflection also apply to scattering by a sphere. In particular, if the incident light is unpolarized, the scattered light is partially linearly polarized with degree of polarization given by an expression of the same form as Eq. (7.108):

$$
-\frac{S_{12}}{S_{11}}.
\tag{7.126}
$$

Despite this similarity there is a difference. If unpolarized light is incident on a planar interface the reflected light is always partially polarized perpendicular to the plane of incidence ($R_\perp \geq R_\parallel$). But because a sphere has an additional degree of freedom, its size relative to the wavelength, light scattered by it can be partially linearly polarized either perpendicular or parallel to the scattering plane. Moreover, this can flip back and forth at different scattering angles, as shown in Fig. 7.12, the degree of polarization of light scattered by water droplets of different sizes.

We wrote Eq. (7.126) with a negative sign so that it reduces to Eq. (7.115) in the small particle limit. Although degree of polarization is strictly a positive quantity, allowing it to take on both positive and negative values is a simple way of showing whether the scattered light is partially polarized perpendicular to the scattering plane (positive) or parallel to this plane (negative).

The degree of polarization of light scattered by molecules or by small particles is nearly independent of wavelength. But this is not true for particles comparable with or larger than the wavelength. Scattering by such particles exhibits *dispersion of polarization*: the degree of polarization at, say, $90°$ may vary considerably over the visible spectrum. This is shown in Fig. 7.13, the degree of polarization of visible light scattered by a water droplet of diameter $0.5\,\mu m$. Dispersion of polarization is not necessarily a consequence of the dispersion of optical

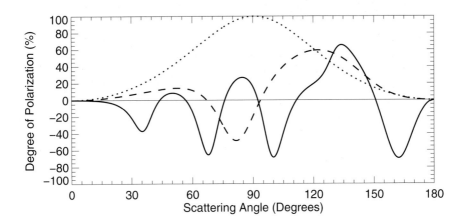

Figure 7.12: Degree of polarization of light scattered by water droplets of different size. The dotted curve is for a droplet of diameter 0.1 μm, the dashed curve for 0.5 μm, the solid curve for 1.0 μm; $\lambda = 0.55$ μm and $n = 1.33$. The incident light is unpolarized.

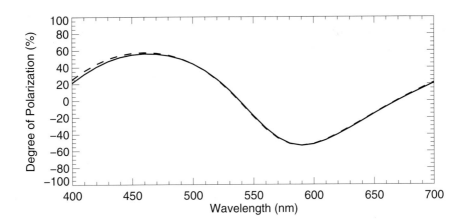

Figure 7.13: Degree of polarization of light scattered at $90°$ by a 0.5 μm diameter water droplet. The dashed line is for a fixed refractive index of 1.33, the solid line for a wavelength-dependent refractive index. The incident light is unpolarized.

constants. For the example shown in Fig. 7.13, the wavelength dependence of polarization is a consequence almost entirely of the dependence of phase differences on size relative to the wavelength.

Particles, unlike molecules, can act as polarizers and retarders. A linear polarizer transforms unpolarized light into partially polarized light by transforming the amplitudes of perpendicular field components differently, whereas a linear retarder transforms polarized light

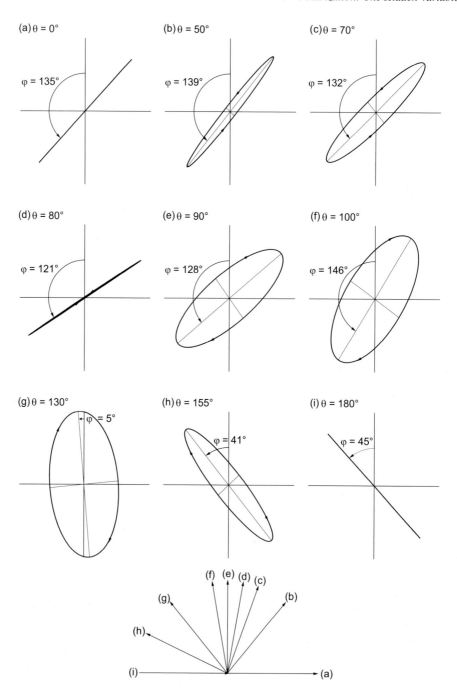

Figure 7.14: Vibration ellipses of visible light (0.55 μm) scattered in different directions by a water droplet of diameter 0.5 μm. The incident light is 100% linearly polarized at 45° to the scattering plane.

of one form into another by transforming the phases of perpendicular field components differently. But to do so a particle has to be sufficiently large relative to the wavelength to cause an appreciable phase shift. A point dipole cannot act as a retarder. If light linearly polarized obliquely to the scattering plane illuminates a spherically symmetric isotropic dipole, the scattered light in all directions is also linearly polarized, although the azimuth of the vibration ellipse (here, a line) varies with scattering angle. But for a sphere comparable with or larger than the wavelength, anything is possible. This can be shown graphically by drawing the vibration ellipses of light scattered in different directions by water droplets illuminated by light linearly polarized at $45°$ to the scattering plane (Fig. 7.14).

Up to this point we have considered only a single particle. Now consider N particles in a small volume illuminated by monodirectional light with irradiance I_i. By small here is meant that the linear dimensions of the volume are small compared with the distance to the observation point, which ensures that the scattering direction for each particle is approximately the same. If there is no fixed phase relation between the waves scattered by the particles (i.e., incoherent array), the Stokes parameters of the light scattered by each of them are additive, and hence the Stokes vector of the total scattered light at the observation point is

$$\mathbf{I_s} = \mathbf{M_1}\mathbf{I_i} + \mathbf{M_2}\mathbf{I_i} + \ldots + \mathbf{M_N}\mathbf{I_i} = \left(\sum_{j=1}^{N}\mathbf{M}_j\right)\mathbf{I_i} = \mathbf{M}\mathbf{I_i} = N\langle\mathbf{M}\rangle\mathbf{I_i}, \qquad (7.127)$$

where $\mathbf{M}_j$ is the scattering matrix for the jth particle and $\langle\mathbf{M}\rangle$ is the average scattering matrix. If the incident light is 100% polarized, the light scattered in any direction by a single particle must be 100% polarized because such a particle is a coherent array. But the light scattered by an incoherent array of particles cannot be 100% polarized if the particles are not all identical. Two particles are different, and hence have different scattering matrices, if they are different in size or shape or composition or, if not spherically symmetric, orientation. Although the light scattered by each particle is 100% polarized, the ellipsometric parameters are, in general, different. Two or more beams with definite but different ellipsometric parameters cannot be superposed to obtain a beam with definite ellipsometric parameters. The elements of each scattering matrix $\mathbf{M}_j$ are such that if the incident light is 100% polarized, the Stokes parameters of the light scattered by each particle satisfy $I_i^2 = Q_i^2 + U_i^2 + V_i^2$ whereas the elements of the average scattering matrix are such that the Stokes parameters of the total scattered light do not. This is a variation on a theme in Section 7.3, where we show that adding beams incoherently can never increase the degree of polarization above that of the beam with the highest degree of polarization.

Although oriented nonspherical particles are not unknown in the atmosphere (e.g., falling ice crystals oriented by aerodynamic forces), we usually are most interested in randomly oriented particles or molecules. The average scattering matrix for a large (large enough for an average to mean something) number of randomly oriented particles that are superposable on their mirror images (no corkscrews, please) has the same block-diagonal symmetry as that for spheres [Eq. (7.125)], although, in general, $S_{22} \neq S_{11}$ and $S_{44} \neq S_{33}$. For such an array of particles, incident unpolarized light gives rise only to partially linearly polarized scattered light perpendicular or parallel to the scattering plane. And incident light partially linearly polarized either perpendicular or parallel to the scattering plane again gives rise only to scattered light similarly polarized. To obtain (partially) circularly polarized light ($V_s \neq 0$) from incident linearly polarized light requires oblique polarization ($U_i \neq 0$). Because of this it would seem that scattering of unpolarized sunlight by atmospheric particles could never yield light with a degree of circular polarization. This would be true if it were not for multiple scattering. Sunlight scattered by one particle or group of particles acquires a degree of linear polarization. This light then becomes incident light for a second particle, and if the scattering direction is such that the second scattering plane is not parallel to the first, this incident light can be obliquely polarized ($U_i \neq 0$).

Let ($E_\parallel, E_\perp$) be the field components of the light scattered by the first particle. These become the field components of the light illuminating the second particle. If the scattering direction is such that the scattering plane for the first particle does not coincide with that of the second, these field components must be transformed to those relative to the second scattering plane:

$$
\begin{pmatrix} E'_\parallel \\ E'_\perp \end{pmatrix} = \begin{pmatrix} \cos\psi & \sin\psi \\ \sin\psi & -\cos\psi \end{pmatrix} \begin{pmatrix} E_\parallel \\ E_\perp \end{pmatrix},
\tag{7.128}
$$

where ψ is the azimuthal angle of the second plane relative to the first (the two coincide if $\psi = \pi/2$) and the prime denotes transformed field components. From Eq. (7.128) it then follows that the Stokes parameters are transformed according to

$$
\begin{pmatrix} I' \\ Q' \\ U' \\ V' \end{pmatrix} = \begin{pmatrix} 1 & 0 & 0 & 0 \\ 0 & -\cos 2\psi & \sin 2\psi & 0 \\ 0 & -\sin 2\psi & -\cos 2\psi & 0 \\ 0 & 0 & 0 & 1 \end{pmatrix} \begin{pmatrix} I \\ Q \\ U \\ V \end{pmatrix}.
\tag{7.129}
$$

This equation also reminds us that the Stokes parameters depend on the coordinate system. Suppose that the Stokes parameters of the light scattered by the first particle, illuminated by unpolarized light, are ($I_s, Q_s, 0, 0$). These then become the Stokes parameters of the incident light for the second particle, which when transformed according to Eq. (7.129) become ($I_s, -\cos 2\psi\, Q_s, -\sin 2\psi\, Q_s, 0$). If $\sin 2\psi \neq 0$ this secondary incident light is partially polarized obliquely to the second scattering plane, and hence the degree of circular polarization of the secondary scattered light is

$$
\frac{S_{34} \sin 2\psi\, Q_s}{S_{11} I_s - S_{12} \cos 2\psi\, Q_s}.
\tag{7.130}
$$

This equation shows how geometry (ψ), linear polarization (Q_s) and the retarding properties of a scatterer (S_{34}) can conspire by way of multiple scattering to yield scattered light with a probably small, but measurable, degree of circular polarization, beginning with unpolarized light.

Now we are better equipped to discuss fine points skirted in previous chapters. We can ignore polarization in the two-stream equations of transfer (Sec. 5.2) because of the symmetry of the media of interest and of the radiation field. We take the media to be isotropic and constrain radiation to only two directions parallel to the normal to planar interfaces. With these restrictions, there is no distinction between light polarized parallel and perpendicular to a reference plane determined by two *different* non-collinear directions. This is evident for specular reflection: at normal incidence, the two reflectivities [Eq. (7.94)] are equal, and polarization, in effect, does not exist. These are examples of a general rule: changes of polarization state always require *asymmetry*: at the molecular level (e.g., birefringence, dichroism), the macroscopic level (e.g., non-spherical, oriented particles), of the illumination (e.g., oblique incidence), or of observation (e.g., scattering directions other than forward or backward).

We continue to assume isotropic media in Chapter 6 but no longer constrain radiation to only two directions. But again polarization is swept under the rug. Why? When polarization is taken into account, the phase *function* in Eq. (6.15) is replaced by a 4×4 phase *matrix* and the scalar radiance by a 4-component Stokes vector. For many applications this added complexity is not worth the effort. Radiative transfer theory in atmospheric science is most often applied to clouds, multiple-scattering media composed of droplets or ice particles distributed in size. In Section 7.3 we show that multiple scattering can only reduce the degree of polarization. And in this section we show that the degree of polarization of light scattered by water droplets varies greatly with their size, scattering angle, and wavelength (Figs. 7.12 and 7.13). Thus several factors conspire to make polarization irrelevant, usually, to radiative transfer in clouds. Cloud light is weakly polarized, which you can verify for yourself with a polarizing filter. But light from the clear sky may have a high degree of polarization (Sec. 7.3). Indeed, this difference between sky and clouds is one reason for polarizing filters on cameras. Such filters can greatly increase contrast (see Sec. 8.2) by reducing skylight more than cloud light. We have seen formless masses smothered in murk suddenly become transformed into clear and distinct clouds by observing them through a rotated polarizing filter. Try this yourself. To explore the consequences of multiple scattering for the polarization state of the clear sky, we cannot avoid going beyond Eq. (6.15) to a radiative transfer equation for the Stokes vector. But if our interest is only in irradiances under a clear sky, ignoring polarization does not lead to serious errors.

At least two charming myths in which the polarization state of cloudlight plays a role are widespread. One is that bees can, without qualification, navigate by polarized skylight. Remarkably clever though they may be, bees cannot do the impossible. The simple wavelength-independent relation between the position of the sun and the direction in which skylight is most highly polarized, which underlies navigation by means of polarized skylight, is obliterated when clouds cover the sky. Indeed this was the key to Karl von Fritsch's solving the puzzle of bee navigation: "Sometimes a cloud would pass across the area of sky... and the bees were unable to indicate the direction to the feeding place. Whatever phenomenon in the blue sky served to orient the dances, this experiment showed that it was seriously disturbed if the blue sky was covered by a cloud." But von Fritsch's words often are forgotten by those eager to spread the word about bee magic to those just as eager to believe what is charming even though untrue.

The other myth, namely that the Vikings could navigate on overcast days by somehow making use of polarized skylight, has been debunked by two modern-day Vikings, Curt Roslund and Claes Beckman. They begin their attack by noting that "Viking exploits in the North Atlantic have aroused people's imagination to a height where responsible judgement of extraordinary claims often seems to be suspended." Even if the Vikings had the means to detect the polarization state of the sky (and there is no evidence that they did), "On most overcast days the Vikings could most certainly not have used the polarization state of light to determine the location of the Sun. Although they might have been able to do so on partly cloudy days, there would have been no need to." The Vikings were indeed remarkable navigators, but not because of their precocity in polarimetry.

References and Suggestions for Further Reading

The treatise on polarization that has influenced us most is William A. Shurcliff, 1962: *Polarized Light: Production and Use*. Harvard University Press. Shurcliff approaches polarization from a fresh point of view. This is a well-written, critical book, mandatory reading for anyone interested in polarization. Another good book on polarization is David Clarke and John F. Grainger, 1971: *Polarized Light and Optical Measurements*. Pergamon. The authors carefully discuss conventions for the handedness of light. A more recent work, also more advanced, is Edward Collett, 1992: *Polarized Light: Fundamentals and Application*. Marcel Dekker.

An advanced treatise on polarization from a statistical point of view is Christian Brosseau, 1998: *Fundamentals of Polarized Light: A Statistical Optics Approach*. John Wiley & Sons. In previous chapters we skirted the entropic properties of radiation but did not meet them head on. For example, Problem 1.40 is a derivation of the form of the Stefan–Boltzmann law by way of the first and *second* laws of thermodynamics. In Section 4.1.6 we referred to specular reflection as reversible. And Eq. (7.124) has the faint odor of entropy about it, being a requirement that the degree of polarization of partially polarized beams can never increase upon superposition. Radiation does indeed possess entropy (see Prob. 7.49). In fact, Brosseau derives a more stringent form of Eq. (7.124) by entropy arguments (Appendix C), includes a section (3.4) on the entropy of radiation, and a brief appendix (E) on the history of the extension of the entropy concept to radiation. Entropy arguments have not been applied extensively to optics problems, possibly because they can be solved by other, more familiar, methods.

For a collection of 75 contributed chapters on various aspects of polarization see Tom Gehrels, Ed., 1974: *Planets, Stars and Nebulae Studied with Polarimetry*. University of Arizona Press.

For popular treatments of polarization, with many color plates, see Günther P. Können, 1980: *Polarized Light in Nature*. Cambridge University Press and David Pye, 2001: *Polarized Light in Science & Nature*, Institute of Physics.

For benchmark papers on polarization see William Swindell, Ed., 1975: *Polarized Light*. Dowden, Hutchinson & Ross. Two other collections edited by Bruce H. Billings are *Selected Papers on Polarization*, Vol. MS 23 (1990) and *Selected Papers on Applications of Polarized Light*, Vol. MS 27 (1992). SPIE Optical Engineering Press.

One of the Arago–Fresnel laws "deduced from direct experiment" long before (1819) the electromagnetic theory of light is "In the same condition in which two rays of ordinary light seem to destroy each other mutually, two rays polarized at right angles or in opposite senses exert on each other no appreciable action" (William Francis Magie, 1965: *A Source Book of Physics*. Harvard University Press, pp. 324–35).

For what Stokes wrote about the polarization parameters that now bear his name see George Gabriel Stokes, 1852: On the composition and resolution of streams of polarized light from different sources. *Transactions Cambridge Philosophical Society*, Vol. 9, pp. 399–416 (reprinted in Stokes's *Mathematical and Physical Papers*. Vol. 3, pp. 233–50 and in the collection edited by Swindell).

A fascinating paper on the development of sheet polarizing filters is Edwin H. Land, 1951: Some aspects of the development of sheet polarizers. *Journal of the Optical Society of America*, Vol. 41, pp. 957–63. This is a paper you must read. It is reprinted in the collections edited by Swindell and by Billings (Vol. MS 23). The October 1994 *Optics & Photonics News* is a special issue devoted to Edwin Land. He is held in such high esteem in the senior author's house that a yellow and brittle obituary of Land is still taped to the refrigerator.

For the discovery of "a strange and wonderful phenomenon. . . according to which objects seen through it appeared not with single images as with other transparent bodies, but with double images" by Erasmus Bartholinus see pp. 280–3 in Magie's *Source Book*. The original Latin version and a partial translation are in the collection edited by Swindell.

For the optics of naturally birefringent media see G. N. Ramachandran and S. Ramaseshan, 1961: Crystal optics, in *Handbuch der Physik*, Springer, Vol. 25/1, pp. 1–217.

For an analysis of the colors seen through airplane windows (with color plates) see Craig F. Bohren, 1991: On the gamut of colors seen through birefringent airplane windows. *Applied Optics*, Vol. 30, pp. 3474–8. These colors are also discussed without mathematics in Craig F. Bohren, 1991: *What Light Through Yonder Window Breaks?*, John Wiley & Sons, Chs. 3 & 4.

For what can be learned about hailstones because ice is naturally birefringent see Charles A. Knight and Nancy C. Knight, 1970: Hailstone embryos. *Journal of the Atmospheric Sciences*, Vol. 27, pp. 659–66; Lobe structure of hailstones, pp. 667–71; The falling behavior of hailstones, pp. 672–81.

For Malus's account of the discovery of polarization upon reflection see Magie's *Source Book*, pp. 315–18. The original French version and a partial translation are in the collection edited by Swindell.

If you dare to spell Snel's name correctly be prepared for a fight with copy editors and those who have misspelled it for a lifetime. Arm yourself with the biographical sketch of Willebrord Snel by his fellow countryman Dirk J. Struik in *Dictionary of Scientific Biography*. Snel's name has been misspelled for centuries because its latinized form is Snellius, which was delatinized (incorrectly) as Snell.

The Fresnel coefficients are derived in many books on optics and electromagnetic theory. See, for example, Julius Adams Stratton, 1941: *Electromagnetic Theory*. McGraw-Hill, pp. 490–511; Max Born and Emil Wolf, 1965: *Principles of Optics*, 3rd rev. ed. Pergamon, pp. 38–51;

John David Jackson, 1975: *Classical Electromagnetic Theory*, 2nd ed. John Wiley & Sons, pp. 278–82; John Lekner, 1987: *Theory of Reflection*. Martinus Nijhoff, pp. 1–10.

For what is now called Brewster's law see David Brewster, 1815: On the laws which regulate the polarisation of light by reflexion from transparent bodies. *Philosophical Transactions of the Royal Society*, Vol. 105, pp. 125–59. Excerpts from this paper are in the collection edited by Swindell. See also Akhlesh Laktakia, 1989: Would Brewster recognize today's Brewster angle? *Optics News*. Vol. 15, pp. 14–17.

For more on polarization of emitted radiation see Oscar Sandus, 1965: A review of emission polarization. *Applied Optics*, Vol. 4, pp. 1634–42. His derivation of the degree of polarization is different from ours but the result is the same. He presents experimental data compared with theory.

For a microscopic (scattering) interpretation of polarization upon reflection, but wrung out of macroscopic physics, see William T. Doyle, 1985: Scattering approach to Fresnel's equations and Brewster's law. *American Journal of Physics*, Vol. 53, pp. 463–68. Doyle's analysis should be included in every textbook derivation of the Fresnel coefficients.

For a proof that backscattering by metallic particles (strictly, ones with infinite refractive index) is 9 times forward scattering, and the physical reason for this, see Hendrik C. van de Hulst, 1957: *Light Scattering by Small Particles*, John Wiley & Sons, pp. 158–61.

Lord Rayleigh's 1871 paper, On the light from the sky, its polarization and colour, originally published in *Philosophical Magazine*, is reprinted in his *Scientific Papers*, Vol. I, Cambridge University Press (1899), pp. 88–103, also in Craig F. Bohren, Ed., 1989: *Selected Papers on Scattering in the Atmosphere*. SPIE Optical Engineering Press.

For an elementary discussion of the polarization of skylight see Craig F. Bohren, 1987: *Clouds in a Glass of Beer*, John Wiley & Sons, Ch. 19.

For a theoretical treatment of the polarization state of the clear sky and comparison with measurements see Subrahmanyan Chandrasekhar and Donna D. Elbert, 1954: The illumination and polarization of the sunlit sky on Rayleigh scattering. *Transactions of the American Philosophical Society*, Vol. 44, pp. 643–54. Reprinted in *Selected Papers on Scattering in the Atmosphere*.

A treatise devoted entirely to polarization of light from the sky is Kinsell L. Coulson, 1988: *Polarization and Intensity of Light in the Atmosphere*, A. Deepak Publishing. Especially recommend for many measurements of the degree of polarization of (clear) sky light (Chapter 5), which indicate that a degree of polarization greater than 85% is all but unattainable.

Section 7.4 is based, in part, on Craig F. Bohren, 1995: Optics, atmospheric, *Encyclopedia of Applied Physics*, Vol. 12, pp. 405–34.

We exclude in Section 7.4 particles that are not superposable on their mirror images. Including such particles (and molecules) opens up a vast field with many important applications, especially in chemistry and biology. For more on this see Akhlesh Lakhtakia, 1990: *Selected Papers on Natural Optical Activity*. SPIE Optical Engineering Press.

It should be evident on physical grounds that ignoring polarization in calculations for clear skies should result in larger radiance than irradiance errors. Irradiance is integrated radiance, the errors in which are about as likely to be positive as negative. This is indeed what is calculated and observed. Andrew A. Lacis, Jacek Chowdhary, Michael I. Mischenko, and Brian Cairns, 1998: Modeling errors in diffuse-sky radiation: Vector vs. scalar treatment. *Geophysical Research Letters*, Vol. 25, pp. 135–8 compare calculations both with and without accounting for polarization. Seiji Kato, Thomas P. Ackerman, Ellsworth G. Dutton, Nels Laulainen, and Nels Larson, 1999: A comparison of models and measured surface shortwave irradiance for a molecular atmosphere. *Journal of Quantitative Spectroscopy and Radiative Transfer*, Vol. 61, pp. 493–502 take a different approach. They compare irradiances calculated using a two-stream theory with measurements made in a very clean environment (Mauna Loa Observatory). They conclude that neglecting polarization introduces negligible errors in irradiance calculations.

For what bees can and, more important, cannot do see Karl von Fritsch, 1971: *Bees: Their Vision, Chemical Senses, and Language*, rev. ed., Cornell University Press, p. 116.

For a delightful debunking of fanciful notions about Viking photopolarimetry see Curt Roslund and Claes Beckman, 1994: Disputing Viking navigation by polarized light, *Applied Optics*, Vol. 33, pp. 4754–5.

Problems

7.1. Show that an unpolarized beam can be considered to be the incoherent superposition of two orthogonally elliptically polarized (100%) beams of equal irradiance.

HINT: This problem is a simple application of the Stokes parameters.

7.2. Show that the electric and magnetic fields of a plane harmonic wave are perpendicular to the direction of propagation only if the wave is homogeneous, and that the electric and magnetic fields are perpendicular to each other only if the medium is nonabsorbing.

7.3. Show by simple arguments that transmission of a beam of unpolarized light by an ideal linear retarder cannot change the state of polarization of the beam. No fancy mathematics is necessary, just simple arguments based on the nature of unpolarized light and the function of a retarder.

7.4. Show that the reflectivities for specular reflection, $R_\parallel$ and $R_\perp$, are both equal to 1 for an angle of incidence of 90° regardless of the refractive index of the medium.

7.5. Show that as m becomes indefinitely large, the degree of polarization of specularly reflected light (given unpolarized incident light) approaches zero for all angles of incidence.

7.6. Many years ago a scheme for reducing glare from the headlights of oncoming automobiles was seriously considered. Linear polarizing filters were to be placed on the headlights and windshields of all automobiles. How should these filters be oriented so that drivers do not see the light from oncoming cars but do see light reflected by the headlights of their own cars? You may assume that the state of polarization of light from headlights is not changed upon reflection. Can you think of at least one reason why this scheme was never adopted?

7.7. Estimate the extent to which the maximum degree of polarization of skylight is reduced because of the finite width (about half a degree) of the sun. You may assume that air molecules are spherically symmetric.

7.8. By how much does an ideal linear polarizing filter reduce the irradiance of an incident unpolarized beam? By how much does the filter reduce the irradiance of an incident beam partially polarized with degree P perpendicular to the transmission axis?

HINT: The first question is easier than the second. You can guess the answer to the second by considering the two limiting cases. To check your guess note that the Mueller matrix for an ideal linear polarizing filter has the same form as the Mueller matrix for reflection in which one of the reflectivities is 1, the other 0.

7.9. It should be evident on physical grounds that incident 100% polarized light yields specularly reflected light that also is 100% polarized although with possibly different degrees of linear and circular polarization. Prove this. That is, show that if $I_i^2 = Q_i^2 + U_i^2 + V_i^2$ it follows that $I_r^2 = Q_r^2 + U_r^2 + V_r^2$. In constructing this proof you will also discover that the Mueller matrix elements for specular reflection are not independent (indeed, this result is an essential part of the proof).

7.10. The transmission axis of a linearly polarizing filter in, for example, polarizing sunglasses, is not perceptible to the human eye. How would you quickly determine the direction of this axis?

7.11. We discussed only the degree of polarization of specularly reflected light, for incident unpolarized light, as a function of angle of incidence for reflection by an infinitely thick medium. Such media are thin on the ground. How do you expect the degree of polarization to vary for light incident on a medium of *finite* thickness? Take it to be a nonabsorbing plate of uniform thickness sufficiently large relative to the wavelength that the consequences of interference need not be taken into account. For sake of visualization, consider the plate to be a microscope slide. Equations in Section 5.1 will help you derive an expression for the degree of polarization of the plate as a function of the reflectivities of the infinite medium. But you should first try to determine by physical reasoning if the degree of polarization increases, decreases, or remains the same (relative to the infinite medium) and how this depends on angle of incidence.

7.12. Find the degree of polarization of light specularly reflected by an incoherent pile of $N = 2, 4, \ldots$ identical plates like those in the previous problem. The plates are sufficiently far apart (relative to the wavelength) that coherence need not be taken into account. Find the degree of polarization of the reflected light and the reflectivity for unpolarized light as a function of the angle of incidence in the limit of indefinitely large N. You can do this by analysis or by physical reasoning. Sketch the reflectivity and degree of polarization as a function of angle in this limit.

7.13. Determine the degree of polarization of light *transmitted* by the pile of plates in the previous problem as a function of the angle of incidence in the limit of an indefinitely large number of plates. Sketch the degree of polarization of the transmitted light and reflectivity for unpolarized light as function of angle of incidence in this limit. First try to make these sketches by only physical reasoning. Compare these sketches with those for reflection. What do you conclude about using a pile of plates as a polarizing filter?

7.14. Show that the surfaces of constant amplitude for an inhomogeneous plane wave that results from illumination of a planar, optically smooth interface between a negligibly absorbing medium and an absorbing medium are planes parallel to the interface. Show that when the imaginary part of the refractive index of the absorbing medium is small compared with the real part, the spatial rate of attenuation of the transmitted field is *approximately* $\exp(-2\pi n_i s/\lambda)$, where n_i is the imaginary part of the complex refractive index of the absorbing medium and s is the distance into this medium *along the direction of refraction*.

7.15. An article in *Science News* (July 3, 2003) about beetles navigating by moonlight begins with the assertion that "an international team of researchers has turned up evidence that the insect aligns its path by detecting the polarization of moonlight." The article goes on to say that the researchers "found that beetles active during the day depend on sunlight polarization patterns." These statements, taken literally (which is the only way we can take them) are incorrect. Why? How would you rewrite them to make them correct?

7.16. On what kind of imaginary but physically allowable planet with what kind of atmosphere illuminated by what kind of sun would skylight be 100% polarized at 90° from the sun?

7.17. Show that in the limit $m \to 1$ Eq. (7.108) for the degree of polarization of specularly reflected light (given incident unpolarized light) approaches Eq. (7.115), the degree of polarization of light scattered by a spherically symmetric electric dipole if by the scattering angle in Eq. (7.108) is meant that between the reflected and transmitted waves.

HINTS: You will need L'Hospital's rule, basic theorems about limits of products and quotients, and trigonometric identities for sums of angles and half angles.

7.18. The angular dependence of scattering (in a given plane) of unpolarized light by a spherically symmetric dipolar scatterer is given by Eq. (7.116). Because this scattering is azimuthally symmetric, you should be able to determine the corresponding (normalized) phase function (the probability of scattering, per unit solid angle, in any direction).

7.19. Show that any optical element that transforms both amplitudes and phases of perpendicular electric field components differently can transform unpolarized light only into partially linearly polarized light. This is a more complicated version of Problem 7.3.

HINT: A proof follows from the definition of the Stokes parameters for light of arbitrary polarization state together with simple trigonometric identities.

7.20. We state without proof in Section 2.2.1 that "for many materials over many wavelength intervals, reflectivity changes hardly at all even with huge increases in absorption coefficient. And if there is a change, it is likely to result in a decrease in absorptivity…" You now have all the ingredients to prove this statement, namely the Fresnel coefficients and the discussion of the complex refractive index (see Sec. 3.5.2).

HINT: Consider specular reflection at normal incidence.

7.21. A plane harmonic wave incident on the interface between two negligibly absorbing media and originating in the (first) medium of higher refractive index is *totally internally reflected* at an angle of incidence, called the *critical angle* ϑ_c, such that the angle of transmission is $\pi/2$ (see Sec. 4.2.1). Show that the reflectivity at the critical angle for incident light polarized parallel or perpendicular to the plane of incidence is 1. Take the second medium to be air with refractive index 1.

HINT: Snel's law [Eq. (7.90)] and the cosine form of the Fresnel coefficients [Eqs. (7.91) and (7.92)] should be helpful.

7.22. This problem is an extension of the previous one. There is no law prohibiting waves from being incident at angles *greater* than the critical angle. What are the reflectivities for incident waves perpendicular and parallel to the plane of incidence at these angles?

HINT: The key to this problem is to recognize that ϑ_t in $\sin \vartheta_t$ and $\cos \vartheta_t$ in Eq. (7.88) does not necessarily correspond to a real angle. That is, $k_t \sin \vartheta_t$ and $k_t \cos \vartheta_t$ are simply the components of what is called the transmitted wavevector, the only requirement being that $\sin^2 \vartheta_t + \cos^2 \vartheta_t = 1$, and Snel's law is a mathematical relation that is always satisfied but not always easy to interpret geometrically. The cosine form of the Fresnel coefficients again should be helpful.

7.23. This problem is an extension of the previous one. Show that for angles of incidence greater than the critical angle, the reflected light is elliptically polarized for incident light linearly polarized obliquely to the plane of incidence.

7.24. We stated in Section 7.1 that the concept of a surface of constant phase is, in general, meaningless. Show this.

HINT: By a surface of constant phase is meant a single such surface. Write down the expression for an arbitrary (complex) vector field and the proof should be obvious.

7.25. Blackbody radiation is unpolarized and isotropic but can be considered the incoherent superposition of two sources of equal magnitude but orthogonally linearly polarized (see Prob. 7.1). From this result, derive an expression for the emissivity of an opaque slab, optically smooth and homogeneous (in air, say), in terms of the reflectivities for incident radiation polarized parallel ($R_\parallel$) and perpendicular ($R_\perp$) to the plane of incidence.

7.26. Derive an expression for the degree of polarization as a function of direction for radiation emitted by the slab in the previous problem. It may help to write the Stokes parameters of the emitted radiation. What is the largest degree of polarization (and in what direction) of $10\,\mu\text{m}$ radiation emitted by a layer of water in air?

HINTS: We chose our words carefully here: "largest degree" rather than maximum degree. The easiest way to do this problem is to use the Fresnel coefficients Eq. (7.91) and Eq. (7.92) to obtain an explicit expression for the degree of polarization as a function of the angles ϑ_i and ϑ_t. Use the form of these equations containing trigonometric functions of the sum and differences of angles. Even though water is absorbing at $10\,\mu\text{m}$, the imaginary part of its refractive index is sufficiently small compared with its real part (about 1.2) that you can ignore the imaginary part. With a bit of algebra and a simple trigonometric identity, you can obtain a fairly simple expression for the degree of polarization.

7.27. Our proof of the inequality Eq. (7.57) was based on the (unproven) assumption that $I_p \leq I$, whereas a rigorous proof must start from Eqs. (7.51)–(7.54). You will need the

Cauchy-Schwarz inequality in the form $\left(\int fg\, dt\right)^2 \le \int f^2\, dt \int g^2\, dt$ and the identity $\cos^2 \delta + \sin^2 \delta = 1$. We confess that we needed the help of two mathematicians, George Greaves and V. I. Burenkov, with a proof, which need not be long.

7.28. Convince yourself that Eq. (7.8) does indeed correspond to two orthogonally polarized waves. All that is needed is a crude sketch.

7.29. We once observed through a polarizing (camera) filter daylight reflected by a polished floor near the Brewster angle. As we rotated the filter, the brightness of the reflection greatly diminished, as expected. But the reflection could not be made to completely disappear. We always observed a dark reflection of a strikingly pure blue. At first we thought this had something to do with illumination by the blue sky. But this hypothesis was quickly discarded when we noticed that the source of illumination was light from an overcast sky. Explain.

HINT: Although it is not necessary to write down the Mueller matrix for an *non-ideal* linear polarizing filter, doing so, or at least thinking about doing so, is likely to help.

7.30. The *Umov effect* or *Umov's law (rule)* is a reciprocal relationship between reflectivity and degree of polarization of light reflected by rough surfaces or granular media (e.g., soils, snow, powders): the higher the reflectivity the lower the degree of polarization and *vice versa*. Although this rule appears to be fairly well known to (planetary) astronomers, a simple, short explanation of it is hard to find. And yet it can be explained adequately in a sentence or two. Provide such an explanation, concise, correct, and easy to understand.

7.31. Obtaining reflectivities from the Fresnel coefficients (ratios of fields, not irradiances) for reflection is straightforward because the incident and reflected waves make the same angle with the normal to the interface and both are in the same medium. Indeed, we presented these reflectivities without fanfare. Transmissivities require a bit more care. Derive expressions for the two transmissivities (ratio of transmitted irradiance to incident irradiance) as a function of angle of incidence for an infinite negligibly absorbing medium in air.

HINTS: You need the Fresnel coefficients for transmission as well as the relation between the Poynting vector and irradiance. We previously ignored a constant of proportionality between the magnitude of the Poynting vector (irradiance) and the square of electric fields. But for this problem we cannot ignore this constant because it is different for different media. For a plane, homogeneous wave in a negligibly absorbing medium $|\mathbf{S}| = n|\mathbf{E} \cdot \mathbf{E}^*|/2Z_o$, where n is the (real) refractive index of the medium and Z_o is a universal constant called the impedance of free space.

7.32. To check the correctness of the result obtained in the previous problem, show that $R + T = 1$, where R is the reflectivity and T is the transmissivity (either polarization state). A further check is the reciprocal relation $T(\vartheta_i, n) = T(\vartheta_t, 1/n)$, which follows from the reversibility of rays.

7.33. In Section 7.2.2 we assert that to obtain elliptically polarized light by specular reflection requires obliquely polarized light incident on a metal (or a material with appreciable absorption at the wavelength of interest) at non-normal incidence. This statement is strictly true only for an *infinite* medium because if it is nonabsorbing $R_{34} = 0$. Show by simple arguments that for oblique incidence R_{34} is not necessarily zero for a *finite* nonabsorbing single layer or many layers (the form of the Mueller matrix is the same as that for an infinite medium). This problem is a variation on the theme that high reflectivity can be obtained either by a metal (appreciably absorbing) or by a multi-layer interference filter (negligibly absorbing).

7.34. Derive the Mueller matrix for an ideal linear polarizing filter with its transmission axis at an arbitrary angle to the reference coordinate system in which the Stokes parameters are defined.

HINT: Begin with Eqs. (7.64) and (7.65).

7.35. Derive the Mueller matrix for an ideal linear retarder with arbitrary retardance and with its fast axis at an arbitrary angle to the reference coordinate system in which the Stokes parameters are defined. This is considerably more difficult than the previous problem. You have to resolve the field components in the reference system along the fast and slow axes of the retarder, introduce different phase shifts, then transform back to the reference system. Trigonometric identities for the cosine and sine of the sum of angles can be used to simplify results.

7.36. As noted at the end of Section 7.1.3, although the Stokes parameters were defined by way of a set of hypothetical measurements not all of which are feasible, once these parameters are defined we can devise ways to measure them. Suppose that you have an irradiance detector, a linear polarizing filter, and a linear retarder with variable retardance. How many measurements and which kind would you have to make in order to determine the Stokes parameters of an arbitrary beam? Do you need both a polarizing filter and a retarder? Try to devise the simplest (easiest to describe) set of measurements. The results of the previous two problems can help you specify in detail what kind of measurements to make.

7.37. Show that if an incoherent suspension of spheres, regardless of their distribution in size and composition but sufficiently thin (optically) that multiple scattering is negligible, is illuminated by light linearly polarized perpendicular (parallel) to the scattering plane, the scattered light in this plane is 100% polarized. You can show this by simple physical arguments, but a mathematical proof (by way of the scattering matrix) leads you into the second part of this problem. By measuring the polarization properties of light scattered by a suspension of particles, how can it be determined if they are nonspherical even if randomly oriented? What scattering matrix elements or combination of elements should be measured and how (we can think of at least two possibilities)?

HINT: Results in Section 7.4 are necessary, and although the solution to Problem 7.34 is not, it might be helpful. For what scattering angles is the quantity measured likely to be most sensitive to departures of the particles from sphericity? You might review Section 3.4.8 before addressing this last question.

7.38. Based on the discussion of fluorescence in the references at the end of Chapter 1 and the discussion in Section 7.4 of the essential role that asymmetry plays in yielding polarized light, what do you expect the state of polarization (at any angle) of fluorescent light excited in gases and liquids to be? What about solids (with other than cubic symmetry)? No mathematical analysis is necessary. This problem tests your physical understanding of polarization and emission by gases and solids.

7.39. Does it seem contradictory that the light scattered at, say, 90°, by randomly oriented nonspherical molecules (for incident unpolarized light) can be partially polarized?

HINT: Consider a spheroidal molecule that differs only slightly from a spherical molecule.

7.40. Show that for negligible absorption, the reflection coefficient Eq. (7.91) or (7.92) reverses sign when the rays are reversed. That is, if we denote $\tilde{r}_{12}$ as the reflection coefficient for light incident at angle ϑ_i from medium 1 onto medium 2, and $\tilde{r}_{21}$ as the reflection coefficient for light incident from medium 2 onto medium 1, where the angles of incidence and refraction are reversed, show that $\tilde{r}_{21} = -\tilde{r}_{12}$. Then show that $1 + \tilde{r}_{12}\tilde{r}_{21} = \tilde{t}_{12}\tilde{t}_{21}$, where $\tilde{t}_{12}$ is the transmission coefficient [Eq. (7.98) or (7.99)], corresponding to $\tilde{r}_{12}$. What is the physical interpretation of this equation?

7.41. We did not derive Eq. (5.24), but you now should be able to do so in a way similar to summing the infinite series in Section 1.4. Although this equation was for normal incidence, no more effort is necessary to derive it for arbitrary incidence.

HINTS: Equation (5.7) suggests the form of the solution. For this problem you have to add fields, taking due account of phase shifts, then take the product of the resultant with its complex conjugate to obtain the reflectivity. Problem 7.40 also is needed for this problem. And you will need the difference in phase between a transmitted wave at $z = 0$ and at $z = h$, which you can obtain from Eq. (7.88).

7.42. With the results of Problem 7.41 you now should be able to explain why different thin film interference colors sometimes are seen in different directions.

7.43. Suppose that the thin film of Problem 7.41 is illuminated by a laser beam of diameter d at oblique incidence. What criterion must be satisfied for the reflectivity obtained in that problem to be applicable to reflection of the laser beam? The purpose of this problem is to underscore the (possible) differences between reflection of an *infinite* plane wave and a *finite* laser beam.

7.44. A single particle is a coherent object, and hence if illuminated by 100% polarized light, the scattered light is also 100% polarized (although not necessarily with all the same ellipsometric parameters). This is not difficult to understand. But we also argue in Section 3.4.2 that a piece of paper is a coherent object, and yet if it is illuminated by 100% polarized light the reflected light is at best weakly polarized. Why the difference? Devise a simple experiment to show that light reflected by a piece of white paper illuminated by 100% polarized light is essentially unpolarized. Try the same experiment with an optically smooth object such as glassware.

7.45. This problem is related to the previous one. If 100% polarized light illuminates a piece of white paper, the reflected light is weakly polarized but not with a degree of polarization of exactly 0%. If 100% polarized light illuminates a piece of glass, the reflected light is not exactly 100% polarized. Explain. Why do we specify "white" paper? What happens when black paper is illuminated by 100% polarized light (see Prob. 7.30)? Do a simple experiment to find out.

7.46. Unlike scattering by a small particle (see Sec. 3.5), reflection because of an interface (i.e., the Fresnel coefficients in Sec. 7.2) does not depend explicitly on wavelength (although it does depend implicitly on wavelength by way of the refractive index). Give a simple explanation why.

HINT: This is not a problem in electromagnetic theory but rather requires invoking a fundamental characteristic of all equations in which the variables are dimensional.

7.47. We note at the end of the chapter that neglecting polarization results in errors in radiance calculations for clear air illuminated by sunlight. Based strictly on physical reasoning you should be able to estimate the (normal) optical thickness for which the error is a maximum.

7.48. In the references for Chapters 2 and 3 we cite (quite favorably) Tony Rothman's *Everything's Relative*. But that doesn't mean that we'll let him get away with the footnote on page 22 of this book: "every time you look through a pair of Polaroid sunglasses you are using Malus' discovery. Polaroids work because reflected light viewed through them loses half its intensity." What's wrong with this (other than the use of intensity for luminance)?

7.49. If you did Problem 1.40 then you already have done almost everything you need to find the specific entropy of blackbody radiation. Do so.

7.50. In what simple, intuitive sense can specular reflection be said to be reversible whereas diffuse reflection is irreversible.

HINT: Consider rays.

7.51. Based strictly on your intuition about entropy determine what happens to the entropy of radiation upon specular reflection and refraction, diffuse reflection, scattering of a beam by a particle, and the incoherent superposition of partially polarized beams.

7.52. Section 7.2 begins with a discussion of reflection and refraction because of illumination of a smooth interface between two different media, the first of which can be taken to be air. Equation (7.86) is the wave vector of the transmitted wave in an arbitrary illuminated medium. Show that if this medium is absorbing, the surfaces of constant amplitude are parallel to the interface and the surfaces of constant phase are not, except for normal incidence.

7.53. Show that for a plane harmonic wave the time-averaged Poynting vector is *not*, in general, parallel to the real part of the complex wave vector. Under what conditions is the Poynting vector parallel to the real part of the wave vector? You may take the constant C in Eq. (7.5) to be real.

7.54. Consider reflection and refraction because of a smooth interface between two dissimilar infinite media. A wave is incident at an arbitrary direction from medium 1, giving rise to a refracted wave in medium 2 and a reflected wave in medium 1. Denote by $\tilde{t}_{12}$ the amplitude of the wave transmitted from 1 to 2 and by $\tilde{r}_{12}$ the amplitude of the reflected wave (for unit incident amplitude). The polarization state is either parallel or perpendicular to the plane of incidence. Now consider the reverse: a wave is incident from medium 2 along the direction of the refracted wave. Denote by $\tilde{t}_{21}$ and $\tilde{r}_{21}$ the corresponding ratios of amplitudes. Show that

$$\tilde{t}_{12}\tilde{t}_{21} = 1 - \tilde{r}_{12}^2,$$

$$\tilde{r}_{12} = -\tilde{r}_{21}.$$

These relations seem to have been derived first by Stokes. You can obtain them by physical arguments about the reversibility of the waves or analytically from the Fresnel equations. A simple diagram is essential. You may find statements in textbooks that these relations are valid only for nonabsorbing media. Show that this is not true.

7.55. With the results of the previous problem derive Eq. (5.24), the normal-incidence reflectivity of a nonabsorbing slab of uniform thickness, by adding all the multiply reflected waves taking account of all phase shifts. The equation you obtain for the reflected amplitude will be valid even for an absorbing medium, illuminated at arbitrary angle of incidence, and for either polarization state (parallel or perpendicular).

7.56. Find the transmitted amplitude for the slab considered in the previous problem. As a check on your result, show that the corresponding transmittivity plus the reflectivity is equal to 1.

7.57. We implicitly assumed in the previous problems that the medium on both sides of the slab is the same. Find the reflectivity of a slab by the series summation method for medium 1 on one side, medium 3 on the other side (all media negligibly absorbing).

7.58. Consider a negligibly absorbing slab between two different infinite media, also negligibly absorbing. Denote the media by subscripts 1, 2, and 3, where 2 denotes the slab. Show that the reflectivity for radiation incident from medium 1 is the same as that for radiation from medium 3. Keep in mind that this *reciprocity principle* requires the angle of incidence for radiation from medium 3 to be the same as the angle of transmission for the slab when illuminated from medium 1.

7.59. Use the results of Problem 7.57 to design an anti-reflection coating. That is, for normal-incidence radiation of wavelength λ, determine the refractive index n_2 and thickness h of a thin layer to be deposited on a substrate with refractive index n_3 such that the reflectivity of the system is zero.

7.60. We state that infinite, monodirectional plane electromagnetic waves do not exist. Prove this.

HINT: About the only physical law that we can always depend upon is conservation of energy.

8 Meteorological Optics: The Reward

The subtitle of this chapter promises that mastering concepts in previous chapters provides the means to understand – even to see in the first place – a rich assortment of atmospheric displays, some strikingly beautiful, some subtle, often remarkably detailed and full of surprises. A day hardly goes by anywhere without something in the sky worth looking, even marveling, at. From sunrise to sunset and beyond, a free show is open to those whose minds have been prepared to see it. Scientific understanding enhances, not diminishes, one's pleasure from scanning the sky. We often have begun a lecture to a large class by asking if anyone has ever seen a sun dog or a halo. Usually, not one student has. But within a week after we have discussed and explained them, excited students come to tell us that what they formerly had been blind to they now have seen. The atmosphere is not different, only the minds of the students.

We use the term *meteorological optics* here, in preference to *atmospheric optics*, to emphasize that our subject does *not* include, say, transmission of laser light through the atmosphere. Here we are interested in what the human observer sees when looking skyward.

Meteorological (or atmospheric) optics is nearly synonymous with light scattering, the only restriction being that the scatterers – molecules or particles – inhabit the atmosphere and the primary source of their illumination is the sun. We know of only a few exceptions in which absorption plays an important role in what is observed. In previous sections, for example that on polarization of skylight (Sec. 7.3), we touch on various topics in meteorological optics. And the chapters on scattering (Ch. 3), radiometry and photometry (Ch. 4), and multiple scattering (Ch. 5) lay the foundations on which we build this chapter.

8.1 Color and Brightness of the Molecular Atmosphere

A few years ago, Spain's leading newspaper, *El País*, published an interview with Manuel Toharia, Director of the Prince Felipe Museum of Sciences, in which he averred that science "is the response to man's curiosity. And this curiosity leads him to raise big and small questions. To me the question of why the sky is blue is not smaller than the question of where we come from and where we are going. Both are big."

Why is the sky blue? is one of the most frequently asked scientific questions. Although not inherently difficult, an answer that is not misleading, incomplete, or outright wrong is almost impossible to find.

One of our former graduate students, Cliff Dungey, taught a game to his daughter when she was three. In a gathering of people he would ask her, "Caryn, Why is the sky blue?" She would answer proudly, "Because of Rayleigh scattering" – and we'd all laugh. This story has

Fundamentals of Atmospheric Radiation: An Introduction with 400 Problems. Craig F. Bohren and Eugene E. Clothiaux
Copyright © 2006 Wiley-VCH Verlag GmbH & Co. KGaA, Weinheim
ISBN: 3-527-40503-8

two morals. One is that invoking the name of a theory doesn't explain anything. And the other is that anything a child can parrot is not a good explanation. Missing from Caryn's childish mantra is the agent responsible for the blue sky. If Lord Rayleigh had never lived, this agent would be the same. Theories come and go, agents endure.

Water is essential for life, truly a vital fluid. Early in our schooling we are told that it is anomalous in its properties. With time, water becomes not merely vital and anomalous but magical. We remind you of two scientific scandals that received worldwide attention, Polywater and Cold Fusion, in which the minds of otherwise sober scientists were clouded by water.

Given the magical status water has acquired, no wonder it often is taken as the cause of the blue sky. Leonardo's explanation was that "the blueness we see in the atmosphere... is caused by warm vapour evaporated in minute and insensible atoms on which the solar rays fall, rendering them luminous against the infinite darkness of the fiery sphere which lies beyond." Newton invoked "Globules of water." Clausius proposed scattering by minute bubbles. All water-based theories – and all wrong. Scattering of sunlight by air molecules is the cause of the blue sky. Which ones? The most numerous: nitrogen and oxygen. Water plays no essential role. Only about one air molecule out of every 100 is a water molecule, and molecule for molecule water vapor scatters *less* visible light than either nitrogen or oxygen. Water vapor does scatter sunlight, and hence does contribute to skylight, but because of its relatively low abundance and smaller scattering per molecule, its contribution is negligible. Nitrogen and oxygen are the agents causing the blue sky because they are selective or wavelength-dependent scatterers. Air is a scattering filter.

Scattering of visible light by air molecules is described to good approximation by Rayleigh's scattering law, according to which scattering is inversely proportional to the wavelength of the illumination to the fourth power (see Sec. 3.2). Reality is a bit more complicated. The power in Rayleigh's law is not on the same footing as, say, that in the law of universal gravitational attraction, which is 2 to a staggering number of digits. Scattering by air molecules over the visible spectrum is described more accurately as inversely proportional to wavelength to the power 4.09. By simple dimensional arguments, Rayleigh captured most but not all of the power law applicable to scatterers small compared with the wavelength of the illumination.

A scattering law with a limited domain of validity was one of Rayleigh's major contributions, but he was not the first to recognize the origins of the blue sky. Indeed, he begins his 1871 paper with the assertion "It is now, I believe, generally admitted that the light which we receive from the clear sky is due in one way or another to small suspended particles which divert the light from its regular course. On this point the experiments of Tyndall with precipitated clouds seem quite decisive." One can go back even further than Tyndall, almost a century, to de Saussure who in 1789 wrote that "air is not perfectly transparent; its elements always reflect some rays of light, especially blue rays. It is these reflected rays which produce the blue color of the sky. The purer the air, and the greater the extent of this pure air, the darker the blue color."

In 1871 Rayleigh confessed that he could not say with certainty which "small suspended particles" are responsible for the blue sky. He opined that they might be common salt, but recognized that his law could neither prove nor disprove this because it is valid for any matter if sufficiently finely divided. This loose end must have bothered him because he returned to

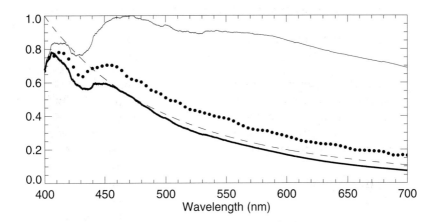

Figure 8.1: Overhead skylight around noon measured in spring in Central Pennsylvania (circles). The thick solid curve is the product of Rayleigh's scattering law (dashes) normalized to one at its greatest value and the solar spectrum (thin solid curve) outside Earth's atmosphere normalized to one at its peak.

the problem of the blue sky 28 years later, beginning his 1899 paper with the assertion "I think that even in the absence of foreign particles we should still have a blue sky."

Are we done? No, we have just begun. In particular, we have to criticize the way Rayleigh's scattering law is almost invariably used to explain the blue of the sky. It is obvious from this law that blue light is scattered more than red, and hence, we are told, the sky is blue. Misusing Rayleigh's law in this way is an example of cutting and bending a theory to fit an observation. It is just as obvious from Rayleigh's law that violet light is scattered even more than blue, and hence by the same logic the sky should be violet. But as we show in Section 4.3, there is no necessary connection between the maximum of a spectrum and the color it evokes.

Evidence for the essential correctness of Rayleigh's explanation is agreement between the product of his scattering law and the solar spectrum outside the atmosphere with the measured zenith skylight spectrum (Fig. 8.1). But again, we are not done. The clear sky is neither uniformly nor inevitably blue.

8.1.1   Variation of Sky Color and Brightness

Selective scattering by molecules is necessary but not sufficient for a blue sky. The atmosphere also must be optically thin, at least for most zenith angles. The blackness of space as a backdrop is taken for granted but also is necessary, as Leonardo recognized. Figure 8.2 shows the normal scattering optical thickness versus wavelength for the Standard Atmosphere.

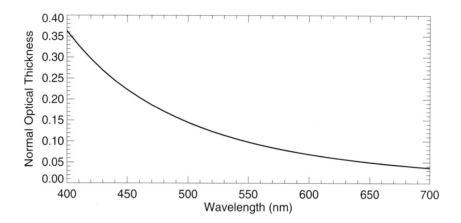

Figure 8.2: Normal scattering optical thickness for the Standard Atmosphere. From Penndorf (1957).

From the two-stream approximation [Eq. (5.66)] with $g = 0$ (molecular scattering) the diffuse downward irradiance $D_\downarrow$ of overhead skylight at the surface is

$$\frac{D_\downarrow}{F_0} = \frac{1}{1 + \tau_n/2} - \exp(-\tau_n), \tag{8.1}$$

where F_0 is the incident irradiance and τ_n is the normal optical thickness of the atmosphere. For $\tau_n \ll 1$ this approximates to

$$\frac{D_\downarrow}{F_0} \approx \frac{\tau_n}{2}. \tag{8.2}$$

Equations (8.1) and (8.2) are for a black underlying surface (zero reflectivity). What about the other extreme, a white underlying surface (a reflectivity of 1)? We can answer this by solving Eq. (5.49) subject to equal downward and upward irradiances at the surface. But it is better for our souls (i.e., our physical intuition) if we guess that it must be approximately twice that given by Eq. (8.2) because the air is illuminated by two approximately equal sources: direct and reflected sunlight. From Fig. 8.2 it follows that the condition for the validity of Eq. (8.2) is satisfied. Thus the spectrum of skylight for a molecular atmosphere should be the solar spectrum modulated by Rayleigh's scattering law, as indeed it is for the overhead sky (Fig. 8.1). What about the other extreme, the horizon sky? To answer this leads us to consider *airlight*.

The only real distinction between airlight and skylight is that the backdrop for airlight can be finite objects at a finite distance, whereas the backdrop for skylight is nearly empty, boundless space. Because of airlight, light scattered by all the molecules and particles along the line of sight from observer to object, even an intrinsically black object is luminous. Consider a horizontally uniform line of sight uniformly illuminated by sunlight (Fig. 8.3). Denote by L_0 the solar radiance illuminating the line of sight. The irradiance in the direction of the sun is

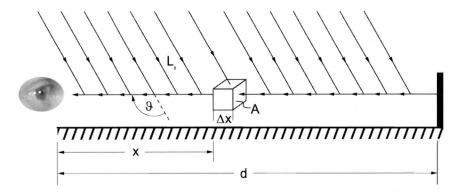

Figure 8.3: An observer looking at a distant black object nevertheless receives some light because of scattering by all the molecules and particles along the line of sight.

$L_0\Omega_s$, where Ω_s is the solid angle subtended by the sun. Assume negligible reflection by the ground. The light scattered toward the observer per unit solid angle by all the molecules and particles in a small volume $A\Delta x$ at a distance x is therefore $L_0\Omega_s\beta\Delta x A p(\vartheta)$, where β is the scattering coefficient and $p(\vartheta)$ is the probability per unit solid angle of scattering in the direction ϑ. Divide by A, which is perpendicular to the line of sight, to obtain the contribution to the radiance from $A\Delta x$. This is the radiance scattered at x *toward* the observer, which has to be multiplied by the transmissivity $\exp(-\beta x)$ to obtain the fraction of this radiance received *by* the observer. The total radiance as a consequence of scattering by everything along a line of sight between the observer and a black object at a distance d is the integral

$$L = L_0\Omega_s p(\vartheta)\beta \int_0^d \exp(-\beta x)\ dx = L_0 G\{1 - \exp(-\tau)\}, \tag{8.3}$$

where $\tau = \beta d$ is the optical thickness along the path d and the two geometrical factors are lumped into a single factor $G = \Omega_s p(\vartheta)$. Underlying Eq. (8.3) is the assumption that light scattered out of the line of sight is not scattered again in this direction, which is a good assumption if the optical thickness in directions lateral to the line of sight is small (which it is for the clear atmosphere but not for fog).

Only in the limit $\tau \to 0$ is $L = 0$ and a black object seen to be black. For $\tau \ll 1$, $L \approx L_0 G\tau$. In a purely molecular atmosphere τ varies with wavelength according to Rayleigh's law, and hence the distant black object is perceived to be bluish. As τ increases so does L but not proportionately: the longer the path, the greater the number of scatterers, but also the greater the attenuation. The limiting value of L ($\beta d \gg 1$) is $L_0 G$, and the radiance spectrum is that of the source illumination on the line of sight *regardless* of the wavelength dependence of β. This result ought to put an end to blather about the white horizon sky infallibly signaling scattering by "big particles."

Although the molecular optical thickness in the visible of Earth's atmosphere is small along a radial path, this is no longer true for paths near or along the horizon. The optical

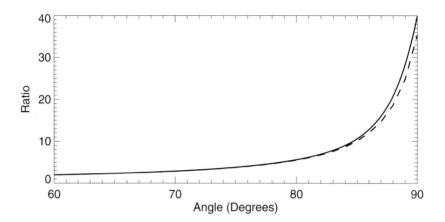

Figure 8.4: Scattering optical thickness of a pure molecular atmosphere with scale height 8 km on Earth relative to the normal optical thickness for a range of zenith angles near the horizon. The solid line is the uniform atmosphere approximation; the dashed line is for an exponentially decreasing scattering coefficient.

thickness along any path is an integral:

$$\tau = \int \beta \, ds. \tag{8.4}$$

For a path from the surface making a constant zenith angle Θ with the vertical direction in an atmosphere with an exponentially decreasing density of scatterers, Eq. (8.4) is

$$\tau = \int_0^\infty \beta_0 \exp\left\{ \frac{R - \sqrt{s^2 + 2sR\cos\Theta + R^2}}{H} \right\} ds, \tag{8.5}$$

where R is Earth's radius, H the scale height for molecular number density (i.e., the rate at which number density decreases exponentially with height), and β_0 the scattering coefficient at sea level. For a radial (normal) path ($\Theta = 0$) Eq. (8.5) can be integrated to obtain

$$\tau_n = \beta_0 H. \tag{8.6}$$

Thus the normal optical thickness of an atmosphere in which the number density of scatterers decreases exponentially with height is the same as that for a uniform atmosphere of finite thickness H.

Although Eq. (8.5) cannot be integrated analytically for arbitrary zenith angle, the uniform, finite atmosphere approximation

$$\frac{\tau}{\tau_n} = \sqrt{\frac{R^2}{H^2}\cos^2\Theta + \frac{2R}{H} + 1} - \frac{R}{H}\cos\Theta \tag{8.7}$$

is surprisingly good right down to the horizon ($\Theta = \pi/2$), as shown in Fig. 8.4. Taking the exponential decrease of molecular number density into account yields an optical thickness at most 10% lower. A flat Earth is one with infinite R, for which Eq. (8.7) yields the expected relation

$$\lim_{R \to \infty} \frac{\tau}{\tau_n} = \frac{1}{\cos \Theta}. \tag{8.8}$$

The tangential (horizon) optical thickness ($\Theta = \pi/2$) from Eq. (8.7) is to good approximation

$$\frac{\tau_t}{\tau_n} = \sqrt{\frac{2R}{H}} \tag{8.9}$$

because $2R/H \gg 1$. For $R = 6400$ km and $H = 8$ km, $\tau_t = 40\tau_n$.

The variation of brightness and color of dark objects with distance was called *aerial perspective* by Leonardo. By means of it we unconsciously estimate distances to objects of unknown size, such as mountains. Aerial perspective is similar to the variations of color and brightness of the sky with zenith angle. Although the optical thickness along a horizon path is not infinite, it is sufficiently large (Figs. 8.2 and 8.4) that GL_0 is a good approximation for the radiance of the horizon sky. For isotropic scattering, a condition almost satisfied by molecules (see Sec. 7.3), G is about 10^{-5}, the ratio of the solid angle subtended by the sun to the solid angle of all directions (4π). Thus the horizon sky is not nearly so bright as direct sunlight.

Unlike in the milk experiment described in Section 5.2, what one observes when looking at the horizon sky is *not* (much) multiply scattered light. Both the whiteness of milk and that of the horizon sky have their origins in multiple scattering but manifested in different ways. Milk is white because it is weakly absorbing and optically thick, and hence all components of incident white light are multiply scattered to the observer even though the violet and blue components traverse a shorter average path in the milk than the orange and red components. White horizon light is that which has *escaped* being multiply scattered, although multiple scattering is why this light is white (strictly, has the spectrum of the source). More light at the short-wavelength end of the spectrum than at the long-wavelength end is scattered *toward* the observer, as evidenced by β in Eq. (8.3). But long-wavelength light has the greater probability of being transmitted to the observer without being scattered *out* of the line of sight, as evidenced by $\exp(-\beta x)$ in Eq. (8.3). For a sufficiently long optical path, these two processes compensate, resulting in a horizon radiance that of the source.

With Eq. (8.3) in hand we can make a stab at estimating the ratio of the horizon radiance to the zenith (overhead) radiance. If we take the incident sunlight to be nearly directly overhead the horizon (tangential) radiance is approximately

$$L_t = L_0 \Omega_s p(90°)\{1 - \exp(-\tau_t)\} \approx L_0 \Omega_s p(90°) \tag{8.10}$$

and the zenith radiance is approximately

$$L_n = L_0 \Omega_s p(0°)\{1 - \exp(-\tau_n)\} \approx L_0 \Omega_s p(0°)\tau_n, \tag{8.11}$$

where p is the phase function for molecular scattering and L_0 is the radiance outside the atmosphere. All we need is the ratio of phase functions for the two scattering directions,

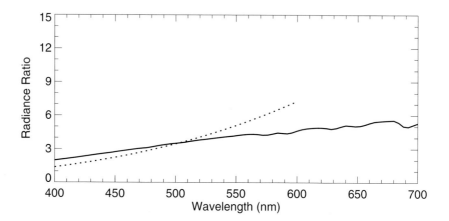

Figure 8.5: Measured ratio (solid line) of the horizon radiance to the radiance directly overhead with the sun high in the sky on a clear day in State College, Pennsylvania. The dotted line is this ratio predicted by simple theory for a pure molecular atmosphere.

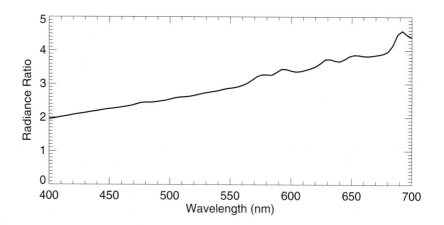

Figure 8.6: Measured ratio of the spectral radiance of magnesium oxide (MgO) powder illuminated by daylight to the spectral radiance of the horizon sky on a clear day in State College, Pennsylvania.

which we get from Eq. (7.117). The result is

$$\frac{L_t}{L_n} \approx \frac{1}{2\tau_n}. \tag{8.12}$$

Although attenuation of sunlight illuminating the line of sight is neglected in Eqs. (8.10) and (8.11), when attenuation is included Eq. (8.12) is unchanged. Also $p(0°)$ does not mean that

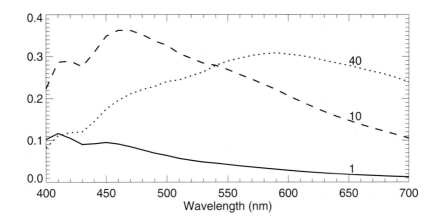

Figure 8.7: Spectra of overhead skylight from the two-stream theory for a molecular atmosphere with the present optical thickness (solid line), 10 times this thickness (dashed line), and 40 times this thickness (dotted line).

the sun is directly overhead and the line of sight is directly toward the sun but rather that the sun is high in the sky and the scattering angle is, say, less than 10–20°. Evidence for the validity of Eq. (8.12) is shown in Fig. 8.5, the ratio of measured radiances of the horizon and overhead skies, with the sun high in the sky, on a clear day. Agreement between measured ratios and those calculated with Eq. (8.12) using the normal optical thickness in Fig. 8.2 is surprisingly good. Moreover, the disagreement at longer wavelengths is in the expected direction: the normal optical thickness is almost always greater than that for a pure molecular atmosphere even in a very clean environment.

The optical thickness through the atmosphere along a horizon path is essentially infinite even in clear air. The source of illumination of this path is sunlight. The optical thickness of a large cumulus cloud is also essentially infinite, the source of illumination for which is also sunlight. Yet the radiance of the brightest cumulus cloud is larger, by roughly a factor of four, than that of the clear horizon sky. This is shown in Fig. 8.6, the ratio of the measured spectral radiance of magnesium oxide powder, which simulates a thick cloud, illuminated by daylight (i.e., direct sunlight and skylight) to the radiance of the horizon.

Although Eq. (8.12) is strictly valid only for small ($\ll 1$) normal optical thicknesses, it does suggest that with increasing optical thickness the gradient (in angle) of skylight radiance should decrease. And, in fact, this is what is observed: on murky days the sky is more nearly uniformly bright. Moreover, Eq. (8.12) also suggests that as one ascends in the atmosphere, and hence τ_n of everything above one's elevation decreases, the gradient of skylight should increase; this can be observed from an airplane.

It follows from the plot of Eq. (8.1) in Fig. 5.13 and the molecular optical thickness spectrum that a blue sky is not inevitable. For optical thicknesses less than about 2.2, skylight irradiance relative to the solar irradiance increases with increasing optical thickness. Because the optical thickness of the molecular atmosphere increases with decreasing wavelength, and over the visible spectrum is less than about 0.36 (Fig. 8.2), skylight irradiance increases with decreasing wavelength. But for optical thicknesses greater than about 2.2, skylight irradiance decreases with increasing optical thickness. The smallest molecular optical thickness in the visible is about 0.04 (at 700 nm). Thus if the atmosphere were about 50 times thicker skylight irradiance would *decrease* with decreasing wavelength. Figure 8.7 shows calculated spectra of the zenith sky over black ground for a molecular atmosphere with the present normal optical thickness as well as for hypothetical atmospheres 10 and 40 times thicker. What we take to be inevitable is accidental: if Earth's atmosphere were much thicker the sky would not only be brighter, its color would be quite different from what it is now.

By showing that the blue sky is not inevitable, we hope to have given you a taste for thinking the unthinkable. We emphasized that the white horizon sky is not a consequence of "big particles" but occurs in a purely molecular atmosphere. Now we are going to turn this on its head and assert that "big particles" not only are not necessary for a white horizon sky, they can make it bluer than it would be otherwise.

Equation (8.3) for airlight has the same form for an atmosphere populated by molecules and particles because scattering coefficients and optical thicknesses are additive. For sufficiently large total optical thickness along a horizon path, the horizon radiance is

$$L = L_0 \Omega_s \overline{p}(\vartheta), \tag{8.13}$$

where $\overline{p}$ is the weighted average phase function for molecules (m) and particles (p):

$$\overline{p}(\vartheta) = \frac{\beta_m}{\beta_m + \beta_p} p_m(\vartheta) + \frac{\beta_p}{\beta_m + \beta_p} p_p(\vartheta). \tag{8.14}$$

To understand the observable consequences of Eqs. (8.13) and (8.14) consider a few limiting cases. Suppose that both scattering toward the observer and total scattering are dominated by particles ($\beta_p p_p \gg \beta_m p_m$ and $\beta_p \gg \beta_m$):

$$L \approx L_0 \Omega_s p_p(\vartheta). \tag{8.15}$$

If the particles are big in the sense that angular scattering by them is independent of wavelength for scattering angles ϑ of interest, the airlight spectrum is white (i.e., that of the source L_0). No surprise here.

Now suppose that both scattering toward the observer and total scattering are dominated by molecules ($\beta_m p_m \gg \beta_p p_p$ and $\beta_m \gg \beta_p$):

$$L \approx L_0 \Omega_s p_m(\vartheta). \tag{8.16}$$

Again the radiance is that of the source because the molecular phase function is to good approximation independent of wavelength. Equation (8.14) also predicts a white horizon if the particles are sufficiently small that scattering by them has the same wavelength dependence as molecular scattering.

If molecules dominate total scattering whereas particles dominate scattering toward the observer ($\beta_m \gg \beta_p$ and $\beta_p p_p \gg \beta_m p_m$)

$$L \approx L_0 \frac{\beta_p p_p(\vartheta)}{\beta_m}. \tag{8.17}$$

Here the horizon radiance is *inversely* related to the molecular scattering coefficient, and hence the airlight is reddish. This is a variation on the theme of distant reddish clouds discussed in the following section on colors at sunrise and sunset. Particles distributed along the entire line of sight play the same role as localized clouds.

When we consider the converse of the previous limiting case, namely total scattering dominated by particles but scattering toward the observer dominated by molecules ($\beta_p \gg \beta_m$ and $\beta_m p_m \gg \beta_p p_p$), we obtain a surprising result:

$$L \approx L_0 \Omega_s \frac{\beta_m p_m(\vartheta)}{\beta_p}. \tag{8.18}$$

For this example, the horizon airlight is bluish. To understand this perhaps contra-intuitive result we return to the example of a pure molecular atmosphere for which the horizon sky is white even though the scatterers are selective. Scattering of sunlight toward the observer favors light at the short wavelength end of the visible spectrum. If this light were transmitted without attenuation, the airlight would be bluish. But it is impossible for molecules to scatter light toward the observer without also scattering some of this light out of the line of sight. This selective attenuation of light scattered toward the observer favors the long wavelength end of the spectrum. For a sufficiently long optical path selective scattering toward the observer is exactly balanced by selective scattering out of the line of sight.

Now we can better understand why big particles can, contrary to what might be expected, make the horizon sky bluer than it would otherwise be. Given the assumptions underlying Eq. (8.18) we can write the airlight radiance as

$$L \approx L_0 \Omega_s \beta_m p_m(\vartheta) \int_0^d \exp\{-(\beta_m + \beta_p)x\} \, dx. \tag{8.19}$$

The factor $\beta_m p_m$ is the wavelength-dependent scattering toward the observer and is the same everywhere along the line of sight. The integral is an attenuation function; the exponential term in the integral is the probability that light scattered at x toward the observer will not be scattered again in traversing this distance. Although light scattered at all points on the line of sight contributes to the radiance, most of the contribution comes from scattering at distances less than about $3/(\beta_m + \beta_p)$, which is approximately $3/\beta_p$ if $\beta_p \gg \beta_m$. Over such distances, however, molecular scattering does not greatly redden the transmitted light (i.e., $\exp\{-\beta_m x\} \approx 1$ for $x < 3/\beta_p$). Thus the color balance is not restored by attenuation as it was for a pure molecular atmosphere.

In his famous book, *The Nature of Light and Color in the Open Air*, Marcel Minnaert notes that "to this day there are scientists who do not consider the problem of the blue sky as being definitively solved...On very exceptional days, occurring perhaps not even once a year, the sky is beautifully blue right down to the horizon. Observations on days like these should be carefully recorded and described... for according to the theory of scattering, such a phenomenon is impossible: with layers of such thickness, the air ought to appear white." Yet the "theory of scattering" [Eq. (8.18)] does show why the sky can be beautifully blue right down to the horizon, although Minnaert was correct in saying that this is "exceptional." The concentration of particles has to be high enough that *total* scattering is dominated by them, but sufficiently low that *differential* scattering is dominated by molecules. This is possible because scattering by molecules does not vary much with scattering angle whereas scattering by particles comparable with or larger than the wavelength is highly peaked in the forward direction and drops by several decades toward the backward direction (see Sec. 3.5).

8.1.2   Sunrise and Sunset

If short-wavelength light is preferentially scattered *out* of direct sunlight, long-wavelength light must be preferentially transmitted *in* the direction of sunlight. Transmission is exponential if multiple scattering is negligible (see Sec. 5.2):

$$L = L_0 \exp(-\tau), \tag{8.20}$$

where L is the radiance in the direction of the sun, L_0 is that of sunlight outside the atmosphere, and τ is the optical thickness along the line of sight. If the wavelength dependence of τ follows Rayeigh's scattering law, transmitted sunlight is reddened, comparatively richer at the long-wavelength end of the visible spectrum than the incident light. But to say that sunlight is reddened is not to say that it is red. The perceived color can be yellow, orange, or red depending on the magnitude of the optical thickness. Equation (8.20) applies to the radiance only in the direction of the sun. Yet oranges and reds can be seen in other directions because reddened sunlight illuminates scatterers that are not on the line of sight to the sun. A striking example of this is a horizon sky tinged with oranges and pinks in the direction *opposite* the sun.

In an atmosphere free of all particles the optical thickness along a path from the sun, even on or below the horizon, is not sufficient to give perceptually red transmitted light. Although selective scattering by molecules yields a blue sky, reds are not possible in a molecular atmosphere, only yellows and oranges. Although this can be proven by the kind of colorimetric analysis in Section 4.3, Nature itself provides the proof. On exceptionally clear days the horizon sky at sunrise or sunset may be tinged with yellow or orange but not red.

The color and brightness of the sun changes as it arcs across the sky because the optical thickness along the line of sight to it changes with solar zenith angle Θ. If Earth were flat, as some still aver, the transmitted solar radiance would be

$$L = L_0 \exp(-\tau_n / \cos \Theta). \tag{8.21}$$

This equation is a good approximation except near the horizon. On a flat Earth, the optical thickness is infinite for horizon paths. On a spherical Earth, all optical thicknesses are finite although much larger for horizon than for vertical paths (Fig. 8.4).

Variations on the theme of reds and oranges at sunrise and sunset can be seen even when the sun is overhead. The radiance at an observer an optical distance τ from a horizon cloud is the sum of transmitted cloudlight and airlight:

$$L = L_0 G\{1 - \exp(-\tau)\} + L_0 G_c \exp(-\tau), \tag{8.22}$$

which is an extension of Eq. (8.3). If the cloud is approximated as an isotropic reflector with reflectivity R and illuminated at an angle Φ from the normal to it, the cloud geometrical factor G_c is $\Omega_s R \cos \Phi$. If $G_c > G$ the observed radiance is redder than the incident radiance, but if

$G_c < G$ the observed radiance is bluer than the incident radiance. Thus distant horizon clouds can be reddish if they are bright or bluish if they are dark.

Underlying Eq. (8.22) is the implicit assumption that the line of sight is uniformly illuminated by sunlight. The first term in this equation is airlight; the second is transmitted cloudlight. Suppose, however, that the line of sight is shadowed from direct sunlight by clouds that do not occlude the distant clouds. This may reduce the first term in Eq. (8.22) so that the second term dominates. Thus under a partly overcast sky, distant horizon clouds may be reddish even when the sun is high in the sky.

Small particles affect the color of the low sun out of proportion to their normal optical thickness because they are concentrated more toward the surface. The scale height for molecules is about 8 km whereas that for particles is typically 1–2 km. Subject to the approximations underlying Eq. (8.7), the ratio of the tangential (horizon) optical thickness for particles τ_{tp} to that for molecules τ_{tm} is

$$\frac{\tau_{tp}}{\tau_{tm}} = \frac{\tau_{np}}{\tau_{nm}} \sqrt{\frac{H_m}{H_p}}, \tag{8.23}$$

where m denotes molecules and p particles. Because of the incoherence of scattering by atmospheric molecules and particles, scattering coefficients are additive, and hence so are optical thicknesses. Even for equal normal optical thicknesses, the tangential optical thickness for particles is more than twice that for molecules. As we noted previously, molecules by themselves cannot give red sunsets and sunrises. Molecules need the help of small particles, and for a fixed normal optical thickness for particles, their tangential optical thickness is greater the more they are concentrated near the surface.

For equal normal optical thicknesses, particles also disproportionately increase the rate at which transmitted radiance decreases with angle near the horizon:

$$\left(-\frac{1}{L}\frac{\partial L}{\partial \Theta}\right)_{\Theta=\pi/2} = \left(\frac{\partial \tau_m}{\partial \Theta} + \frac{\partial \tau_p}{\partial \Theta}\right)_{\Theta=\pi/2} = \tau_{np}\frac{R}{H_p} + \tau_{nm}\frac{R}{H_m}, \tag{8.24}$$

which follows from Eq. (8.7) and Eq. (8.20) with $\tau = \tau_m + \tau_p$. The rate of decrease because of particles can be so great that the color of the setting sun varies across its diameter, from yellow at its top, to red at its bottom.

8.1.3   Ozone and the Twilight Sky

No theory ought to be accepted until it has been pushed to its limits. The theory of the blue sky discussed in previous sections gives cause for doubts when we consider the overhead sky at or around sunset. Even without doing any calculations we ought to suspect a discrepancy between theory and observations. The source of light illuminating molecules and particles along a vertical path at sunset is sunlight that has been appreciably reddened (i.e., its spectrum skewed toward the long-wavelength end of the spectrum) over long tangential atmospheric paths. Selective scattering by molecules does not create blue light: if red light illuminates molecules the scattered light is red. So how can the overhead sky be blue when the sun is low on the horizon? Solely by scattering it cannot. To show this quantitatively we appeal to the uniform, finite atmosphere approximation (Fig. 8.8).

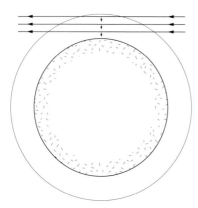

Figure 8.8: With the sun on the horizon, an observer looking straight overhead sees light scattered along a vertical path, where the source of this light is sunlight attenuated by different amounts along horizontal paths. The scale here is greatly distorted.

If we neglect multiple scattering, the total overhead radiance as a consequence of scattering of attenuated sunlight is

$$L = L_0 G \int_0^H \beta \exp\{-\beta(d+z)\} \, dz = L_0 G \int_0^H \beta \exp\{-\tau(z)\} \, dz, \qquad (8.25)$$

where G is a geometrical factor of no concern here, d is the path length of sunlight incident at height z above the surface, and τ is the optical thickness. Because the radius R of Earth is much greater than the scale height H, to good approximation

$$d \approx \sqrt{2R(H-z)}. \qquad (8.26)$$

Even without evaluating Eq. (8.25) it should be evident that L does not correspond to bluish light. Although the scattering coefficient β is greater at the short-wavelength end of the spectrum, the wavelength dependence of the exponential function is just the reverse. And because the optical thickness lies between τ_n (normal) and τ_t (tangential), the exponential function dominates. If the variable of integration is transformed to $H - u$, H neglected in comparison with R, the variable of integration transformed to u/H and finally to u^2, Eq. (8.25) becomes

$$L \approx 2L_0 G \exp\{-\tau_n\}\tau_n \int_0^1 \exp\{-\tau_t u\} \, u \, du. \qquad (8.27)$$

The integral in this equation is approximately $1/\tau_t^2$ because $\exp\{-\tau_t\} \ll 1$. The final result is the tidy expression

$$L \approx L_0 G \frac{H}{R} \frac{\exp\{-\tau_n\}}{\tau_n}. \qquad (8.28)$$

As a check on the correctness of this result note that $L \to 0$ as $H/R \to 0$. That is, when the atmosphere doesn't exist ($H = 0$) or Earth is flat, and hence the horizon path length is

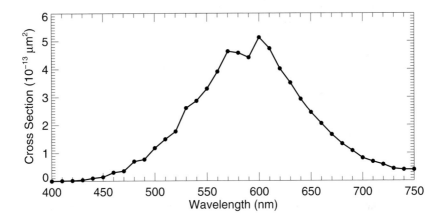

Figure 8.9: Absorption cross section of ozone for 238 K and 10 mb total pressure obtained from the line-by-line code developed by Clough *et al.* (1992) and cited at the end of Chapter 2. The Chappuis bands of ozone, originating from electronic transitions, do not depend strongly on temperature and pressure.

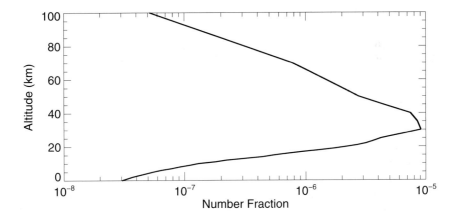

Figure 8.10: Mid-latitude summer ozone profile taken from McClatchey *et al.* (1972) cited at the end of Chapter 2.

infinite, the overhead radiance must vanish. Equation (8.28) is just the inverse of the blue sky: wavelength to the fourth power (approximately) instead of the inverse fourth power.

What is missing from this analysis is absorption by ozone. We briefly mention the Chappuis bands of ozone in Section 2.8.2, which are shown in Fig. 2.12. Figure 8.9 shows these bands in more detail on a linear wavelength scale. Note the broad absorption peak in the middle of the visible spectrum. Because photo-dissociation of oxygen into atomic oxygen

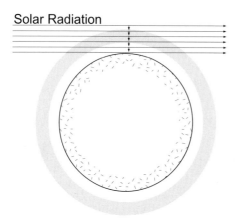

Figure 8.11: Path lengths, and hence absorption optical thicknesses, through an ozone layer concentrated mostly between about 20 km and 40 km (shaded) are much greater when the sun is on the horizon than when overhead. The scale of this figure is greatly distorted

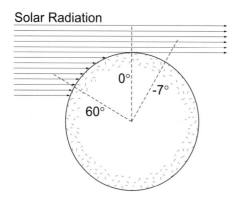

Figure 8.12: Dashed lines intersect Earth at points for which the sun elevation is 60°, on the horizon, and below the horizon. The angle below the sun is exaggerated (calculations were done for −7°) to show that much of the observer's line of sight is not illuminated by direct sunlight.

and subsequent combination of atomic and molecular oxygen is the source of ozone in the atmosphere, excluding lightning and anthropogenic sources, its concentration is sharply peaked between about 20 km and 40 km (Fig. 8.10). Figure 8.11 depicts this layer of ozone, path lengths through which depend on the solar elevation angle: the lower the sun, the greater the path lengths and hence absorption optical thicknesses.

Calculations of the radiance spectrum of the overhead sky for a molecular atmosphere with and without ozone are shown in Fig. 8.13 for solar elevation angles 60°, 0° (sun on the horizon), and −7° (sun below the horizon). For these calculations the uniform atmosphere approximation was not used because most ozone is well above the scale height $H \approx 8$ km. Figure 8.12 depicts the position of the observer for these three solar elevations.

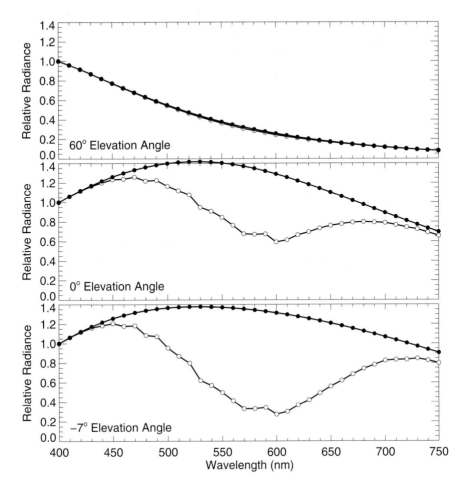

Figure 8.13: Calculated radiance spectra (normalized to 1 at 400 nm) of the overhead sky for a molecular atmosphere with (open circles) and without (closed circles) absorption by ozone for three different solar elevation angles. The spectrum of the illumination is a 6000 K blackbody, which approximates the solar spectrum outside Earth's atmosphere.

When the sun is 60° above the horizon, radiance spectra with and without absorption by ozone are indistinguishable (Fig. 8.13). And this is more or less true even when the sun is as low as 10° above the horizon. Thus over much of the day ozone plays no essential role in the blue of the sky. But when the sun is near or below the horizon, the overhead sky would not be blue without ozone. Absorption by ozone takes a big bite out of the middle of the radiance spectrum. But there is more to this story than just the *amount* of ozone: the radiance spectrum depends on *where* the ozone resides. Calculated radiance spectra (Fig. 8.14) for the sun on the horizon and for a fixed integrated amount of ozone but uniformly distributed between the surface and 15 km, between 20 km and 35 km, and between 85 km and 100 km

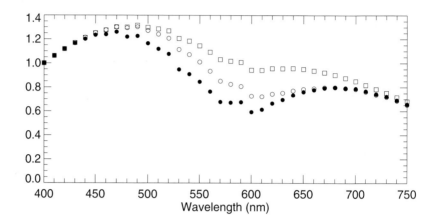

Figure 8.14: Calculated radiance spectrum (normalized to 1 at 400 nm) of the overhead sky with the sun on the horizon for a fixed total amount of ozone but distribted differently: a uniform layer between 20 km and 35 km (solid circles), a uniform layer from the surface to 15 km (open circles), and a uniform layer from 85 km to 100 km (squares). The spectrum of the illumination is a 6000 K blackbody, which approximates the solar spectrum outside Earth's atmosphere.

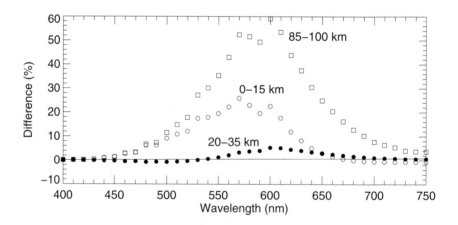

Figure 8.15: Relative difference between the radiance spectrum of the overhead sky with the sun on the horizon for a fixed total amount of ozone in a uniform layer between 20 km and 35 km (solid circles), between the surface and 15 km (open circles), and between 85 km and 100 km (squares), and the radiance spectrum corresponding to the non-uniform distribution shown in Fig. 8.10. The spectrum of the illumination is a 6000 K blackbody, which approximates the solar spectrum outside Earth's atmosphere.

are different than that for ozone distributed as shown in Fig. 8.10. The difference is greatest for the uniform 85–100 km layer, least for the 20–35 km layer, which roughly corresponds to the actual non-uniform distribution (Fig. 8.15).

This result strengthens our assertion that the blue sky is accidental rather than inevitable even though scattering by individual air molecules obeys Rayleigh's scattering law. A blue overhead sky for Earth's atmosphere from dawn to dusk requires not only a sufficiently small optical thickness but ozone in just the right amount and in the right place.

8.2 Atmospheric Visual Range

On a clear day can we really see forever? If not, how far can we see? To answer this question requires qualifying it by restricting viewing to more or less horizontal paths during daylight. Stars at staggering distances can be seen at night, partly because there is little skylight to reduce contrast, partly because stars overhead are seen in directions for which attenuation by the atmosphere is least.

We are careful to distinguish between visibility, a *quality*, and visual range, a *quantity*. One can say that visibility is good or poor but not that it is 10 km or 100 km. W. E. Knowles Middleton, whose book *Vision in the Atmosphere* is the standard work on atmospheric visibility, inveighed against the careless confutation of these two terms. He didn't have much effect, and neither will we, so all we can do is express our scorn for folks, especially learned doctors of science, so devoid of linguistic sense that they can't distinguish a quality from a quantity.

The radiance in the direction of a black object is not zero because of airlight (Sec. 8.1). At sufficiently large distances, this airlight is indistinguishable from the horizon sky. An example is a phalanx of parallel dark ridges, each ridge brighter than those in front of it (Fig. 8.16). The farthest ridges blend into the horizon sky. Beyond some distance we cannot see ridges because of insufficient contrast.

Figure 8.16: The brightness of each of these ridges, all covered with the same dark vegetation, increases with increasing distance, and hence their contrast with the horizon sky decreases.

Equation (8.3) gives the airlight radiance L, from which we obtain the airlight luminance B, which is what humans sense, by integrating over the visible spectrum:

$$B = K \int V(\lambda) L(\lambda) \, d\lambda, \tag{8.29}$$

where V is the luminous efficiency of the human eye and K is a constant of no concern here (see Sec. 4.1). The *contrast* C between any object, with luminance B, and the horizon sky is

$$C = \frac{B - B_\infty}{B_\infty}, \tag{8.30}$$

where B_∞ is the horizon luminance. For a uniformly illuminated line of sight of length d, uniform in its scattering properties, and over black ground, the contrast obtained from Eq. (8.3) is

$$C = -\frac{\int GV L_0 \exp(-\beta d) \, d\lambda}{\int GV L_0 \, d\lambda}, \tag{8.31}$$

where we assume an infinite horizon optical thickness. This ratio of integrals defines an average (over the visible spectrum) optical thickness

$$\langle \tau \rangle = -\ln |C| \tag{8.32}$$

and a corresponding scattering coefficient

$$\langle \beta \rangle = \frac{\langle \tau \rangle}{d}. \tag{8.33}$$

Equation (8.32) for contrast reduction with optical thickness is formally, but not physically, identical to the expression for exponential attenuation at any wavelength of radiance [Eq. (8.20)], which perhaps is responsible for the misconception that atmospheric visibility is reduced because of attenuation. But if there is no light from a black object to be attenuated, its finite visual range cannot be a consequence of attenuation.

The distance beyond which a black object cannot be distinguished from the horizon sky depends on the *contrast threshold*, the smallest absolute value of contrast detectable by a human observer. Although this depends on the particular observer, the angular size of the object observed, the presence of nearby objects, and the absolute luminance, a contrast threshold of 0.02 is often taken as a typical value. To find the visual range for this threshold we have to evaluate the integrals in Eq. (8.31) numerically for various values of d to find the one for which the right side of this equation is -0.02. But if β is independent of wavelength the solution is simply

$$\beta d = 3.9. \tag{8.34}$$

This equation is often called Koschmieder's law, although W. E. Knowles Middleton notes that "there can be no doubt that Bouguer was quite clear about the main factors determining the horizontal visual range, and that he effectively stated the law which has lately been called Koschmieder's law."

The scattering coefficient for molecules, however, is not independent of wavelength, although the geometrical factor G is. The function V is fairly sharply, and symmetrically, peaked around 550 nm. Because the molecular scattering coefficient decreases, and hence the exponential function in Eq. (8.31) increases with increasing wavelength, the average molecular scattering coefficient must correspond to wavelengths somewhat greater than 550 nm. We numerically solved Eq. (8.31) for a pure molecular atmosphere at standard pressure (1013 mb) and temperature (0 °C) to obtain $d = 330$ km, which corresponds to a wavelength of about 560 nm. Because the wavelength dependence of scattering is steepest for molecules and particles small compared with the wavelength, we usually can estimate the visual range from Eq. (8.34) using the scattering coefficient at or around 560 nm.

According to this analysis, therefore, "forever" is around 330 km, the maximum visual range at which a black object at sea level can be distinguished from the horizon sky for a purely molecular atmosphere, assuming a contrast threshold of 0.02 and also ignoring the curvature of the Earth. This maximum is more than twice the visual range considered exceptionally high. From this we conclude that almost always visual range is limited by particles.

We also observe contrast between elements of the same scene, a hillside mottled with stands of trees and forest clearings, for example. The extent to which we can discern details in such a scene depends on sun angle as well as distance. The airlight radiance for a black object is given by Eq. (8.3), whereas that for a reflecting object is given by Eq. (8.22). These two equations can be combined to give the contrast between adjacent reflecting and non-reflecting objects

$$C = \frac{-\Gamma \exp(-\tau)}{1 + (\Gamma - 1)\exp(-\tau)}, \tag{8.35}$$

where

$$\Gamma = \frac{R \cos \Phi}{\pi p(\Theta)}. \tag{8.36}$$

Although Eq. (8.35) specifies the contrast of radiance, this equation is a good approximation to the luminance contrast if we take τ to be in the middle of the visible spectrum.

All else being equal, contrast decreases as p increases. And as we show in Section 3.5, p is more sharply peaked in the forward direction the larger the scatterer. Thus we expect the details of a distant scene to be less distinct when looking toward than away from the sun if the optical thickness of the line of sight has an appreciable component contributed by particles comparable with or larger than the wavelength. Indeed, on many occasions we have observed marked improvements in contrast on a distant ridge or mountain to the east from morning to late afternoon despite no obvious change in particle concentration.

The misconception that water vapor is a powerful scatterer of sunlight is probably largely a consequence of the common observation that on humid, hazy days, visibility is often depressingly poor. But haze is not water vapor, rather water that has *ceased* to be vapor. At high relative humidities, but still well below 100%, small soluble particles in the atmosphere accrete liquid water to become solution droplets. Although these droplets are much smaller than cloud droplets, they markedly diminish visual range because of the sharp increase in scattering with particle size (see Fig. 3.11). Because of coherence, the same number of water molecules when aggregated in haze scatter vastly more than when apart, increasing the scattering coefficient and therefore decreasing visual range.

8.3   Atmospheric Refraction

Atmospheric refraction is a consequence of molecular scattering, which is rarely stated given the historical accident that before light and matter were well understood refraction and scattering were locked in separate compartments and subsequently have been sequestered more rigidly than monks and nuns in neighboring cloisters.

Consider a beam of light propagating in an optically homogeneous medium. Light is scattered laterally to the beam, weakly but observably, and more strongly in the same direction as the beam (i.e., the forward direction). The observed beam is a coherent superposition of incident light and forward-scattered light excited by the incident light. Although real refractive indices are often defined by ratios of phase velocities (see Sec. 3.5.1), we may also look upon the real refractive index as a parameter that specifies the phase shift between an incident beam and forward-scattered light. The connection between incoherent scattering and refraction, coherent scattering, can be divined from the expression for the refractive index n of a gas and that for the scattering cross section σ_s of a gas molecule:

$$n = 1 + \frac{1}{2}\alpha N, \tag{8.37}$$

$$\sigma_s = \frac{k^4}{6\pi}|\alpha|^2, \tag{8.38}$$

where N is the number (not mass) density of gas molecules, k is the wavenumber of the incident light, and α is the polarizability of the molecule (i.e., induced dipole moment per unit incident inducing electric field). The appearance of the polarizability in Eq. (8.37) but its square in Eq. (8.38) is the clue that refraction is associated with electric fields whereas scattering is associated with electric fields squared. Scattering without qualification often means incoherent scattering in all directions. Refraction, in a nutshell, is coherent scattering in a particular direction.

8.3.1   Terrestrial Mirages

Mirages are not illusions, any more so than are reflections in a pond. Reflections of plants growing at its edge are not interpreted as plants growing into the water. If the water is ruffled by wind, the reflected images may be so distorted that they are no longer recognizable as those of plants. Yet we would not call such distorted images illusions. And so it is with mirages. They are images noticeably different from what they would be in the absence of atmospheric refraction, creations of the atmosphere, not of the mind. An example of a true illusion is the *moon illusion*, a moon seen to be larger on the horizon than overhead. This seemingly enlarged moon is a creation of the mind, not the atmosphere. And yet the moon illusion is still often attributed to atmospheric refraction even though this has been known not to be true for at least 1000, possibly 2000 years, and can be verified by simple measurements of the angular size of the moon at different elevations.

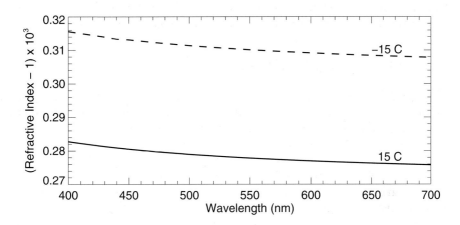

Figure 8.17: Refractive index of dry air at a pressure of one atmosphere and for the two temperatures noted. From the compilation by Penndorf (1957).

Mirages are vastly more common than is realized. Look and you shall see them. Contrary to popular opinion, they are not unique to deserts. Mirages can be seen frequently even over ice-covered landscapes and highways flanked by deep snowbanks. Temperature *per se* is not what produces mirages but rather temperature *gradients*.

Because air is a mixture of gases, the polarizability for air in Eq. (8.37) is an average over all its molecular constituents, although their individual polarizabilities are about the same at visible and near-visible wavelengths. The vertical refractive index gradient can be written so as to show its dependence on pressure p and absolute temperature T by way of the ideal gas law $p = Nk_{B}T$ and Eq. (8.37):

$$\frac{d}{dz}\ln(n-1) = \frac{1}{p}\frac{dp}{dz} - \frac{1}{T}\frac{dT}{dz}. \tag{8.39}$$

Pressure decreases approximately exponentially with height [i.e., $\exp(-z/H)$], where the scale height H is about 8 km. The first term on the right side of Eq. (8.39) is therefore about $0.1 \, \text{km}^{-1}$. Temperature usually decreases with height in the atmosphere. An average lapse rate of temperature (i.e., its decrease with height) is about $6 \, ^{\circ}\text{C} \, \text{km}^{-1}$. A characteristic temperature in the troposphere, within about 15 km of the surface, is 280 K. Thus the magnitude of the second term in Eq. (8.39) is about $0.02 \, \text{km}^{-1}$. On average, therefore, the refractive index gradient is dominated by the vertical pressure gradient. But within a few meters of the surface, conditions are far from average. On a sun-baked highway your feet may be touching asphalt at $50 \, ^{\circ}\text{C}$ while your nose is breathing air at $35 \, ^{\circ}\text{C}$, which corresponds to a lapse rate thousands of times the average. Moreover, temperature near the surface can increase with height. In shallow surface layers, in which pressure is nearly constant, the temperature gradient dominates the refractive index gradient. In such shallow layers mirages, which are caused by refractive index gradients, are seen.

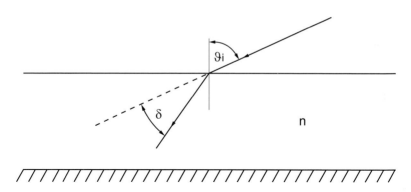

Figure 8.18: Deviation of incident light because of refraction by a uniform slab with refractive index n.

Cartoonists with fertile imaginations unfettered by science and careless textbook writers have engendered the notion that atmospheric refraction can work wonders, lifting images of ships, for example, from the sea high into the sky. A back-of-the-envelope calculation dispels such notions. The refractive index of air at sea level is about 1.0003 (Fig. 8.17). Light from free space incident on a uniform slab (Fig. 8.18) with this refractive index is displaced from where it would have been in the absence of refraction by an angle δ given by Snel's law

$$\sin \vartheta_i = n \sin \vartheta_t = n \sin(\vartheta_i - \delta) = n(\sin \vartheta_i \cos \delta - \sin \delta \cos \vartheta_i), \tag{8.40}$$

which at glancing incidence ($\vartheta_i = 90\,°$) yields

$$\cos \delta = \frac{1}{n}. \tag{8.41}$$

Because $n \approx 1$, and hence $\delta \ll 1$, we can approximate Eq. (8.41) as

$$\delta \approx \sqrt{2(n-1)}. \tag{8.42}$$

For $n - 1 = 0.0003$, Eq. (8.42) gives an angular displacement of about $1.4°$, which is a rough upper limit.

Trajectories of light rays in nonuniform media can be expressed in different ways. According to Fermat's principle of least time, which ought to be *extreme* time, the actual path taken by a ray between two points is such that the path integral

$$\int_1^2 n \, ds \tag{8.43}$$

is an extremum; strictly, this integral is *stationary*, which includes the possibility of a point of inflection. That is, of all possible paths between 1 and 2, that taken by a light ray is such that Eq. (8.43) is either a minimum (least time) or a maximum (greatest time). Why time?

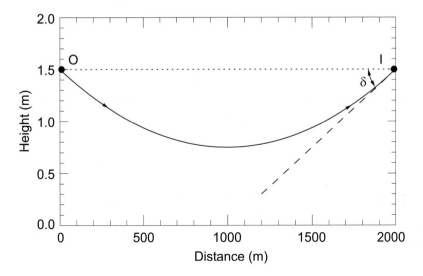

Figure 8.19: Ray trajectory from object point O to image point I in air with a temperature decreasing at a rate more than 100 times the average rate in Earth's atmosphere. To an observer at I it is as if the light from O comes from an object displaced downward from the line of sight OI by an angle δ. Note that the horizontal and vertical scales differ by a factor of about 600, which creates the impression that δ can be much larger than it is in reality ($\sim 1°$).

Because n is the ratio c/v, where c is a universal constant (the free-space speed of light) and v is the phase speed, and hence except for the constant factor c, Eq. (8.43) has the dimensions of time. But this time is *not* the time it would take a signal to propagate from 1 to 2 except in a non-dispersive medium (see Sec. 3.5). The principle of least time has inspired piffle about the alleged efficiency of nature, which directs light over routes that minimize travel time, presumably giving light more time to attend to important business at its destination.

The scale of terrestrial mirages is such that in analyzing them we may pretend that Earth is flat. On such a planet, with an atmosphere in which the refractive index varies only in the vertical, Fermat's principle yields a generalization of Snel's law:

$$n \sin \vartheta = \text{constant} = C, \tag{8.44}$$

where ϑ is the angle between the ray and the vertical direction. We could have bypassed Fermat's principle to obtain this result.

A ray in such a medium is a curve $z = f(y)$, where the yz-plane is the plane of incidence and z is the vertical coordinate. The slope of this curve is

$$\frac{dz}{dy} = \tan(\pi/2 - \vartheta) = \frac{1}{\tan \vartheta}. \tag{8.45}$$

Square Eqs. (8.44) and (8.45) and combine them to obtain

$$\left(\frac{dz}{dy}\right)^2 = \frac{n^2 - C^2}{C^2}. \tag{8.46}$$

Take the derivative with respect to y of both sides:

$$\frac{d^2 z}{d^2 y} = n\frac{dn}{dz}\frac{1}{C^2}. \tag{8.47}$$

Here $n \approx 1$, and if we restrict ourselves to nearly horizontal rays (i.e., $\vartheta \approx \pi/2$), we can set both n and C equal to 1 in Eq. (8.47) to obtain the approximate differential equation satisfied by nearly horizontal rays:

$$\frac{d^2 z}{d^2 y} = \frac{dn}{dz}. \tag{8.48}$$

This equation shows that terrestrial mirages are a consequence of vertical refractive index gradients: if this gradient is zero, ray paths are straight lines.

For a constant refractive index gradient, which to good approximation occurs for a constant temperature gradient, the solution to Eq. (8.48) is a parabola. One such parabola, for a constant lapse rate more than 100 times the average, is shown in Fig. 8.19. Note the greatly different horizontal and vertical scales. If we had plotted the parabola to uniform scale its curvature would not have been noticeable. The image is displaced downward from what it would be in the absence of the atmosphere, strictly in the absence of a vertical refractive index gradient. That is, an observer at I sees light that originated from O coming from a direction below (in angle) the straight line between O and I: hence the designation *inferior mirage*. This is the familiar mirage seen over highways warmer than the air above them. The downward angular displacement is

$$\delta = \frac{1}{2}s\frac{dn}{dz}. \tag{8.49}$$

This was obtained by solving Eq. (8.48), then determining the two constants of integration by requiring the ray to go through the points $(h, 0)$ and (h, s), where h is the height and s the horizontal distance between object (O) and image (I). The displacement is then

$$\tan\delta = \left(\frac{dz}{dy}\right)_{y=s} \approx \delta. \tag{8.50}$$

Even for temperature gradients 1000 times the average lapse rate, angular displacements of mirages are less than a degree at distances of a few kilometers.

If temperature increases with height, as it might, for example, in air over a colder sea, the resulting mirage is called a *superior mirage*. The refractive index gradient in Eq. (8.49) changes sign, as does δ. Inferior and superior are not designations of lower and higher castes but rather of displacements downward and upward.

For a constant temperature gradient, one and only one parabolic ray trajectory connects an object point to an image point. Multiple images therefore are not possible. But temperature gradients close to the ground are rarely linear. The upward transport of energy from a hot surface occurs by molecular conduction through a stagnant boundary layer of air. Somewhat above the surface, however, energy is transported by air in motion. As a consequence the temperature gradient steepens near the ground if the energy flux is constant. This variable gradient can lead to two consequences: magnification and multiple images.

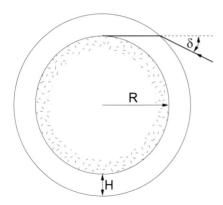

Figure 8.20: When the sun is at an angle δ (a fraction of a degree) below the horizon it still can be seen because of atmospheric refraction. The relative value of the scale height H to Earth's radius R is greatly exaggerated in this figure.

According to Eq. (8.49) all image points at a given horizontal distance are displaced downward by the same amount proportional to the constant refractive index gradient. This suggests that the closer an object point is to a surface, where the temperature is greatest, the greater the downward displacement of the corresponding image point. Thus nonlinear vertical temperature profiles may magnify images. Magnification in the optical sense is an increase in angular size, which is all that human observers directly perceive. Transforming angular sizes into linear sizes (lengths) is a complicated perceptual process. There is far more to seeing than the formation of images on retinas.

Multiple images are seen frequently on highways. What often appears to be water on a highway but evaporates before the water is reached is the inverted image of either the horizon sky or horizon objects brighter than the highway.

8.3.2   Extraterrestrial Mirages

When we turn from mirages of terrestrial objects to those of extraterrestrial bodies, most notably the sun and moon, we can no longer pretend that Earth is flat. But we can pretend that its atmosphere is uniform and bounded with a constant refractive index equal to the surface value n_0. The integrated refractive index of a vertical ray from the surface to infinity is the same in an atmosphere with an exponentially decreasing molecular number density as in a hypothetical atmosphere with a uniform density equal to the surface value up to the scale height H.

A ray refracted along a horizon path by this hypothetical atmosphere and originating from outside it (Fig. 8.20) must have been incident on it from an angle δ below the horizon. From Snel's law we have

$$\sin \vartheta_i = \sin(\vartheta_t + \delta) = n_0 \sin \vartheta_t = \sin \vartheta_t \cos \delta + \cos \vartheta_t \sin \delta, \tag{8.51}$$

from which it follows that

$$\sin \delta = \tan \vartheta_t (n_0 - \cos \delta), \qquad (8.52)$$

where

$$\tan \vartheta_t = \frac{R}{\sqrt{2HR + H^2}} \approx \sqrt{\frac{R}{2H}}. \qquad (8.53)$$

The radius R of Earth is much greater than the scale height of its atmosphere. Given that we expect $\delta \ll 1$ because $n \approx 1$, we can approximate $\sin \delta$ by δ and $\cos \delta$ by 1 in Eq. (8.51) to obtain

$$\delta \approx (n_0 - 1)\sqrt{\frac{R}{2H}}. \qquad (8.54)$$

A slightly more accurate (about 10%) but more cumbersome expression can be obtained from Eq. (8.51) by truncating the series expansion for the cosine after the second term rather than the first.

According to Eq. (8.54) when the sun or moon is seen to be on the horizon it is actually more than halfway below it, δ being about 0.34°, whereas the angular width of the sun or moon is about 0.5°.

Extraterrestrial bodies seen near the horizon also are vertically compressed. The simplest way to estimate the amount of compression is from the rate of change of angle of refraction with angle of incidence for a uniform slab, which from Eq. (8.51) is

$$\frac{d\vartheta_t}{d\vartheta_i} = \frac{\cos \vartheta_i}{\sqrt{n_0^2 - \sin^2 \vartheta_i}} = \sqrt{\frac{1 - \sin^2 \vartheta_i}{n_0^2 - \sin^2 \vartheta_i}}, \qquad (8.55)$$

where the angle of incidence is taken to be that for a curved but uniform atmosphere such that the refracted ray is horizontal:

$$\sin \vartheta_t = \frac{R}{R + H}. \qquad (8.56)$$

Equations (8.55) and (8.56) combined yield

$$\frac{d\vartheta_t}{d\vartheta_i} \approx \sqrt{1 - \frac{R}{H}(n_0 - 1)} \qquad (8.57)$$

if we neglect terms of order $(H/R)^2$ and approximate $n_0 + 1$ as 2 and n_0 as 1 when it is a multiplicative factor. According to Eq. (8.57) the sun or moon near the horizon is distorted into an ellipse with aspect ratio about 0.87. We are unlikely to notice this distortion, however, because we expect the sun and moon to be circular, and hence we see them that way. But if we compare two photographs of a low sun or moon taken at the same moment, one rotated by 90° relative to the other, the elliptical shape may become obvious.

Our conclusions about the downward displacement and distortion of the sun were based on a refractive-index profile determined mostly by the pressure gradient. That is, the average

Figure 8.21: Atmospheric refraction transformed this low sun into nearly a triangle. The serrations are a consequence of horizontal variations in the atmospheric refractive index.

refractive index gradient for a uniform slab of thickness H is $(1 - n_0)/H$, which is the same as Eq. (8.39) with $n = n_0$ if the temperature gradient term is negligible. But as we noted, near the surface the temperature gradient is the prime determinant of the refractive-index gradient, as a consequence of which the horizon sun can take on shapes more striking than a mere ellipse. For example, Fig. 8.21 shows a nearly triangular sun with serrated edges. Assigning a cause to these serrations provides a lesson in the perils of jumping to conclusions. Obviously, the serrations are the result of sharp changes in the temperature gradient – or so one might think. Setting aside how such changes could be produced and maintained in a real atmosphere, a theorem by Alistair Fraser gives pause for thought: "In a horizontally (spherically) homogeneous atmosphere it is impossible for more than one image of an extraterrestrial object (sun) to be seen above the astronomical horizon [horizontal direction determined by a bubble level]." These serrations on the sun are multiple images. But if the refractive index varies only vertically (i.e., along a radius), no matter how sharply, multiple images are not possible. Thus the serrations must owe their existence to *horizontal* variations of the refractive index, which Fraser attributes to gravity waves propagating along a temperature inversion.

8.3.3 The Green Flash

Compared to the rainbow, the green flash is not a rare phenomenon. Before you dismiss this assertion as the ravings of lunatics, consider that rainbows require raindrops as well as sunlight to illuminate them, and yet the clouds that are the source of these raindrops often completely obscure the sun. Moreover, the sun must be below about $42°$ (see Sec. 8.4). As a consequence, rainbows do not occur frequently (at least not in many parts of the world), but when they do occur, they are difficult *not* to see. And they are seen often enough to be considered the paragon of color variation ("all the colors of the rainbow" is a cliché). Yet tinges of green on the upper rim of the sun can be seen every day at sunrise or sunset given a sufficiently low horizon and a cloudless sky. Thus the conditions for seeing a green flash are met more often than those for seeing a rainbow. Why then is the green flash considered to be so rare ("the rare green flash" is another cliché)? The distinction here is that between a rarely *observed* phenomenon (the green flash) and a rarely *observable* phenomenon (the rainbow). To see the green flash requires knowing when and where and how to look whereas even people who go through life in a daze do occasionally trip over rainbows.

The green flash is not without its commercial uses, although much fewer than rainbows. Several green flash restaurants and bars can be found near beaches, including in the Caribbean, and even a Green Flash Brewing Company in California.

To understand the origins of the green flash we may consider the sun to be an infinite set of overlapping discs, one for each visible wavelength. When the sun is overhead, all these discs coincide and we see the sun as white. But as it descends in the sky, atmospheric refraction displaces the discs in angle by slightly different amounts, the red less than the violet (see Fig. 8.17). Most of each disc overlaps all the others except for the discs at the extremes of the visible spectrum. As a consequence, the upper rim of the low sun is violet or blue, its lower rim red, whereas its interior, the region in which all discs overlap, is still white.

At least this is what would happen in the absence of lateral scattering of sunlight. But refraction and lateral scattering go hand in hand; one cannot occur without the other even in an atmosphere completely free of particles. Selective incoherent scattering by atmospheric molecules and particles causes the spectrum of transmitted sunlight to shift toward longer wavelengths, and hence the perceived color of the sun to change. In particular, the violet-bluish upper rim can be transformed to green.

According to Eq. (8.55) and the refractive indices in Fig. 8.17 the angular separation between the violet and red solar discs when the sun is on the horizon is about $0.01°$, which is too small to be resolved by the human eye. You can verify this yourself by drawing two black parallel lines, say 4–6 mm apart, on a white piece of paper and observing them at increasingly greater distances. At a certain distance, they will merge into one, and this distance corresponds to an angular separation of around 1 minute of arc (0.3 mrad). You can make one of the lines red and the other green to convince yourself that different colors do not change the resolving power of the human eye. Thus in order to see the upper green rim of the sun requires binoculars or a telescope. But depending on the temperature profile, the atmosphere itself can magnify this rim and yield a second image of it, thereby enabling it to be seen by the naked eye. Green rims, which require artificial magnification, can be seen more frequently than green flashes, which require natural magnification. Yet both can be seen often by those who know what to look for and when and are willing to look.

Although the green flash is objectively real, Andrew Young argues that there is "compelling evidence that adaptation in the visual system strongly affects the perceived color of most green flashes." He notes that "photography shows that there is a real green flash in some sunsets. Green flashes are not afterimages. Nevertheless... physiological effects in the visual system must usually make the preceding yellow stage of a sunset flash appear green to an attentive observer." But green flashes at sunrise are a horse of a different color: "green flashes are also seen at sunrise, when the eye has not been previously exposed to bright light... Sunrise flashes are therefore seen more nearly in their intrinsic colors."

8.4 Scattering by Single Water Droplets

All the colored atmospheric displays that result when water droplets or ice crystals are illuminated by sunlight have the same underlying cause: light is scattered in different amounts in different directions by particles larger than the wavelength, and the directions in which scattering is greatest depend on wavelength. Thus when such particles are illuminated by sunlight, the result can be angular separation of colors even if scattering integrated over all directions is independent of wavelength, as it is for droplets and ice crystals. This description, although correct, is too general to be completely satisfying. We need something more specific, more quantitative, which requires theories of scattering.

Because superficially different theories have been used to describe different optical phenomena, the notion has become widespread that they are caused by these theories. For example, coronas are said to be caused by diffraction and rainbows by refraction. Yet both the corona and rainbow can be described quantitatively to high accuracy with Mie theory (Sec. 3.5.2) in which diffraction and refraction do not appear explicitly. As we noted in Section 3.1, no impenetrable barrier separates scattering from specular reflection, refraction, and diffraction. Because these terms came into general use and were entombed in textbooks before the nature of light and matter were well understood, we are stuck with them. But if we insist that diffraction, for example, is somehow different from scattering, we do so at the expense of shattering the unity of the seemingly disparate observable phenomena that result when light interacts with matter. What is observed depends on the composition and disposition of matter, not on which approximate theory in a hierarchy is used for quantitative description.

Atmospheric optical phenomena are best classified by the direction in which they are seen and the agents that cause them. Accordingly, the following sections are arranged in order of scattering direction, from forward to backward.

When a *single* water droplet is illuminated by white light and the scattered light projected onto a screen, the result is a set of colored rings. But this same set of rings in the sky is a mosaic to which a thin cloud of *many* droplets contributes. Light from each direction is that scattered by a different set of droplets in each patch of sky. Thus for a complete mosaic droplets must be present in sufficient number and illuminated by sunlight, and the cloud must be sufficiently thin that multiple scattering does not wash out the mosaic.

8.4.1 Coronas and Iridescent Clouds

A cloud of droplets narrowly distributed in size and thinly veiling the sun or moon can yield a striking series of concentric colored rings around it. This *corona* is most easily described quantitatively by the Fraunhofer diffraction theory, a simple approximation valid for particles large compared with the wavelength for scattering near the forward direction. According to this approximation, which must break down because it is oblivious to the polarization state of the incident light and the composition of the scatterer, the differential scattering cross section of a spherical droplet of radius a illuminated by light of wavenumber k is $|S|^2/k^2$, where the scattering amplitude is

$$S = x^2 \frac{1 + \cos \vartheta}{2} \frac{J_1(x \sin \vartheta)}{x \sin \vartheta}. \tag{8.58}$$

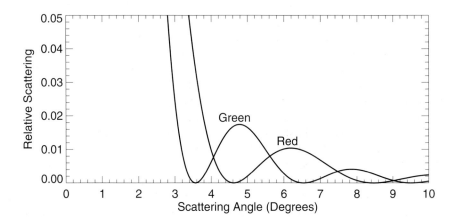

Figure 8.22: Differential scattering cross section calculated by the Fraunhofer approximation for a sphere of diameter 10 μm. Green corresponds to 510 nm, red to 660 nm. Both cross sections are normalized to the value for green at $0°$.

J_1 is the Bessel function of first order and the size parameter $x = ka$. The quantity $(1 + \cos \vartheta)/2$ usually is approximated as 1 because only near-forward scattering angles ϑ are of interest.

This differential scattering cross section, which determines the angular distribution of scattered light, has maxima for

$$x \sin \vartheta = 5.137,\ 8.417,\ 11.62, \ldots \tag{8.59}$$

Thus the dispersion in the position of the first maximum is

$$\frac{d\vartheta}{d\lambda} \approx \frac{0.817}{a}, \tag{8.60}$$

and is greater for higher-order maxima. This dispersion determines the upper limit on droplet size such that a corona can be observed. For the total angular dispersion over the visible spectrum to be greater than the angular width of the sun ($0.5°$), the droplets cannot be larger than about 60 μm in diameter. Drops in rain, even drizzle, are appreciably larger, which is why coronas are not seen through rainshafts. But scattering by a droplet of diameter 10 μm (Fig. 8.22), a typical cloud droplet size, gives sufficient dispersion to yield colored coronas.

Suppose that the first angular maximum for a droplet of radius a occurs at a wavelength λ. For a droplet of radius $a + \Delta a$, the position of this maximum is the same at a wavelength $\lambda + \Delta \lambda$, where from Eq. (8.59)

$$\frac{\Delta a}{a} \approx \frac{\Delta \lambda}{\lambda}. \tag{8.61}$$

If we take $\Delta \lambda$ to be half the width of the visible spectrum (about 0.15 μm) and λ to be in the middle of the spectrum (0.55 μm), $\Delta a/a \approx 0.3$. Thus coronas require fairly narrow size distributions: if droplets are distributed in size with a relative variance much greater than 30–40%, color separation is not possible.

Because of the stringent requirements for complete coronas (a thin veil of droplets narrowly distributed in size extending within 10–20° of the sun), they are not observed often in the sky, although you might see them in clouds formed on your breath, the source of illumination street lights or automobile lights. Or you might see them as a result of scattering by water droplets condensed onto a window or the windshield of a car. Of greater occurrence are the corona's cousins, iridescent clouds, which display patches of colors but usually not arranged in any obvious geometrical pattern. You may miss iridescence because it is seen toward the sun and so may be dazzled by sunlight. And yet on many partly cloudy days the thin edges of even thick clouds may be tinged with red and green. To enhance your ability to see these colored patches you can reduce the luminance with sunglasses or by reflection by a window or a puddle. Alistair Fraser used to give students in his observing class black glass tiles, about 10 cm on a side, to be used for scanning the sky for iridescence. When you know where and how to look, you can see it frequently.

Coronas are not the unique signature of spherical scatterers. Randomly oriented ice columns and plates give similar near-forward scattering patterns according to Fraunhofer theory. As a practical matter, however, most coronas probably are caused by droplets. Many clouds at temperatures well below freezing contain subcooled water droplets. Only if a corona were seen through a cloud at a temperature known to be lower than $-40\,°C$ could one assert with confidence that it must be an ice-crystal corona. Ken Sassen and his collaborators have made what appear to be incontrovertible observations of occasional coronas and, even more rarely, iridescence in clouds composed of ice crystals. But the rarity of this is evident from the title of one of Sassen's papers, "Cirrus cloud iridescence: a rare case study".

Another exception to the assertion about most coronas being caused by droplets is elliptical coronas. These have been observed, photographed, simulated, and attributed to scattering by more or less spheroidal pollen grains oriented by aerodynamic forces as they fall through air.

8.4.2 Rainbows

In contrast with coronas, which are seen toward the sun, rainbows are seen away from it, and caused by water drops much larger than those that give coronas. To describe the rainbow quantitatively we pretend that light incident on a transparent sphere much larger than the wavelength is composed of individual rays, each of which suffers a different fate determined only by the laws of specular reflection and refraction. Each incident ray splinters into an infinite number of scattered rays: reflected at the first interface, transmitted without internal reflection, transmitted after one, two, and so on internal reflections. For any scattering angle ϑ, each splinter contributes to the scattered light and hence to the differential scattering cross section (see Sec. 3.5). Consider a small area $a^2 \sin \vartheta_i \Delta\vartheta_i \Delta\varphi_i$, defined by co-latitudes between ϑ_i and $\vartheta_i + \Delta\vartheta_i$ and azimuthal angles between φ_i and $\varphi_i + \Delta\varphi_i$, on a sphere of radius a. If ϑ_i is the angle an incident ray makes with this small area, the radiant energy intercepted by it is proportional to $a^2 \sin \vartheta_i \cos \vartheta_i \Delta\vartheta_i \Delta\varphi_i$, a fraction of which is reflected, a fraction transmitted without internal reflection, and so on. The solid angle of the corresponding set of scattered rays is $\sin \vartheta \Delta\vartheta \Delta\varphi$, and hence the contribution to the differential scattering cross section by a splinter has the form

$$\mathcal{T}\frac{a^2 \sin \vartheta_i \cos \vartheta_i \Delta \vartheta_i \Delta \varphi_i}{\sin \vartheta \Delta \vartheta \Delta \varphi}, \tag{8.62}$$

where $\mathcal{T}$ is a transmissivity (in general, a product of various transmissivities and reflectivities obtained from the Fresnel coefficients discussed in Sec. 7.2). For our purposes here all we need to know about $\mathcal{T}$ is that it is finite and nonzero and hence can be ignored.

Because of the azimuthal symmetry of a sphere, $\Delta \varphi = \Delta \varphi_i$, whereas ϑ is a *different* function of ϑ_i, or conversely, for every splinter. In the limit $\Delta \vartheta \to 0$ Eq. (8.62) becomes

$$a^2 \frac{\cos \vartheta_i \sin \vartheta_i}{\sin \vartheta} \frac{d\vartheta_i}{d\vartheta}, \tag{8.63}$$

where we omit $\mathcal{T}$. This can be written more compactly by way of the *impact parameter* $b = a \sin \vartheta_i$:

$$\frac{b}{\sin \vartheta} \frac{db}{d\vartheta}. \tag{8.64}$$

The differential scattering cross section is an infinite series of terms of the form of Eq. (8.63) or Eq. (8.64). Singularities, or caustics, in the differential scattering cross section occur at scattering angles for which

$$\frac{d\vartheta_i}{d\vartheta} \to \infty, \quad \frac{\cos \vartheta_i \sin \vartheta_i}{\sin \vartheta} \neq 0, \tag{8.65}$$

or, equivalently,

$$\frac{db}{d\vartheta} \to \infty, \quad \frac{b}{\sin \vartheta} \neq 0. \tag{8.66}$$

According to geometrical (i.e., ray) optics, at a caustic the differential scattering cross section is infinite. In reality, it is finite for all scattering angles, so geometrical optics can at best only point out the approximate positions of spikes. For our purposes, Eq. (8.65) is more convenient for determining caustics and the condition on the derivative is written more conveniently as

$$\frac{d\vartheta}{d\vartheta_i} = 0. \tag{8.67}$$

Equation (8.67) defines *rainbow angles*. No rainbow angle is associated with splinters externally reflected and transmitted without internal reflection, but there may be rainbow angles for transmission after one internal reflection (*primary* rainbow), after two internal reflections (*secondary* rainbow) and so on. Consider first splinters transmitted after one internal reflection (Fig. 8.23). The scattering or deviation angle is

$$\vartheta = (\vartheta_i - \vartheta_t) + (\pi - 2\vartheta_t) + (\vartheta_i - \vartheta_t) = 2\vartheta_i - 4\vartheta_t + \pi. \tag{8.68}$$

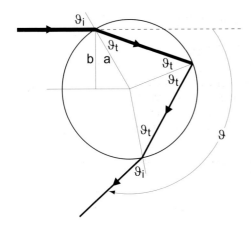

Figure 8.23: Any of the infinitude of rays imagined to be incident on a large, compared with the wavelength, transparent water sphere can be transmitted into it and reflected once before being transmitted out. Only one ray is shown here, its angle of incidence ϑ_i.

The curve of ϑ as a function of ϑ_i for a water sphere exhibits a minimum (Fig. 8.24). The angle of incidence corresponding to this minimum follows from Eqs. (8.67) and (8.68) and Snel's law:

$$\sin \vartheta_i = \sqrt{\frac{4 - n^2}{3}}, \tag{8.69}$$

where n is the real refractive index. Because $\sin \vartheta_i$ must be a real number less than or equal to 1, primary rainbows require drops with refractive index less than 2.

In the middle of the visible spectrum (550 nm) the primary rainbow angle is 137.7°, but this angle is better expressed as its complement (42.3°) because the primary rainbow is seen at about 42° from the *antisolar point*, the direction directly opposite the sun (Fig. 8.25). Because n varies with wavelength, so does the rainbow angle, the angular spread from violet (425 nm) to red (650 nm) about 2.3°. This is appreciably greater than the angular width of the sun, which is why color separation can be seen in natural rainbows. Because the refractive index of water is least in the red, so is the corresponding rainbow angle, and hence the red bow is highest above the antisolar point.

Contrary to appearances, rainbows are not palpable objects lying in a vertical plane. Nor are they even caused solely by raindrops in a plane. All raindrops in space lying on a cone with its apex at the observer and an apex angle of about 84° contribute to a rainbow (Fig. 8.25). Moreover, because the rainbow angle is an angle of minimum deviation, light is scattered in all directions within this cone. Thus the rainbow is the bright outer edge of a luminous disc. But why do we see only part of this disc? The facile answer is that the ground gets in the way, but a more satisfying answer is that the optical thickness along paths that intersect the ground is in general much less than along paths that do not. We show this schematically in Fig. 8.25.

To our surprise and delight, the noted adventure photographer Galen Rowell responded to a challenge to photograph a complete rainbow. He writes on page 52 of *Galen Rowell's Vision:*

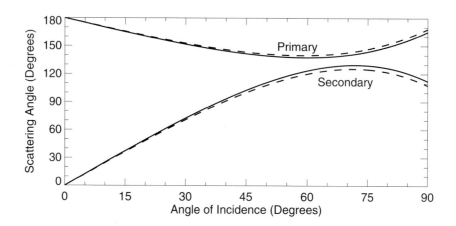

Figure 8.24: The scattering angle for incident rays undergoing one internal reflection by a large, transparent sphere is a minimum for a primary rainbow, whereas that for incident rays undergoing two internal reflections is a maximum for a secondary rainbow. The solid curve is for a wavelength of 650 nm, the dashed curve for 425 nm.

The Art of Adventure Photography: "When my wife, Barbara flew her single-engine Cessna to Patagonia, I was able to photograph…the extremely rare 360-degree double rainbow…that appeared in a sunlit rainstorm along the coast of Mexico. Atmospheric physicist Craig F. Bohren says in his 1987 book, *Clouds in a Glass of Beer*, 'to the best of my knowledge, one has never been photographed'." He continues with "a suggestion for anyone who would like a bit of fame and fortune: photograph a complete rainbow. You will need an airplane…You will also have to persuade the pilot to fly in stormy weather. If you survive your flight you will have acquired something rare indeed."

But what is this "double rainbow" that Rowell mentions? This is a term for the primary and secondary rainbows seen together. The secondary rainbow often is missed because it is less bright than the primary, but if you know where to look you often can see at least parts of the secondary bow. The deviation angle for splinters that have been reflected internally twice before transmission (Fig. 8.26) is

$$\vartheta = 2\vartheta_i - 6\vartheta_t + 2\pi. \tag{8.70}$$

Note that this is the *total* deviation angle. To obtain from it the corresponding scattering angle, which by definition lies between 0 and π, requires subtracting Eq. (8.70) from 2π. This scattering angle exhibits a *maximum*, the total deviation a minimum, for a water drop (Fig. 8.24). The corresponding angle of incidence follows from Eqs. (8.67) and (8.70) and Snel's law:

$$\sin \vartheta_i = \sqrt{\frac{9 - n^2}{8}}. \tag{8.71}$$

Figure 8.25: The primary rainbow is a consequence of strong scattering by all illuminated drops lying on a cone with its apex at the eye of the observer and with apex angle approximately 84°.

If rain were composed of titanium dioxide ($n \approx 2.7$), a common constituent of paints, primary rainbows would be absent from the sky and we would have to be content with secondary rainbows.

In the middle of the visible spectrum the secondary rainbow angle is 128.3°, but again its complement (51.7°) is more relevant to an observer: the secondary rainbow lies about 52° above the antisolar point, and hence about 9° above the primary rainbow. Indeed, we drew

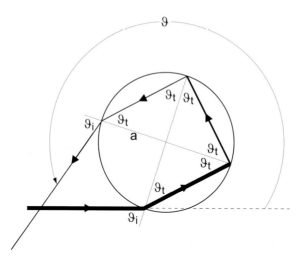

Figure 8.26: Any of the infinitude of rays imagined to be incident on a large, compared with the wavelength, transparent water sphere can be transmitted into it and reflected twice before being transmitted out. Only one ray is shown here, its angle of incidence ϑ_i corresponding to the secondary rainbow angle for which the scattering angle ϑ is a maximum.

Figs. 8.23 and 8.26 differently in order to make this clear. The angular spread in the secondary rainbow from violet to red is about $4.1°$. Because the scattering angle corresponding to the deviation angle determined by Eqs. (8.70) and (8.71) is greatest in the red, the angular color order in the secondary bow is reversed. It is sometimes said that this is because the second internal reflection turns the rainbow inside out. But the different color order of the two bows is an accident of the magnitude of the refractive index of water. Because the primary rainbow angle is a minimum scattering angle and the secondary rainbow angle is a maximum scattering angle, geometrical optics predicts a dark band, called *Alexander's dark band*, between the primary and secondary bow, which indeed is observed.

The colors of rainbows are a consequence of sufficient dispersion of the refractive index of water over the visible spectrum to give a spread of rainbow angles that appreciably exceeds the angular width of the sun. As we noted, the width of the primary bow is about $2.3°$ and that of the secondary bow about $4.1°$, although these values depend on the choice of end points of the visible spectrum and the values of the refractive index. Because of its band of colors arcing across the sky, the rainbow has become the paragon of colors, the standard against which all other colors are compared. Raymond Lee and Alistair Fraser, however, have challenged this status of the rainbow, pointing out that even the most vivid rainbows are colorimetrically far from pure (see Sec. 4.3). Rainbows are almost invariably discussed as if they occurred literally in a vacuum. But real rainbows, as opposed to the pencil-and-paper variety, are necessarily observed in an atmosphere, the molecules and particles of which scatter sunlight that adds to the light from rainbows but subtracts from their purity of color.

According to geometrical optics, the differential scattering cross section is infinite at all rainbow angles. But in reality, scattering by raindrops is everywhere finite, and the scattered light at each successive rainbow angle diminishes because of ever decreasing transmission.

Secondary rainbows are less bright than primary rainbows. Tertiary and higher-order rainbows are drowned in natural environments by background illumination (e.g., skylight). In a darkened laboratory, with a laser illuminating a single droplet, caustics of remarkably high order can be observed, but it is a stretch to call them rainbows.

Although geometrical optics yields the positions, widths, and color separation of rainbows, it yields little else. For example, it is blind to *supernumerary bows*, a series of narrow bands sometimes seen below the primary bow. These bows cannot be described without *explicitly* invoking interference. (As noted in Secs. 3.1 and 7.2, specular reflection and refraction are implicit interference patterns.) Except at the rainbow angle, a horizontal line intersects the curve of scattering angle versus angle of incidence in two points (Fig. 8.24). The corresponding two rays therefore have the same direction but follow different trajectories in a drop. If we look upon ray trajectories as specifying (approximately) the transmission paths of waves, the two waves corresponding to the two rays in the same direction are different in phase and hence interfere. Moreover, this interference depends on drop size unlike the positions of rainbow angles, which according to geometrical optics are independent of size. The interference interpretation of supernumerary bows has been seized upon despite the embarrassing fact that it is puzzling to anyone who knows about rain. Raindrops are widely distributed in size (see Prob. 1.3) so how can supernumerary bows be seen in rain showers? In a nice piece of detective work, Alistair Fraser answered this question.

Raindrops falling in a vacuum are spherical. Those falling in air are distorted by aerodynamic forces, not, despite the depictions of countless artists, into teardrops but rather into more nearly oblate spheroids with their axes approximately vertical. Fraser argues that supernumerary bows are caused by drops of diameter about 0.5 mm, for which the angular position of the first and second supernumerary bow is a minimum. Interference causes the angular position of the supernumerary bow to increase with decreasing size whereas drop distortion causes it to increase with increasing size. Supernumerary patterns contributed by drops on either side of the minimum cancel, leaving only the contribution from drops at the minimum. This cancellation occurs only near the tops of rainbows, where supernumerary bows are seen. In the vertical parts of a rainbow, a horizontal slice through a distorted drop is more or less circular, and hence these drops do not exhibit a minimum supernumerary angle.

According to geometrical optics, all spherical drops, regardless of size, yield the same rainbow. But it is not necessary for a drop to be spherical for it to yield rainbows independent of its size. This merely requires that the plane defined by the incident and scattered rays intersect the drop in a circle. Even distorted drops satisfy this condition in the vertical part of a bow. As a consequence, the absence of supernumerary bows there is compensated for by more vivid colors of the primary and secondary rainbows. Smaller drops are more likely to be spherical, but the smaller the drop, the less light it scatters. Thus the dominant contribution to the luminance of rainbows is from the larger drops. At the top of the bow, the plane defined by the incident and scattered rays intersects the large, distorted drops in an ellipse, yielding a range of rainbow angles varying with the amount of distortion, and hence a pastel rainbow. To the careful observer rainbows are no more uniform in color and brightness than is the sky.

Although geometrical optics predicts that all rainbows are equal, neglecting background light, real rainbows do not slavishly follow the dictates of this approximate theory. Rainbows in nature range from nearly colorless fog bows, or cloud bows, to the vividly colorful vertical portions of rainbows likely to have inspired myths about pots of gold.

8.4.3 The Glory

Continuing our sweep of scattering directions, from forward to backward, we arrive at the end of the arc to the *glory*. Because it is most easily seen from airplanes it sometimes is called the *pilot's bow*. Another name is *anticorona*, a corona around the antisolar point. Although glories and coronas share some common characteristics, there are differences between them other than direction of observation. Unlike coronas, which may be caused by nonspherical ice crystals, glories require spherical cloud droplets (but see the references at the end of the chapter for a possible exception). And a greater number of colored rings may be seen in glories because the decrease in luminance away from the backward direction is not as steep as that away from the forward direction. To see a glory from an airplane, look for colored rings around its shadow cast on clouds below. This shadow is not an essential part of the glory, merely a way of finding the antisolar point.

Like the rainbow, the glory may be looked upon as a singularity in the differential scattering cross section. Equation (8.65) gives one set of conditions for a singularity; the second set is

$$\sin \vartheta = 0, \ b(\vartheta) \neq 0. \tag{8.72}$$

That is, the differential scattering cross section is infinite for nonzero impact parameters, corresponding to incident rays that do not intersect the center of the sphere, that give forward ($0°$) or backward ($180°$) scattering. The forward direction is excluded because intense scattering in this direction is accounted for by the Fraunhofer theory.

For one internal reflection, Eqs. (8.68) and (8.72) and Snel's law yield the condition

$$\sin \vartheta_i = \frac{n}{2} \sqrt{4 - n^2}, \tag{8.73}$$

which is satisfied only for refractive indices between 1.414 and 2, the lower refractive index corresponding to a grazing-incidence ray. The refractive index of water lies outside this range. Although a condition similar to Eq. (8.73) is satisfied by four or more internal reflections, insufficient radiant energy is associated with such rays. Thus it seems that we have reached an impasse: the theoretical condition for a glory cannot be met by water droplets. Not so says Henk van de Hulst. He argues that 1.414 is close enough to 1.33 given that geometrical optics is, after all, an approximation. Cloud droplets are large compared with the wavelengths of visible light, but not so large that geometrical optics is an infallible guide to their optical behavior. Support for the van de Hulstian interpretation of glories was provided by Bryant and Cox, who showed that the dominant contribution to the glory is from the last term in the exact series for scattering by a sphere. Each successive term in this series is associated with ever-larger impact parameters. Thus the terms that give the glory are indeed those corresponding to grazing rays. Further unraveling of the glory and vindication of van de Hulst's conjecture about the glory were provided by Nussenzveig.

It sometimes is said that geometrical optics is incapable of treating the glory. Yet the same can be said about the rainbow. Geometrical optics explains rainbows only in the limited sense that it predicts singularities for scattering in certain directions (i.e., rainbow angles). But it can predict only the angles of intense scattering, not the amount of light scattered. Indeed, the error is infinite. Geometrical optics also predicts a singularity in the backward direction but is powerless to predict more. Results from geometrical optics for both rainbows and glories are not the end but rather the beginning, an invitation to take a closer look with more powerful magnifying glasses.

8.5   Scattering by Single Ice Crystals

Scattering by spherical particles in the atmosphere can result in three distinct displays: coronas, rainbows, and glories. If the particles depart somewhat from perfect spheres, the displays are not greatly changed. An entirely new and more varied set of displays arises when the particles are ice crystals, which are far from spherical and, because of their shape, can be oriented by aerodynamic forces. As with rainbows the gross features of ice-crystal displays can be described simply, but approximately, by following the various trajectories of rays incident on crystals. Colorless displays (e.g., the subsuns touched on in Sec. 4.1) are generally associated with reflected rays, colored displays (e.g., sun dogs and halos) with refracted rays. Because of the wealth of ice-crystal displays, we cannot treat all of them, but one example should point the way toward understanding many of them.

8.5.1   Sun Dogs and Halos

Because of its hexagonal crystalline structure, ice can form as hexagonal plates in the atmosphere. The stable position of a plate falling in air is with its face more or less horizontal, which can be demonstrated with an ordinary business card. When dropped with its face vertical, the supposedly aerodynamic position that many people choose instinctively, the card somersaults in a helter-skelter path to the ground. But when dropped with its face horizontal, the card gently rocks back and forth in descent.

   A hexagonal ice plate falling through air and illuminated by a low sun is like a $60°$ prism illuminated normally to its sides (Fig. 8.27). Because there is no mechanism for orienting a plate within a horizontal plane, all plate orientations in this plane are equally probable. Stated another way, all angles of incidence on a fixed plate are equally probable. Yet all deviation (i.e., scattering) angles of rays refracted into and out of the plate are not equally probable. Let $p(\vartheta_i)$ be the uniform probability distribution for incident angles ϑ_i and $P(\vartheta)$ that for deviation angles ϑ, where ϑ is a function of ϑ_i. As with all probability distributions, the integral of $P(\vartheta)$ over any interval is the probability that the deviation angle lies in that interval. From the theorem in Section 1.2 for transforming from one distribution to another (i.e., transforming variables of integration)

$$P(\vartheta)\left|\frac{d\vartheta}{d\vartheta_i}\right| = p(\vartheta_i), \tag{8.74}$$

which is more illuminating when written as

$$P(\vartheta) = \frac{p(\vartheta_i)}{|d\vartheta/d\vartheta_i|}. \tag{8.75}$$

Note that Eq. (8.75) does not give the radiant energy in the twice-refracted radiation. To obtain this would require including the Fresnel transmission coefficients for the two interfaces.

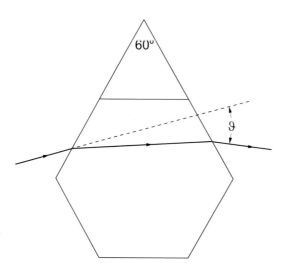

Figure 8.27: Deviation (i.e., scattering) of incident light by an angle ϑ because of refraction by a 60° prism that is part of a hexagonal plate.

Figure 8.28 shows deviation angle as a function of incidence angle for a 60° ice prism that is part of a hexagonal plate. For angles of incidence less than about 13° the transmitted ray is totally internally reflected at the second interface; for angles of incidence greater than about 70° reflection rises sharply (e.g., see Fig. 7.6) and hence transmission plunges. Thus most incident rays of consequence lie in this range. According to Eq. (8.75) the probability density $P(\vartheta)$ becomes infinite if the deviation angle has a minimum, which it does for $\vartheta_i \approx 40°$; the corresponding deviation angle is about 22°. The observable manifestation of this singularity, or caustic, at the angle of minimum deviation for a 60° ice prism is a bright spot about 22° from either or both sides of a sun low in the sky. These bright spots are called *sun dogs*, because they accompany the sun, or *parhelia* or *mock suns*.

The minimum deviation angle ϑ_m, and hence the angular position of sun dogs, depends on the prism angle Δ and refractive index n (see Prob. 8.20):

$$\vartheta_m = 2\sin^{-1}\left(n\sin\frac{\Delta}{2}\right) - \Delta. \tag{8.76}$$

Because n varies with wavelength, the separation between the angles of minimum deviation for red (650 nm) and violet (430 nm) light is about 0.7° (Fig. 8.28), slightly greater than the angular width of the sun. As a consequence, sun dogs may be tinged with color, most noticeably toward the sun. Because the refractive index of ice is least at the red end of the spectrum, the red components of a sun dog are closest to the sun. Moreover, for any wavelength, except that corresponding to red, a horizontal line tangent at the minimum angle to the curve of deviation angle versus incident angle intersects curves for other wavelengths. Because of this overlap of deviation angles, red is the purest color in sun dogs, which fade into white away from their red inner edges.

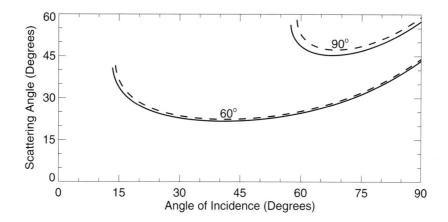

Figure 8.28: Deviation (i.e., scattering) angle versus angle of incidence for a 60° ice prism and a 90° ice prism. The solid line is for a wavelength of 650 nm, the dashed line for 430 nm.

With increasing solar elevation, sun dogs move away from the sun. A falling ice plate is approximately equivalent to a prism the angle of which increases with solar elevation. From Eq. (8.76) it follows that the angle of minimum deviation, hence the position of the sun dog, also increases. Why is only the 60° prism portion of a hexagonal plate singled out for attention? According to Fig. 8.27, a hexagonal plate could be considered to be made up of 120° prisms. For a ray to be refracted twice, its angle of incidence at the second interface must be less than the critical angle [see Eq. (4.63)], which imposes limitations on the prism angle. For $n \approx 1.31$ (ice at visible wavelengths), all incident rays are totally internally reflected by prisms with angles greater than about 99.5°.

A close relative of the sun dog is the 22° halo, a bright ring approximately 22° from the sun. Lunar halos and moon dogs are also possible, but because of their low brightness may not be noticed as frequently as their solar counterparts. Until Alistair Fraser analyzed halos the conventional wisdom had been that they obviously were the result of randomly oriented crystals, yet another example of jumping to conclusions. By combining optics and aerodynamics, he showed that ice crystals small enough to be randomly oriented by Brownian motion are too small to yield sharp scattering patterns.

But partially oriented larger plates can produce halos, especially ones of non-uniform brightness. Each part of a halo is contributed to by a plate with a different tip angle, angle between the normal to the plate and the vertical. The transition from oriented plates with zero tip angle to randomly oriented plates occurs over a narrow range of sizes. In the transition region, plates can be small enough to be partially oriented yet large enough to give a distinct contribution to the halo. Moreover, the mapping between tip angles and azimuthal angles on the halo depends on solar elevation. When the sun is near the horizon, plates can give a distinct halo over much of its azimuth. When the sun is high in the sky, hexagonal plates cannot give a sharp halo but hexagonal columns – another possible form of atmospheric ice particles – can. The stable position of a falling column is with its long axis horizontal. When the sun is directly overhead, such columns can give a uniform halo even if they all lie in the horizontal plane. When the sun is not overhead but well above the horizon, columns also can give halos.

A corollary of Fraser's analysis is that halos are caused by crystals with sizes in the range 12–40 µm. Larger crystals are oriented, smaller crystals too small to yield distinct scattering patterns. More or less uniformly bright halos with the sun neither high nor low in the sky could be caused by mixtures of hexagonal plates and columns or by clusters of bullets (rosettes). Fraser opines that the latter is more likely.

One of the by-products of his analysis is an understanding of the relative rarity of the 46° halo. Rays can be transmitted through two sides of a hexagonal column ($\Delta = 60°$) or through one side and an end ($\Delta = 90°$). For $n = 1.31$ and $\Delta = 90°$ Eq. (8.76) yields a minimum deviation angle of 46° (see Fig. 8.28). Although 46° halos are possible, they are seen much less frequently than 22° halos. Plates cannot give distinct 46° halos although columns can. But they must be solid and most columns have hollow ends. Moreover, the range of sun elevations is restricted.

Like the green flash, ice-crystal phenomena are not intrinsically rare. Halos and sun dogs can be seen frequently once you know what to look for, where, and when. Hans Neuberger reports that halos were observed in State College, Pennsylvania an average of 74 days a year over a 16-year period, with extremes of 29 and 152 halos a year. Although the 22° halo was by far the most frequently seen display, ice-crystal displays of all kinds were seen, on average, more often than once every four days at a location not especially blessed with clear skies. Although thin clouds are necessary for ice-crystal displays, clouds thick enough to obscure the sun are their bane.

8.6   Clouds as Givers and Takers of Light

Despite their apparent solidity, clouds are so flimsy as to be almost nonexistent – except optically. The fraction of cloud volume occupied by water substance, liquid or solid, is about 10^{-6} or less, and hence the mass density of clouds is a small fraction of the density of sea-level air. And yet their optical thickness per unit physical thickness is much greater because scattering by a water molecule when part of a coherent array (e.g., water droplet) is vastly greater than that by a single, isolated molecule (see Fig. 3.11).

Clouds seen by passengers in an airplane flying above cloud can be dazzling, but if the airplane were to descend through the cloud these passengers might describe the sky overhead as gloomy. Clouds are both givers and takers of light. Their dual role is exemplified in Fig. 5.13, which shows the calculated diffuse downward irradiance below clouds of varying optical thickness. On an airless planet the sky would be black in all directions except directly toward the sun. But if the sky were to be filled from horizon to horizon with thin clouds, the brightness overhead would markedly increase. As so often happens, however, more is not always better. Beyond a certain cloud optical thickness, the diffuse irradiance decreases, and for a sufficiently thick cloud the sky overhead can be darker than the clear sky (see Probs. 5.25 and 5.26).

Why are clouds bright? Why are they dark? No inclusive one-line answer can be given to these questions. Better to ask, Why is that particular cloud bright? Why is that particular cloud dark? Each observation must be treated individually; generalizations are risky. Moreover, we must keep in mind the difference between brightness and luminance when addressing the queries of human observers (see Sec. 4.1). If the luminance of an object is appreciably greater than that of its surroundings, we call the object bright; if appreciably less, we call it dark. But these are relative, not absolute terms. Two clouds, identical in all respects, including illumination, still may appear different because they are seen against different backgrounds, a cloud seen against the horizon sky appearing darker than when seen against the zenith sky. Of two clouds under identical illumination, the optically smaller will be less bright. If an even larger cloud were to appear, the cloud formerly described as white might be demoted to gray. With the sun below the horizon, two identical clouds at markedly different elevations might be quite different in brightness, the lower cloud being shadowed from direct illumination by sunlight. A striking example of dark clouds sometimes can be seen well after sunset. Low-lying clouds not illuminated by direct sunlight may be inky blotches staining the faint twilight sky.

Because the dark objects of our everyday lives usually owe their darkness to absorption, nonsense about dark clouds is rife: they are filled with pollution. Yet of all the reasons why clouds sometimes are seen to be dark or even black, absorption is not among them. But there is at least one example in which absorption by clouds of pure water droplets or ice crystals can result in observable consequences, which we discuss in the following section.

8.6.1   Green Thunderstorms

Blue or red skies, clear or cloudy, are unremarkable, but green is the color of ground-level objects: grass and trees. A green sky therefore gets your attention, a source of wonder, even fear, an indication that the world has gone topsy-turvy. And yet green thunderstorms are seen from time to time. Their existence is not in doubt, although explanations of them are. Where tornadoes occur frequently, they are said to be the cause of green thunderstorms, although the mechanism is not specified. Where hail occurs, it is said to be the cause of green thunderstorms. Both of these so-called explanations exemplify one of the most widespread forms of defective reasoning: if two events occur in succession or nearly simultaneously, hail and green thunderstorms, for example, one causes the other. We go beyond folklore and offer two rational explanations, variations on previous themes, of green thunderstorms.

Equation (8.3) is a simple expression for the airlight radiance L seen in the direction of a black object at an optical distance τ:

$$L = L_0 G\{1 - \exp(-\tau)\}, \tag{8.77}$$

where G is a geometrical factor of no importance here and L_0 is the radiance on the uniformly illuminated line of sight. If $\tau \ll 1$, L is approximately proportional to τ, which, if that for scattering by molecules and small particles, results in bluish airlight. But underlying this assertion, which is consistent with many observations, is an implicit assumption: L_0 is for sunlight that is not greatly attenuated. For sufficiently long optical paths through the atmosphere, we cannot ignore attenuation of the source illumination. So let us rewrite Eq. (8.77) more carefully:

$$L = L_s \exp(-\tau_s)G\{1 - \exp(-\tau)\}, \tag{8.78}$$

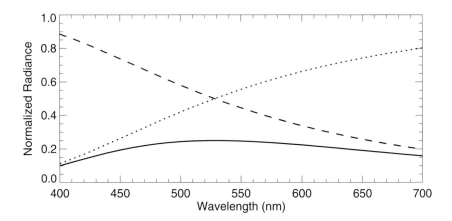

Figure 8.29: Airlight radiance due to an unattenuated source (dashed curve) and the radiance of attenuated solar radiation (dotted curve), both normalized by the solar radiance outside Earth's atmosphere, for an optical thickness 6 times the normal optical thickness of the molecular atmosphere. The product of these two curves is the solid curve.

where L_s is the solar radiance above Earth's atmosphere and τ_s is the optical thickness through the atmosphere to the line of sight. Thus the airlight radiance is a product of two spectral functions. One function, $1 - \exp(-\tau)$, *decreases* steadily from violet to red over the visible if τ is that for scattering by molecules and small particles; the other function, $\exp(-\tau_s)$, *increases* steadily from violet to red. The product of a function that yields red attenuated sunlight and a function that yields blue airlight can yield the intermediate green. Figure 8.29 shows $1 - \exp(-6\tau_n)$, $\exp(-6\tau_n)$, and their product, where τ_n is the normal molecular optical thickness. This corresponds to a black object some tens of kilometers away and the sun low in the sky. The result is an airlight spectrum with a broad peak in the green. This figure underscores a point made in our discussion of Fig. 8.7: for an atmosphere composed only of nonabsorbing molecules over black ground, blue is possible, as is red and everything in between.

Suppose that the distant black object is a thunderstorm, so thick that little sunlight is transmitted by it. For τ_s sufficiently large, the corresponding airlight can be green. The thunderstorm is not the source of this green light, merely the dark backdrop against which it is seen. But it is possible that the thunderstorm itself is green, which we turn to next.

We stated in Section 5.3 that the bottoms of very thick clouds can have a bluish cast because water, liquid or solid, has an absorption minimum in the blue. But we omitted the implicit assumption that the source of illumination is sunlight that has not been greatly attenuated. Sunlight is reddened by atmospheric attenuation, with the result that its chromaticity coordinates (see Sec. 4.3) move to the right of the achromatic point, which we may take to be that of unattenuated sunlight. If reddened sunlight, by which we do not necessarily mean perceptually red, illuminates an intrinsically blue object, such as a cloud, the transmitted light can be perceptually green. An almost childishly simple demonstration shows this (see Prob. 4.44).

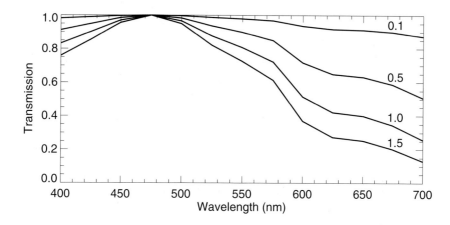

Figure 8.30: Transmission normalized to the maximum at around 470 nm for water droplet clouds with liquid water paths ranging from 0.1 cm to 1.5 cm.

Add some yellow food coloring to a glass of water, some blue food coloring to another glass of water. Then put the glass of bluish water in front of the glass of yellowish water and view the latter through the former. You will see green. This demonstration serves more than one purpose. It shows very simply a mechanism for green thunderstorms and also puts more nails in the coffin containing nonsense about yellow objects transmitting only yellow light, blue objects only blue light, and so on. If this were true, the demonstration would be impossible because the yellow water would be a source of only yellow light, whatever that is, and hence the blue water could not transmit green.

Green thunderstorms are associated with exceedingly thick clouds and seem to be observed mostly at sundown, when the illumination is reddened sunlight. Can this combination of red illumination of an intrinsically blue object be the cause of green thunderstorms? To answer this we have to be more precise about what is meant by "very thick clouds." By solving the two-stream Eq. (5.69) subject to the boundary conditions $F_\downarrow(0) = F_0$ and $F_\uparrow(\bar\tau) = 0$ we obtain the transmissivity of a normally illuminated finite cloud:

$$\frac{F_\downarrow(\bar\tau)}{F_0} = \frac{(1 - R_\infty^2)\exp(-K\bar\tau)}{1 - R_\infty^2 \exp(-2K\bar\tau)},\tag{8.79}$$

where $\bar\tau$ is the total optical thickness, K is given by Eq. (5.70), and R_∞ is the reflectivity of the corresponding infinite medium. From Eqs. (5.73), (5.83), and (5.86) we have the approximation

$$K\bar\tau \approx h_w \sqrt{\frac{3\kappa_w(1 - g)}{d}},\tag{8.80}$$

where h_w is the liquid water path of the cloud, κ_w is the bulk absorption coefficient of water, g is the asymmetry parameter and d is a mean droplet diameter. Because R_∞ is more or less flat across the visible spectrum (Fig. 5.15) and we are interested in clouds that are thick in

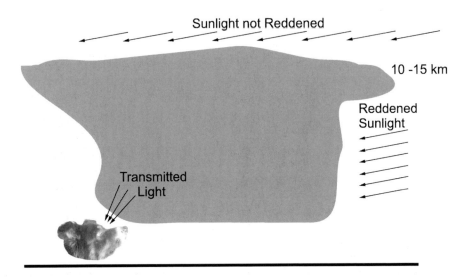

Figure 8.31: With the sun low in the sky the top of a thick cloud is illuminated by sunlight that has not been reddened, whereas the side of the cloud is illuminated by reddened sunlight. An observer on the ground receives transmitted light from both sources (top and side illumination).

the sense that $K\bar{\tau} > 1$, we plot $\exp(-K\bar{\tau})$ relative to its peak value versus wavelength in Fig. 8.30 for various liquid water paths instead of the transmissivity [Eq. (8.79)]. We take the asymmetry parameter $g = 0.85$ and mean droplet diameter $d = 10\ \mu\text{m}$, although their precise values are irrelevant given that they are factors of h_w in Eq. (8.80). Only when the liquid water path of a cloud exceeds about 0.5 cm does transmission dip markedly in the red. Such a value is high but not unrealistically so, although it does correspond to clouds 10–15 km thick.

If a sufficiently thick cloud is illuminated by reddened sunlight, the result can be perceptually green transmitted light. But herein lies a problem. The cloud to which Eq. (8.79) applies is illuminated from above, and if thick clouds are needed, their tops must be high. Above altitudes of 10 km to 15 km, there is insufficient atmosphere to appreciably redden sunlight even with the sun low in the sky. So the cloud top illumination is essentially unattenuated sunlight. What is needed is reddened light illuminating the sides of two-dimensional, perhaps even three-dimensional clouds (Fig. 8.31). The one-dimensional analysis therefore does not prove that thick clouds illuminated at their tops give green thunderstorms but rather that long paths traversed by reddened sunlight in clouds can yield green light. Such clouds must be thick both vertically and horizontally.

Both explanations of green thunderstorms, which are not mutually exclusive, require illumination by reddened light, which seems to demand a low sun such as at sunrise or sunset. And, in fact, most green thunderstorms seem to be seen late in the day. But this by itself is not definitive because late in the day is when most thunderstorms occur. If a midday green thunderstorm is ever reported, we have a possible explanation for it. Suppose that sunlight shines through a gap in exceedingly thick clouds (Fig. 8.32). Illuminated air and particles at the bottom of the gap are the source of light scattered toward the observer, but only that frac-

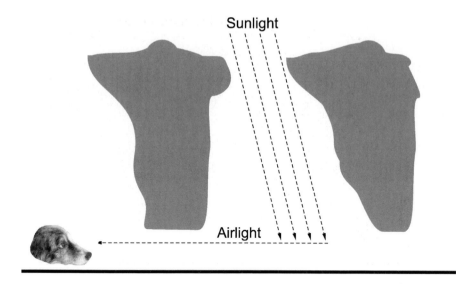

Figure 8.32: Scattered light results mostly from illumination by sunlight of an open patch of air between the two thick clouds whereas this scattered light is attenuated along the entire line of sight to an observer. Depending on the relative lengths of the paths below cloud and in the clear air, the spectrum of the transmitted light can be perceptually green: scattering shifts the spectrum toward shorter wavelengths, transmission shifts it toward longer wavelengths.

tion of it not scattered again along a path beneath the cloud is seen by an observer. Depending on the size of the gap and its distance from the observer, the transmitted light can be perceptually green. The mathematical analysis is identical to Eq. (8.78): the light within the gap is proportional to $1 - \exp(-\tau_g)$, where τ_g is the horizontal optical thickness of the gap, whereas the fraction of this light transmitted beneath the cloud is $\exp(-\tau_c)$, where τ_c is the optical thickness of the path beneath the cloud to the observer. If the cloud is thick, this path receives little illumination. Again, green light is possible with just the right combination of bluing by scattering (τ_g) and reddening by transmission (τ_c); it makes no difference in which order this occurs. We would not call this green light seen under thick clouds a green thunderstorm but rather green light seen in conjunction with a thunderstorm. A thunderstorm might be called green simply because it occurs simultaneously with green light.

Green in the twilight sky is a consequence of a similar mechanism but different geometry. Space is the black backdrop against which light at the short wavelength end of reddened sunlight is preferentially scattered toward the observer. What about reflection by grass as the source of green thunderstorms? Although theory [Eq. (5.68)] and measurements cast doubts on this explanation, the best refutation of it is observational: green thunderstorms have been reported over water and over parched ground nearly devoid of vegetation.

References and Suggestions for Further Reading

This chapter is based on Craig F. Bohren, 1995: Optics, atmospheric, *Encyclopedia of Applied Physics*, Vol. 12, pp. 405–34, subsequently reprinted (with minor changes and the title changed to atmospheric optics) in two other encyclopedic works: 2003: *Handbook of Weather, Climate, and Water*, Thomas D. Potter and Bradley R. Coleman, Eds., Wiley-Interscience, pp. 453–500, and 2004: *The Optics Encyclopedia* Vol. 1, Th. G. Brown, K, Creath, H. Kogelnik, M. A. Kriss, J. Schmit, M. J. Weber, Eds., Wiley-VCH, pp. 53–91.

Many of the seminal papers in atmospheric optics are more readily accessible in the collection compiled by Craig F. Bohren, Ed., 1989: *Selected Papers on Scattering in the Atmosphere*, SPIE Optical Engineering Press than in the original journals. An asterisk (*) before the name of authors in the citations that follow indicates that they are reprinted in this compendium.

Over the years special issues of Optical Society of America journals have been devoted to papers presented at Topical Meetings on Meteorological Optics (later changed to Topical Meetings on Light and Color in the Open Air): *Journal of the Optical Society of America*, Vol. 69, pp. 1051–198 (1979); Vol. 73, pp. 1622–64 (1983), Vol. A4, pp. 558–620 (1987); *Applied Optics*, Vol. 30, pp. 3381–552 (1991); Vol. 33, pp. 4535–760 (1994); Vol. 37, pp. 1425–1588 (1998); and Vol. 42, pp. 307–525 (2003).

The title of these Topical Meetings was inspired by Marcel Minnaert, 1954: *The Nature of Light and Color in the Open Air*. Dover. This is the bible of atmospheric optics. Minnaert inspired a few successors: Robert Greenler, 1980: *Rainbows, Halos, and Glories*. Cambridge University Press; R. A. R. Tricker, 1970: *Meteorological Optics*, Elsevier; Earl J. McCartney, 1976: *Optics of the Atmosphere*, Wiley; David K. Lynch and William Livingston, 1995: *Color and Light in Nature*. Cambridge University Press. The books by Greenler and by Lynch and Livingston are especially recommended for their many beautiful color plates.

Although not devoted exclusively to atmospheric optics William Jackson Humphreys, 1964: *Physics of the Air*. Dover contains a few chapters on the subject.

For an excellent history of polywater see Felix Franks, 1981: *Polywater*, MIT Press; for cold fusion see Frank Close, 1991: *Too Hot to Handle: The Race for Cold Fusion*, Princeton University Press.

Lord Rayleigh's On the light from the sky, its polarization and colour (1871) and On the transmission of light through an atmosphere containing small particles in suspension and the origin of the blue of the sky (1899) were originally published in *Philosophical Magazine*, but are more accessible in his *Scientific Papers*, Cambridge University Press, Vol. I, 1899, pp. 88–103 and Vol. III, 1902, pp. 397–405, respectively, or in the compendium *Selected Papers on Scattering in the Atmosphere*.

Not only does invoking Rayleigh scattering explain nothing, the term itself is fraught with different meanings. See *Andrew T. Young, 1982: Rayleigh scattering. *Physics Today*, January, pp. 2–8.

For a good history of the blue sky see Pedro Lilienfeld, 2004: A blue sky history. *Optics and Photonics News*, June, pp. 32–9. Evidence that he takes the history of science seriously is that he is one of the few authors to correctly credit Bouguer with the law of exponential attenuation (see Sec. 2.1).

For a detailed discussion of how the eye processes skylight containing more light that is not blue than is blue and yet yields the sensation blue, see Glenn S. Smith, 2005: Human color vision and the unsaturated blue color of the daytime sky. *American Journal of Physics*, Vol. 73, pp. 590–7.

Problem 4.10 asks for an estimate of the exposure time necessary to photograph the moonlit blue night sky. For a photograph of the night sky together with exposure times for various speed films and apertures see Joseph A. Shaw: 1996: What color is the night sky? *Optics and Photonics News*, November, pp. 54–5. Joe subsequently photographed the night sky with a digital camera and sent us some striking photos. He tells us that there are some "interesting differences" (e.g., shorter exposure times with digital) between digital and film night-sky photography, which he hopes to elucidate in a future article.

For more on daylight, especially the history of daylight spectroscopy, see S. T. Henderson, 1977: *Daylight and its Spectrum*, 2nd ed. John Wiley & Sons.

The (molecular) normal optical thickness (Fig. 8.2) and refractive index of air (Fig. 8.9) were taken from the compilation by *Rudolf Penndorf, 1957: Tables of the refractive index for standard air and the Rayleigh scattering coefficient for the spectral region between 0.2 and 20.0 μ and their application to atmospheric optics. *Journal of the Optical Society of America*, Vol. 47, pp. 176–82.

The heretical if not outright dangerous idea that big particles are not only unnecessary for a white horizon, they can make the horizon sky bluer is put forth by Craig F. Bohren and Clifton E. Dungey, 1992: Colors of the sky: can big particles make the sky bluer? *Contributions to Atmospheric Physics (Beiträge zur Physik der Atmosphäre)*, Vol. 64, pp. 329–34.

For an expository article on sky colors, including a proof that red sunsets and sunrises are not possible in a purely molecular atmosphere, see Craig F. Bohren and Alistair B. Fraser, 1985: Colors of the sky. *The Physics Teacher*, Vol. 23, pp. 267–72.

It seems that the first to recognize that molecular scattering is insufficient to explain the blue of the zenith sky when the sun is low was *E. O. Hulburt, 1953: Explanation of the brightness and color of the sky, particularly the twilight sky. *Journal of the Optical Society of America*, Vol. 42, pp. 113–18.

For a detailed treatment of the twilight sky see Georgii Vladimirovich Rozenberg, 1965: *Twilight: A Study in Atmospheric Optics*. Plenum.

The standard work on atmospheric visibility is by William Edgar Knowles Middleton, 1952: *Vision Through the Atmosphere*. University of Toronto Press.

For a simple demonstration that the contrast threshold depends on angular size, see Craig F. Bohren, 1987: *Clouds in a Glass of Beer*. John Wiley & Sons, Fig. 16.1.

With a bit of effort Eq. (8.34) can be extracted from Eq. (5.33) in *Harald Koschmieder, 1924: Theorie der horizontalen Sichtweite. *Beiträge zur Physik der freien Atmosphäre*, Vol. 12, pp. 33–54.

The quoted statement in Section 8.2 about Bouguer is in *William Edgar Knowles Middleton, 1960: Random reflections on the history of atmospheric optics. *Journal of the Optical Society of America*, Vol. 50, pp. 97–100.

There is more to contrast than visual range. Coming to grips with contrast may save your life. The senior author always has had white vehicles. In *Vision and Highway Safety* (Chilton Book Company, 1970) Merrill J. Allen asserts (p. 138) that "Studies have shown that up to ten times as many accidents happen to black cars as to white. Other data indicate white cars are up to forty times more visible than black." This is backed up by measurements of "relative visibility" of automobiles of difference colors.

Mirages, no matter how complicated, are readily explicable by fairly simple physics. This cannot be said of the moon illusion, a full understanding of which awaits (perhaps forever) a thorough understanding of how the human brain constructs a world out of raw visual data. The compendium by Maurice Hershenson, Ed., 1989: *The Moon Illusion*. Lawrence Erlbaum, is the result of its editor soliciting contributions from authors who have proposed (since 1962) explanations – there are at least nine – of the moon illusion, and commentaries from researchers in visual perception. The multiplicity of opinions in this compendium is evidence of how difficult it is to state *the* cause of the moon illusion. For a good treatise on the moon illusion see Helen E. Ross and Cornelis Plug, 2002: *The Mystery of the Moon Illusion: Exploring Size Perception*. Oxford University Press.

One of the best ever expository articles on mirages is by Alistair B. Fraser and William H. Mach, 1976: Mirages. *Scientific American*, Vol. 234(1), pp. 102–11. A mathematician's view of mirages, but without equations, is by Walter Tape, 1985: The topology of mirages. *Scientific America*. June, pp. 120–9.

For a putative explanation of the sun standing still because of a superior mirage, see Dario Camuffo, 1990: A meteorological anomaly in Palestine 33 centuries ago: How did the sun stop? *Theoretical and Applied Climatology*, Vol. 41, pp. 81–5.

The theorem about the impossibility of multiple images of astronomical objects above the horizon is in *Alistair B. Fraser, 1975: The green flash and clear air turbulence. *Atmosphere*, Vol. 13, pp. 1–10.

For a discussion of controversy and confusion over the role of the air temperature profile near the surface in astronomical refraction (angular displacement of astronomical objects) see Andrew T. Young, 2004: Sunset science IV: Low-altitude refraction. *The Astronomical*

Journal, Vol. 127, pp. 3622–37. He concludes that "At and below the astronomical horizon, the refraction depends primarily on atmospheric structure *below* the observer and varies so much (tens of minutes, or even several degrees) that only very crude predictions can be made. The observed time of sunset at a sea horizon often varies by a few minutes from day to day, and the variations increase with height above the sea."

For many color photographs of the low sun and references to papers on the green flash (but not much in the way of explanation) see D. J. K. O'Connell, 1958: *The Green Flash and Other Low Sun Phenomena*. North Holland.

A detailed colorimetric analysis of the rim of the horizon sun was done by *Glenn E. Shaw, 1973: Observations and theoretical reconstruction of the green flash. *Pure and Applied Geophysics*, Vol. 102, pp. 223–35. His calculations include attenuation of sunlight by molecular scattering, scattering by particles, and absorption by ozone. Unfortunately, he did not do calculations in which ozone or particles were omitted, thus leaving unanswered questions about the degree to which they contribute to the green flash. We leave it as a problem (Prob. 8.61) for you to determine if absorption by ozone is an essential ingredient in the green flash.

A simple demonstration of the green flash (strictly, the green rim of the sun) using a slide projector as the source of illumination and a pan of slightly milky water with a mirror under the water is in Craig F. Bohren, 1982: The green flash. *Weatherwise*, Vol. 35, pp. 271–5. A rewritten version was published as Chapter 13 in Craig F. Bohren, 1987: *Clouds in a Glass of Beer*. John Wiley & Sons without, alas, the color plates in the original article.

For a discussion of the role of physiology in the perception of the green flash see Andrew T. Young, 2000: Sunset science. III. Visual adaptation and green flashes. *Journal of the Optical Society of America A*, Vol. 17, pp. 2129–39.

One of the classic papers on coronas is by *George C. Simpson, 1912: Coronae and iridescent clouds. *Quarterly Journal of the Royal Meteorological Society*, Vol. 38, pp. 291–301. Many years later Simpson "developed glaucoma and used his extensive knowledge of atmospheric haloes to analyze ocular haloes." These "haloes" are what we would call coronas, which can originate from scattering within the eye. For more on this see David Miller and George Benedek, 1973: *Intraocular Light Scattering: Theory and Clinical Applications*. Charles C. Thomas, Ch. IV.

Near-forward scattering by randomly oriented hexagonal columns and plates is shown to be similar to that for spheres by Y. Takano and S. Asano, 1983: Fraunhofer diffraction by ice crystals suspended in the atmosphere. *Journal of the Meteorological Society of Japan*, Vol. 61, pp. 281–300. A figure showing the similarity between near-forward scattering by randomly oriented circular cylinders and spheres, according to the Fraunhofer approximation, is in Hendrik C. van de Hulst, 1957: *Light Scattering by Small Particles*. John Wiley & Sons, Fig. 16.

For observational evidence that coronas and iridescence are on rare occasions the result of scattering by nonspherical ice crystals see Kenneth Sassen, 2003: Cirrus cloud iridescence:

a rare case study. *Applied Optics*, Vol. 42, pp. 486–91; Kenneth Sassen, Gerald G. Mace, John Hallett, and Michael R. Poellet, 1998: Corona-producing ice clouds: a case study of a cold mid-latitude cirrus layer. *Applied Optics*, Vol. 37, pp. 1477–85; Kenneth Sassen, 1991: Corona-producing cirrus cloud properties derived from polarization lidar and photographic analysis. *Applied Optics*, Vol. 30, pp. 3421–8. Evidence for coronas and iridescence caused by scattering by ice particles in mountain wave clouds is put forward by Joseph A. Shaw and Paul J. Neiman, 2003: Coronas and iridescence in mountain wave clouds. *Applied Optics*, Vol. 42, pp. 476–85. They argue that ice-particle iridescence is more common than is believed, although only in wave clouds, not cirrus clouds.

For more about elliptical coronas see Pekka Parviainen, Craig F. Bohren, and Veikko Mäkelä, 1994: Vertical elliptical coronas caused by pollen. *Applied Optics*, Vol. 33, pp. 4548–51; Eberhard Tränkle and Bernd Mielke, 1994: Simulation and analysis of pollen coronas. *Applied Optics*, Vol. 33, pp. 4552–62.

The rainbow book to end all rainbow books is by Raymond Lee, Jr. and Alistair B. Fraser: 2001: *The Rainbow Bridge: Rainbows in Art, Myth, and Science*. The Pennsylvania State University Press and SPIE Press.

For a detailed discussion of supernumerary bows see *Alistair B. Fraser, 1983: Why can supernumerary bows be seen in a rain shower? *Journal of the Optical Society of America*, Vol. 73, pp. 1626–8. For a popular account, accompanied by superb photographs and illustrations, see Alistair B. Fraser, 1983: Chasing rainbows: numerous supernumeraries are super. *Weatherwise*, Vol. 36, pp. 280–9.

Investigation of reports of tertiary and higher-order rainbows in nature almost always reveal that they are something else, especially given that to many people any splash of color in the sky is a rainbow just as to other than bird watchers any small brown bird is a sparrow. One of the few exceptions is the observation by David E. Pedgley, 1986: A tertiary rainbow. *Weather*, Vol. 41, p. 401. Pedgley is a sufficiently careful observer to be believed. But to date no one seems to have photographed a tertiary rainbow in nature. In the laboratory is another matter. See Jearl D. Walker, 1976: Multiple rainbows from single drops of water and other liquids. *American Journal of Physics*, Vol. 44, pp. 421–33. For a photograph of an infrared rainbow see *Robert G. Greenler, 1971: Infrared rainbow. *Science*, Vol. 173, pp. 1231–2.

For a theory of the glory and its confirmation by detailed calculations see *Hendrik C. van de Hulst, 1947: A theory of the anti-coronae. *Journal of the Optical Society of America*, Vol. 37, pp. 16–22, *H. C. Bryant and A. J. Cox, 1966: Mie theory and the glory. *Journal of the Optical Society of America*, Vol. 56, pp. 1529–32, and *H. Moysés Nussenzveig, 1979: Complex angular momentum theory of the rainbow and the glory. *Journal of the Optical Society of America*, Vol. 69, pp. 1068–79.

For evidence of a possible ice-crystal glory see Kenneth Sassen, W. Patrick Arnott, Jennifer M. Barnett, and Steve Aulenbach, 1998: Can cirrus clouds produce glories? *Applied Optics*, Vol. 37, pp. 1427–33.

The backscattering direction is a source of delights for the eye and mind: the glory, coherent backscattering (see Probs. 6.28 and 6.29), the heiligenschein, and its fairly recently discovered cousin the *sylvanshine* (Alistair B. Fraser, 1994: The sylvanshine: retroreflection from dew-covered trees. *Applied Optics*, Vol. 33, pp. 4539–47). Keep in mind that these are distinguished by the agents responsible for them, not the theories used to describe them: no theory causes anything.

For a collection of stunning ice-crystal displays see Walter Tape, 1994: *Atmospheric Haloes*. American Geophysical Union. The display on the cover takes your breath away. Tape photographed many displays, some at the South Pole, and at the same time collected the crystals responsible for them. Several of the displays he photographed are compared with computer simulations. For an earlier treatise on haloes but without color photographs see R. A. R. Tricker, 1979: *Ice Crystal Haloes*. Optical Society of America.

Arguments that halos need not be caused by randomly oriented crystals are set forth by *Alistair B. Fraser, 1979: What size of ice crystals causes the halos? *Journal of the Optical Society of America*, Vol. 69, pp. 1112–18.

For a discussion of the consequences of multiple scattering to halos see *Eberhard Tränkle and Robert G. Greenler, 1987: Multiple-scattering effects in halo phenomena. *Journal of the Optical Society of America A*, Vol. 4, pp. 591–9.

The frequency of ice-crystal phenomena seen in Central Pennsylvania during a 16-year period is reported by Hans Neuberger, 1951: *Introduction to Physical Meteorology*. School of Mineral Industries, Pennsylvania State College, p. 174. (The School of Mineral Industries is now The College of Earth and Mineral Sciences, and Pennsylvania State College is now Pennsylvania State University).

It is unusual for proponents of different theories of the same phenomenon to publish them jointly in the same paper, but this was done by Craig F. Bohren and Alistair B. Fraser, 1993: Green thunderstorms. *Bulletin of the American Meteorological Society*, Vol. 74, pp. 2185–93. Observational evidence, by way of spectral measurements, that green thunderstorms are not "optical illusions" (whatever they are) is in Frank W. Gallagher III, William H. Beasley, and Craig F. Bohren, 1996: Green thunderstorms observed. *Bulletin of the American Meteorological Society*, Vol. 77, pp. 2889–97.

For evidence that green thunderstorms are not a consequence of reflection by green ground see Frank W. Gallagher III, 2001: Ground reflection and green thunderstorms. *Journal of Applied Meteorology*, Vol. 40, pp. 776–82.

Understanding atmospheric optical phenomena is enhanced by at least some knowledge of the particles responsible for them. To this end, the following are recommended: Hans R. Pruppacher and James D. Klett, 1980: *Microphysics of Clouds and Precipitation*. Dordrecht; Sean A. Twomey, 1977: *Atmospheric Aerosols*. Elsevier.

Problems

8.1. How might it be possible to determine the spectral solar irradiance at the top of the atmosphere and the normal optical thickness (clear sky) of the atmosphere by making spectral irradiance measurements at the surface of Earth? Devise a scheme for doing so.

HINT: Make measurements on a clear day. Assume exponential attenuation of direct solar radiation. Assume that your instrument is capable of measuring absolute irradiance of the (attenuated) sun.

HINT: Consider the variable $1/\cos\vartheta$, where ϑ is the solar zenith angle, to vary from 0 to infinity even though you can measure this quantity only for values greater than 1.

HINT: If you plot your data in the right way, it should be evident how to obtain the optical thickness and the irradiance at the top of the atmosphere.

HINT: The normal optical thickness varies from day to day and from place to place.

8.2. The clear sky is blue even though violet light is scattered more than blue light and the peak of the skylight spectrum is in the violet. Yet we sometimes see violet light in rainbows. Explain this apparent contradiction.

8.3. Estimate the largest zenith angle (smallest elevation angle) of the moon such that it cannot be seen against the daylight sky. You may take the reflectivity of the moon to be 0.06 over the visible spectrum. You may take the contrast threshold to be 0.02. You may take the normal optical thickness of the atmosphere (molecules and particles) to be 0.3 in the middle of the visible spectrum.

HINTS: In determining the contrast between the moon and the surrounding sky, be careful to ask yourself what light you see when you look at the moon through the atmosphere. For this problem you may assume that scattering by the molecules and particles is isotropic (i.e., the probability of scattering per unit solid angle in any direction is $1/4\pi$). Be sure that your answer makes sense in light of your own observations of the moon.

8.4. We noted in Section 8.1.1 that the radiance of the brightest cumulus cloud is larger, by roughly a factor of four, than that of the clear horizon sky. Please explain why. The phase function of cloud particles is largely irrelevant to the radiance of a cloud (if it is optically thick); scattering by air molecules is roughly isotropic.

Although we say "roughly a factor of four", this figure can be made more precise by using the result of Problem 7.18.

8.5. Although the blue sky still is sometimes attributed to water vapor it is not difficult to show that scattering of visible light by water vapor is less and by how much, per molecule, than that by the dominant components of air. You'll need the following data, taken from the 47th edition of the *Handbook of Chemistry and Physics*, for refractive indices at the sodium D line (about in the middle of the visible spectrum) at 0 °C and pressure of one atmosphere: oxygen, 1.000271–1.000272; nitrogen, 1.000296–1.000298; water vapor, 1.000249–1.000259.

8.6. This problem is related to the previous one. Suppose that Earth's atmosphere were composed entirely of helium. What would the clear sky look like? How would the maximum degree of polarization of skylight change? To answer the second question requires knowing just a bit about the helium atom. The refractive index of helium at the sodium D line at 0 °C and one atmosphere is 1.000036.

8.7. David E. H. Jones used to write a regular column for *New Scientist*, under the pen name Daedalus, in which he would float highly imaginative and original scientific schemes, ostensibly plausible, sometimes outrageous, always interesting. These columns have been collected in two compendiums *The Inventions of Daedalus* and *The Further Inventions of Daedalus*. One such "plausible scheme" is the "optically flat Earth." Daedalus notes that if the refractive index gradient were such that the curvature of rays matched that of Earth it would be optically flat. Daedalus implicitly assumed an isothermal atmosphere, or, at the very least, that the refractive index gradient is dominated by the pressure gradient. In all that follows you may make this same assumption even though it is not true near the surface.

First, determine the condition for an optically flat Earth. Strictly speaking, this problem should be done by deriving the equation for ray trajectories in spherical coordinates, but you can use Eq. (8.48) to approximate the amount a ray initially horizontal at the surface descends vertically for a given (small) horizontal distance. Then match this with the same distance Earth's surface is below a horizontal plane. From this expression, determine the condition that the refractive index gradient must satisfy, then the radius of Earth such that a ray will follow its curvature. A quick approach is to simply guess, by dimensional analysis, the condition on the refractive index gradient. Although we obtained essentially the same expression as Daedalus, he claims that the radius of the optically flat Earth is only 13 km less than its actual value, whereas we obtain a radius almost 5 times larger. So we conclude that Daedalus made a computational error.

After doing this problem we noticed an inconsistency in the analysis. We (and Daedalus) assume a fixed scale height, whereas it depends on g, the acceleration due to gravity at the surface, which in turn depends on Earth's radius (assuming constant density). Given that the scale height is inversely proportional to g, determine the radius of the optically flat Earth (you'll need to know or find out how g depends on radius for fixed density).

But after doing this we noted another inconsistency. We assume a fixed refractive index at the surface. But if the total number of molecules in the atmosphere is fixed, the surface number density changes with scale height and radius. You can determine this dependence by integrating the number density (assumed exponential) over a spherical atmosphere or, to save effort, you can guess the correct form by dimensional analysis. If number density is not independent, neither is refractive index. Take this into account and determine the radius of the optically flat Earth. The result may surprise you.

Daedalus claimed that if Earth were optically flat, "people would not have realized the Earth was round until they discovered that, with a good telescope, you could see the back of your head." He ends with the assertion that "if the Earth were only 13 km smaller in radius, we could see round it." Discuss the physical correctness of this assertion, paying no heed to the fact that 13 km is incorrect.

8.8. The diffuse downward irradiance at the surface [Eq. (8.2)] for a molecular atmosphere was obtained from the two-stream approximation. But it can be obtained by direct integration assuming only single scattering. Take the incident sunlight to be directly overhead. You'll need the phase function for molecular scattering (see Prob. 7.18).

HINTS: You'll need to integrate over a laterally infinite slab of physical thickness h. This integration is done most easily in cylindrical polar coordinates. Ignore attenuation of light along the path from the point it is scattered to the point at which the irradiance is to be determined.

8.9. What is the normal (radial) optical thickness of the molecular atmosphere above the altitude at which commercial airlines fly (relative to the value from the surface to infinity)? Suppose that you were to determine this by using the uniform, finite atmosphere approximation. How much error would you make? What does this tell you about the limitations of this approximation?

8.10. Many years ago the niece of one of the authors pointed to the sunset sky and asked him why low scattered clouds were tinged with red whereas high clouds were white. He gave her an answer, which satisfied her. A few days later they were driving together in the early morning. This time she noted that low clouds were gray whereas those higher were reddish. She therefore challenged the previous explanation. Answer both of her questions.

HINT: It might be possible to answer this question without diagrams but it would be foolish to try.

8.11. We calculated "forever" to be about 330 km. This is the greatest distance at which a black object can be seen against the horizon sky at sea level in a molecular atmosphere for a contrast threshold of 0.02 (ignoring Earth's curvature). This result was based on the assumption that the horizon optical thickness is essentially infinite (i.e., the horizon radiance is negligibly different from its asymptotic value). Although this assumption is a good one for even a clean atmosphere (for which particles increase the optical thickness), it is not quite true for an atmosphere *completely* free of particles. Estimate by how much "forever" is changed for a finite horizon optical thickness.

8.12. The parents of one of our students made the interesting observation that crescent moons are rarely, if ever, red or orange. Explain

8.13. If you look at distant clouds through a polarizing filter on a clear day, with the sun high in the sky, you are likely to notice a perceptible change in color of the clouds, from white to red, as you rotate the filter. Explain.

8.14. By how much is the visual range (at sea level) in an atmosphere free of particles expected to vary because of (realistic) variations in surface air temperature? Where we live, in Central Pennsylvania, we have noticed that exceptionally clear days often are also cold. Does this exceptional clarity have anything to do with temperature *per se*? Support your answer quantitatively.

8.15. In deriving an expression for the visual range we implicitly assumed that the ground was everywhere black. That is, the source illuminating the line of sight is only sunlight. Suppose that the ground were everywhere white (diffusely reflecting with a reflectivity close to 1). All else being equal, would the visual range increase, decrease, or remain the same? Explain your answer.

8.16. This problem is related to the previous one. Consider a distant black object, a line of hills, for example. Suppose that the ground were black everywhere from the observer to the object, then white everywhere beyond the object. Would the contrast between the object and the horizon sky go up, down, or stay the same? Now suppose that the ground were white everywhere from the observer to the object, then black everywhere beyond the object. Would the contrast go up, down, or stay the same?

HINT: You can answer these questions without doing detailed and complicated derivations of the airlight.

8.17. Solve Eq. (5.49) to verify our assertion that the diffuse downward irradiance (clear sky) for an atmosphere overlying a white surface should be approximately twice that for an atmosphere overlying a black surface [Eq. (8.2)].

8.18. Derive Eq. (8.8).

HINT: A binomial expansion is helpful.

8.19. We derived an expression for visual range in an atmosphere containing scatterers but not absorbers. Does the visual range change, and if so does it increase or decrease, when the (uniform) line of sight is characterized by an absorption coefficient κ as well as a scattering coefficient β? Before tackling a detailed solution, try to obtain it by simple physical reasoning.

8.20. Derive Eq. (8.76), the angle of minimum deviation for an arbitrary prism. First show that deviation is least when the incident ray and exit rays make the same angle with the prism face.

HINTS: You need Snel's law, simple trigonometry, and implicit differentiation

$$\frac{d}{dx}f\{y(x)\} = \frac{df}{dy}\frac{dy}{dx}$$

is useful.

8.21. Show that all rays incident at less than about $13°$ on a $60°$ ice prism are totally internally reflected (see Sec. 8.5.1).

8.22. Show that all incident rays are total internally reflected by ice prisms with prism angle greater than about $99.5°$ (see Sec. 8.5.1).

HINT: First convince yourself that the internal angle of incidence is *least* for an external angle of incidence of $90°$.

8.23. Misconceptions about what Newton did are widespread. For example, Newton is said to have broken up a beam of white light into a spectrum using a prism, then passed this spectrum through another prism to obtain the original beam of white light. Show that in general it is not possible to use two and only two identical prisms to transform a beam of white light back into a beam of white light. Show that this is possible with two identical slabs. Show that there is one special instance in which two prisms can restore a beam of white light. Find out what Newton really did by reading his *Optiks*. This problem was inspired by two notes in *American Journal Physics*. The first by Henry Perkins (Common misunderstanding of Newton's synthesis of light, Vol. 9, 1941, pp. 188–9), the second (The synthesis of light, Vol. 12, 1944, p. 232), in which Albert E. Hennings asserts that Perkins misunderstood Newton's synthesis of light. We think that they both are wrong, although Hennings not as wrong as Perkins.

HINT: Consider a single ray of white light to be a superposition of rays corresponding to a continuous set of wavelengths, all propagating along the same line. We can say that such a ray is broken up into its spectral components if interaction with an optical element results in rays separated in space or direction or both according to wavelength. To restore the original ray of white light requires that all these separate rays again propagate together in the same direction along the same line. The reversibility of refraction also should be helpful.

8.24. Estimate the degree of polarization of a sun dog. You need the Fresnel coefficients for transmission (Sec. 7.2) and the Mueller matrix for transmission by a prism. Ignore the birefringence of ice and take the (visible) refractive index of ice to be 1.31.

8.25. Estimate the degree of polarization of the primary rainbow. This is not a complicated problem. A sketch and a few simple calculations are sufficient. Concentrate on essentials, don't get bogged down in details. Figure 7.5 is helpful. Before tackling this problem ask yourself why the primary rainbow might be partially polarized.

8.26. Estimate the solar elevation above which a sun dog is not possible. You can do this in one line. Lengthy calculations are not needed.

8.27. Show that a zeroth-order rainbow is not possible. By zeroth-order is meant associated with rays transmitted after zero internal reflections.

8.28. Under what (plausible) conditions might it be possible to see a secondary but not a primary rainbow in nature?

8.29. We note in Section 8.4.2 that the reversed (angular) color order in the primary and secondary rainbows is an accident of the refractive index of water, not a necessary consequence of the different number of internal reflections. Show this.

HINT: The easiest way is by combining a bit of analysis with trial and error.

8.30. Prove the laws of specular reflection and refraction for a planar interface between two optically homogeneous negligibly absorbing media using Fermat's principle. That is, show that of all possible ray trajectories connecting any two points above the interface, subject to the constraint that a trajectory intersect it, the trajectory that *minimizes* Eq. (8.43) is the one for which the law of specular reflection holds. For the law of refraction, the ray trajectories must intersect an arbitrary fixed point in the medium from which they are incident and a fixed point inside the second medium.

   To show that Fermat's principle is an extremum principle, not a minimum principle, consider a spherical concave mirror. Take the mirror to be a hemisphere. Show that of all possible ray trajectories that connect two points above the mirror and intersect it, the actual trajectory is the one that *maximizes* Eq. (8.43). For simplicity, take the two points to lie in the plane that intersects the hemisphere in a circle with radius equal to that of the hemisphere and equidistant from the center.

8.31. We showed that "forever" is about 330 km, the maximum distance at which a black object can be distinguished from the horizon sky, in a molecular atmosphere at sea level, assuming a contrast threshold of 0.02. Suppose, however, that you lived on the Tibetan Plateau at an altitude of about 4 km. By how much would "forever" be increased? You may take the scale height of the molecular atmosphere to be 8 km.

8.32. What is the maximum error in the optical thickness of a slant path through the molecular atmosphere, for zenith angles less than $85°$ (i.e., at least $5°$ above the horizon), as a consequence of assuming a flat Earth?

8.33. Derive Eq. (8.79) and make a stab at giving it a physical interpretation.

8.34. Show quantitatively that hail is irrelevant to green thunderstorms.

HINT: Consider the origin of hailstones and their size relative to cloud droplets.

8.35. Equation (8.80) does not include scattering by air, and yet clouds are not suspended in a vacuum. How does the air within a cloud affect any conclusions drawn from this equation used in Eq. (8.79)? Could scattering by air within a cloud result in reddening that can yield green light transmitted by a sufficiently thick cloud?

HINT: Before trying to answer this equation by modifying Eq. (8.80) so that it includes scattering by air, try to answer these questions by simple physical reasoning.

8.36. People who have witnessed green thunderstorms sometimes assert that the sky suddenly glowed bright green. Is this consistent with the two explanations of green thunderstorms discussed in Section 8.6.1? If not, what do you make of such assertions?

HINT: This is as much a problem in psychology and physiology as in physics.

8.37. Although the sky is blue, it is far from being pure blue, and atmospheric particles small compared with the wavelengths of sunlight can only make the sky a less pure blue. But because of such particles, sunsets can be orange or even red of much higher purity than the blueness of the sky. Explain this difference in purity.

8.38. Why was no mention made in this chapter of supernumerary sun dogs, the counterpart of supernumerary rainbows? Both sun dog and rainbow angles are caustics, and the curves of deviation angle versus angle of incidence are similar.

HINT: To understand why supernumerary sun dogs are conspicuous by their absence you have to understand why supernumerary rainbows are seen in rain showers.

8.39. Several years ago we were at a scientific meeting devoted to meteorological optics. While saying not a word, one speaker showed a slide that stunned the audience. Most of us understood the slide immediately and gasped. It showed a rainbow caused partly by rain, partly by spray from ocean waves crashing against a rocky coast. What caused us to gasp was a kink in the rainbow. Explain. Devise a simple demonstration of a kinky rainbow.

HINT: This has nothing to do with drop size. One of our students did this demonstration at the cost of a few dollars spent in the garden section of a hardware store.

8.40. The Old Testament contains an account of Joshua (10, 12–14) commanding the sun to stand still so that he would have more light for killing his enemies: "So the sun stood still in the midst of heaven, and hasted not to go down about a whole day. And there was no day like that before it or after it". This account has been attributed to a superior mirage as a consequence of hailstones on the ground (mentioned in this passage). Discuss. In particular, address such questions as, Could atmospheric refraction raise the sun to "the midst of heaven"? It if could, how would the sun appear? How much longer is the day because of atmospheric refraction?; Could this increase have been measured, or likely even been noticed, 33 centuries ago? And so on. This is an open–ended question.

8.41. Show that the generalization of Eq. (8.44) to a radially homogeneous atmosphere on a spherical Earth is $n \sin \vartheta / r = \mathrm{const.}$, where ϑ is the angle between a ray and the normal to a spherical surface at r.

HINT: Consider two adjacent, concentric spherical shells, sufficiently thin that the refractive index within each can be taken to be a (different) constant, and use Snel's law and the law of sines.

8.42. In deriving Eqs. (8.10) and (8.11) the source of radiation illuminating each point of the line of sight was taken to be unattenuated sunlight. How are these equations modified if attenuation is taken into account? Show that doing so does not affect Eq. (8.12).

HINT: Use the uniform, finite atmosphere approximation.

8.43. Do Problem 8.42 again but this time account for the variable number density of molecules with height. That is, take the scattering coefficient to be $\beta_0 \exp(-z/H)$, where β_0 is the scattering coefficient at $z = 0$ and H is the scale height.

HINT: A first step is to determine the optical thickness of the atmosphere from z to infinity.

8.44. Estimate the distance d from a black object such that the airlight spectral radiance (horizontal line of sight) is approximately the same as that of the overhead sky for a molecular atmosphere. How does your estimate change if the atmosphere also contains particles that are small compared with the wavelength? How does your estimate change if the atmosphere also contains particles that are comparable with or larger than the wavelength?

8.45. Leonardo introduced a fine spray of water into a darkened chamber illuminated by sunlight. The result was "blue rays", as a consequence of which he attributed the blue of the sky to the "particles of moisture which catch the rays of the sun." The only problem with this interpretation is that it is impossible to make a spray of water droplets (i.e., using a nozzle) much smaller than the wavelengths of visible light. If Leonardo did indeed see blue in his darkened chamber when he sprayed water into it, what is a physically plausible explanation for his observation?

HINT: To answer this question you need to know the rudiments of cloud formation, especially the distinction between cloud and haze droplets.

8.46. If a rainbow really did exhibit "all the colors of the rainbow", what would the curve of colorimetric coordinates on a transect through a bow (points along a radial line perpendicular to the bow) look like on the CIE chromaticity diagram (see Sec. 4.3.1)?

8.47. Problem 3.14 addressed the *existence* of the refractive index of a cloud (at sufficiently long wavelengths). Estimate its *value*.

HINTS: You will need Eq. (8.37) and Problem 8.5. For simplicity assume that the polarizability of a water droplet at the wavelengths determined in the previous problem is the polarizability of a single water molecule at visible wavelengths times the number of water molecules in the droplet. This is a crude approximation which can be made better by digging up the refractive index of water vapor at the wavelengths calculated in Problem 3.14. Also estimate the refractive index of the snowpack in Problem 3.14.

8.48. A correspondent wrote to us that "water in motion, when it freezes, is clear rather than ice cubes (or in clouds) which are cloudy (white)." This statement embodies an understandable misconception, namely, that the individual particles in clouds, water droplets or ice, have the same optical properties as the clouds themselves. What visual observations can you adduce to argue that cloud particles are not cloudy? What fairly simple laboratory experiment might you do in support of the observational evidence?

HINT: For your experiment a laser would be handy, although not absolutely necessary, and an eyedropper or hypodermic syringe.

8.49. One clear and dry evening many years ago the senior author and Sean Twomey emerged from a pub in Tucson, Arizona into the parking lot behind the pub. As they did their eyes were drawn toward an intense green lamp on a pole perhaps a dozen meters distant, the only source of illumination in the lot. The following conversation ensued. "Do you see it?" "Yep." "It's not real is it?" "Nope." And then both nearly simultaneously made a simple test to confirm their supposition. The "it" here was a green halo around the lamp, perhaps a few degrees from it. What simple test did they make and what was the source of this halo?

8.50. We once received a letter from a retired Canadian commercial airline pilot who told us that in all his years of flying he had never once seen the horizon from cruising altitude. Do a rough calculation to show that this is exactly what is to be expected even in a very clean atmosphere.

HINT: The uniform, finite atmosphere approximation is adequate here. For more on this problem see Craig F. Bohren and Alistair B. Fraser, 1986: At what altitude does the horizon cease to be visible? *American Journal of Physics*, Vol. 54, pp. 222–7.

8.51. Why does the agreement between the measured radiance ratio in Fig. 8.5 and that calculated using Eq. (8.12) with a molecular optical thickness get worse for longer wavelengths?

8.52. In 1899 Lord Rayleigh published a derivation of the scattering coefficient of air based on the assumption that the radiant power scattered by N molecules is N times scattering by one molecule ("phases are entirely distributed at random"). The wavelength dependence of this scattering coefficient (inverse fourth power) is in accord with the observed color of the clear sky. But this observation is qualitative. There remains the question of the correct magnitude of the scattering coefficient. What observations (without using a radiometer) might Rayleigh have made to verify the correct magnitude of his scattering coefficient? Or, stated another way, what observations would you make?

8.53. Approximately how much (what fraction) of the total mass density of a cloud is contributed by cloud particles? What is the ratio of this contribution to that by water vapor only?

8.54. Over what range of wavelengths are rainbows possible?

8.55. Attenuation of beams of radiation (also sound) is often expressed in decibels (dB), defined as 10 times the logarithm (base 10) of the ratio of the radiance at any distance to that at the source. What is the attenuation of a beam of near-infrared radiation (say 1 μm) that travels completely around Earth (at sea-level) in an atmosphere free of particles? What thickness of fog gives the same attenuation? Fog liquid water contents are typically $0.1\,\mathrm{g\,m^{-3}}$ or less, and a representative fog droplet diameter might be 2 μm. Ignore multiple scattering.

8.56. For some purposes one can take the refractive index of air to be identically equal to 1 without introducing appreciable error. But for other purposes this would result in infinite error. Give examples. You might want to review Sections 3.5 and 7.2.

8.57. Although it is difficult to make a complete indoor rainbow, you can make indoor sun-dogs with a $60°$ prism mounted with its vertical axis perpendicular to a base that can be rotated. A slide projector and a black slide with a hole in it serves as a monodirectional source of white light. Illuminate the prism while it is rapidly rotating and two colored spots will be projected onto a wall behind the prism. The only difference between these artificial sun dogs and those in nature is the angle. What is the angle for the artificial sun dogs? Suppose that you were intent on duplicating the exact $22°$ sundog with a glass prism. What would be the required prism angle?

8.58. Strictly, Eq. (8.37) applies to an ideal gas composed of a single species. Stated another way, the polarizability in this equation is an average. Generalize this equation to a mixture of ideal gases, and hence verify the assertion made in the references at the end of Chapter 3 that the refractive index of air is the volume-weighted average of its gaseous components.

HINT: You have to be careful to specify exactly what you mean by a volume-weighted average.

8.59. Students of the atmosphere who understand the conditions under which Rayleigh's scattering law is valid should also be able to state why scattering by tiny (pure) water droplets cannot be the origin of the blue sky. Please do so.

8.60. Estimate the absorption optical thickness through the ozone layer, at the peak of the Chappuis bands, for a horizontal path (sun on the horizon) tangential to Earth. Make any reasonable approximations.

8.61. With the results of the previous problem, determine if the Chappuis bands of ozone markedly affect the green flash.

8.62. In Section 4.3.1 we assert that colors in our everyday lives are far from pure and support this with measurements (Fig. 4.22). But there is at least one exception: the often high purity of orange and red sunsets and sunrises. Explain.

8.63. A student once showed us a puzzling photograph of a low sun several degrees above a lake. The sun was yellowish but its specular reflection in the lake was reddish. As evidenced by Fig. 3.8, specular reflection by water does not depend on wavelength over the visible spectrum. Moreover, anything dissolved in the water will not change this. Explain the difference in the photographed color of the sun and its reflection. You may take the water to be pure and free of any suspended matter. What observation would you try to make in order to support your explanation?

8.64. We state in Section 8.4.2 that light scattered by atmospheric molecules and particles reduces the purity of the colors of rainbows. Suppose that by some magic the atmosphere were to disappear. Would the rainbow colors now be pure? Assume that geometrical optics is exact. That is, this question is not about the appropriateness of a theory but about one of the fundamental reasons why rainbow colors are not pure.

8.65. Steven Greenberg, a graduate student in meteorology at Penn State University, asked us the following question: "The attached picture shows a mid-visible (660 nm) satellite image of the northern Alaska area. During the period of this satellite image, the lidar (532 nm) is unable to penetrate through the persistent liquid-topped boundary layer clouds (as expected). My question is…if both the satellite and lidar operate at visible wavelengths, then why are we able to see the geography beneath the cloud layer? Wouldn't the cloud be opaque for both instruments?" Answer his question. Point Barrow, the northernmost tip of Alaska, is in the middle of the image. The Brooks Range is the rugged topography in the lower right. Sea ice can be seen through cloud cover at the top right. The ground is covered with snow, but the Arctic Ocean to the north and west has yet to freeze.

Index

Fundamentals of Atmospheric Radiation: An Introduction with 400 Problems. Craig F. Bohren and Eugene E. Clothiaux
Copyright © 2006 Wiley-VCH Verlag GmbH & Co. KGaA, Weinheim
ISBN: 3-527-40503-8

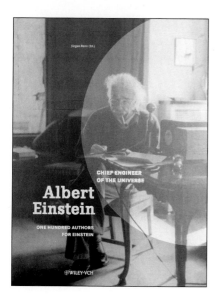

Edited by *JÜRGEN RENN*
Max-Planck-Institut Berlin

Albert Einstein – Chief Engineer of the Universe

One Hundred Authors for Einstein

2005. 472 pages. Hardcover.
ISBN 3-527-40574-7

In 1905, Albert Einstein published five scientific articles that fundamentally changed the world-view of physics: The Special Theory of Reativity revolutionized our concept of space and time, E=mc² became the best-known equation in physics.

In this book, 100 authors explain the historical background of Einstein's life and work, shed light on many different aspects of his biography, and on the scientific fields and topics that are connected to Einstein's work. The authors are some of the most renowned Einstein researchers in the world, such as Jürgen Ehlers, Peter Galison, Zeev Rosenkranz, John Stachel and Robert Schulmann.
The essays form a bridge between scientific and cultural history, opening up a perspective on Einstein's biography which goes beyond the traditional picture of the exceptional science genius.

Wiley-VCH
P.O. Box 10 11 61 • D-69451 Weinheim, Germany
Fax: +49 (0)6201 606 184
e-mail: service@wiley-vch.de • www.wiley-vch.de